OXFORD SERIES ON NEUTRON SCATTERING IN CONDENSED MATTER

General Editors

S. W. Lovesey, E. W. J. Mitchell

OXFORD SERIES ON NEUTRON SCATTERING IN CONDENSED MATTER

---

1. W. G. Williams: *Polarized neutrons*
2. E. Balcar and S. W. Lovesey: *Theory of magnetic neutron and photon scattering*
3. V. F. Sears: *Neutron optics*
4. M. F. Collins: *Magnetic critical scattering*
5. V. K. Ignatovich: *The physics of ultracold neutrons*
6. Yu. A. Alexandrov: *Fundamental properties of the neutron*
7. P. A. Egelstaff: *An introduction to the liquid state*
8. J. S. Higgins and H. C. Benoît: *Polymers and neutron scattering*

Polymers and Neutron Scattering

Julia S. Higgins

Department of Chemical Engineering and Chemical Technology
Imperial College, London

and

Henri C. Benoît

Université Louis Pasteur
Strasbourg

CLARENDON PRESS · OXFORD

Oxford University Press, Great Clarendon Street, Oxford OX2 6DP

Oxford New York
Athens Auckland Bangkok Bombay
Calcutta Cape Town Dar es Salaam Delhi
Florence Hong Kong Istanbul Karachi
Kuala Lumpur Madras Madrid Melbourne
Mexico City Nairobi Paris Singapore
Taipei Tokyo Toronto

Oxford is a trade mark of Oxford University Press

Published in the United States
by Oxford University Press Inc., New York

First published 1994
Reprinted (with corrections) 1996

A catalogue record for this book is available from the British Library

Library of Congress Cataloging in Publication Data
Higgins, Julia S.
Polymers and neutron scattering / Julia S. Higgins and Henri C. Benoît.
p. cm.—(Oxford series on neutron scattering in condensed matter)
1. Polymers—Use of radiation to study. 2. Neutrons—Scattering.
I. Benoît, Henri, 1921- . II. Title. III. Series.
QC173.4.P65H54 1994 547.7—dc20 93-15616
ISBN 0 19 850063 7

Printed in Great Britain by
Bookcraft Ltd., Midsomer Norton, Avon

Preface

Neutron sources which provide beams for experimental use are available in institutions in many different countries. Almost all of these invite experimenters from universities and from research laboratories to come and use their neutron scattering facilities. In the last 20 years or so these invitations have been taken up with increasing frequency by polymer scientists so that anyone who surveys the current journals covering polymer science and technology will find papers involving the application of neutron scattering; yet there has been until now no basic text to which newcomers can refer. The problem of how to introduce new research students to the field (including and especially her own) eventually drove one of us to attempt to fill the gap and to persuade the other to join her in the endeavour.

This background and the perceived requirements have led us to write an introductory text. We could not attempt to cover in depth the whole field or even review the whole literature. This would have necessitated a very large volume indeed. In this sense we are bound to disappoint our specialist colleagues, and to them we apologize but insist that our aims were otherwise. Our hope has been to provide an introductory work which will help the newcomer both to find his or her feet among the experimental techniques and to penetrate the complexities of the theoretical formulae needed to interpret the results. We have tried to demonstrate how neutron scattering is used, for which problems it is suitable, and how the results might be interpreted.

In attempting to achieve our aims and provide a truly introductory text we have introduced certain peculiarities in the structure of our book. Firstly, the chapters are to a fairly large extent independent and can be read independently, which, of course, means that there is a certain amount of repetition. Then, we have developed a number of points in detail in three of the appendices—the first devoted to the Fourier transform, the second to the application of thermodynamics to polymers, and the third to the relationship between neutron, light, and X-ray scattering formulae. This latter may appear at first sight to be trivial but in the experience of the authors the exact use of the theoretical formulae in interpreting data can cause a lot of wasted time and effort for novices. As far as possible we have shown derivations for all the theoretical formulae commonly used to interpret neutron scattering data from polymers but we have been more interested in the way the formulae are based on physical models than in mathematical rigour. In this context, and after some hesitation, we have shown how the formula introduced by de Gennes under the name random phase approximation may be derived from the formalism of Ornstein and Zernicke. This demonstration may not be any simpler, but it has the advantage of introducing a different point of view and it has not often been presented in the literature.

One of our biggest problems was the heroic task of choosing just a few examples from the huge and growing literature. With apologies to the many of our colleagues whose beautiful data we could not present, we reiterate that we could only choose a few examples to illustrate the use of the techniques and the appropriate theoretical interpretation. Because of this our reference list is much shorter than the literature would warrant. To cover these omissions we have indicated for the reader a number of excellent literature reviews where further information on the wide applications of neutron scattering in the polymer field can be found.

Here then, is the fruit of a collaboration between two researchers not only of different nationality but each with a different method of working and a different conception of the subject. Our frequent meetings and long discussions, face to face, by telephone, fax, and post and our strenuous efforts to speak the same language (both in words and in mathematical formalism) have, we hope, resulted in a volume which will attract new researchers to the field and contribute to new developments in this branch of science. There are still many problems in polymer science where neutron scattering can make valuable contributions.

Finally, both authors wish to thank all their colleagues and friends who have contributed both directly and indirectly to the production of this volume. We hope they will forgive us the omissions and imperfections of our text and for not thanking them all personally. Without their support and their encouragement the book would never have been completed.

London J. S. H.
Strasbourg H. C. B.
March 1993

Contents

Glossary of symbols

α Rouse unit in a polymer chain
$A(q)$ Amplitude scattered by an ensemble of scattering points where the origin of phases is at $q = 0$.
A_2 Second virial coefficient in a polymer solution
$\mathcal{B}$ Total scattering length of the sample (Chapters 2, 3, 7)
$\overline{\Delta \mathcal{B}^2}$ Square of the fluctuations of the total scattering length
$\overline{\Delta b^2}$ The incoherent scattering cross section divided by 4π
ℓ_i Scattering length of a polymer molecule of species i
b_i Scattering length of the 'monomer' of species i
b_i' Scattering length per unit volume of species i
$\langle b \rangle$ Average scattering length in a mixture of deuterated and ordinary polymers $\langle b \rangle = x b_{\mathrm{D}} + (1 - x) b_{\mathrm{H}}$
b_{coh} or b Coherent scattering length
b_{app} Apparent scattering length in a solution; difference between the scattering length of the dissolved molecule and the same volume of solvent $= \ell_1 - b_0 \dfrac{V_1}{v_0}$
$\bar{b}_i$ Contrast factor in the presence of a solvent

$$\bar{b}_i = \frac{\ell_i}{z_i} - b_0$$

$b_i(t)$ Time-dependent position vector for rotational motion
b_{v} Contrast factor per unit volume $b_{\mathrm{v}} = b_i' - b_0' = \dfrac{b_i}{v_i} - \dfrac{b_0}{v_0}$
$[\bar{\ell}]$ Column vector of the $\bar{b}_{\mathrm{i}}$
c Concentration in unit mass per unit volume
$c_i(t)$ Time-dependent position vector for centre of mass motion
D Translational diffusion coefficient
D_{R} Tube diameter in the reptation model
$G(, p, T, N_1, N_2 \ldots)$ Gibbs free energy of mixing
$G(r, t)$ Van Hove correlation function
$G_{\mathrm{s}}(r, t)$ Autocorrelation function
$g(r)$ Radial distribution function
$g_{\mathrm{c}}(T, p, \varphi_1, \varphi_2, \ldots)$ Gibbs free energy of mixing per unit cell
$g_{\mathrm{c}}'' = \dfrac{\partial^2 g_{\mathrm{c}}}{\partial \varphi^2}$
$g(\omega)$ Amplitude weighted density of states

h Planck constant
$\hbar = (1/2\pi)h$
$I(q) = \dfrac{\partial\sigma}{\partial\Omega_{\text{coh}}}$, scattering by a sample of volume V
$i(q) = I(q)/N_{\text{T}}$, scattering by the volume v_0
k_{B} Boltzmann constant
$\boldsymbol{k}$ Wavevector of a neutron beam $\left(\dfrac{2\pi}{\lambda}\boldsymbol{u}\right)$ where $\boldsymbol{u}$ is a unit vector in the direction of propagation
$\mathcal{L}$ Total length of a linear molecular chain often called the contour length or the chemical length
$\mathbb{L}^*$ Persistence length of a rigid molecule
l Length of one segment of a chain of z segments
$\overline{l^2}$ Root mean square value of the length of an elementary step
$\overline{L^2}$ Mean square distance between the centre of mass of the blocks of a copolymer
$\overline{L_p^2}$ Root mean square distance between two points separated by p segments
m Mass of a scattering unit (in Chapters 1 and 4 it is the neutron mass)
M Molecular weight
M_{w} Weight averaged molecular weight
M_{n} Number averaged molecular weight
N Number of scattering molecules
N_i Number of molecules of species i
$\overline{N}_i$ Average number of molecules of species i in a given volume
N_{T} Ratio of the volume of the sample to the volume of the scattering unit
$n = \dfrac{N}{V}$ Number of scatterers per unit volume = density of scattering units
$n(\boldsymbol{r})$ Local density of scatterers, $\Delta n(\boldsymbol{r}) = n(\boldsymbol{r}) - n$
$\hat{n}(\boldsymbol{q})$ Fourier transform of $n(\boldsymbol{r})$
P_z Net polarization of a neutron beam
$P(q)$ Form factor for one molecule made of z subunits

$$P(q) = \frac{1}{z^2}\sum_{i_1=1}^{z}\sum_{j_1=1}^{z}\langle\exp(-\mathrm{i}\boldsymbol{q}\cdot\boldsymbol{r}_{ij})\rangle$$

$Q(q)$ Normalized intermolecular interference term:

$$Q(q) = \frac{1}{z^2}\sum_{i_1=1}^{z}\sum_{j_2=1}^{z}\langle\exp(\mathrm{i}\boldsymbol{q}\cdot\boldsymbol{r}_{i_1j_2})\rangle$$

q Modulus of the scattering vector $\left(\dfrac{4\pi}{\lambda}\sin\dfrac{\theta}{2}\right)$

q Scattering vector
R Gas constant, (in Chapter 10 it is the reflectivity)
$\mathcal{R}$ Rayleigh ratio, ratio of the scattered intensity per unit volume to the brightness of the illuminated scattering volume
$R(q)$ Growth rate of concentration fluctuations in a phase separating polymer blend
$\overline{R^2}$ Average value of the square of the radius of gyration
R_g Radius of gyration $\sqrt{\overline{R^2}}$
R_H Hydrodynamic radius
$\langle r_E^2 \rangle$ Root mean square distance of the entanglements in a melt
$\boldsymbol{r}$ Vector characterizing the position of a scattering point—its modulus is r.
S Entropy of the system
$S(q)$ Coherent structure factor $= \frac{1}{b^2}$ or $\frac{1}{\bar{b}^2}$ times $\left(\frac{\partial \sigma}{\partial \Omega}\right)_{\text{coh}}$

$s(q) = \frac{S(q)}{N_T}$, where N_T is the volume of the system divided by the volume of the solvent or the basic monomer
$S(q, t)$ Time-dependent intermediate structure factor (inverse Fourier transform of $S(q, \omega)$
$S(q, \omega)$ Energy-dependent structure factor. It can be split into its coherent $S_{\text{coh}}(q, \omega)$ and its incoherent part $S_{\text{inc}}(q, \omega)$
$S_{ab}(q)$ Partial structure factor describing the scattering by pairs a and b:

$$\sum_{i_a}^{N_a} \sum_{j_b}^{N_b} \langle \exp(-\mathrm{i}\boldsymbol{q}\cdot\boldsymbol{r}_{i_a j_b}) \rangle$$

T_g Glass transition temperature of a polymer
U Energy of the system
$\boldsymbol{u}_i(t)$ Time-dependent position vector for vibrational motion
V Scattering volume
V_i Volume of the polymer molecule i
v_i Volume of the 'monomer' of polymer i
$\mathfrak{v}_{ij}$ Excluded volume characterizing the interactions between molecules of type i and j
v_{ij} Excluded volume parameter for the interaction between molecules of type i and j, $v_{ij} = \frac{\mathfrak{v}_{ij}}{v_0}$
v_0 Volume of one molecule of solvent or of an arbitrarily chosen reference unit
v_i^s Specific volume of the species i
$w(\boldsymbol{r})$; $w(r)$ Probability of finding a point at the position $\boldsymbol{r}$ (or at the distance r from the origin)

x Volume fraction of one species of molecule in a binary mixture of identical molecules

z_i Number of subunits ('monomers') in a macromolecule $\left(z_i = \frac{V_i}{v_0}\right)$

$z(\omega)$ Density of states

Greek letters

β Isothermal compressibility $\left(\frac{-1}{v}\frac{\mathrm{d}v}{\mathrm{d}p}\right)_T$

$\delta(x)$ The improper Dirac delta function, x can be a number or a vector

$\Gamma(r, t)$ Normalized correlation function $= \frac{1}{n^2} G(r, t)$

Γ Inverse correlation time

Γ_{R} Inverse correlation time, Rouse model

Γ_{Z} Inverse correlation time, Zimm model

$\frac{\partial^2 \sigma}{\partial \Omega \partial \omega}(q)$ Scattering intensity (for an incident beam of energy unity and for a given q value in a given range of energy $\omega\hbar$). It can be can be split into its coherent part $\left(\frac{\partial^2 \sigma}{\partial \Omega \partial \omega}\right)_{\mathrm{coh}}$ and its incoherent part $\left(\frac{\partial^2 \sigma}{\partial \Omega \partial \omega}\right)_{\mathrm{inc}}$

$\frac{\partial \nu}{\partial c}$ Refractive index increment

$\frac{\partial \sigma}{\partial \Omega}(q)$ Ratio of the intensity scattered in a given solid angle Ω, characterized by the direction q, to the incident intensity.

ε Dielectic permittivity of the scattering medium

η_0 Solvent viscosity

θ Angle between the incident and the scattered beam

κ Phonon wavevector in a crystal

λ Wavelength of neutron (or light *in vacuo*). In Chapter 8: ratio of the actual length to the initial length in a stretched rubber

μ_i Chemical potential $\left(\frac{\partial G}{\partial N_i}\right)_{p, T, \mu_j}, \ldots$

$\bar{\mu}_i$ Exchange chemical potential $\left(\frac{\partial g}{\partial \varphi_i}\right)_{p, T, \varphi_j}, \ldots = \frac{\mu_i}{z_i} - \mu_0$

ν Refractive index of the scattering medium

ξ Correlation length or screening length; the interaction between two molecules decreases as a function of the distance r following the law $r^{-1} \exp\left(-\frac{r}{\xi}\right)$

ξ_0 Friction factor per segment in a polymer melt
Ξ Grand partition function
ρ_a Absorption cross section per unit volume
ρ_b Scattering length per unit volume
ρ_l Scattering length per unit length.
ρ_s Scattering length per unit surface.
ρ_x Mass per unit volume of the constituent x in the system
σ Scattering cross section
σ Reciprocal lattice vector in a crystal (Chapters 4 and 9)
σ^2 Rigidity parameter $\sigma^2 = \overline{R^2}/\overline{R_\mathrm{f}^2}$ where $\bar{R}_\mathrm{f}^2$ is the radius of gyration of a freely rotating chain with constant valence angle
τ Neutron time of flight
φ_i Volume fraction occupied by the constituent i
χ Flory–Huggins interaction parameter
Ω Solid angle, dΩ elementary scattering angle in the scattering direction
ω Circular (angular) frequency

1

Introduction

1.1 INTRODUCTION

When the relationship between their physical properties and their molecular structure is considered, the most obvious characteristic of macromolecules is their length. For example, a polyethylene molecule with 10^6 repeat units stretched out in its planar zigzag form is 0.2 mm long. The organization of these enormously long molecules, even in dilute solution, let alone in bulk samples, poses fascinating questions. When molecular motion is added too, it is easy to see why research groups, or indeed whole laboratories are set up for the study of molecular properties of polymers.

In order to study molecular organization and molecular motion, physicists and chemists are used to turning to various parts of the electromagnetic spectrum: X-rays for crystal structure, infrared spectroscopy for molecular vibrations, quasielastic light scattering for diffusional motion of molecules or aggregates. For each of these there is an analogous technique involving scattering of neutrons. Unfortunately, sources of neutrons, unlike lasers, do not come in 'pocket' sizes for laboratory experiments. Until recently most were in fact nuclear reactors; now we have in addition pulsed accelerator sources. What they all have in common is largeness of scale and relative inaccessibility to the scientist who has just prepared an interesting sample. Why then do it? Why undertake experiments at distant reactor centres, using complex equipment shared with many other scientists, often at antisocial times (reactors run continuously with no respect for meal times, weekends, or holidays) and necessitating endless writing of proposals and reports. That many of us do face these difficulties and succeed is evidenced by the steadily growing number of publications in this field. In 1973 when the UK became the third partner with France and Germany in the then recently commissioned high flux reactor at the Institut Laue–Langevin in Grenoble in the French Alps, the techniques were known only to a few specialists. Now, nearly 20 years later no book on the physics or physical chemistry of polymers is published without reference to the results of neutron experiments. So what is it about this small particle that makes it such a useful tool? This introductory chapter is devoted to answering that question with a survey of the types of questions about polymeric molecules which can be addressed using neutron techniques. In this chapter everything is kept simple. The aim is to give a basic idea of what the technique is all about without getting involved with the complexities. All of the ideas introduced here are developed subsequently in more detail in later chapters. As an appetizer for these chapters, several examples of the unique information available are briefly described.

1.2 POSSIBILITIES–DEUTERIUM LABELLING

In Fig. 1.1 a schematic polymer molecule in a solid sample is sketched at the centre of the page. It has some sort of side group structure–for example a phenyl group as in polystyrene or a methyl group as in polypropylene,

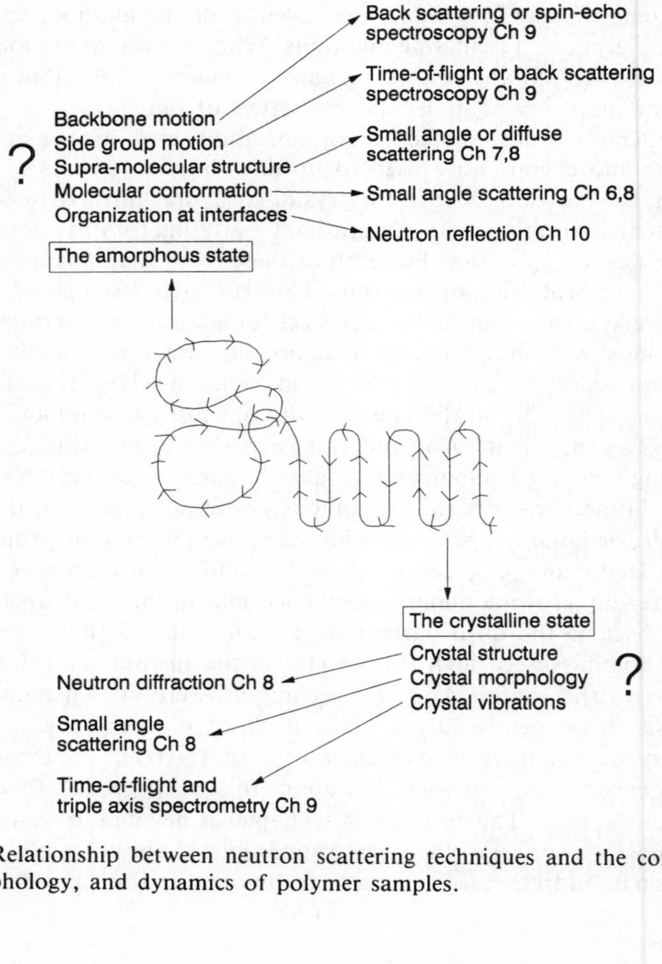

Fig. 1.1 Relationship between neutron scattering techniques and the conformation, morphology, and dynamics of polymer samples.

or something much more complex as in the comb-like liquid crystal polymers. Half of the molecule is randomly coiled as in the amorphous, rubber or melt state. The remaining half is folded into parallel segments as in the lamellae typical of the polymeric crystalline state. Each type of structure raises a series of questions. These questions might be quite specific to the chemistry of one polymer molecule: for example 'What is the frequency of rotation of the side group and its associated temperature dependence?' Alternatively there may be much wider implications. The molecule in Fig. 1.1 has been drawn with consecutive segments in the 'crystalline region' next to each other, but the whole question of 'adjacent re-entry' in the crystalline polymeric state is one that has exercised polymer scientists for several decades. Associated with each question in the diagram is one or more neutron scattering technique which could be used to study the problem. For each neutron technique there is usually a more conventional equivalent which involves use of various parts of the electromagnetic spectrum. A useful way to introduce the special advantages of neutrons is to contrast a neutron experiment with an analogous, and perhaps more familiar measurement using photons.

In structural studies neutron diffraction and neutron small angle scattering cover the same ranges of spatial resolution as X-ray diffraction and small angle scattering. The major difference between the techniques lies in the interaction between the probing radiation and the sample. X-rays are scattered by electrons, neutrons are scattered by nuclei. In this simple statement lies the basis of the special advantage of the neutron. (At this point apologies should be made to those physicists who study magnetic structures and who do observe effects due to electronic magnetism. However, to date, studies of polymers have not extended to the field of magnetism so that it is perfectly proper to confine attention in this book to nuclear scattering.) The details of the neutron–nuclear interaction are not yet of interest. What is of vital importance is that the strength of this interaction, characterized by a property called 'scattering length', is not a function of atomic number as is the case for X-rays where the heavier the element the more it scatters. The neutron–nuclear interaction varies much more randomly throughout the periodic table, as can be seen in Fig. 1.2. (There are theories based on nuclear interactions which model this variation but these are beyond the scope of this work.) As a consequence, lighter elements, and in particular, hydrogen, are no longer under the same disadvantage they suffer in X-ray diffraction. Figure 1.2 shows one further very interesting detail—hydrogen and deuterium differ markedly in the values of their neutron scattering lengths. This observation is of fundamental importance. Almost without exception, the use of neutron technique for studying polymers has relied on this difference in order to highlight molecules or sections of molecules. Deuterium labelling is much less intrusive than labelling with heavy metals

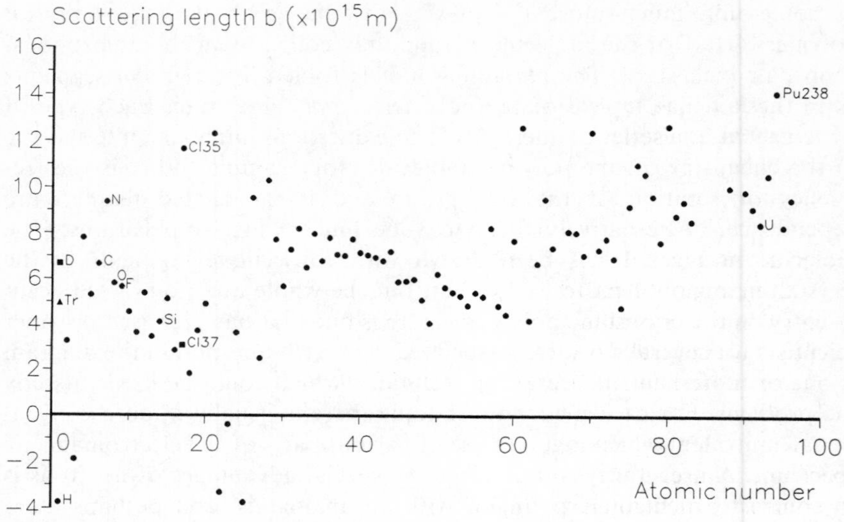

Fig. 1.2 Variation of neutron scattering length, *b*, with atomic number. Unless indicated the mean value has been used for elements with several isotopes.

(as is done for X-ray investigations) so that labelled molecules can be studied in a neutron experiment surrounded by other molecules almost in their 'natural' state. It is for this reason that the example chosen for Fig. 1.1 was a molecule in a solid sample. There are, though, situations where investigation of molecules in solution can benefit from labelling—for example when supramolecular structures such as aggregates or micelles are formed and the molecular conformations within these are required.

In relaxation studies or in spectroscopy the neutron does not hold quite such an ace card over other techniques. In NMR or in infrared spectroscopy, for example, deuterium labelling is also important, and for similar reasons. The extra mass of deuterium reduces its vibrational frequency compared to hydrogen so that infrared spectra are shifted. In NMR it is the different nuclear spin of deuterium that shifts the resonance lines. In Chapter 9 it will be seen that the study of polymer dynamics using neutron spectroscopies has been much less extensive than the structural studies. It can be argued that this will always be the case, and that such applications of neutron scattering will be limited to very specific examples. In this category falls the study of backbone motion in melts using neutron spin echo (Chapter 9). On the other hand the applications of small angle scattering show no signs of limitations beyond those of access to the neutron sources themselves.

1.3 THE COMPONENTS OF A NEUTRON EXPERIMENT

In any spectrometer which uses radiation to investigate organization or motion at a molecular level a number of general characteristics can be identified. The experiment allows observation of changes in properties of the scattering radiation and the experimenter interprets these in terms of structure or motion in the scattering sample. Figure 1.3 represents diagramatically such an experiment in which various components are indicated. Firstly, a source of radiation is required. This apparently trivial statement is in fact quite important since the nature of the source will determine many of the characteristics of the incident radiation just as, for example the use of lasers rather than mercury arc lamps has had a profound effect on optical spectroscopy. Neutron sources are discussed briefly in Chapter 2. There are only two types of interest—reactors or pulsed sources. In the former a nuclear chain reaction produces a continuous flux of neutrons. In the latter sharp bursts of high energy protons or electrons from an accelerator hit a heavy metal target and 'boil' off neutrons, also in sharp bursts. Both share one characteristic: the neutrons have to be moderated. When the neutrons emerge from the nuclei that produce them they have very large energies and are in a very real sense 'too hot to handle'. In a moderator the neutrons are scattered many times within a suitable medium, exchanging energy at each collision until they achieve approximate thermal equilibrium with the moderator material (Fig. 1.4). As a result the emerging beam contains a wide spread of energies and thus has more in common with the mercury arc lamp rather than with the laser. In every scattering experiment changes

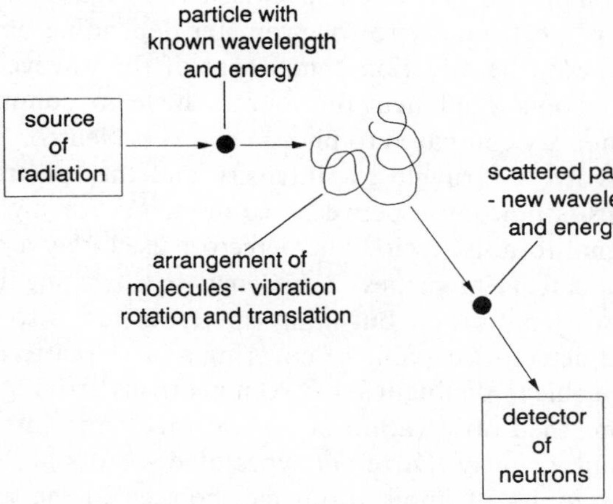

Fig. 1.3 A schematic neutron scattering experiment.

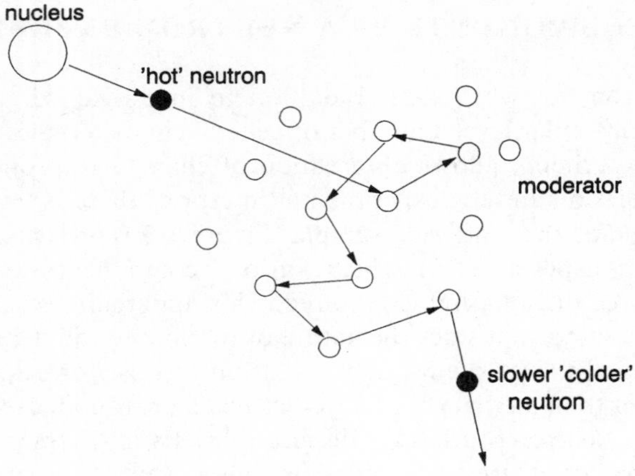

Fig. 1.4 Moderation of the neutron energy.

involving direction and/or energy must be detected. It is thus necessary to define these properties before the radiation reaches the sample. The energy (and hence wavelength) spread in the emerging beam has to be defined and in many cases a much narrower spread of values selected. This process is called monochromation and the method chosen will depend on the type of experiment in progress. Various examples will be seen in the different spectrometers described in Chapter 3.

After scattering a similar procedure is required in order to define the changes in direction and energy which have occurred. Again the procedures adopted vary from spectrometer to spectrometer depending on the resolution required in space (the direction component of the wavevector) and in energy. Another property all neutron sources have in common is their relative lack of intensity compared to photon sources. Neutron experiments almost always involve a struggle for intensity and the spectrometers are often an ingenious compromise between requirements for high resolution and a sensible signal-to-noise ratio. The detectors used play a role in combating the low incident intensities. Efficiency in detecting the arriving neutrons is obviously important but many spectrometers also incorporate spatially extended detectors capable of covering a large scattering angle. If these detectors are able to distinguish between neutrons arriving at different positions on them, then observation at several different scattering angles can be made simultaneously. Different types of detectors are discussed in Chapter 2 and examples of applications can be seen in the various spectrometers in Chapter 3.

1.4 BACKGROUND—A LITTLE HISTORY

All of the theory necessary for the design of spectrometers was available almost as soon as the neutron was discovered by Chadwick in 1932. The first reactors were built in the forties and fifties and the first scattering experiments were carried out soon afterwards. On searching the literature for applications of neutron scattering in polymer science, however, very few references will be found prior to the seventies, while subsequently what was a trickle has become a steady stream of publications. The reason for this comparatively late application to polymer science are several. There was firstly the question of know-how and access. The first scientists to use neutrons to study the structure and dynamics of condensed matter were solid state physicists who were employed at the reactor centres themselves. It took some time for information about the possibilities of this technique to filter through to other interested communities. There were also the problems of source intensities and detector efficiencies already mentioned. Most of the techniques of particular interest in polymer science make heavy demands on resolution and hence on neutron flux and detector design. High flux reactors were first designed and built in the USA and France during the late sixties. Both the technology for building large area detectors and revolutionary new ideas for achieving high resolution in spectrometers were developed in the early seventies most especially at the Institut Laue–Langevin. This institute was specially open to the wider scientific communities in the three member countries (France, Germany, and the UK) and welcomed visitors from many others. There was, thus, at the beginning of the seventies in Grenoble a unique combination of technical advances and scientific expertise which blossomed into an ever increasing use of neutron scattering in many fields, among them polymer science. This use of neutrons in polymer science continues extensively at Grenoble but has also now spread to many other laboratories worldwide.

1.5 SCATTERING—THE BASICS: INTRODUCING q

It is not possible to proceed much further in the discussion of neutron scattering without looking at the processes in which information about the sample is converted into changes in wavevector and energy of the neutrons. In this section the question of nomenclature inevitably rears its head and in particular the choice of symbols. Much of the theory of neutron scattering parallels if not reproduces that for scattering of light or X-rays. Given the limitations imposed by the English and Greek alphabets together with separate developments by different scientific communities it is not surprising to find duplicate uses of the same symbol or duplicate symbols for the same parameter. Throughout this book an attempt has been made to achieve self-consistency and to comply with current usage in the neutron scattering community. To avoid total confusion for those already familiar with nomenclature in one of the parallel fields an index of symbols has been included. It can be found at the beginning of the book.

When considering the production and detection of neutrons, it is their particulate nature that tends to be foremost in one's thinking. It is not possible to understand the scattering process itself without remembering that these particles also have wave properties. Louis de Broglie first formulated the relationship between the wave properties and particulate properties in 1924 by expressing the wavelength, λ, associated with a particle in terms of its mass m and velocity v

$$\lambda = h/mv \tag{1.1}$$

(h is the Planck constant).

The veracity of this relationship is usually demonstrated experimentally by diffracting electrons since these are relatively easy particles to obtain with variable velocity and to detect. From thermal reactors, typical neutron energies will be of the order of $k_B T$ (where k_B is the Boltzmann constant). Remembering that the reactor core and moderator are at temperatures not far above room temperature (at least for water cooled reactors which includes all the research beam reactors)—i.e. $k_B T$ is around 4 to 5×10^{-21} J—then the corresponding value of λ from eqn (1.1) is around 10^{-10} m. This is ideal for investigating interatomic distances. (If longer or shorter wavelengths are required these are obtained by cooling or heating a small subsidiary moderator.)

The first experiment considered therefore is a classical Bragg diffraction experiment in which a beam of neutrons is scattered by the layers of a crystal lattice. This is shown diagramatically in Fig. 1.5. The scattering angle is θ. The beam of particles is replaced by a wavefront impinging on the crystal and being scattered. The scattered wave from each nucleus spreads out over all directions—it is isotropic. The detector collects a combination of all the waves at that point. The part of the wavefront scattered at B travels further to the detector than that scattered at E. Following the geometry in Fig. 1.5 the extra distance travelled is $2d\sin(\theta/2)$. When the wavefront is collected in the detector the two parts will only be in phase with each other if the extra distance travelled is a whole number of wavelengths. This requirement is shown diagramatically below Fig. 1.5. On the left two waves of equal amplitude are exactly in phase. On combination a wave of equal wavelength but double the amplitude will be obtained. On the right the two waves are exactly one half wavelength out of phase. On addition zero amplitude would be obtained. These two situations corres-

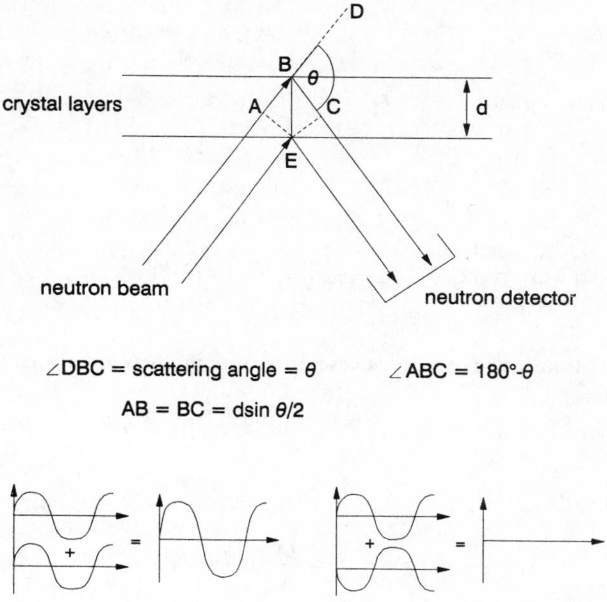

Fig. 1.5 Diffraction of neutrons by two layers, e.g. in a crystalline sample. Constructive and destructive interference of waves.

pond to constructive and destructive interference respectively. Thus for constructive interference—i.e. a finite amplitude of the recombined wave fronts—it is necessary that

$$n\lambda = 2d\sin(\theta/2) \tag{1.2}$$

where n is any integer. Immediately, a relationship has been obtained between the neutron wavevector (wavelength and direction) and a property of the sample—in this case the crystal lattice spacing.

In Fig. 1.6 the same scattering event as in Fig. 1.5 is drawn in terms of the neutron wavevectors. These are vectors of magnitude $2\pi/\lambda$ pointing in the direction of travel of the neutrons. In the scattering event under consideration the neutron does not exchange energy with the crystal lattice, so its wavelength remains unaltered. Only its direction changes. Thus the magnitude of $\boldsymbol{k}$, the initial wavevector, which expressed as $\boldsymbol{k}_i$ is the same as that of $\boldsymbol{k}_f$, the final wavevector. The change, $\boldsymbol{q} = \boldsymbol{k}_f - \boldsymbol{k}_i$ is shown in the vector diagram in Fig. 1.6. It will be useful here to state the convention to be used about vector and scalar quantities. If $\boldsymbol{k}$ is the vector, then k will be the equivalent scalar where $k = |\boldsymbol{k}|$.

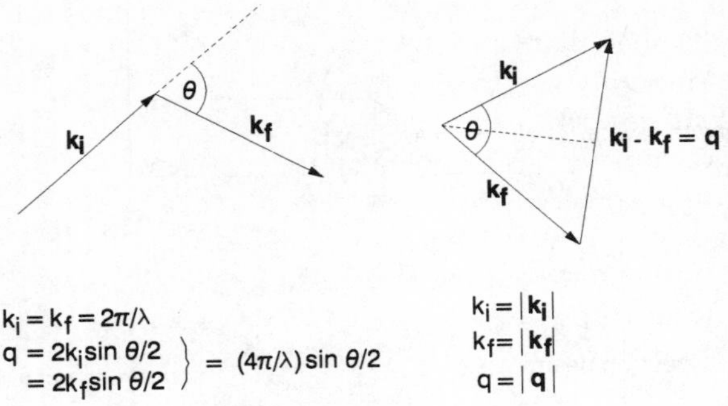

Fig. 1.6 Relationship between wavevectors and momentum transfer for elastic scattering.

In magnitude

$$|q| = q = \frac{4\pi}{\lambda}\sin(\theta/2) \tag{1.3}$$

and eqn (1.2) can now be expressed in terms of q as

$$2\pi/q = d/n. \tag{1.4}$$

The wavevector change, $\boldsymbol{q}$, may be referred to as the scattering vector and is often also called the momentum transfer. This can be understood by referring back to eqn (1.1). Momentum $mv = h/\lambda = hk/2\pi = \hbar k$ (where $\hbar$ is $h/2\pi$). Thus the momentum change in the scattering when the neutron is considered as a particle is just $\hbar q$.

All expressions relating the neutron scattering event to the spatial properties of the scattering sample in real space are expressed in terms of q—as for example in eqn (1.4). (In experiments where the incident wavelength is fixed, as is usually the case in reactor spectrometers, q is directly related to the angle of scatter measured. Spectrometers on pulsed sources, however, may use fixed angles and variable incident wavelength to scan q. The dual dependence of q on angle and wavelength must be carried through in the development of the analysis therefore.) The inverse relationship between distance and q seen in eqn (1.4) is fundamental. Large-scale structures within the sample require very small values of q for their exploration and conversely small distance scales are observed via large q values. Since q depends on two parameters, wavelength and scattering angle, it can be varied by changing either or both values.

In Chapters 4–7 all the calculations will start from spatial arrangements of the scattering nuclei and convert these into a structure factor $S(q)$ which predicts the probability of neutrons being scattered as a function of q. For the example in Fig. 1.5 $S(q)$ would consist of sharp peaks in probability at scattering angles such that the q values satisfy eqn (1.4) but more generally $S(q)$ will be a smoothly varying function of q. Other factors will of course come into play in governing the actual intensity of neutrons detected. These include the spectrometer design (incident flux, detector size, and efficiency, for example) and the strength of the neutron–nuclear interaction for the nuclei constituting the sample. Lumping all the factors due to the spectrometer into a function $C(q)$, and expressing the neutron–nuclear interaction as a function, $f(\sigma)$, of the scattering cross section, σ, then the number of neutrons observed $I(q)$ is given by

$$I(q) = f(\sigma)C(q)S(q). \tag{1.5}$$

The factors that govern $C(q)$ are considered in Chapter 2 and 3.

1.6 RELATIONSHIP BETWEEN SCATTERING AND STRUCTURE

The derivation of $S(q)$ from specific models of the molecular structure will be described in Chapter 4, 6, and 7. What can be done here though is to take a more general case than that of Bragg diffraction and look at the mathematical bridge between the positions of the scattering nuclei, the wave nature of the neutron and the shape of $S(q)$. Initially the time variation of the wave will be considered, as well as its spatial variation since this is needed in order to consider inelastic scattering in the next section.

In one dimension a plane wave travelling along the x-axis is written as

$$A(x, t) = A_0\cos(\omega t - 2\pi x/\lambda) \tag{1.6}$$

where λ is the wavelength, ω is the angular frequency and t is the time. At constant t the amplitude, $A(x, t)$, is a cosine function of x, while at constant x it is a cosine function of t. The term $2\pi x/\lambda$ gives the phase of the wave at a point a distance x from an arbitrary origin. As discussed for the case of Bragg diffraction, if two waves initially in phase (i.e. sharing the same origin of phase) travel over different paths to a detector then their relative phases will depend on the factor $2\pi d/\lambda$ where d is the difference in path travelled. In Fig. 1.7 a plane wave strikes a nucleus at the point O. It will

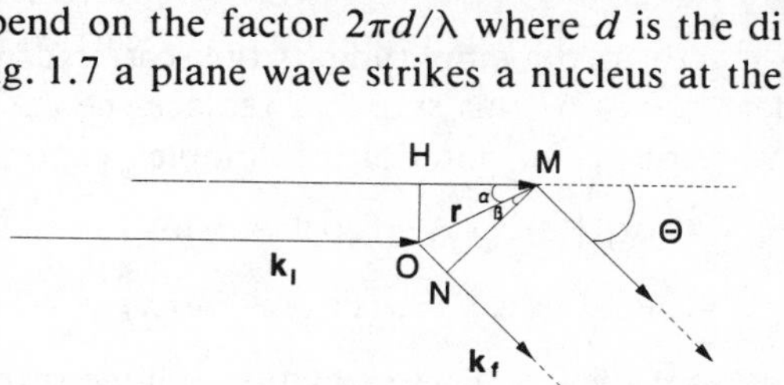

Fig. 1.7 Scattering of a plane wave.

be scattered as a spherical wave, but far enough away from O the wave front will be indistinguishable from a plane wave (the amplitude will of course have decreased as R^{-1}). Given the relative sizes of internuclear distances and sample detector distances the scattered wave will always appear locally like a plane wave. The incident wave is defined by the vector $\boldsymbol{k}_i$ and that from the sample to the detector by the vector $\boldsymbol{k}_f$. The angle between $\boldsymbol{k}_i$ and $\boldsymbol{k}_f$ is the scattering angle, θ and as before $2\pi/\lambda = k$.

The plane wave will strike many nuclei within the sample, each the source of a new spherical wave. Consider just one of them at M, and compare the phases of the two waves, one scattered from O and one from M when they arrive at the detector. Assuming that there is no change in phase on scattering, the phase difference will just be given by the difference in distance travelled, d.

$$\text{phase shift} = (\text{HM} - \text{ON})2\pi/\lambda. \tag{1.7}$$

If we let O be the origin of coordinates and of phase, then OM is the vector $\boldsymbol{r}$ and $\text{HM}(2\pi/\lambda) = (\text{OM}\cos\alpha)(2\pi/\lambda) = \boldsymbol{r}\cdot\boldsymbol{k}_i$. We have arbitrarily put an origin on one of the nuclei and specified the positions of all the others with respect to this arbitrary origin. Since it will soon become evident that only differences in position of the nuclei are important (in the same way that only differences of phase are important) this arbitrariness really does not matter. Now $2\pi/\lambda\ \text{ON} = (2\pi/\lambda)(\text{OM}\sin\beta) = \boldsymbol{r}\cdot\boldsymbol{k}_f$ and

$$\begin{aligned}\text{phase shift} &= \frac{2\pi}{\lambda}(\boldsymbol{r}\cdot\boldsymbol{k}_f - \boldsymbol{r}\cdot\boldsymbol{k}_i) \\ &= \boldsymbol{r}\cdot(\boldsymbol{k}_f - \boldsymbol{k}_i)\end{aligned} \tag{1.8}$$

and remembering the definition of $\boldsymbol{q}$,

$$= \boldsymbol{r}\cdot\boldsymbol{q}.$$

Allowing for changes in amplitude on scattering (but not changes in phase) the wave scattered at M can be written relative to that scattered at O as

$$A(x,t) = b\cos(\omega t - \boldsymbol{q}\cdot\boldsymbol{r}) \tag{1.9}$$

b, the scattering length, is the amplitude of the scattered wave relative to that of the incident wave. At this stage we replace the cosine in eqn (1.9) with the much more powerful notation of complex exponentials where

$$\begin{aligned} a\exp(+\mathrm{i}x) &= a(\cos x + \mathrm{i}\sin x) \\ a\exp(-\mathrm{i}x) &= a(\cos x - \mathrm{i}\sin x)\end{aligned} \tag{1.10}$$

and i is the square root of -1. These equations allow us to define a complex amplitude which takes into account the phase of the wave. The real part is the 'in-phase' amplitude and the term in i, the so-called imaginary part, is the 'out-of-phase' amplitude. The wave scattered at M is written as

$$A(x,t) = b\exp(\mathrm{i}(\omega t - \boldsymbol{q}\cdot\boldsymbol{r})) \tag{1.11}$$
$$= b\exp(\mathrm{i}\omega t)\exp(-\mathrm{i}\boldsymbol{q}\cdot\boldsymbol{r})$$

The frequency term in ωt will only change if there is a change in the energy of the neutron. When considering elastic scattering which just depends on the relative positions of the nuclei and not their motion this term can be ignored. The relative amplitude of waves scattered by any of the nuclei in the sample just depends on the terms $b\exp(-\mathrm{i}\boldsymbol{q}\cdot\boldsymbol{r})$. We can easily express the total amplitude in the detector in this complex notation—it is the sum of all these complex amplitudes. This total amplitude will be called $A(q)$ and it is given by

$$A(q) = b\sum_{i=1}^{N} \exp(-\mathrm{i}\boldsymbol{q}\cdot\boldsymbol{r}_i). \tag{1.12}$$

The intensity is the square of the amplitude, and when A is complex the intensity is expressed as the product of A by its complex conjugate, A^* i.e. if $A = (a + \mathrm{i}b)$ then $AA^* = (a + \mathrm{i}b)(a - \mathrm{i}b) = (a^2 + b^2)$ which is always real and positive, as is necessary for an intensity. For two superimposed waves the amplitude will be

$$A(q) = b_1\exp(-\mathrm{i}\boldsymbol{q}\cdot\boldsymbol{r}_1) + b_2\exp(-\mathrm{i}\boldsymbol{q}\cdot\boldsymbol{r}_2) \tag{1.13}$$

and

$$I(q) = A(q)A(q)^* = b_1^2 + b_2^2 + 2b_1b_2\cos[\boldsymbol{q}\cdot(\boldsymbol{r}_1 - \boldsymbol{r}_2)]. \tag{1.14}$$

The last step back to cosines is made by referring back to equation (1.10). It is interesting to note that by considering what happens when $\cos x$ is unity (so that I is a maximum) the condition for constructive interference is recovered $(q = 2\pi/(r_1 - r_2)) = 2\pi/d$. (When the cosine $= -1$, and the waves are exactly out of phase we have destructive interference.) Using eqn (1.12) it is now possible to cover the many cases where the nuclei are not arranged on a periodic lattice—i.e. non-Bragg scattering or diffuse scattering.

Returning to a system of many nuclei and using eqn (1.12), the intensity is given by

$$I(q) = A(q)A^*(q) = \sum_{i=1}^{N} b_i\exp(-\mathrm{i}\boldsymbol{q}\cdot\boldsymbol{r}_i)\sum_{j=1}^{N} b_j\exp(+\mathrm{i}\boldsymbol{q}\cdot\boldsymbol{r}_j) \tag{1.15}$$

$$I(q) = \sum_{i=1}^{N}\sum_{j=1}^{N} b_ib_j\exp(-\mathrm{i}\boldsymbol{q}\cdot(\boldsymbol{r}_i - \boldsymbol{r}_j)). \tag{1.16}$$

It is now clear that the result is independent of the choice of origin, O, since only the differences $r_i - r_j$ enter into the final formula.

Equation (1.16) is the basis for calculating all the model scattering laws, $S(q)$, for elastic scattering. It depends, as can now be seen, on the relative positions of every pair of atoms within the sample. For the moment any changes in phase on scattering have been ignored, and the scattering is supposed to be elastic—i.e. no change in energy.

1.7 ELASTIC AND INELASTIC SCATTERING

It is easier to envisage the use of neutrons to detect motion in the sample by considering them as particles rather than the waves we have just been discussing. If a particle is scattered by a nucleus in a vibrating, rotating or translating molecule, then there is a finite probability that an exchange of energy may occur. If this does happen the neutron is said to be inelastically scattered. (It is not necessary to consider exchanges with nuclear energies because the moderated neutrons now have energy many orders of magnitude less than the nuclear binding energy.) The inelastically scattered neutron may gain energy from or lose energy to the molecular motion. From quantum mechanics it is known that the energies associated with these motions are quantized, i.e. only discrete quanta of energy can be taken up or lost by the molecules. For vibrational motion the spacings between the quantized energy states are large compared to thermal energies ($k_B T$) and thus also compared to the neutron energy. Gains or losses of energy thus appear as distinct shifts in the neutron energy, as represented in Fig. 1.8.

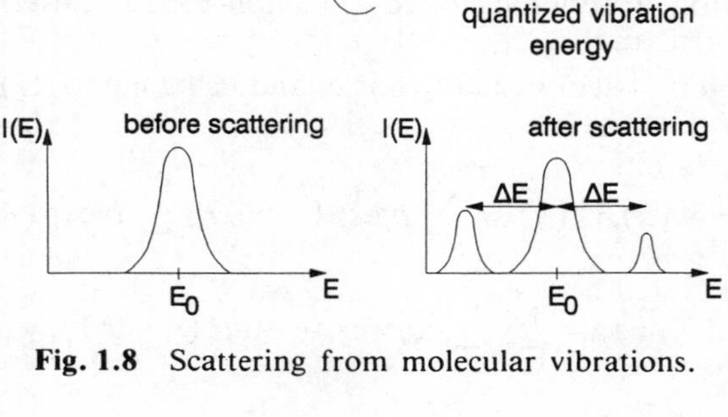

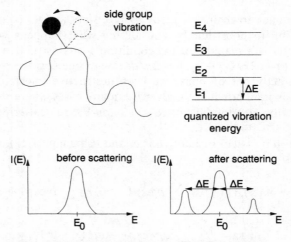

Fig. 1.8 Scattering from molecular vibrations.

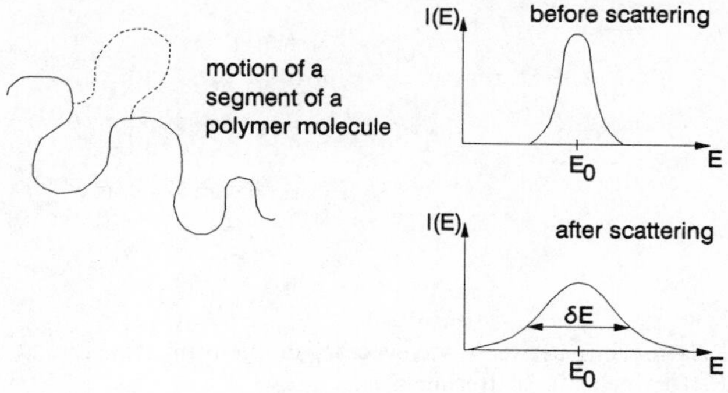

Fig. 1.9 Scattering from rotation or translation of molecules.

There will be fewer neutrons gaining energy from the molecular vibrations—as indicated by the lower intensity of the energy gain peak at $E_0 + \Delta E$ because this depends on occupation of the excited states E_1, E_2 etc. which is governed by the Boltzmann factor. Quantization of rotational and translational energies on the other hand is usually negligibly small compared to $k_B T$ and thus to the neutron energy. The effect of these motions on the neutron energy is to broaden the initially sharp distribution. The process is sometimes called Doppler broadening in analogy with the frequency shift of a sound wave when it is emitted from a moving source. Such a scattering experiment is represented schematically in Fig. 1.9. The width $\delta(E)$ of the broadening is related to the energy of the molecular motion involved. It is now no longer sufficient to consider only the spatial properties of the sample in $S(\boldsymbol{q})$. Dynamic properties must be included by considering the dynamic structure factor, $S(\boldsymbol{q}, \Delta E)$, which now relates the structure and dynamics simultaneously to the probability of a neutron being scattered with energy change ΔE and momentum change $\boldsymbol{q}$.

Now clearly if such energy exchanges can occur then they must be observed in any experiment where molecular motion is possible—i.e. at any temperature above absolute zero. How then do the spectrometers designed for structural studies differentiate between elastic and inelastic scattering events? This is essentially a question of spectrometer design. There are two alternatives in performing an elastic scattering experiment, i.e. one designed only to investigate the structure of the sample via $S(\boldsymbol{q})$. The inelastically scattered neutrons can be discriminated out, thus giving a purely elastic experiment or the inelasticity can be ignored and the inelastically scattered neutrons counted along with the elastically scattered ones. If E_i is the initial and E_f the final enrgy, then in the former experiment $S(\boldsymbol{q}, \Delta E)\,[E_i = E_f]$ is observed, in the second $\int S(\boldsymbol{q}, E)\mathrm{d}E$. The first

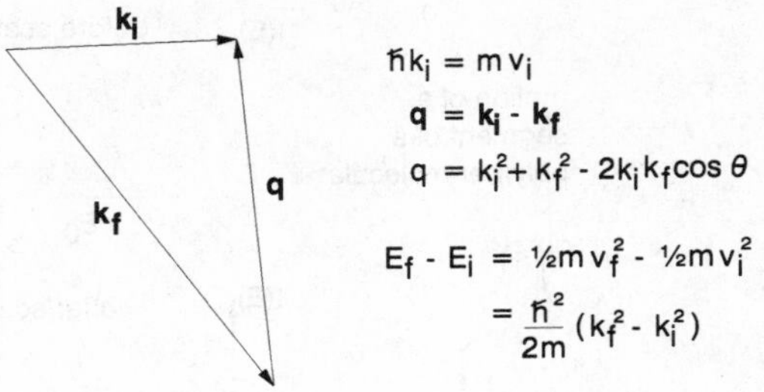

Fig. 1.10 Relationship between wavevector, momentum transfer, q, and energy transfer, ΔE, for inelastic scattering.

case is clearly more technically demanding since it requires measurement of the neutron energy after scattering. It is thus clear why many diffractometers are designed simply not to distinguish changes in energy, i.e. they assume constant wavelength. How much error is this likely to introduce? Well, the probability of a neutron exchanging energy with the scattering nucleus will be calculated in Chapter 4. Anticipating these calculations it will be found that this probability is very much less than the probability that the neutron will be scattered elastically so that at worst the inelastically scattered neutrons will form some sort of a background to the structural study. There are, moreover, strict relationships arising from the geometry of the scattering experiment between the possible values of q and ΔE as seen in Fig. 1.10. In Chapter 4 a further restriction on possible values of q and ΔE in $S(q, \Delta E)$ arising from the spatial and dynamic properties of the sample will be introduced. When all these factors are taken into account it will be seen that for most diffraction experiments, ignoring the effects of inelastic scattering does not threaten the interpretation of the data. For certain experiments, special care to discriminate for truly elastic scattering has been found necessary.

1.8 CROSS SECTIONS

In the discussion so far, the neutron–nucleus interaction has been dealt with in the most general of terms and two parameters have been mentioned in this context—the neutron scattering length b and the cross section σ. In order to understand these quantities the question of wave particle duality must be recalled and the way in which the two sets of properties are reconciled. The wave nature of the particle is related to the probability that it will be found at some point in space. In fact it is the square of the amplitude that determines this probability. In the experiment each scattering nucleus is considered to be the source of a new scattered wave. The neutron scattering length, b, introduced in Section 1.6 determines the amplitude of this scattered wave with respect to the amplitude of the incident wave. A negative value as shown for hydrogen in Fig. 1.2 then signifies a change of phase on scattering (i.e. the scattered wave is one-half wavelength shifted with respect to the incident wave as in the right hand diagram below Fig. 1.11). The square of the amplitude at a point in space determines the probability that the neutron will be found at that point and thus b^2 determines the probability that a neutron from the incident beam will be found in the scattered beam. The dimensions of b are chosen so that b^2 is the probability that a neutron will be scattered per nucleus, per incident neutron, per solid angle (solid angle is normally expressed in steradians. There are 4π steradians in a sphere.). The probability of the neutron being scattered somewhere in all space is then just $4\pi b^2$ and it is this quantity which is given the name cross section and represented by the symbol σ. A negative value of b still gives rise to a positive value of σ and thus has no significance when considering a single neutron.

The probability of a beam of neutrons being scattered by an assembly of nuclei will be governed by the sum of the wavelengths scattered by individual nuclei. Remembering that the net signal will be a sum of all the wave amplitudes, if there are nuclei present with different values of b then these must be averaged before squaring them to obtain the net probability of scatter. (Now the significance of the negative values of b becomes clear – it will be important in calculating cross sections in an assembly of different nuclei.) If there are spatial correlations between the nuclei with the same scattering length – for example, if some polymer molecules among many others are deliberately labelled by deuterating them – then the correlations in scattering lengths over each of these molecules give rise to structure factors $S(q)$ characteristic of their shape. Such scattering, carrying information about structural arrangement within the sample, is called coherent scattering. If, however, there are no correlations between the position in the sample of a nucleus and the value of its scattering length, as for example with naturally occurring isotopes (which are randomly distributed), then there will be a random or incoherent part of the scattering probability giving rise to incoherent scattering. The terms coherent and incoherent often cause confusion so it is worth considering the source of the two types of scattering in more detail. If, in a sample, all the nuclei have the same value of b then at any angle of scatter the amplitude, and hence the probability of finding neutrons, will be predictable in terms of the sample structure (as in Fig. 1.5). If there are randomly placed nuclei in the sample with different scattering lengths then in the scattered beam at any angle there will be a mean amplitude determined by the sample structure – i.e. a mean probability of finding neutrons – together with random fluctuations away from

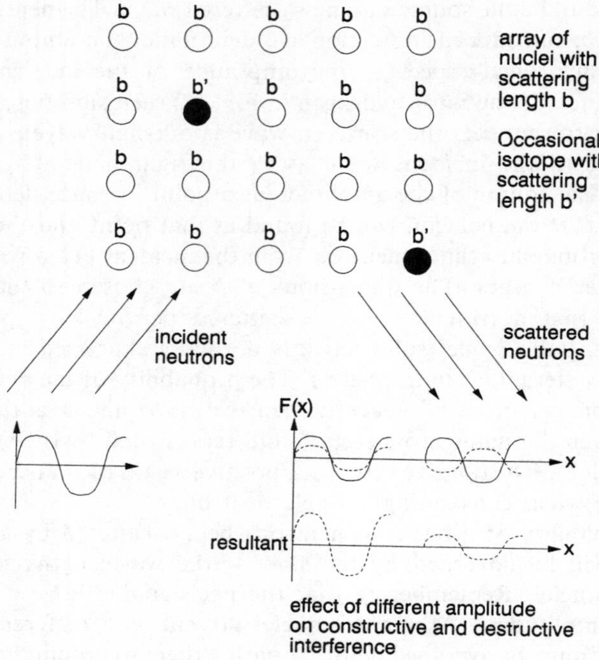

Fig. 1.11 Coherent and incoherent scattering from an array of nuclei with scattering lengths, b or b'. The neutron waves scattered from b' produce neither complete destruction nor addition when combined with those from b. They produce an angularly independent contribution to the scattering.

this mean. These random fluctuations, since they arise from random points in the sample, cannot contribute to constructive interference as in Fig. 1.5. The scattering probability calculated from these fluctuations in amplitude is therefore isotropic since the scattering from any individual nucleus is isotropic. Because of the random positions of the nuclei which are its source incoherent scattering can carry no structural information about the sample. These effects are shown diagramatically in Fig. 1.11.

The incoherent scattering arising from isotopes is not the only, or even the most important incoherent scattering from polymeric samples. Nuclei have spins and so does the neutron. When the interactions between these spins are taken into account further differences in scattering lengths may arise. In particular, hydrogen has two very different scattering lengths for neutrons depending on its spin state. Since the spin of a nucleus is generally uncorrelated with its position in the sample these scattering lengths also give rise to spin incoherent scattering. From the calculations in Chapter 4 it will be seen that the spin-incoherent cross section of hydrogen is very large. Not only does this give rise to large backgrounds to the spin coherent scattering from hydrogen-containing samples but it may contain useful information. Although the incoherent scattering carries no information about the spatial arrangements within the sample, it will carry information about the dynamics via any changes of energy incurred in the scattering event. Incoherent scattering from hydrogen is very intense and has been used to observe the dynamics of polymer molecules. In experiments aimed at determining structure the incoherent scattering forms a background which has to be removed. Values of all these cross sections and scattering lengths will be found in Chapter 2 (Table 2.1)

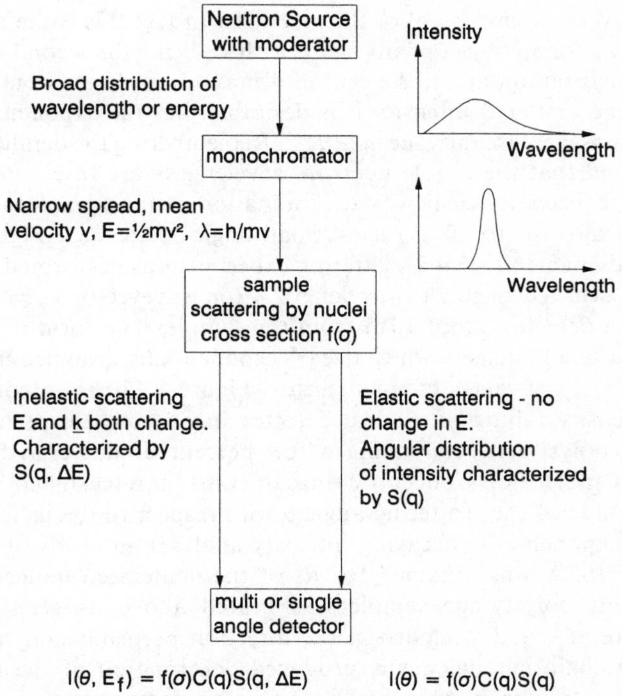

Fig. 1.12 Scheme of elastic and inelastic scattering experiment. The intensity $I(\theta, E_f)$ or $I(\theta)$ depends on the neutron–nuclear interaction, σ, the spectrometer geometry $C(q)$ and the sample structure and dynamics $S(q, \Delta E)$ or $S(q)$.

1.9 SUMMARY

Figure 1.12 summarizes the ideas introduced in this chapter by showing diagramatically the components of neutron scattering experiments. In order to put flesh on these dry bones the rest of the chapter is devoted to outlining a few applications to polymeric systems.

The first two follow the right-hand side of Fig. 1.12, involving elastic, coherent scattering; the third uses inelastic incoherent scattering.

1.9.1 Molecular shape

In Chapter 6 (eqn (6.105)) the scattering from any polymer molecule) is shown to have a very simple form when $q \ll R_g^{-1}$

$$I(q) = I_0 \exp(-(q^2 R_g^2)/3). \tag{1.17}$$

I_0 depends on the size and number of molecules and R_g is the radius of gyration (average dimension) of the molecule. In fact this is the asymptotic form at low q for an object of any shape with R_g being the second moment of the mass distribution about the centre of mass. In order to obtain sensible values of the scattered intensity it is desirable that the exponential in eqn (1.17) remains close to unity, i.e. $q \simeq R_g^{-1}$. Remembering the definition of q in eqn (1.3) and that the usable neutron wavelengths are rarely much larger than 10 Å it becomes clear why determination of macromolecular dimensions (typically 10^2 to 10^3 Å) uses 'small angle scattering'. Figure 1.13(a) indicates a scheme of such a scattering experiment as performed on a reactor with a selected incident 'wavelength λ (or wavevector $\boldsymbol{k}_o = 2\pi/\lambda$) and a large area detector about 10 m from the sample. The form of eqn (1.17) leads to scattered intensity which decays exponentially away from zero scattering angle at the centre of the detector. Figure 1.13(b) is a contour map of the intensity falling on such a detector in an actual experiment on a sample of polystyrene containing a few percent of deuterated (labelled) polystyrene molecules. Concentric rings of constant intensity correspond to the same value of the scattering angle θ with respect to the incident beam. From the exponentially decaying intensity analysed in terms of eqn (1.17) a value of 105 Å was obtained for R_g of the deuterated molecules.

When this polystyrene sample was heated above its glass transition temperature, T_g, and stretched in the direction perpendicular to the incident beam chain entanglements produced deformation of the molecules. The sample was quickly quenched back to room temperature, before stress relaxation occurred so that anisotropy of the molecular dimensions was frozen in. Figure 1.13(c) shows the contour map of intensity on the detector scattered by such a stretched-quenched sample. The molecular anisotropy is clearly seen, with a steeper curve (larger R_g) parallel to the sample stretch direction. The rate of stress relaxation in the sample was followed by waiting for sequentially longer times at elevated temperature after stretching and before quenching to room temperature. Figure 1.13(d) shows R_g parallel and perpendicular to the stretch direction as stress relaxation proceeds. Experiments such as these have been used to test the theories of affine molecular deformation, and of viscoelasticity (such as reptation). Further discussion will be found at the end of Chapter 8.

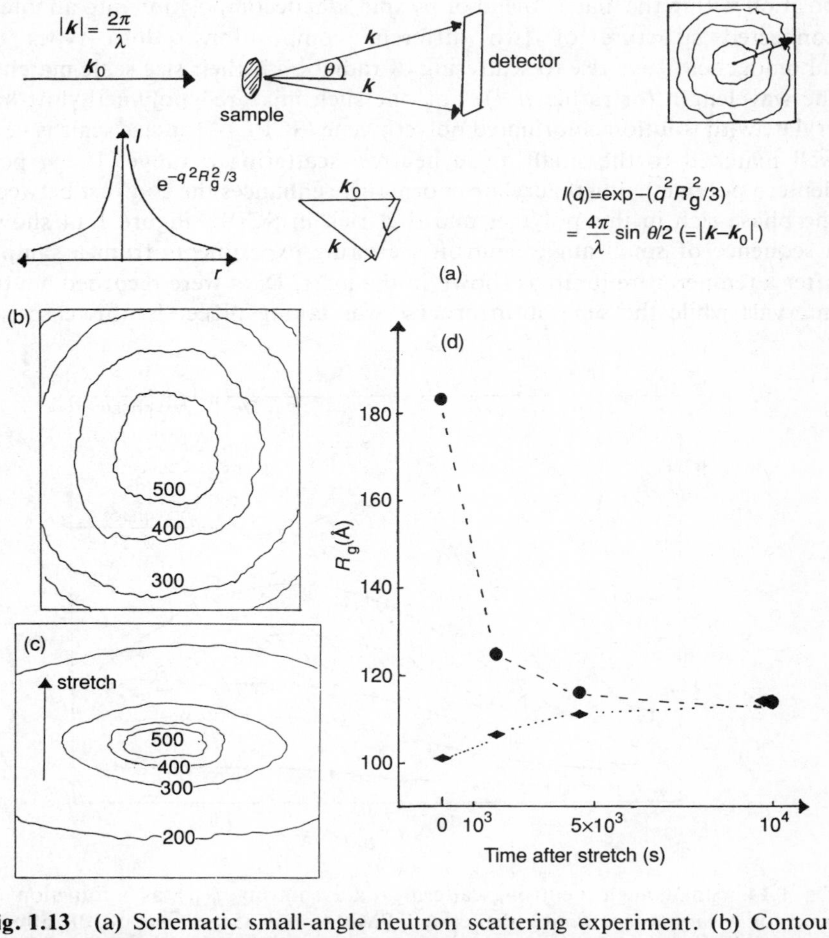

Fig. 1.13 (a) Schematic small-angle neutron scattering experiment. (b) Contour plot of the intensity distribution about the beam scattered by a sample of polystyrene containing deuterated polystyrene molecules. (c) The sample in (b) after stretching. (d) The dimensions parallel (●) and perpendicular (♦) to the direction of stretch as a function of the time delay before freezing for the sample in (b) and (c). (Data from Maconnachie *et al.* (1981)).

1.9.2 Sample morphology

Another type of molecular motion that can be followed by a series of 'snapshots' is the separation that takes place in a binary polymer mixture when this is heated from the one-phase to the two-phase region of the temperature–composition phase diagram (Fig. 1.14). In the two-phase region, which is usually reached by a temperature jump as shown in the inset of Fig. 1.14 the originally homogeneous mixture develops small scale heterogeneities, either via a nucleation and growth process of a new composition within the matrix blend or by spinodal decomposition into an interconnected structure of two different compositions. Both types of inhomogeneity give rise to scattering of radiation if their size scale matches the wavelength (or rather q^{-1}). For one such mixture—polymethylmethacrylate with solution chlorinated polyethylene (SCPE)—the size scale is very well matched to the small angle neutron scattering q range. Using perdeutero polymethylmethacrylate enormously enhances the contrast between the phase rich in this polymer and that rich in SCPE. Figure 1.14 shows a sequence of small angle neutron scattering experiments from a sample after a temperature jump as shown in the inset. Data were recorded at 10 s intervals while the separation process was taking place. In this case the

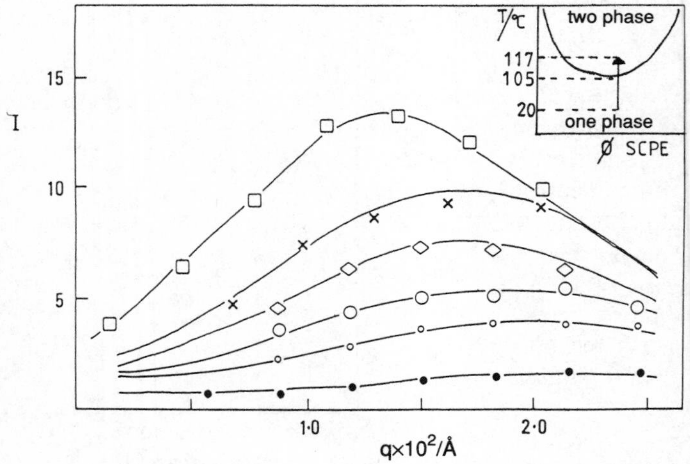

Fig. 1.14 Small angle neutron scattering data showing $I(q)$ as a function of time for a sample containing deuterated PMMA and 55 per cent SCPE after a fast temperature jump from 105 to 117°C. (●) 30 s, (○) 80 s, (○) 110 s, (◇)150 s, (×) 200 s, (□)290 s. Most of the data points have been omitted for clarity. Inset is the phase diagram for this system showing the temperature jump. (Data from Higgins *et al.* (1989*a*).)

signal, being isotropic, was radially averaged over the area detector, and this factor together with the enhanced contrast and the fact that scattering arises from whole regions rather than single molecules gave a high enough counting rate to allow these 'real time experiments' to be undertaken and to eliminate the necessity of quenching samples for subsequent measurement. The position of the broad peak in the scattering curves in Fig. 1.14 gives the size of the developing structure via the Bragg condition in eqn (1.4) (in this case $d \sim 1000$ Å). Following the separation process, as in these experiments, tests the current theories, provides thermodynamic information about the blend and opens the possibility of producing materials with novel morphologies and mechanical properties by allowing phase separation to proceed to a desired level and then freezing in the structure. Further discussion of the phase-separation in polymer blends will be found in Chapter 8 (Section 8.3.4).

1.9.3 Main chain segmental motion

It is easy to forget that a rubbery polymer is in fact a liquid. The effects of the entanglements of these long molecules on the physical properties such as viscosity or diffusion are obvious but such properties involve large scale molecular motion. On a very local scale neutron experiments show that the molecules are wriggling at the same sort of frequency as the Brownian motion of low molecular weight liquids. The energy of segmental motion is quite small compared to $k_B T$ and thus the energy gained or lost by neutrons scattered from these molecules gives a broadened energy spread as described in Fig. 1.9. If this experiment is carried out on a sample containing only hydrogenous molecules then the incoherent scattering of hydrogen swamps everything else. There is no spatial information carried by the scattered neutrons which are isotropically distributed over all scattering angles. (This statement will require some modification in detail in later chapters but remains essentially true.) The energy spread in the scattered beam, however, is found to be dependent on the q value (i.e. for fixed incident energy, the observed spread varies with scattering angle). The spread in energy is often expressed as the full width at half maximum, δE, as shown in Fig. 1.9. For liquids in general it is found that

$$\delta E = 2\hbar D q^2 \tag{1.18}$$

where D is the self-diffusion coefficient of the molecules.

This Q dependence of the broadening can be understood by remembering the inverse relationship between q and spatial dimensions. The smaller the q value the longer the corresponding distance scale explored. Now, because we are considering incoherent scattering, only the correlations between a scattering nucleus and itself can be detected. Nevertheless this relationship between distance and q^{-1} can be extended to the motion of individual nuclei. Thus at small q values the scattering nucleus explores longer distances. These large excursions of the nucleus are slow and therefore correspond to low energies. Thus the broadening, δE as the neutron gains or loses energy from the motion is less for lower q values. A segment of a polymer molecule can be imagined as dragging along other segments as it wriggles. The net effect is to slow the motion down as longer and longer distances are explored, i.e. as observation is made at smaller and smaller q values. The result is that the diffusion coefficient in eqn (1.18) behaves as though it, too, is q dependent. For molecules in a polymer melt it is predicted that δE varies as q^4 for small q values. Figure 1.15 shows the Doppler broadening δE of the incident beam (which itself does not have infinitely good resolution of course) observed in the scattered beam at increasing q for a melt of polydimethyl siloxane molecules ($M_w = 1.74 \times 10^5$).

In Fig. 1.16 the values of δE for this polymer at room temperature are compared with those for water molecules at the same temperature. The difference between the q^2 variation of δE for water and the much faster variation for the polymer is clearly seen. At higher q when only short excur-

Fig. 1.15 The scattered intensity as a function of neutron energy for several q values (scattering angles) for a sample of polydimethylsiloxane. (Data from Allen *et al.* (1974).)

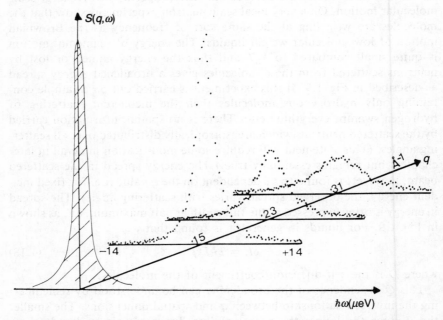

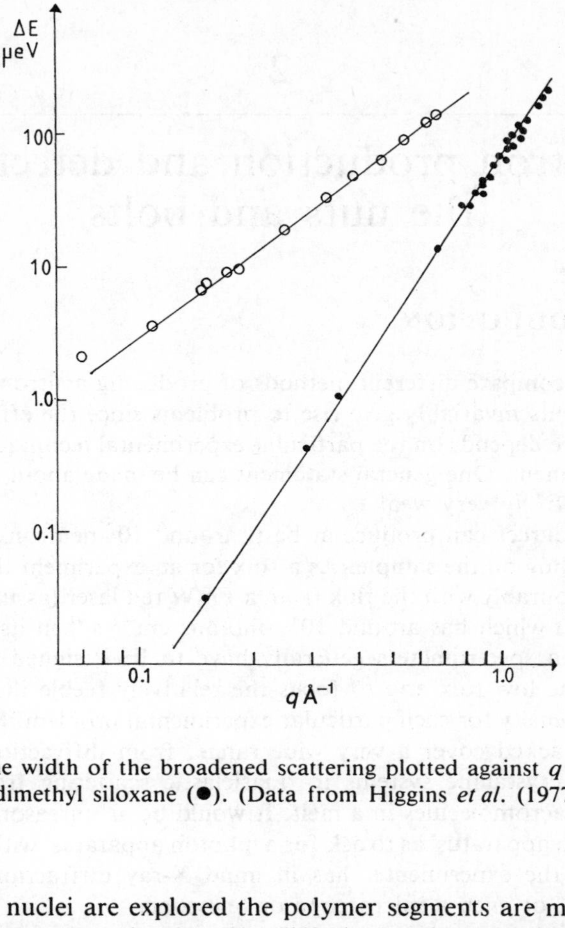

Fig. 1.16 The width of the broadened scattering plotted against q for water (○) and for polydimethyl siloxane (●). (Data from Higgins *et al.* (1977).)

sions of the nuclei are explored the polymer segments are moving almost as fast as the water molecules. The diffusive motion of small molecules is described theoretically in Chapter 4, that of large molecules in Chapter 9. Experimental examples for small molecules and polymers are given in Chapter 9.

SUGGESTED FURTHER READING

Higgins and Maconnachie (1987).
Squires (1978).
Willis (1973).
Wignall (1987).

2

Neutron production and detection: the nuts and bolts

2.1 INTRODUCTION

Attempts to compare different methods of producing neutrons for scattering experiments invariably give rise to problems since the effectiveness of a given source depends on the particular experimental technique being used in the experiment. One general statement can be made about all sources—they are relatively very weak.

Neutron sources can produce at best, around 10^7 neutrons cm^{-2} s^{-1} in the beam falling on the sample. As a flux for an experiment this compares rather unfavourably with the flux from a 1 mW red laser (assuming a 2 mm square beam) which has around 10^{17} photons cm^{-2} s^{-1} in its beam. As a result neutron spectrometers generally have to be designed to maximize the use of the low flux and to focus the relatively feeble illumination to maximize intensity for each particular experimental problem. Moreover the problems attacked cover a very wide range, from diffraction studies of structure in crystalline systems to quasielastic scattering from the slow motion of macromolecules in a melt. It would be as unreasonable to look for 'a neutron apparatus' as to ask for a 'photon apparatus' without specifying whether the experimenter has in mind X-ray diffraction or Raman spectroscopy, or even a microscope or telescope.

This preamble is aimed at emphasizing the wide diversity of neutron spectrometers, whose designs depend crucially on that aspect of structure or motion in the scattering systems which is to be studied. A description of specific neutron scattering techniques should then necessarily be closely associated with a discussion of the scattering systems and model structure factors. This will be undertaken systematically in later chapters. Nevertheless since neutron sources are very expensive, they are multi-user facilities, and there are many factors common to all spectrometers which will be considered here. A number of ways of efficiently detecting neutrons have been developed and there are only limited possibilities for selecting and measuring neutron energies. The discussion in this chapter then, will briefly address the questions 'where do neutrons come from' and 'how do I handle them?' It should be seen more as a reference base for questions posed by later chapters than as a section to be swallowed whole at one sitting.

2.2 NEUTRON SOURCES

From the point of view of the experimenter what lies behind the hole from which the neutron beams emerge may be to all intents and purposes a black box. The energy and intensity of the neutrons in the beams and the time structure of the flux are the important factors, as well as any background radiation, for example γ-rays. The energy characteristics determine the type of structural or dynamic investigation suited to the source while the time structure of the flux—continuous or pulsed—determines the design of the spectrometers. These characteristics are determined both by the type of source, which may be a nuclear reactor or a particle accelerator, and by the environment through which the neutrons pass immediately after production—the process which is called moderation.

2.2.1 Reactor sources

A reactor designed for use as a source of neutrons is very different from one in use as a power generator. The designer of the latter is concerned with increasing efficiency by keeping all the neutrons trapped, while by definition the former has beam tubes to allow the neutrons to escape. In the reactor core fast neutrons (1–2 MeV) are produced in a fission chain reaction often from enriched uranium—^{235}U. The reactor core is surrounded by a moderator, often D_2O or H_2O, sometimes graphite in which the emitted neutrons are scattered many times, losing energy at each collision until they have an average thermal energy which is characteristic of the moderator temperature. Given the nature of the usual moderator (water) it is hardly surprising that these unpressurized reactors usually function at around 50°C. A Maxwell–Boltzmann velocity distribution for neutrons at these temperatures has a most probable velocity of $\sim 3000\,\mathrm{m\,s^{-1}}$ which translates via the de Broglie relationship ($\lambda = h/mv$) to a most probable neutron wavelength around 1.4 Å. The maximum flux of neutrons is usually designed to be just outside the core near the entrance to the beam tubes and may be as high as 10^{15} neutrons $\mathrm{cm^{-2}\,s^{-1}}$. Unfortunately, only a small fraction of these neutrons will be travelling in the right direction to pass down the beam tubes through the biological shielding. The peak flux in these beam tubes may be reduced by as much as a factor of 10^5 from that in the core. Subsequent 'tailoring' of the beam reduces intensity still further: 10^7 neutrons $\mathrm{cm^{-2}\,s^{-1}}$ is a typical value at a sample position. Figure 2.1 is a schematic diagram of a research reactor showing the

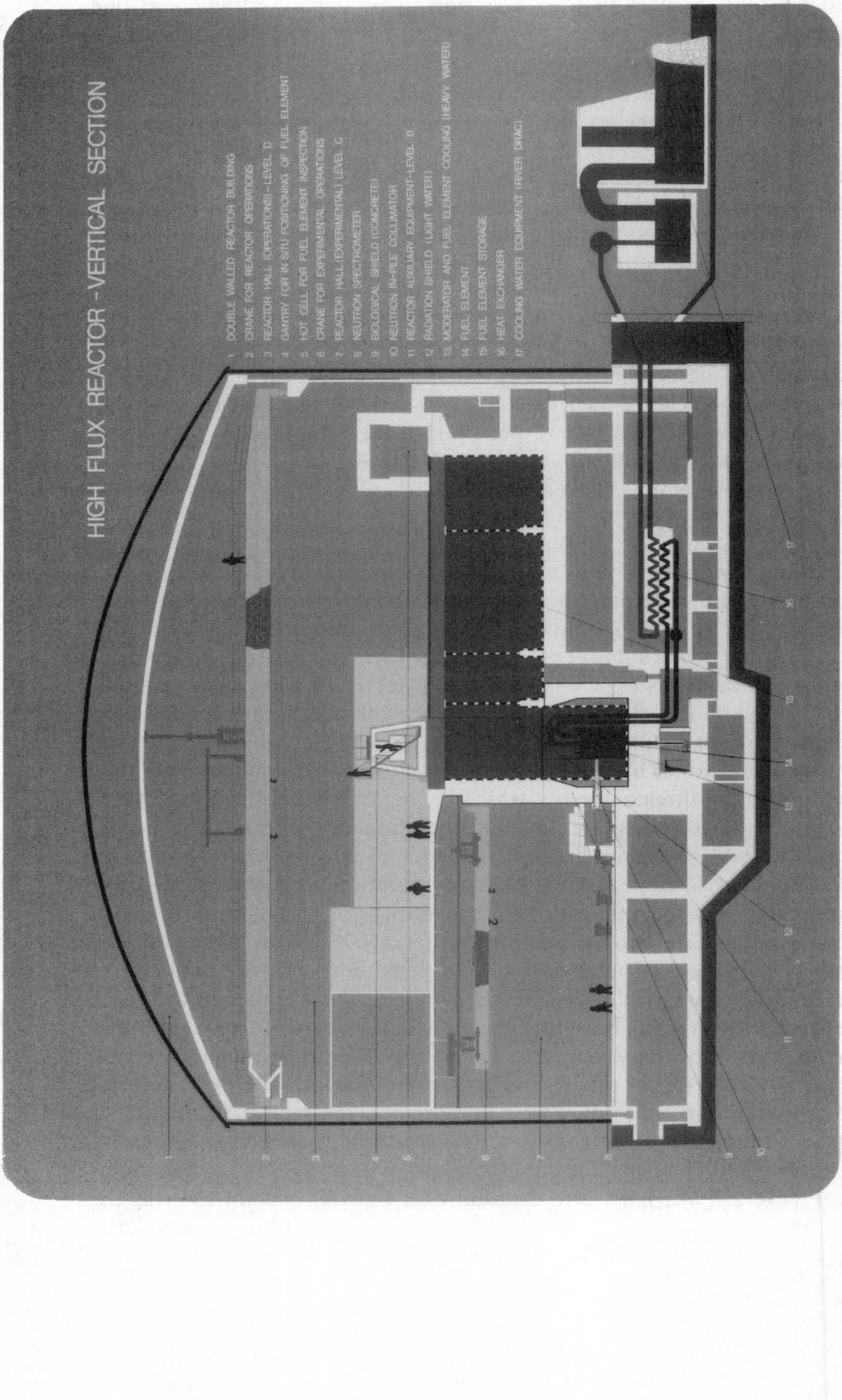

Fig. 2.1 Cut-away diagram of a research reactor at the Institut Laue–Langevin at Grenoble.

essential features while Fig. 2.2 is a photograph of the inside of the experimental hall, i.e. what the experimenter is likely to see when arriving to do an experiment.

Almost all reactors act as continuous sources of neutrons. They start up, run for a period of weeks and then close down for fuel replacement, maintenance (and holidays). Most run 24 hours in 24. In Russia, however, a pulsed reactor has been developed. The fixed core of the so-called IBR reactor is designed to be just subcritical. In the earlier model a second section of fissile material is attached to a large wheel which rotates through the core, bringing the two sections into close proximity for a short period. During this time the reactor goes critical and neutrons are produced. In a more recent version of these pulsed reactors, the core is stationary and a nonfissile reflector rotates, bringing the core into and out of criticality. Since the flux limitation of a steady state reactor is usually the cooling rate, which is related to the average flux, during the short pulse of criticality in these reactors much higher fluxes can be achieved. Clearly, considerable effort has to go into the safe control of such a reactor, since if the wheel stops while the assembly is in the critical state the cooling system (geared to the average power only) is unable to cope. Within the pulse production rates as high as 10^{17} neutrons s^{-1} can be achieved for periods of around 100 μs. As will be seen, for certain inelastic scattering experiments such high intensity pulses may allow considerable gain in the counting rates over steady state sources.

2.2.2 Electron accelerator sources

Electron linear accelerators (linacs) are relatively common and the technology has a solid industrial base. This is because electron linacs have a wide range of users from nuclear physics to medicine. Some of the neutron sources based on linacs are 'parasites', often at nuclear physics installations, while others have been built as dedicated neutron sources, though not necessarily only for neutron scattering experiments. Conversion from high-energy electrons to neutrons takes two stages. Firstly, interactions between the electrons and atoms in a heavy element target produce bremsstrahlung gamma radiation. The gamma rays then interact with the nuclei to produce neutrons by the photo fission reaction. Linacs are pulsed sources, the duration of the neutron pulse being determined by factors such as the target dimensions, as well as the length of the electron burst. The neutrons produced have a very high effective temperature and thus have extremely short wavelengths. For neutron scattering experiments these energies have to be reduced by suitable moderators placed around the primary source. These moderators contain a high concentration of light atoms (water or polythene are typical materials) so that in each collision

Fig. 2.2 Photography of inside of the reactor at the Institut Laue–Langevin at Grenoble.

the neutrons exchange a large amount of energy and eventually emerge with a Maxwellian distribution of energies. Because the moderators are small, the effective temperature of the distribution is much higher than the physical temperature of the moderators. The neutrons are said to be 'undermoderated' and there is still a large component of epithermal neutrons. A typical linac may have peak production rates around 10^{17} neutrons s^{-1} with a mean value of about 10^{14} neutrons s^{-1} or a flux of thermal neutrons around 10^9 neutrons cm^{-2} s^{-1}. Though the mean value is much lower than for even medium flux reactor sources, exploitation of the pulsed nature of the source when designing spectrometers allows the high peak flux to play its part. Electron sources are very flexible since both pulse width and frequency can be tuned to suit particular experiments—unlike the synchrotons described in the next section. On the other hand their background of γ-rays is high due to the bremsstrahlung radiation.

2.2.3 Spallation neutron sources

Unlike the processes in reactor cores and electron linacs the spallation reaction (from the verb to 'spall') does not involve the disintegration of the nuclei in the target. Heavy energetic particles such as 800 MeV protons chip or splinter neutrons from heavy nuclei. The yield is high—around 30 neutrons per proton, but so is the cost of accelerating the protons to these energies. On the other hand, the heat dissipated per neutron produced is quite low—about 20 times less than that in the electron linac.

The high energy protons used for spallation sources can be produced in a number of ways. A linear proton accelerator is in principle adequate, but the cost of achieving very high proton energies escalates very quickly since the accelerator must be extremely long. A proton synchroton achieves these energies more economically, but can reach much higher currents (and therefore neutron fluxes) if the protons are injected at high energies. Several spallation sources therefore use a combination of a proton linear accelerator followed by a proton synchroton. The neutrons produced have extremely high energies which have to be reduced by moderators to suitable values for scattering experiments. Figure 2.3 shows the ISIS spallation neutron source at the Rutherford-Appleton Laboratory. The linear accelerator, the synchroton ring and the proton beam line to the target station can be seen.

As with electron linacs, the methods of acceleration tend to produce short intense bursts of high energy protons, and hence pulses of neutrons. The pulse at ISIS is 4×10^{16} n_f s^{-1} (where n_f means fast neutrons), and the thermal flux after moderation is an order of magnitude brighter than a reactor at its peak. It is possible to create a continuous spallation source by feeding the accelerated protons into a storage ring and deliberately

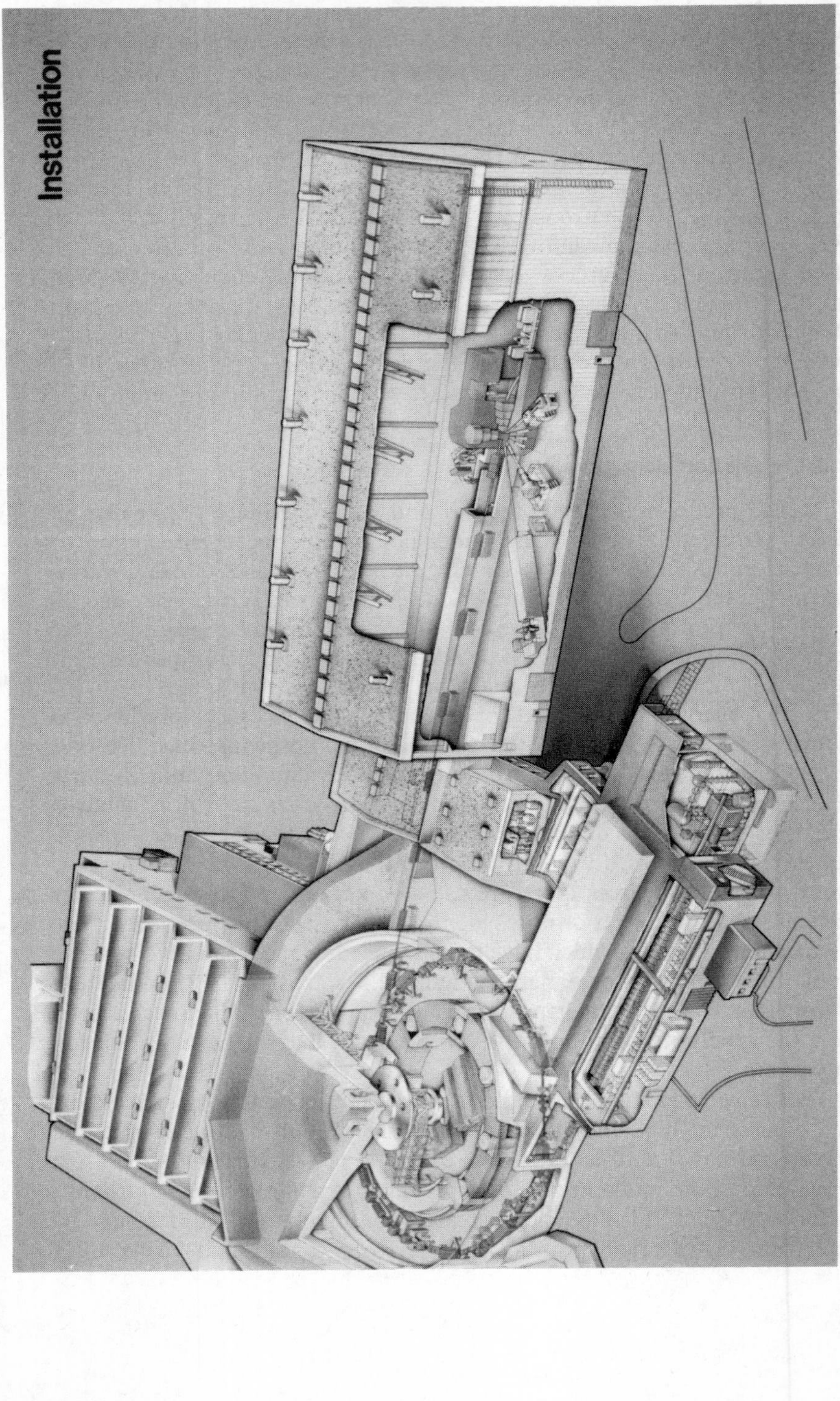

Fig. 2.3 Cut-away diagram of the ISIS. Spallation Neutron Source at Rutherford–Appleton Laboratory.

smearing out the pulse structure. Such a continuous source was proposed by the German scientific community (the Spallation Neutronen Quelle project) but was in the end not constructed. A new accelerator source is being constructed in Switzerland and is planned to be quasicontinuous. It is thus a first approximation only to assume that sources using accelerated particles are pulsed sources while reactors (except in Russia!) are continuous. Later in this chapter the fundamental differences in spectrometer design for pulsed or continuous sources will be discussed.

2.2.4 Relative advantages of different sources

The question frequently arises as to the relative advantages of pulsed or continuous neutron sources. Only by comparing results for the same scientific problem studied by spectrometers designed for the purpose on each type of source can a true comparison be made between them. Nevertheless, it is sometimes necessary to discuss the question in general terms, in order, for example, to plan disposition of national and international resources.

An early discussion of the factors which might need to be taken into consideration was given by Windsor (1981) but at that time there was relatively little experience in running experiments on intense pulsed neutron sources. This experience is now much more extensive, so that it is now possible to outline some more general points on the basis of results obtained. Given the current high level of activity in both designing and carrying out experiments at pulsed sources, even the discussion which follows below is unlikely to remain fully up-to-date for long.

Pulsed sources produce many fewer neutrons than continuous ones, but provided the peak intensity in the pulse is comparable with the steady state flux on the continuous source, and provided this peak flux can be used for as much of the period between the pulses as possible, techniques can be designed at the pulsed source which are competitive with those at the continuous one. To achieve these provisos white beam techniques using time-of-flight analysis had to be developed for the pulsed sources. As a consequence there is much less flexibility in spectrometer design and location (very long guides are required in order to achieve high resolution) and more demands are made on the source, or moderator, specification than for a continuous source such as a reactor. Although the net neutron production is much lower, pulsed sources produce many more high energy, short wavelength neutrons than do reactors. For example the ratio of the intensity in the pulse to the average intensity from the reactor is a factor of 300 for 1 eV neutrons and is still as high as unity for 5 meV neutrons when the ISIS spallation source (Rutherford Laboratory) is compared to the ILL reactor (Grenoble). There is thus a very useful flux of cold neutrons from a pulsed neutron source, and a distinct advantage over the reactor at very short wavelengths.

The debate about whether pulsed sources or continuous ones are better depends on many circumstances (including politics!) and we can do no more in this section than to summarize what appear to be important pros and cons.

1. Pulsed sources are better if high energy neutrons are required.
2. Reactors allow more flexibility in spectrometer design and location.
3. Pulsed sources have a low background between pulses.
4. Reactors have fewer very fast neutrons which can cause shielding problems on pulsed sources.
5. Pulsed source instruments often have fewer moving parts and fixed sample geometry.
6. The short pulses mean ultrahigh resolution is easy to obtain on pulsed sources, but moderate resolution high intensity spectrometers are much easier to design for reactor sources.
7. It seems likely that the targets on pulsed sources have not reached their cooling limits (much lower integrated neutron flux and therefore much lower energy released). Pulsed sources may offer therefore more potential for increased intensity in the future than reactors which are quite close to their cooling limits already.

More often than not, the decision facing an experimenter is not whether to build a pulsed source or a reactor, but simply the best place to carry out a given experiment. Such decisions are frequently made on the basis of non-scientific considerations such as finance or geography (how near are the sources to the home laboratory and who will pay the expenses?) All these factors being equal, however, the net result of the above pros and cons is that reactors are better for long wavelength experiments than short wavelength ones, and pulsed sources are better for short wavelength experiments than for long wavelength ones. There is a crossover value which depends on the power of particular sources under review. For the ISIS spallation source and the ILL reactor this crossover is 5 meV.

2.3. MODERATORS: WAVELENGTHS AND PULSE LENGTH

Several times in the preceding sections, moderators have been mentioned. These are important because the beam emerging from the 'back box' reactor or accelerator target is characteristic both of the type of source and of the moderator. A moderator is composed of material rich in light nuclei. In this material neutrons suffer many collisions, and following Newton's laws, transfer a large fraction of their momentum (losing energy) to the nucleus at each collision. If the moderator is large enough they eventually achieve, approximate equilibrium as a 'gas' at the temperature of the moderator. In a reactor therefore the temperature of the moderator determines the spectrum of thermal energies, and hence (eqn (1.1)) the distribution of wavelengths. As mentioned earlier, for a water-moderated reactor the temperature is around 50°C and the corresponding peak wavelength is about 1.4 Å. This is not ideal for many experiments (especially structural and dynamic studies of large, rather slowly moving macromolecules).

Figure 2.4 shows the wavelength distribution from a thermal reactor.

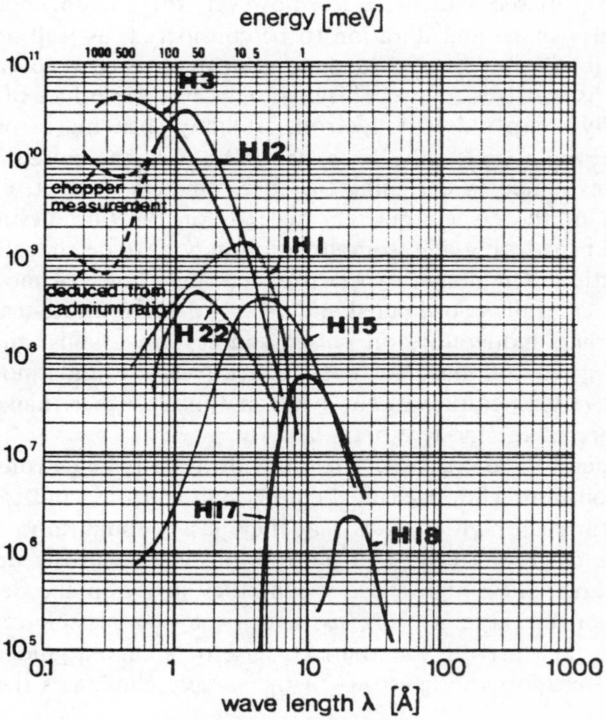

Fig. 2.4 The wavelength distribution in various beams at the Institut Laue-Langevin, Grenoble. H12, 1H1, and H3 are all beam tubes, looking at the thermal core, at a liquid D_2 'cold source', and at graphite 'hot source' respectively. H22, H15, H17, and H18 are all curved guides with radii of curvature 27 000, 2700, 150, and 25 m respectively. H22 looks at the thermal core, the other three at the cold source.

As can be seen the flux of neutrons at a modest 5 Å is already two orders of magnitude less than the peak and beyond this the decrease is very rapid. The solution is to provide a local 'cold spot' in the moderator—a cold source. In practice this is around 20 l of liquid hydrogen or deuterium at 25 K. A new Maxwellian distribution around 25 K results with a peak wavelength around 6 Å and a considerable enhancement in intensity of long wavelength neutrons compared to the thermal moderator. The effect of a cold source on the wavelength distribution is also shown in Fig. 2.4. For certain experiments enhanced fluxes at very short wavelengths are desirable, and a corresponding 'hot spot' is included. The effect of such a hot source (a 20 cm cube of graphite at 2400 K) on the wavelength is included in Fig. 2.4.

Moderators are also an essential part of the production of useful neutron beams from pulsed sources. Here, however, there is an important effect on the pulse shape and duration to be considered, as well as the desired wavelength distribution. Reduction in neutron energy to produce long wavelengths involves many collisions. This has the effect of lengthening the initially sharp pulse of neutrons, roughly speaking in proportion to the lengthening wavelength. Since, as will be seen, the pulsed nature of the source is exploited in designing the spectrometers using it, and since the resolution of these spectrometers depends on the time definition of the pulse, the broadening due to moderation may have detrimental effects on the resolution. The design of spectrometer and choice of moderator have to be very carefully coordinated and usually a compromise made. As mentioned earlier, moderators on pulsed sources are usually small, in order to obtain good coupling, so that the neutrons are undermoderated, i.e. the effective temperature of the neutron flux is higher than the physical temperature of the moderator.

No moderators are perfect. There are always a few neutrons that escape without collision. They are called epithermal neutrons, and, since they are travelling at such high energies that they pass through most samples, the main problem they cause is in background counts in the detector. In a pulsed source, their high speed means they arrive in the detector much sooner than the thermal neutrons, and it may be possible to stop counting during their arrival. Alternatively, use of a curved guide (see Section 2.4.1) to transport the neutrons to the sample eliminates them from the beam.

2.4 NEUTRON TRANSPORTATION

2.4.1 Guides

The maximum flux of a thermal reactor is just outside the core and beam tubes penetrating through the 3 m or so of shielding to this point are effectively holes which allow neutrons from the source to arrive at a spectrometer. Since the neutron flux per square centimetre will decrease as the square of the distance from the source it is obviously advantageous to construct the spectrometers as close to the outside of the shielding as possible. Figures 2.1 and 2.2 show how crowded this region becomes. It is clearly a severe constraint on the number of spectrometers which can be built around a given source. Both reactors and pulsed sources solve this crowding problem by using neutron guides. Although proposed quite early (Maier-Leibnitz and Springer 1963) their full scale exploitation was first demonstrated when the high flux reactor at the Institut Laue–Langevin was built in the early seventies. Neutron guides are analogous to light guides. Pipes—usually rectangular in shape and several centimetres in dimension are coated internally with a suitable reflecting material. 'Suitable' means such that neutrons will be totally internally reflected just as light is when it impinges at angles less than the critical angle on a medium with higher refractive index—for example travelling from glass to air the critical angle for light is about 45°. Neutron refractive indices are calculated from the nuclear scattering lengths introduced in Chapters 1 and 4. For non-absorbing materials, the refractive index can be expressed as

$$\nu = 1 - \lambda^2 \rho_b / 2\pi \tag{2.1}$$

(this expression will be derived and discussed in detail in Chapter 10).

ρ_b is the scattering length density. Typical refractive indices are very close to, and usually slightly less than unity, $1 - \nu$ being of order 10^{-6} so

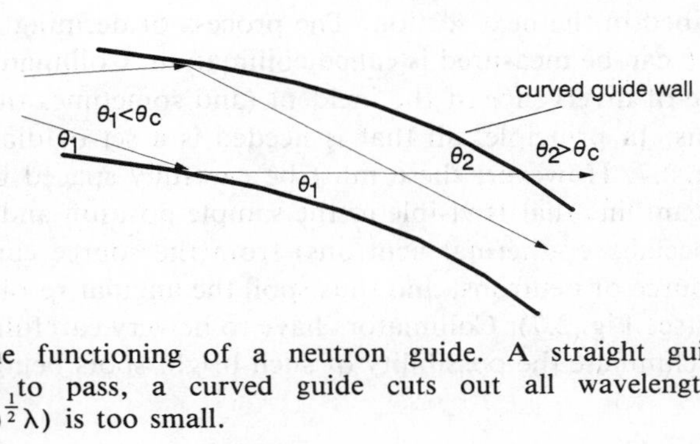

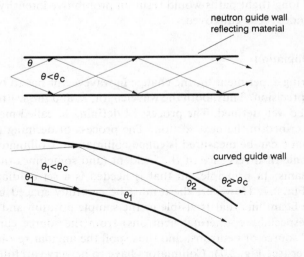

Fig. 2.5 The functioning of a neutron guide. A straight guide allows all wavelengths to pass, a curved guide cuts out all wavelengths for which $\theta_c (= (\rho_b/\pi)^{\frac{1}{2}} \lambda)$ is too small.

that the critical angle is very small. ν for air is effectively unity and neutrons are totally externally reflected at close to glancing angles. A common material for neutron guides is nickel-coated glass with critical angle $1.73 \times 10^{-3}\lambda$ mrad (i.e. for 10 Å neutrons, $\theta_c \simeq 0.5°$. Guide surfaces must be very flat, and the guides themselves very well aligned). The 'acceptance angle' at the entrance to a guide is fairly small, and different for each wavelength. Equally, the beam divergence at the the guide exit is also wavelength dependent. Figure 2.5 demonstrates diagrammatically the functioning of a neutron guide.

The primary function of guides at reactors is to increase the possibilities of exploitation by allowing more spectrometers to 'see' the reactor. Figure 2.6 shows the large number of spectrometers which can be accommodated at the high flux reactor at the Institut Laue–Langevin in the adjacent guide hall. There are however other advantages arising from the use of guides. They allow spectrometers to be constructed far from the reactor in regions where the background radiation is low. They can be curved, thus further spreading the beams and relaxing spatial constraints on spectrometers, and, by changing angles of incidence, cutting off short wavelength neutrons since θ_c depends on λ (see Fig. 2.5). The flux for H15, H17, and H18 in Fig. 2.4 show progressively more of the short wavelengths are lost as the guides become more curved. In spectrometers on pulsed sources, resolution in wavelength often depends on measuring the velocity (or flight time) of the neutrons. Since different wavelength neutrons in the initial sharp pulse will become more spread out in time the further the pulse travels, the high resolution spectrometers at pulsed sources are often designed to be very long. Such long flight paths would result in prohibitive intensity loss unless neutron guides were employed.

2.4.2 Collimation

In a scattering experiment, $\boldsymbol{q}$, the change in wavevector has to be measured with good precision. Thus, both the wavelength, λ, and the scattering angle, θ, have to be well defined. The process of defining λ, called monochromation, is described in the next section. The process of defining the neutron beam so that θ can be measured is called collimation. Collimators limit the size and angle of divergence of the incident (and sometimes the scattered) neutron beams. In principle, all that is needed is a set of diaphragms as shown in Fig. 2.7. However, these must be carefully spaced because any part of the beam line that is visible to the sample position and exposed to neutrons (especially epithermal neutrons) from the source can become a 'secondary' source of neutrons, and thus spoil the angular resolution in the spectrometer (see Fig. 2.7). Collimators have to be very carefully designed, therefore, to eliminate the possibility of such bright spots being seen from

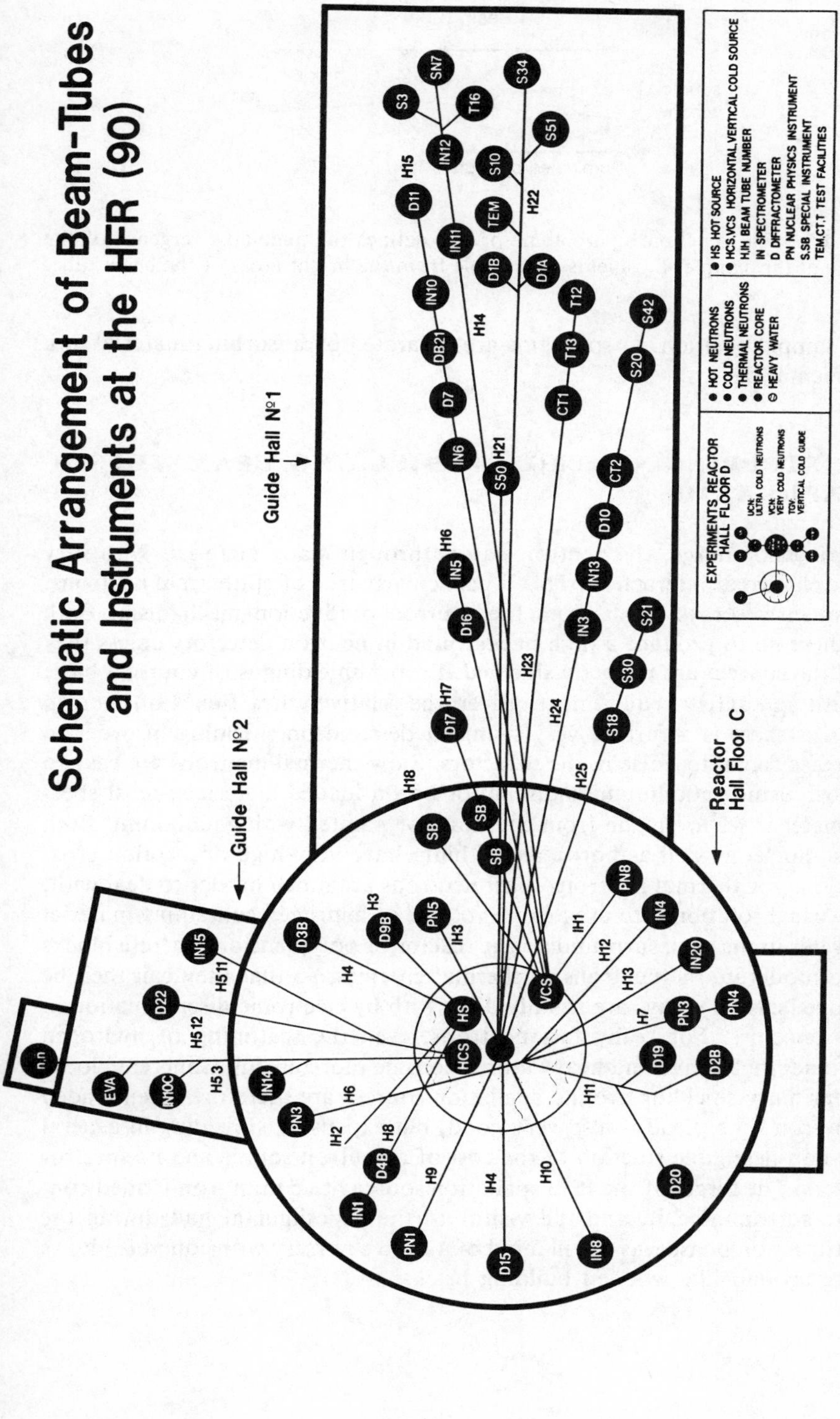

Fig. 2.6 Diagram of the arrangement of the instruments in the neutron guide hall at the Institut Laue–Langevin, Grenoble.

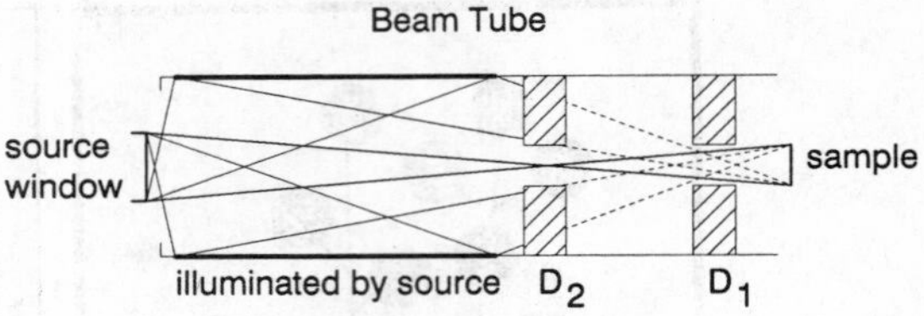

Fig. 2.7 The function of a collimator. D_1 defines the angular divergence of the beam at the sample. D_2 shields the sample from the bright areas of the beam tube.

the sample position. Diaphragms are separated by absorbing material in a segmental design.

2.5 STOPPING NEUTRONS–SHIELDING, BEAMSTOPS, DIAPHRAGMS

Bearing no charge, the neutron passes through many materials with very little chance of interaction. This is particularly true of epithermal neutrons. Frequently γ-rays result from the neutron production mechanism. Both of these go to produce a high background in neutron detectors unless they and the sources are properly shielded. Proper shielding is of course a basic health and safety requirement. Given the relatively low fluxes of neutron sources there is a further very stringent demand on shielding in order to increase signal to noise in the detectors. Slow thermal neutrons are easy to absorb using gadolinium, cadmium or boron-loaded materials: most spectrometer irises are made from cadmium or painted with gadolinium. Both these nuclei as well as boron and lithium have very high absorption cross sections for thermal neutrons. Fast neutrons are much harder to deal with. The usual solution is to use a large volume of hydrogen containing material –water in the tank surrounding the reactor or polythene or concrete blocks –to moderate the neutrons to thermal energies, so that they can then be stopped easily. γ-rays are usually dealt with by electronic discrimination in the detectors. For really fast neutrons, even the scattering of hydrogen becomes ineffective and heavy atoms become more useful. Concrete blocks for primary shielding around spallation sources are therefore often loaded with iron shot. Lead is also very good, but expensive. Shielding in general is a non-negligible fraction of the cost of a neutron source and its spectrometers. The target of the ISIS spallation source has 5 m of iron-loaded concrete surrounding it, and the visitor to the experimental hall during the construction phase was put in mind of a giant's nursery with concrete blocks lying around like wooden building bricks.

2.6 MONOCHROMATION – CHOOSING WAVELENGTHS

The term monochromation refers to the selection of a particular wavelength from the spread of wavelengths or 'white' beam produced by the source. For the original reactor sources this process was essential if the parameters, q, and ΔE were to be clearly defined in the spectrometers. Pulsed sources give the option of using the whole of (or a wide spread from) this white beam and measuring the wavelength distribution carefully.

First, wavelength selection; there are basically only two ways of choosing a given wavelength. The first uses the wave nature of the neutron via Bragg diffraction from a suitable crystal. The second uses the particle nature and mechanical choppers to select a narrow velocity range.

2.6.1 Crystal monochromators

Crystals for selecting thermal wavelengths must have not only a suitable crystal lattice spacing (in eqn (1.2) λ and d must be matched to given a reasonable value for θ) but also high reflectivity and a good mosaic spread. This last property refers to how much of the crystal lattice is slightly misaligned. This misalignment is important if reasonable beam intensities are to be diffracted. If the crystal were perfect, all the neutrons of the Bragg wavelength which were diffracted in one region of the crystal would have exactly the correct wavelength to be diffracted again by an adjacent region and would finish travelling back along the beam tube. This process is called extinction. By choosing crystals in which blocks or 'mosaics' are perfect but each very slightly misaligned, extinction is avoided, but of course the spread in wavelength diffracted at a given angle is increased (Fig. 2.8). Obviously wavelength resolution is decreased, but the intensity is increased. Choice of the mosaic spread in monochromator crystals is a balancing act between requirements of resolution and of intensity. The nuclei in the chosen crystals need to have a high coherent cross section and negligible absorption and incoherent cross sections. Metallic crystals have been found not only to satisfy these cross section requirements but also to have mosaic spreads of the correct order which can be controlled while the crystals are grown. Typical materials are copper, for short wavelengths, germanium or aluminium, for thermal neutrons, and pyrolytic graphite for longer wavelengths. One limitation of the use of crystal monochromators is the absence of suitable materials with long lattice spacings. The longest wavelength diffracted from pyrolytic graphite is 4.8 Å at 45°.

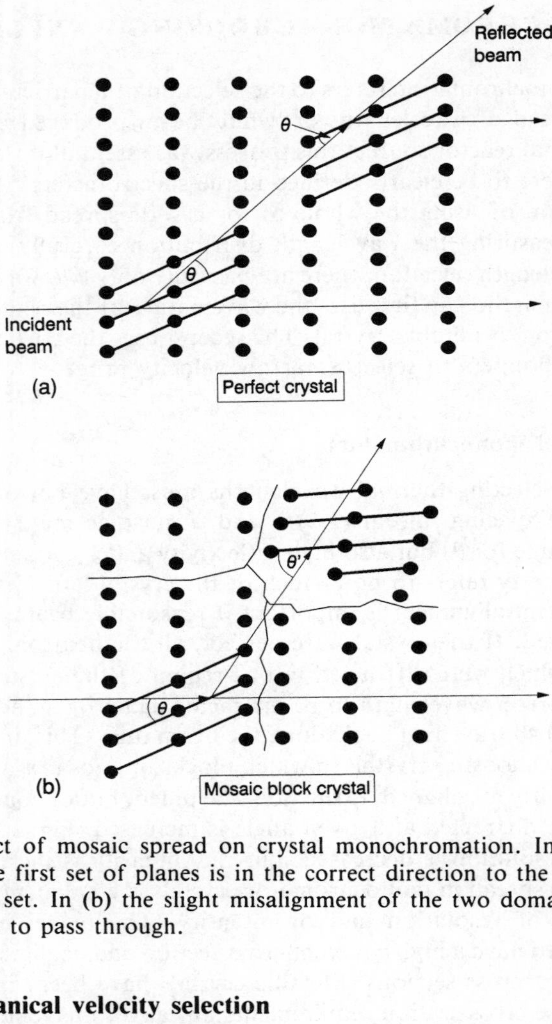

Fig. 2.8 Effect of mosaic spread on crystal monochromation. In (a) the beam reflected in the first set of planes is in the correct direction to the reflected back by the second set. In (b) the slight misalignment of the two domains allows the reflected beam to pass through.

2.6.2 Mechanical velocity selection

Thermal neutrons travel relatively slowly. The velocity of a 6 Å neutron is about 600 ms^{-1} or 0.6 metre per millisecond. Such velocities lend themselves to mechanical selection devices. An example is shown in Fig. 2.9(a). Two gates which open periodically are arranged sequentially. The first allows a sharp polychromatic pulse of neutrons to pass. Neutrons with different velocities arrive at the second gate at different times, so that by suitable phasing of the opening and shutting of the gates a narrow wavelength *pulse* of neutrons is produced. The opening and shutting is conveniently arranged by making each gate a transparent window on a rotating

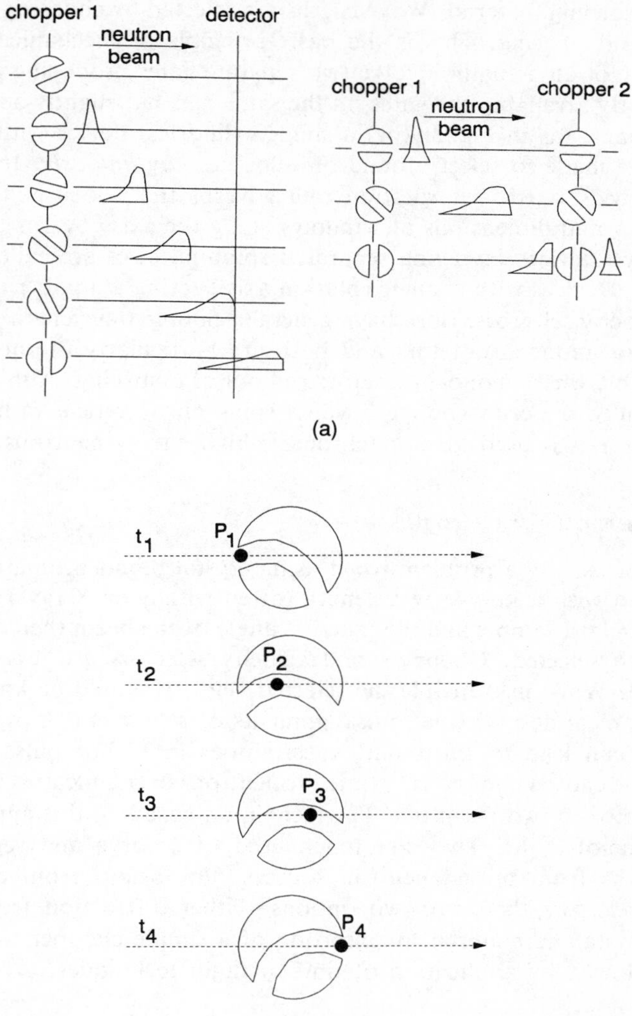

Fig. 2.9 (a) The function of mechanical velocity selection by phased choppers. The second chopper selects a narrow band of velocities from the distribution transmitted by the first. (b) A single 'helical slot' velocity selector. Only those neutrons travelling to be at P 1, 2, 3, 4 at t, 1, 2, 3, 4 pass through the chopper.

disc of absorbing material. Wavelengths are selected by choosing speeds of rotation and phasing. This is the basic principle of mechanical velocity selectors. Not all actually involve two separate gates producing a pulse of neutrons. By arranging the gates on the same axis but slightly advanced in rotation angle (as in Fig. 2.9(b)) a single cylindrical device with a helical slot can be made to select a band of velocities *continuously*. In this case the phasing is fixed, but rotation rate governs the velocity chosen, and the number and dimensions of windows along the axis governs the spread in velocity. Devices involving separated spinning discs are called 'phased choppers'. Devices with a single rotation axis bearing a number of slits are called velocity selectors. Both have generally poorer wavelength resolution than crystal monochromators and both are particularly useful at longer wavelengths, where monochromators run out of *d*-spacing. A third type of device, called a Fermi chopper, which spins on a vertical rather than a horizontal axis is used to monochromate high energy neutrons.

2.6.3 Measuring wavelengths

This is not usually a problem when using crystal monochromators. These are aligned with a known wavelength (often with γ or X-rays) calibrated with a standard sample and the take-off angle of the beam then defines the wavelength selected. Choppers and velocity selectors are less obviously calibrated. While in principle the selected velocity should be known from the mechanical design, small misalignments of selectors or misphasing of choppers can lead to shifts and uncertainties in λ. For pulsed chopper systems the chosen method is to time the neutrons over a measured distance, usually between two monitors. The techniques based on this approach are called time-of-flight. They are much used to observe and separate the wavelengths from pulsed neutron sources. For beams from continuous velocity selectors there are two options—either diffraction from a well-known crystalline material, or insertion of a simple chopper to pulse the beam followed by application of time-of-flight techniques.

2.7 DETECTING AND MONITORING NEUTRON BEAMS

Neutron detection involves a process of absorption by a suitable nucleus followed by detection of the charged particles produced by these events. The most important reactions involve the light nuclei ^{3}He, ^{6}Li, and ^{10}B.

(a) $^3\text{He} + \text{n} \rightarrow {}^3\text{H} + {}^1\text{H} + 0.77\,\text{MeV}$
in which a neutron is absorbed by helium to produce a proton and a tritium nucleus.

(b) $^{10}B + n \rightarrow {}^{7}Li + {}^{4}He + 2.3\,MeV$
in which a neutron is absorbed by boron to produce lithium and helium.

(c) $^{6}Li + n \rightarrow {}^{4}He + {}^{3}He + 4.79\,MeV$
in which the neutron meets a lithium nucleus and produces two helium nuclei.

Two things are obvious from the above nuclear reactions—the atoms produced are ionized, and they are produced with considerable energy in exothermic reactions. It is these fast ions which are used for the actual detection process. In gas detectors they produce a trail of secondary ionization which travels towards an electrode and is detected as an electronic pulse. Typical gas detectors contain ^{10}B enriched BF_3 or ^{3}He-enriched helium gas. Conventional gas detectors are cylindrical in shape with diameters of order of a few centimetres and lengths from 10 to a few tens of centimetres. The collecting anode runs along the central axis. ^{3}He has a higher absorption cross section than ^{10}B and higher pressures can be used at moderate voltages so that the ^{3}He detectors can be made more efficient and compact. They are also more expensive! Scintillation detectors are in principle even more compact. Such a detector uses a solid, and therefore much denser, neutron absorber—typically ^{6}Li in a glass containing a phosphor such as ZnS (Ag). The phosphor emits a flash of light as the ionizing particles produced by the (n, α) reaction pass through. The light is detected by photomultipliers. In principle the light flash is very intense so that high counting efficiency can be achieved, but careful optical coupling to the photomultiplier is required.

An efficient detector is required to react to as many of the neutrons impinging on it as possible. By contrast a monitor, placed in the direct beam falling on the sample, should be able to react to a very few neutrons but in a way that is directly proportional to the total beam intensity. A typical monitor relies on the reaction of neutrons with ^{235}U to give a nuclear fission. A layer of a few atoms of ^{235}U on one glass surface of a flat, gas containing detector produces enough ionizing products to be easily detected.

While there have been notable improvements in the efficiency of detector design, the major advances were associated with the development during the 1970s of large, area detectors. Clearly an 'in principle' neutron experiment can be carried out by moving a single detector to each scattering angle, counting and then moving on to the next angle. This was done in the early days of neutron scattering and is very slow. Arranging banks of detectors to cover large ranges of scattering angles improves counting times enormously. The size of the detector housing, however, still leaves large dead spots and limits the possible resolution. Area detectors avoid both these problems and replace them by problems purely of design, cost, and data handling! An area detector based on glass scintillators is a relatively simple prospect if a light guide leads from each small area of the detector to a dedicated photomultiplier. For covering a square metre with a resolution of 1 cm^2 this requires 10^4 multipliers! A solution is to use a pattern of three light guides from each area, leading in bunches to many fewer photomultipliers, and then to search electronically for threefold coincidences to pinpoint the neutron arrival position. In this way 10 000 elements can be coded using about 100 phototubes.

Gas area detectors usually depend on coincidences in two sets of cathode wires forming a 2-D grid. This grid could be Cartesian (x and y) or radial (circles and spokes). Gas-filled area detectors for neutrons have typical resolution of 1 cm and areas up to a square metre. An alternative to detection of coincidences is to arrange that different positions of arrival of the ion pulse on a detection wire give rise to recognizable differences in the electronic pulses detected. If the wire is resistive, then the amount of charge arriving at each end will depend on the distance from the end in a direct ratio to the resistance (i.e. the length) between the detection point and the end. If the total charge is Q and the position of detection x, then the ratio of charge collected at the two ends is Q/x: $Q/(1 - x)$ and hence x can be determined. This technique is most usually used for linear detectors, but it has been adapted to area detection by folding a single wire in a zigzag to cover the required area. An adaptation of the process to use the arrival times at each end of the wire rather than the charge levels has also been produced.

2.8 POLARIZED NEUTRON BEAMS

One of the techniques described in Chapter 3 and used for observing the slow motion of polymer molecules relies on following what happens to the neutron spin during the scattering process. Understanding this technique therefore requires an introduction to methods of handling polarized neutron beams. Moreover, when questions arise about separating contributions from coherent and incoherent scattering, polarization analysis is potentially invaluable, though not as yet widely applied. In this section therefore a brief introduction is given to methods of preparing polarized beams, and of analysing the net remaining polarization after scattering.

In a polarized beam, the neutron magnetic moments are preferentially aligned. Since the spin is 1/2, there are only two directions the spin can take with respect to a given direction – up or down (+ or −). Polarization works by *selecting* one of these directions and rejecting the other. This, at its most efficient, reduces the flux in a 100 per cent polarized beam by a factor of two. If the fraction of neutrons with polarization in the (+) state is f, then the degree of polarization of the beam $P_z = 2f - 1$ (so if 100 per cent are in the (+) state the beam is perfectly polarized and, $P_z = 1$ while if 50 per cent are in the (+) state and 50 per cent in the (−) state it is unpolarized and $P_z = 0$).

Polarized spectrometers will involve some or all of the following components:

1. A method of polarizing the beam by selecting (or rejecting) half of the spin states.
2. A magnetic field to define a z-direction and preserve the polarization once selected.
3. A method for changing the spin direction—usually exploiting the Larmor precession of the spins about a magnetic field. The Larmor precession frequency $\omega_L = \gamma H$ where γ is the magnetic moment and H is the field. γ is 1.83×10^4 rad s^{-1} Oe^{-1} and for a small field of 10 Oe the frequency is 10^6 Hz. The spin is reversed in 5×10^{-5} s, i.e. for a 5 Å neutron travelling at 500 ms^{-1} the spins are reversed over a distance of about 4 cm.
4. An analyser system to determine polarization after scattering by the sample.

Polarizers and analysers may be either magnetic crystals which diffract only one polarization state, reflecting surfaces or 'supermirrors' which again reflect only one state, or filters in which one polarization state is either resonantly absorbed or preferentially scattered. Typical polarizing crystals are Co_{92} Fe_8 (the (2 0 0) reflection with spacing 1.7725 Å) or the so-called Heusler alloys Cu_2 Mn Al (the (1 1 1) reflection with spacing 3.9864 Å). The problem with this method is that it necessarily monochromates the beam—which is undesirable for applications on pulsed sources where wide wavelength bands are employed. Moreover the range of wavelengths available may not be suitable.

The use of scattering or absorbing filters suffers from problems in achieving high polarization together with high transmission and also involves cryogenic operations. The best absorbing filter material is samarium-149. For best effect it has to be diluted with other nuclei, cooled to 0.15 K and polarized. Under these conditions high polarization of short wavelength neutrons can be achieved with a transmission around 20 per cent. For scattering filters the practical problems are even greater. The best scattering filters use polarized protons: neutrons with spins parallel to the polarization of the filter suffer much less scattering than those with antiparallel spins. Polarizing the protons involves working at 1 K and transferring energy to the protons via electrons! Needless to say at the present time such proton scattering filters are not in wide application, though simpler, cheaper versions such as magnetized iron can produce $P_z \sim 0.5$ at wavelengths around 3.6 Å.

The polarization method that seems to overcome nearly all these problems is the polarizing reflector or 'supermirror'. In the earlier section on guide tubes it was mentioned that total reflection of the neutron beam occurs for angles less than the critical angle where $\theta_c = (\rho_b/\pi)^{\frac{1}{2}}$. If the surface material is a ferromagnet then the nuclear scattering length b is increased to $b + p$ where p is the magnetic scattering length. There are now two critical angles, and in the range of angles between the old and new critical angles, i.e.

$$\left(\frac{\rho_b}{\pi}\right)^{\frac{1}{2}} \lambda < \theta < \left(\frac{\rho_{b+p}}{\pi}\right)^{\frac{1}{2}} \lambda$$

only one spin component in the neutron beam is reflected. The critical angle is very small so the beam divergences tend to be small. This is overcome firstly by depositing the magnetic layer onto thin plastic films, which can be curved to increase illumination. The supermirrors consist of many layers of magnetic and non-magnetic material alternatively laid down in an artificial crystal with a layer thickness decreasing as $n^{\frac{1}{4}}$ where n is the number order of the deposited layers.

2.9 SPECTROMETERS

The scientist arriving at a neutron spectrometer is likely to be faced by large blocks of (possibly attractively painted) shielding and a control room containing computers and racks of electronics. Rarely may he or she actually see all the parts of the spectrometer. It is helpful therefore to have in mind what the computer and electronics are actually controlling. Various spectrometers will be described in detail in Chapter 3. In order to summarize the information in this chapter, one will be described schematically here.

Figure 2.10 shows diagrammatically a typical reactor spectrometer for observing elastic scattering at small angles. (Small angle neutron scattering is one of the most important tools for studying macromolecules.) The beam tube is directed at a cold source, allowing long wavelength neutrons to be used. From within the beam port a section of guide tube transports the neutrons with low loss to a region with more space and low background. The neutron wavelength is selected using a velocity selector. The control of this is likely to be via the computer, or perhaps a separate unit nearby in the control room. Next there is a monitor which will certainly be read into the electronic array and recorded in the computer. It will be used for checks on sample transmission as well as most importantly for controlling

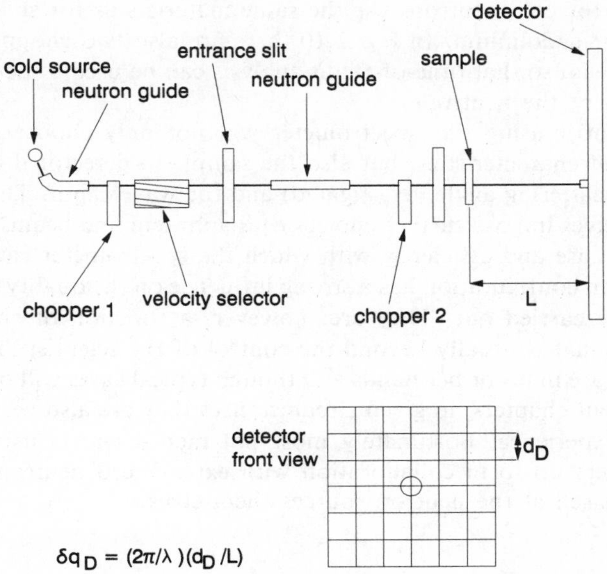

Fig. 2.10 Typical spectrometer for a reactor source.

the length of experiments. Neutron sources may not provide constant flux over short periods so that time is not always a sensible control parameter. Total neutron intensity, monitored continuously gives the best method of control. The next stage of the spectrometer is the collimator. In this spectrometer it is just provided by a diaphragm immediately before the sample position. The diverging beam from the end of the last guide section is cut down to a narrow angle by this diaphragm. For this spectrometer the sample position is fixed. The divergence is changed via the end position of the guide which is varied by moving in and out movable sections. The divergence is chosen so that the spot falling on the detector, also movable, is small enough for the required spatial resolution. The sample may be heated or cooled and surrounded by a controlled atmosphere or vacuum. Control of these parameters may also be via the computer. The long flight paths are evacuated so there have to be neutron transparent windows at several points along the spectrometer. This evacuation is essential for such a long (up to 80 m) spectrometer or the scattering and absorption by the air would remove most neutrons before they arrived at the detector. Finally there is the area detector whose output will be collected via a rack of electronics into the computer. In front of the detector there will be a beam-stop to protect it from the full intensity of the unscattered beam. Typical beamstops for cold neutrons use the same materials as for shielding—i.e. cadmium or gadolinium. In Fig. 2.10 there are also two choppers used to pulse the beam so that time-of-flight analysis can be used to determine the wavelength of the neutrons.

The scientist using this spectrometer will not only choose the sample environment characteristics, but also the sample to detector distance (and hence the scattering angle investigated) and the wavelength. These choices will themselves impose further choices of diaphragm and beamstop dimensions. The ease and efficiency with which the spectrometer can be set up in the chosen configuration has a strong influence on the quality of research that can be carried out. They are, however, a function of spectrometer design and that is usually beyond the control of the scientist. The choices themselves are in his or her hands and though typical cases will be discussed in subsequent chapters, in given circumstances they are also very much the result of experience. Fortunately most polymer scientists using neutron spectrometers do so in collaboration with experienced neutron scattering scientists based at the neutron sources themselves.

SUGGESTED FURTHER READING

Egelstaff (1965).
Willis (1973).
Windsor (1981).

3

Spectrometers and what they measure

3.1 INTRODUCTION

Any reader who has just ploughed through Chapter 2 and now reads the title of this chapter could be forgiven for wondering when we are going to get to the 'meat' and discuss scattering from polymers. It is true that the previous chapter could have its place in any work on neutron scattering. On the other hand, omission of the material it contains could result in an irritating lack of completeness. In this chapter we will narrow the field of view and concentrate attention on the relatively few types of spectrometer with significant numbers of applications to polymer systems. Without this limitation, the sheer variety of neutron sources and applications would lead to a very extensive volume.

All experiments are attempts to measure and to interpret, for a sample of interest, the scattering laws to be introduced in Chapter 4. While in principle a careful measurement of $S(\boldsymbol{q}, \omega)$ for all values $\boldsymbol{q}$ and ω contains all the information required, in practice resolution requirements have meant that compromises have to be made. For example, energy changes are not measured if high resolution structural information is required via $S(\boldsymbol{q})$ and spatial information is often sacrificed for high energy resolution, especially from amorphous materials or solutions. Even from the most powerful sources the neutron beams are relatively feeble, and the spectrometers are designed to maximize the illumination of the particular problem in hand. The different types of spectrometer are classified below in order of complexity. Some will later be described in further detail because they have important polymer applications. The experiments all depend on preparing samples with suitable cross sections for neutrons. In Chapter 1 we showed (Fig. 1.2) how the neutron scattering lengths, b, vary through the periodic table. In Table 3.1 are listed values of b, and the various cross sections for elements commonly found in polymeric samples. For much more comprehensive tables the reader is referred to the literature (for example Bée 1988) or the original lists in Koester *et al.* (1991) or Sears (1992)).

Table 3.1 Scattering lengths and cross section for some common isotopes

	$b \times 10^{12}$ (cm)	$\sigma_{coh} \times 10^{24}$ (cm^2)	$\sigma_{inc} \times 10^{24}$ (cm^2)	$\sigma_{abs} \times 10^{24}$ (cm^2 at 1.798 Å)
^{1}H	−0.374	1.76	79.7	0.33
^{2}D	0.667	5.59	2	0.0005
^{12}C	0.665	5.55	0	0.003
^{14}N	0.94	11.01	0.5	1.9
$^{16}0$	0.58	4.23	0	0.0001
^{19}F	0.57	4.02	0	0.0096
$^{ave\ 28.06}Si$	0.415	2.16	0	0.17
^{32}S	0.28	1.02	0	0.53
$^{ave\ 35.5}Cl$	0.96	11.53	5.3	33.5

3.1.1 Diffractometers for liquids and amorphous materials

In order to observe $S(\boldsymbol{q})$, the scattered intensity must be measured as a function of a defined wavelength and scattering angle. In reactor instruments it is normal to select λ with a monochromator, and to vary θ. In principle, a single detector, shifted sequentially to different angles, would suffice, but in order to increase the counting rate it is usual to include an array of detectors or a large position sensitive detector. For pulsed sources it would be sufficient to place a single detector at a fixed angle, and allow a broad wavelength band pulse to fall on the sample. Subsequent time-of-flight analysis defines the wavelengths and hence $\boldsymbol{q}$ values. In practice such spectrometers often combine the wavelength scan with an angular scan by incorporating multidetectors as well, and thus increase the count rate dramatically. In Fig. 3.1 the design of typical spectrometers of this type for a reactor source and for a pulsed source are compared.

Amorphous or liquid samples usually have no inherent orientation and $S(\boldsymbol{q})$ is a rather slowly varying function of $\boldsymbol{q}$, so that high $\boldsymbol{q}$ resolution is not required. Small angle scattering which falls in this general category of techniques poses special resolution demands by requiring very small $\boldsymbol{q}$ values. Since $\boldsymbol{q} = (4\pi/\lambda)(\sin(\theta/2))$ and since λ is rarely available with values longer than about 15 Å, a value of $\boldsymbol{q}$ of 0.01 Å requires $\theta < 1.5°$. Thus specially designed detectors are required since, while $\Delta\boldsymbol{q}/\boldsymbol{q}$ of about 10 per cent is satisfactory, the absolute value of $\Delta\boldsymbol{q}$ has now become very small.

The overwhelming importance of *small angle neutron scattering* for studying polymer systems justifies describing in some detail in the next section the design and use of the relevant spectrometers. Although there are applications for classical diffractometers in polymeric systems in the

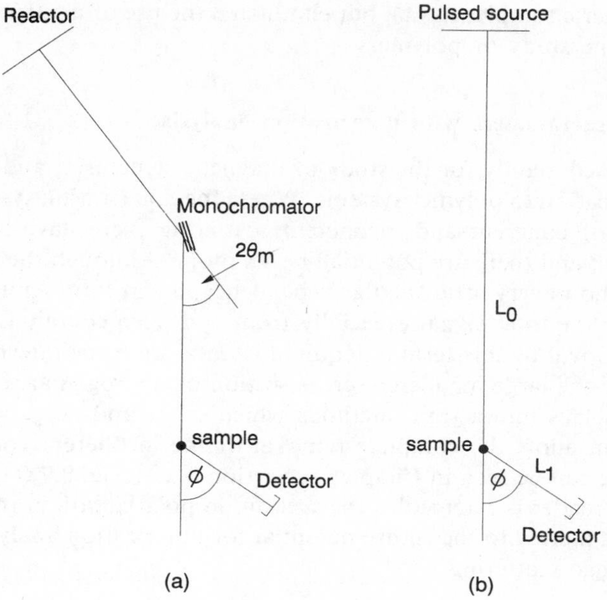

Fig. 3.1 Comparison of spectrometer design on a pulsed or a continuous source for wide angle elastic scattering. (a) The incident wavelength is defined by the angle of diffraction θ_m at the crystal monochromator. (b) The incident wavelength is not selected but is measured from the flight time to the detector over the flight path $L_0 + L_1$. A 'white' pulse falls on the sample.

study of local structures, these applications are much more limited in scope and, we believe, can be understood as an extension of the principles learned in small angle scattering. No further detail of their design will be included.

Another subclass of diffractometer is the recently developed family of *neutron reflectometers*. Since these have a fast expanding range of applications in the study of surfaces and interfaces, these will be described in detail in a subsequent section.

3.1.2 Diffractometers for crystalline powders

These differ from the previous class by requiring much higher $\boldsymbol{q}$-resolution. The wavelength and the scattering angle have to be more carefully defined, with a consequent drop in flux which is compensated for by the very high intensity in many of the diffraction peaks. The potential applications to polymeric systems are already covered by the more widely available X-ray techniques, and this factor, together with the relatively low crystallinity of most polymeric samples has all but eliminated the use of neutron crystallography in the study of polymers.

3.1.3 Diffractometers with polarization analysis

These are used mainly for the study of magnetic structures, and are not of direct application to polymer systems. Where the polarization system allows separation of coherent and incoherent scattering there have been a few applications, and there are potentially even more. Although the incoherent scattering shows very little angular dependence, it can form an intense contribution to the total signal especially from hydrogen containing samples. Correct removal of this term is required if data are to be interpreted properly. Since the large incoherent cross section of hydrogen arises from the neutron–nucleus interaction, methods which select and keep track of the neutron spin allow the complete removal of the incoherent contribution. An example can be seen in Chapter 8, Section 8.2.1, Fig. 8.7 (Gabryś *et al.* 1986). The reader is referred to the section on polarization in the previous chapter and alerted to the future potential for polarization analysis options in small angle scattering.

3.1.4 Single crystal diffractometers

More complex geometry of the relative positions of monochromator, sample, and detector now has to be accessible since sample orientation as well as scattering angle has to be done independently. Given the impossibility of growing larger polymer single crystals, it is hardly surprising that such diffractometers do not feature in applications of neutron scattering to polymers. The principles and design of spectrometers are similar to those for X-ray diffraction.

3.1.5 Spectrometers for measuring $S_{inc}(q, \omega)$ (inelastic incoherent scattering) with modest resolution and high intensity

Most spectrometers in this category use time-of-flight techniques to determine the neutron energy. The beam therefore needs to be pulsed. From reactors this is achieved either by combinations of choppers or choppers and crystal monochromators, or by a rotating single crystal monochromator. For a pulsed source, the design geometry is often ‘inverted’—that is a ‘white’ pulse is incident on the sample but the analyser system will only accept neutrons of a fixed energy. A filter of beryllium polycrystal, for example, will only transmit neutrons with wavelengths longer than 4.2 Å, i.e. with energies between 0 and 5 meV. In combination with a graphite crystal this can produce a spectrometer with good resolution over a wide energy range.

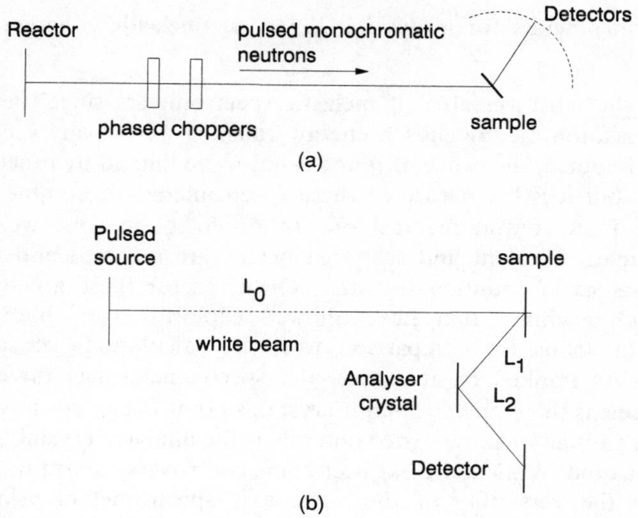

Fig. 3.2 Comparison of spectrometer design on a pulsed or a continuous source for inelastic scattering. (a) A time-of-flight spectrometer on a reactor. The incident energy is *selected* by the phase of the choppers, the final energy is *measured* from the time of arrival in the detectors and the flight path, L. (b) An inverted geometry spectrometer. The incident energy is *measured* by the flight time and the flight paths (L_0, L_1 ,L_2), the final energy is *selected* by the angle of diffraction.

In Fig. 3.2 the design of general purpose spectrometers such as these for reactor and pulsed neutron sources are compared. A characteristic they share is limited and fixed scattering angles and hence limited $\boldsymbol{q}$ ranges. Early neutron scattering from polymers involved uses of such spectrometers to observe vibrational bands in the far infrared range often exploiting the differences in incoherent cross section between hydrogen and deuterium. Such applications—never very widespread—have become even less fashionable in recent years, mainly because of the difficulty of interpreting any but the grossest features of the inelastic scattering from amorphous samples.

The inter- or intramolecular forces acting on side groups of polymer chains can also be explored by observing the very small energy transfers involved in changes in their rotational states (as opposed to vibrational transitions). This is called quasielastic scattering and requires rather good energy resolution. For this reason, much of the experimental work is now undertaken on spectrometers purpose-built to have high resolution in the near-elastic region, but sacrificing for this advantage the wide energy range of the time-of-flight spectrometers. Some of these high resolution spectrometers will be described in detail in subsequent sections.

3.1.6 Spectrometers for measuring $S_{coh}(q, \omega)$ (inelastic coherent scattering)

These are the most versatile of inelastic spectrometers since they should allow observation of *any* chosen energy transfer, $\hbar\omega$, at any vector $\boldsymbol{q}$. In practice, of course, the range of $\boldsymbol{q}$ and ω has to be limited by practical considerations but it is the nature of these spectrometers to be able to cover q–ω space in a continuous fashion. In order to do this, well-defined monochromatic incident and scattered beams are required and there are severe penalties in counting statistics. On a reactor there are the classic 'triple axis' machines that have allowed exploration of phonons and magnons in solids for comparison with the calculations of solid state physicists. As implied by its name, the spectrometer has three axes of rotation – one is the monochromator crystal, so that E_i can be chosen, one is the sample so that θ can be varied and one is the analyser crystal, so that E_f can be measured. While there has been some controversy about the best way to achieve the versatility of the triple axis spectrometers using pulsed neutron sources, satisfactory analogues are now being found. Unfortunately, these thoroughbreds of the neutron scattering technique have had almost no applications in polymer science because they are designed for single crystal samples.

3.1.7 High resolution quasielastic scattering spectrometers

In order to explore the slow motion of the polymer backbone, or of large side groups with frequencies in the range below 10^9 Hz, energy transfer of the order of microelectronvolts (μ eV) have to be measured. Given that the lowest available neutron energies are about 1 to 5 meV this requires that E_i and E_f should be defined to better than one part in 10^3. Even if this were possible via the classical time-of-flight methods, the reduction in the counting rate would be unacceptable. Two types of spectrometer have been developed to access this range of $\hbar\omega$ and both achieve their ends by sacrificing energy range and $\boldsymbol{q}$ resolution. The *back scattering spectrometers* are modified triple axis spectrometers – exploiting the fact that Bragg reflections at 180° from a single crystal have an extremely narrow wavelength or energy spread. The *spin-echo spectrometers* exploit the precession of the neutron spin in a magnetic field and use it as a sort of clock to measure changes in time-of-flight thus entirely avoiding defining E_i and E_f separately (except, however, as they define $\boldsymbol{q}$) and measuring merely changes in energy. Both of these high resolution machines are applied very successfully to study polymer motion and will be described in detail below.

GLE SCATTERING

ting there are at least a dozen small angle neutron spec-
ope, about half that number in the USA and others
ide, each designed to suit a particular neutron source, and
emand. It is clearly not practicable to describe all these
s. A typical design has already been described at the end
A description of their spectrometers is always provided by
urces whose facilities are open to experimental scientists and
ons can provide up-to-date detail much better than a general
n would quickly become obsolete. In this, and the following
refore, it is the general principles of the technique which will be discussed. Armed with these principles, the potential user should be able to approach a new spectrometer with a lot of pertinent questions and quickly devise a *modus operandi*. The method of attack can be summarized under the following headings, which will be expanded below:

(i) setting up the spectrometer—usually involving choosing wavelength, sample to detector distance, diaphragms etc.;

(ii) samples—their environment, concentration, dimensions, containers etc.;

(iii) what to measure—background, resolution, transmission;

(iv) data reduction—i.e. $I(\theta, \lambda)$ to $S(q)$.

In polymer science the SANS technique is likely to be the most widely used of all the neutron techniques so these practical points will be dealt with in some detail. Then for the other techniques in subsequent sections only special points of difference need to be drawn out.

3.2.1 Setting up the spectrometer

All small angle machines these days are furnished with area detectors, although the construction varies from installation to installation. A very likely question then will be the choice of sample to detector distance. In some cases this distance is continuously variable as the detector moves inside an evacuated tube, in others only fixed choices can be made by removing or adding sections of the evacuated tube between the detector and the sample. This evacuated flight path is necessary because of the large distances involved, over which the absorption of neutrons by air would be prohibitive. Replacing air with helium is a possible, though expensive alternative. The choice of sample–detector distance effectively determines the scattering angle θ and hence q. Figure 3.3 shows an area detector, at a

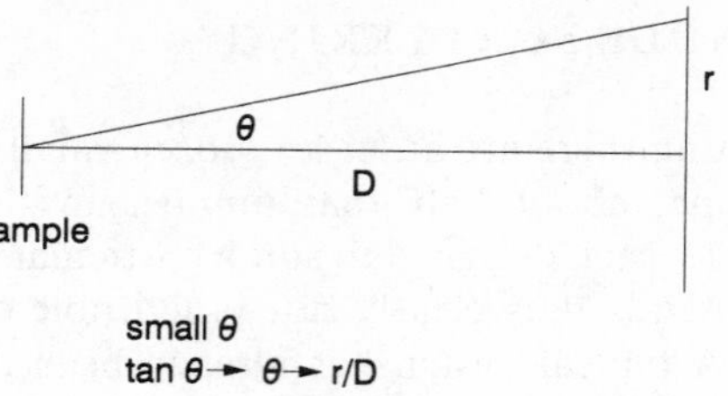

Fig. 3.3 The scattering angle for an area detector a long distance, D, from the sample.

distance D from the sample. At a distance r from the centre of the detector, the scattering angle θ is defined by $\tan\theta = r/D$. Since θ is very small, however, $\tan\theta \to \theta$ and $q = 4\pi/\lambda \sin\theta/2 \to (2\pi/\lambda)(r/D)$.

The choice of D is, of course, tied up with the available wavelength range. However, given that the largest area detectors are 1 m in diameter, and that a typical value for λ might be 5 Å, for $D = 1$ m the maximum value of q, q_{max} would be 0.6 $Å^{-1}$ and for $D = 10$ m, 0.06 $Å^{-1}$. The maximum value of q (for a single incident wavelength) will be determined by the resolution of the detector elements (typically 1 cm) and the collimation of the beam. Even for strongly scattering samples, most of the beam is transmitted and would fall on the centre of the detector. To avoid the damage to the detector which such a relatively intense flux of neutrons would cause, the beam is caught by a 'beamstop' made of neutron-absorbing material, such as cadmium, in front of the detector. The size of this central mask then determines how close to $\theta = 0$ the intensity can be measured, i.e. the minimum scattering angle. The collimation is determined by the distance from the sample to the reactor, and by the number of diaphragms or slits interposed. Typically it is not possible to measure closer than 5 cm to the centre of the detector. For $D = 10$ m, $\lambda = 5$ Å and $q_{min} = 0.06$ $Å^{-1}$. For a detector 1 m across therefore the q range available for a single λ and D value is just one order of magnitude. For many detectors, the range is even smaller. If (as in many cases) a wide q range is important then several measurements at different λ or D values have to be overlapped. There is an interesting balance of choices here. Increasing λ always results in a strong decrease in intensity (see Fig. 2.4). On the other hand, changing the detector distance by a factor x, results in a change in the solid angle subtended by a factor x^2. The collimation also has to be tightened up, usually resulting in another reduction in intensity by a factor of order x to x^2. Thus, halving q_{min}, by doubling the detector distance would result in a factor between 8 and 16 reduction in counting rate.

Pulsed sources have a great advantage when it comes to consideratior

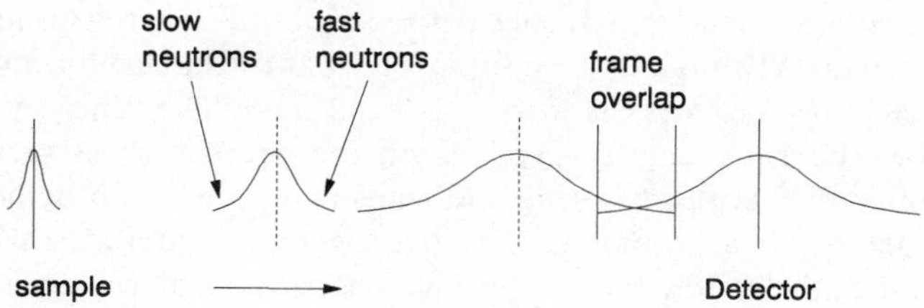

Fig. 3.4 The frame-overlap problem. If a sequence of 'white' pulses travel over a long enough distance the fast neutrons eventually 'catch up' with the slow ones in the previous pulse.

of the q range available as a single measurement. By combining the angular range of an area detector with a considerable wavelength range, they can easily double or treble the q range of a reactor instrument. There are penalties. Firstly, the computing back-up is complicated by having to label each neutron with its arrival time as well as its position. It is not possible to calculate immediately a q value and store this simplified label since many of the corrections to the data are wavelength dependent. The separate velocity (i.e. λ) and position information has to be preserved at least until the first stages of data reduction have been carried out. Secondly, wide wavelength spreads cannot be used for long sample to detector distances because of frame overlap. In order to determine λ unambiguously from flight-times the pulsed nature must be preserved. However, at increasing distance from the detector, the pulse is spreading out, with the fast neutrons arriving much earlier. Eventually these fast neutrons will start to catch up the slower ones in the previous pulse (Fig. 3.4). At this point it is not possible to distinguish these fast and slow neutrons and ambiguity in the assignment of λ creeps in. To overcome this there are two options. Either some of the pulses can be removed by a 'chopper' thus giving the others more space to spread out, or the range of wavelengths in the pulse can be limited, also using phased choppers.

3.2.2 Samples and their environment

Clearly the detailed preparation of the sample is determined by the nature of the scientific problem. The question of the exact deuterated hydrogenous mixture to use in SANS experiments is discussed in detail in Chapters 5–8. Some examples of contrast factors are given in Appendix 3. There are, however, a number of general points that can be made.

The area of sample 'in view' is usually defined by a diaphragm of absorbing material (cadmium) just in front of it. Since this diaphragm is part of the collimation system, and hence helps to define the resolution of the system, it is usual to make its area similar to that of the resolution elements on the detector, i.e. of the order 1 to 2 cm^2. The thickness of the sample—i.e. the path length—depends on the neutron cross section. The aim is to maximize both scattering and transmission which is achieved with about 50 per cent transmissions. For hydrogenous material, this requires about 1 mm path length, but for nonhydrogenous materials path lengths as long as 1 cm may be feasible. The correct thickness may be calculated from the Beer-Lambert law

$$I/I_0 = \exp\ (-n\sigma_T t) \tag{3.1}$$

where I is the transmitted and I_0, the incident neutron flux, σ_T is the total cross section (both coherent and incoherent) for some convenient molecular unit, n is the number of such units per unit volume and t is the thickness. To illustrate this point consider the transmission of a sample of water 1 mm thick.

The total cross section per molecule using the values given in Table 3.1 and ignoring the small absorption cross section is

$$\sigma_T = \{2 \times (79.7 + 1.76) + 4.23\} \times 10^{-24}\,cm^2 = 167.15 \times 10^{-24}\,cm^2.$$

Assuming a density of about 1 g cm^{-3} and a molar mass of 18, the number of molecules per unit volume is Avogadro's number/18 = 3.36×10^{22}. From eqn (3.1) then

$$I/I_0 = \exp - (3.36 \times 10^{22} \times 167.15 \times 10^{-24} \times 10^{-1}) = 0.57.$$

The equivalent transmitted fraction for D_2O would however be 0.95 and for this material to have a transmission of 0.5, the thickness would be as much as 11.7 mm. Table 3.1 shows that samples containing chlorine have a high absorption cross section and these must be thinner than usual to obtain good transmission. Clearly the water discussed above, and, in general, polymer solutions and melts, would have to be held within a container. Moreover, as mentioned earlier the flight path between the sample and the detector (and also, of course, the collimation system) will almost certainly be evacuated. Either the sample container itself must be capable of sitting in a vacuum, or windows will close the flight paths immediately before and after the sample. The requirements of the sample container and window materials are that they have a low cross section and no large scale morphological features which would give rise to small angle scattering. Aluminium has a very low cross section, but normal sheet aluminium is polycrystalline and gives rise to quite strong small angle scattering. Single crystal aluminium has been used successfully for window materials especially when extreme sample conditions (high or low temperatures or pressures) demand extra strength. More common polymer applications use silica (quartz) for liquid-containing cells. Sapphire is frequently used for window material.

The advantages of eliminating windows and placing the samples in the same vacuum as the detector are most evident when very weak scatterers are involved. Most polymer samples give rise to reasonably strong signals and for solution samples it is clearly then much easier to work outside the evacuated area. Even for solid samples there is considerable inconvenience involved in breaking into the vacuum system every time a sample change is required. Solid samples can be free standing if they are large enough, or pressed into aluminium rings which will be masked by cadmium diaphragms, or even wrapped in a little aluminium cooking foil (a surprisingly good material for such uses being very thin and pure).

Since spectra from polymer samples are often obtained in relatively short times, and neutron sources run 24 hours in 24, the experimenter will be thankful if a sample changing device is available, allowing him or her to load a number of samples and obtain nourishment, sleep, and other basic necessities. Heaters, cryostats, and more sophisticated sample environments such as devices for applying high pressure or sample deformation are often also available.

3.2.3 Ancilliary measurements – background, resolution, transmission

Any experiment will probably involve in practice a sequence of measurements which have subsequently to be normalized one to the other. As a first step the counts on the detector must be correlated to the incident neutron flux. This is usually achieved by simultaneously recording the counts from a monitor (see Chapter 2) permanently placed in the incident beam. Often the length of an experiment is set for a certain number of monitor counts, rather than a certain length of time.

When subtracting the scattering from a container, for example, the transmission of the sample in the container with respect to the empty container is required. Transmissions are measured by comparing the counts recorded by a detector placed after the sample position. This detector might be a second monitor although this option has the potential to cause small angle scattering. Alternatively, a detector may be placed in the centre of the beam stop, or the main detector itself might be used with the beamstop removed, and an attenuator introduced upstream to reduce the overall intensity on the detector.

Background subtraction means different things to different scientists. A scientist talking about background subtraction might be referring to any one of a number of processes. In essence, however, they all boil down to removing signal intensity which the scientist is unable (or unwilling) to consider when modelling $S(q)$.

These contributions might include:

(a) electronic noise in the detector;

(b) neutrons arriving from surrounding spectrometers or the neutron source itself;

(c) neutrons scattered by the cell and/or sample support;

(d) neutrons scattered by parts of the sample not of interest, for example the solvent;

(e) neutrons scattered by the molecules of interest, but incoherently.

Background (a) is independent of the neutron flux and usually proportional to the time the experiment runs. All the other possible contributions arise from the neutron source but only (c), (d), and (e) depend directly on the flux falling on the sample. The precise methods adopted for removing the various backgrounds vary from laboratory to laboratory, but the general principles can be outlined and a few *caveats* indicated. (a) and (b) are often measured together, with the beam shut-off somewhere before the sample (c) and (d) can be measured directly by recording the counts from an empty sample cell or support and then from one containing pure solvent.

An important point to bear in mind is that the signal from the cell containing solvent will not be just the direct addition of scattering from cell and from solvent, i.e. $I(\text{solvent}) \neq I(\text{solvent} + \text{cell}) - I(\text{cell})$. This is illustrated in Fig. 3.5 and arises because the neutron flux scattered by the cell when it contains solvent is again reduced by the scattering of the solvent. Any subtraction procedure must therefore take account of the transmissions of the various contributors to the scattering. One way of performing the subtraction mentioned above is to calculate $I(\text{solvent}) = I(\text{solvent} + \text{cell}) - (Trs/c)\, I(\text{cell})$ where Trs/c is the transmission of the cell + solvent with respect to the cell alone. The remaining scattering now comes only from the solvent, but is reduced by the transmission of Trs/c.

A special case of background removal is separation and/or removal of the incoherent signal from the sample itself—especially a hydrogen-containing sample. The procedures for this depend on the data analysis programme undertaken, and discussion of this problem is, therefore, postponed until the whole question of data analysis is broached below.

Resolution measurements are not usually a problem with small angle scattering except in special cases. The sources of uncertainty in the q values are, firstly, the spread $\Delta\lambda/\lambda$ in the incident beam (or, on pulsed sources, the uncertainty $\Delta\lambda/\lambda$ in determining the wavelengths), secondly, the angular spread, or divergence of the neutron beam, thirdly, the diaphragm defining the sample, and fourthly, the spatial resolution of the detector. Each of these contributions can be separately determined, or the effect of the com-

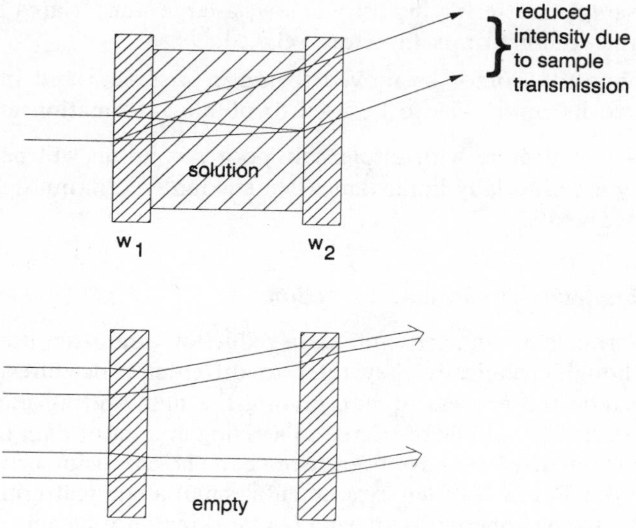

Fig. 3.5 The effect on scattering from an empty cell by the solution it contains. The scattered intensity from the first wall, and the incident intensity on the second wall are both reduced, by the scattering in the solution itself.

bination may be observed by scattering from a sample with a sharp Bragg reflection in the q range of interest. This method requires a crystal with a d spacing of 10^2–10^3 Å, a rather unusual sample, but available in, for example collagen or lamellar copolymers.

Since the spatial resolution is likely to be fixed, its effect on $\Delta q/q$ via $\Delta\theta/\theta$ may well vary from the centre of the multidetector outwards—for example for a 1 m detector with 1 cm resolution elements, 10 cm out from the centre $\Delta\theta/\theta$ is 10 per cent while at the outer edge it is 2 per cent. In usual practice, since the detector resolution is a fixed quantity, the wavelength spread and collimation are designed to be of the same order of magnitude. (There is little point in losing flux by choosing a better than 1 per cent wavelength resolution, if all this effort will effectively be wasted in a much poorer spatial resolution on the detector.)

If the wavelength resolution is of the order of 10 per cent, its effects in smearing the scattering data are small (< 5 per cent) and can safely be ignored (Wignall *et al.* 1988). The time to consider resolution effects is if, for some reason, poorer resolution has to be tolerated. Possible cases are:

(a) The flux is low and the wavelength spread $\Delta\lambda/\lambda$ is increased to compensate for this.

(b) The sample scattering intensity is low, a large sample area is required and the collimation is therefore relaxed.

(c) The detector cannot be moved to longer distances and information close to its centre has to be used to obtain information at low q.

Methods for dealing with intolerably poor resolution will be described later. They are especially important when encountered in quasielastic scattering experiments.

3.2.4 Data analysis and data reduction

The two terms 'data analysis' and 'data reduction' are often used as alternatives, though technically they refer to different procedures. The first should include the process of interpreting the data, fitting models, etc., while the second should be restricted to arriving at a set of data from which all instrumental artefacts have been removed. The confusion arises because the true reduction of, say, an experimental small angle scattering observation $I(\theta, \lambda)$, to the coherent scattering law $S(\boldsymbol{q})$ may not be achievable. For example removal of the resolution effects as described above may prove impossible. The procedure adopted is often to measure the effects of the instrumental resolution on what should be an infinitely sharp scattering peak (for example, Bragg diffraction, though of course infinitely sharp here only means 'very sharp compared to the resolution') or to calculate a resolution function, and then to apply it to the theoretical model of $S(\boldsymbol{q})$ before fitting this to the data. Analysis and reduction are inextricably entwined.

Two steps towards proper data reduction, which can (and indeed often must) be undertaken before a model $S(\boldsymbol{q})$ is chosen, are normalization and removal of incoherent background. The measured scattered intensity $I_{\mathrm{m}}(\boldsymbol{q})$ is related to the differential cross section per unit sample volume $(\partial\sigma/\partial\Omega)_{\mathrm{v}}$.

$$I_{\mathrm{m}}(\boldsymbol{q}) = I_0\Omega_0\varepsilon A d T(\partial\sigma/\partial\Omega)_{\mathrm{v}} \tag{3.2}$$

where I_0 is the incident neutron flux, Ω_0 is the solid angle of acceptance of the detector, ε is the detector efficiency, A is the area of the beam, d is the sample thickness, T is the transmission of the sample. (In this section the transmission will include effects of absorption as well as scattering of neutrons.) The terms I_0 Ω_0, ε, and A are normally constant for a series of measurements. In order to be able to correct for detector efficiency, it is necessary to measure the scattering from a substance that scatters entirely incoherently in the q range of the experiment: that is, there must be no angular dependence of the scattering from such a substance. Common standards that are used are water and vanadium.

The scattering from a standard incoherent scatterer can be written as

$$I_m(\boldsymbol{q})_S = I_0\Omega_0\varepsilon A d_S T_S(\partial\sigma/\partial\Omega)_{i,S} \tag{3.3}$$

where the subscript S refers to the standard. For an incoherent scatterer, the scattered intensity is spread uniformly over 4π steradians (except for a small decrease with angle due to the Debye–Waller factor discussed in Chapter 4, Section 4.9.3)

$$d_S T_S(\partial\sigma/\partial\Omega)_{i,S} = (1 - T_S)/4\pi \tag{3.4}$$

and eqn (3.3) becomes

$$I_m(\boldsymbol{q})_S = I_0\Omega_0\varepsilon A(1 - T_S)/4\pi \tag{3.5}$$

substituting for $I_0\Omega_0\varepsilon A$ in eqn (3.2) gives

$$I_m(\boldsymbol{q}) = \frac{I_m(\boldsymbol{q})_S 4\pi}{(1 - T_S)} \mathrm{d}T \left(\frac{\partial\sigma}{\partial\Omega}\right)_V \tag{3.6}$$

$$I_m(\boldsymbol{q}) = D\left(\frac{\partial\sigma}{\partial\Omega}\right)_V \tag{3.7}$$

where $D = I_m(q)_S(4\pi)\mathrm{d}T/(1 - T_S)$.

Note that intensity is a dimensionless number, D has units of length (d) and $\partial\sigma/\partial\Omega$ has units of inverse length.

Unfortunately, there is one drawback, especially with the use of water as a standard and that is that its scattering is not entirely angularly independent due to inelastic scattering effects. If water is used as a standard it is necessary to take account of these effects when normalizing the data. The scattering from water tends to peak in the forward direction, and a factor $g(\sigma, \lambda)$ is used to account for this. The term $g(\sigma, \lambda)$ is a function of cross section and wavelength and, therefore, varies with the temperature of the water sample. Values of $g(\sigma, \lambda)$ have been measured for a 1 mm sample of water on one SANS spectrometer (May *et al.* 1982). Jacrot (1976) has suggested a formula that can be used to calculate $g(\sigma, \lambda)$, but the values are not in agreement with the measured values.

To take account of this effect when water is being used as a standard, D becomes

$$D = I_m(\boldsymbol{q})_S[4\pi d T/(1 - T_S)]g(\sigma, \lambda). \tag{3.8}$$

Another method used to normalize the data is to use the intensity at $q = 0$ from a sample of known molecular weight or known scattered intensities. In this case, water is used to correct for detector efficiency etc. but not to normalize the data. A detailed discussion of the use of secondary standards for absolute calibration has been given by Wignall and Bates (1987).

If models of $S(\boldsymbol{q})$ are to be fitted then it must be assumed that the incoherent scattering has been subtracted, a procedure that is not always straightforward. For a dilute mixture of a deuterated polymer in a hydrogenous matrix, the incoherent scattering is almost entirely due to that from the hydrogenous matrix. In this case the scattering from a 100 per cent hydrogenous sample can be subtracted with virtually no error because of the very small amount of incoherent scattering from the deuterium. However, as the amount of deuterium increases, the incoherent scattering from the hydrogenous matrix reduces and that from the deuterium increases. In this case it is no longer correct to subtract only the scattering from the pure hydrogenous matrix as this would be too large—account must be taken of the relative amounts of deuterium and hydrogen present in the sample. For a polymer mixture in which some of the chains of one of the polymers are deuterated, the subtraction is more complicated. In this case, it is necessary to take account of the incoherent scattering from three different components. If there are no changes in the incoherent cross sections of the polymers when they are mixed together, then the background can be estimated from the scattering from the pure components. However, it has recently been observed that the cross section for polymers varies with temperature and wavelength due to changes in the incoherent inelastic cross section arising from increased molecular motion at higher temperatures (Maconnachie 1984). When a polymer chain is mixed with a different polymer, the local environment around the chains may well be different from that in the bulk homopolymer, thus affecting the dynamics of the chains and the incoherent cross sections. Caution should, therefore, be exercised when estimating incoherent backgrounds for polymer blends (or concentrated solutions) from the pure components.

A number of methods have been used either to estimate or to measure the incoherent scattering. For a mixture of x deuterated chains and $(1 - x)$ hydrogenous chains the incoherent scattering can be written in the form

$$I_i = xI_i^{\mathrm{D}} + (1 - x)I_i^{\mathrm{H}} \tag{3.9}$$

where I_i^{D} and I_i^{H} are the normalized incoherent scattering from pure deuterated and pure hydrogenous chains, respectively. The term I_i can thus be estimated from the scattering from the pure components if the proportion of incoherent to coherent scattering is known.

A method of calculating this scattering from such a mixture has been proposed by Hayashi *et al.* (1983). This method assumes that the cross section for hydrogen is the same as that of the bound atom, which is a reasonable approximation for $\lambda \sim 5$ Å or longer, and also that the incoherent scattering is equally distributed over 4π steradians. The expression that is used is

$$I_i = \frac{1 - \exp(-N\tau_i d)}{1 - \exp(-N_{\mathrm{H}}\tau_i d_{\mathrm{H}})} I_i^{\mathrm{H}} \tag{3.10}$$

in which I_i^{H} is the incoherent scattering from the pure hydrogenous polymer; N_{H} and N are the number of protons in the hydrogenous polymer and the mixture per unit volume respectively; τ_i is the incoherent cross section of hydrogen; and d_{H} and d are the thickness of the hydrogenous polymer and the mixture, respectively. The incoherent scattering from the pure hydrogeneous polymer is also adjusted for the small amount of coherent scattering present as follows. Assuming that the incoherent scattering from deuterium is negligible and that the coherent scattering is proportional to the coherent cross sections, then

$$I_i^{\mathrm{H}} = I^{\mathrm{H}} I^{\mathrm{D}} - \left(\frac{d_{\mathrm{H}} T_{\mathrm{H}} \tau_{c,\mathrm{H}}}{d_{\mathrm{D}} T_{\mathrm{D}} \tau_{c,\mathrm{D}}}\right) \tag{3.11}$$

where T_{H}, $\tau_{c,\mathrm{H}}$, and T_{D}, $\tau_{c,\mathrm{D}}$ are the transmission and coherent cross section of the hydrogenous polymer and deuterated polymer, respectively, I^{H} and I^{D} are the total scattering from the pure components. Equation (3.11) is then used with eqn (3.10) to obtain the level of incoherent scattering in the system.

Gawrisch *et al.* (1981) have used another method for analysing the data that does not involve the actual subtraction of the incoherent background. In this method, the total scattering as a function of c, the concentration of deuterated chains, is fitted to an expression that is quadratic in c. The coefficients of this expression as a function of $\boldsymbol{q}$ give the single-chain scattering, the incoherent scattering, and the density fluctuations. The drawback of this method is that a number of concentrations must be used to obtain the single-chain scattering. It is also not clear whether this method would be applicable to the scattering from polymer blends or concentrated solutions where concentration fluctuations are present. The incoherent scattering from a random copolymer that has the same proportion of H and D monomers in the chain as in the mixture should be identical to that from the mixture. This method has been used (Schelten *et al.* 1973) but it does rely on the fact that the proportions of H and D have to be exactly the same as in the mixture and that their distribution is random, both of which can be difficult to achieve.

At high q the scattering from a polymer is mainly incoherent, so that it is possible to measure the level of scattering at high q and use this value at low q. There are two drawbacks with this method: (1) there may still be some coherent scattering present and (2) it is assumed that the incoherent scattering is independent of q (see comments on the use of water, p. 65).

Finally, there is one method that has not been widely used so far in SANS and which can measure directly the actual level of incoherent scattering: polarization analysis. It is to be hoped that, in the future, SANS spectrometers will have polarization analysis as a standard option, thus obviating the need to estimate the level of incoherent scattering.

One final set of corrections that has to be considered arises from the possibility that a neutron may be scattered twice by the nuclei in the sample – multiple scattering. This is one occasion where the low cross section of most materials for neutrons is a positive advantage. Although multiple scattering corrections have to be taken into account when fitting quasielastic and inelastic scattering, in small angle scattering they have been shown by Monte Carlo simulation, to have, in general, a very small effect (less than 2 per cent as, shown by Goyal *et al.* (1983)).

After the data have been *reduced* towards $S(\boldsymbol{q})$ as far as possible, analysis can proceed by interpreting the experimental $S(\boldsymbol{q})$ in terms of that calculated for one of the available models some of which are described in subsequent chapters. Two methods of approach are possible – plot the data in such a way that linearity is expected and the slope and intercept can be interpreted, or, fit a calculated $S(\boldsymbol{q})$ to the experimental one. As mentioned earlier, it is at this point that decisions have to be made about whether, and if so how, to deal with the resolution.

An example of a possible model to fit was described at the end of Chapter 1. For a polymer molecule at low q the Guinier function applies and –

$$S(q) = N \exp\ (-q^2 R_g^2/3) \tag{3.12}$$

thus plotting ln $S(q)$ vs q^2 yields the molecular weight, N (or rather ln$N$) from the intercept at low q and the radius of gyration, R_g from the slope. Clearly if resolution effects are important, it may not be easy to remove them during this procedure. However, if the function $F(\lambda)$ representing the wavelength distribution has an analytical form, then this can be mathematically convoluted with $S(q)$ and the effect on such quantities as slope and intercept determined.

An example of this is the Guinier curve in eqn (3.13) below. For $qR_g < 1$ only the leading terms in the exponential need to be considered. Then, writing q explicitly in terms of θ and λ, and remembering that for small θ, $\sin\theta \rightarrow \theta$

$$S(\theta, \lambda) = N\left(1 - \left(\frac{2\pi\theta}{\lambda}\right)^2 \frac{R_g^2}{3}\right). \tag{3.13}$$

If there is a wavelength spread, $F(\lambda)$ then

$$S(\theta, \lambda) = N\left(1 - \frac{2\pi\theta^2 R_g^2}{3}\left\langle\frac{1}{\lambda^2}\right\rangle\right) \tag{3.14}$$

where

$$\left\langle\frac{1}{\lambda^2}\right\rangle = \int \frac{1}{\lambda^2} F(\lambda) \Big/ \int F(\lambda)\mathrm{d}\lambda = \frac{1}{\lambda_0^2} \tag{3.15}$$

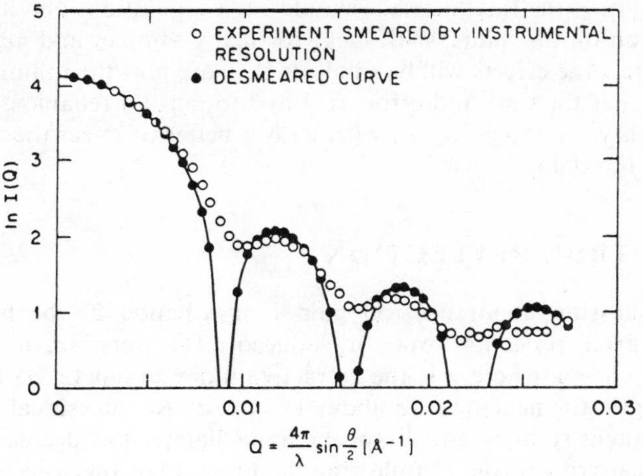

Fig. 3.6 Scattering from a 3.98 vol per cent solution of monodisperse polymethylmethacrylate spheres in D_2O/H_2O mixture. (○) experimental data, (●) desmeared data. $Q \equiv q$. Reprinted with permission from Wignall *et al.* (1988).

and the correct R_g is obtained from the slope of the logarithmic plot, if the correct value of λ_0 (i.e. the inverse second moment of the wavelength distribution) is employed. In an early neutron scattering experiment, the effect of two different velocity selectors on λ_0 was examined (Cotton *et al.* 1974). For a selector with $\Delta\lambda/\lambda$ of 10 per cent, $\lambda_0 = 0.99\lambda$ while even for $\Delta\lambda/\lambda = 40$ per cent, $\lambda_0 = 0.95\lambda$. As mentioned earlier and demonstrated by Wignall *et al.* (1988) the effect is very small unless $F(\lambda)$ is extremely broad.

More serious is the effect of finite λ and θ resolution on curves showing complex shapes. Figure 3.6 shows the scattering curve from the set of uniform hard spheres in dilute solution. The shape of this curve, and the effect of interparticle interference on it will be discussed in Chapter 6 (Section 6.3.2). Here it forms a useful example of the effect of finite resolution on a scattering curve with sharp features. The example is taken from Wignall *et al.* (1988) and included in the figure is a set of points resulting from the desmearing process, i.e. removal of resolution effects by mathematical procedures. In this case the process appears to have been quite successful, but it would be more difficult in a situation where the form of $S(q)$ is not known *a priori*, or if there is also polydispersity in the sample. A rather more detailed discussion of resolution and desmearing (or deconvolution) and the many pitfalls and hazards will be necessary when discussing quasielastic scattering in Sections 3.4 and 3.5. In the context of small angle scattering, the final message should be 'ask questions about the effect of resolution on the data'. With luck, for many samples and many SANS spectrometers the effects will be small. If they are not, then think carefully about whether the time and effort required to make a reliable desmearing procedure is worth investing, or whether it is better to smear the model and fit this to the data.

3.3 NEUTRON REFLECTION

In the discussion about neutron guides, in Chapter 2, the basic ideas about neutron reflection were introduced. The very small values of $\lambda^2\rho_b/2\pi(=1-\nu$ where ν is the refractive index in eqn (2.5)) mean that critical angles for neutrons are about 1° or less. At the critical angle, the reflected intensity does not disappear immediately, but decreases over a range of scattering angle, θ, following the Fresnel law (derived for electromagnetic radiation).

$$R = \left|\frac{\nu_0 \sin\theta_0 - \nu_1 \sin\theta_1}{\nu_0 \sin\theta_0 + \nu_1 \sin\theta_1}\right|^2. \tag{3.16}$$

The subscripts 0 and 1 refer to the two media on either side of the reflecting surface. The reflectivity, R, is a function of the scattering angle and the refractive index, and hence, of the wavelength and also the scattering length density, ρ_b. θ and λ are the experimental variables, and the reflectivity then contains information about the variation of ρ_b, i.e. the chemical composition, normal to the reflecting surface. For a thin film between two bulk media (for example, air and a substrate) interference occurs between the waves reflected at each surface, leading to the observation of interference fringes at angles greater than the critical angle, and hence to a value of the film thickness.

Clearly, small angle scattering spectrometers would allow the necessary range of θ to be explored. Remember, however, that these are *reflection* experiments, so that for fixed λ, on a reactor source, the angle of incidence must be varied as θ is scanned, or the wavelength must be changed for fixed θ. For fixed collimation $\Delta\theta$ and fixed wavelength resolution, $\Delta\lambda$, the resolution varies as $\cot\theta\Delta\theta$. Moreover, the illuminated area of the sample varies as θ varies. Thus a better alternative in many cases is to use fixed angular geometry and scan λ. Time-of-flight analysis on a pulsed source is thus well suited to neutron reflection studies. Liquid surfaces can only be easily explored in horizontal geometry. Thus, the incident neutron beam needs to approach the sample slightly off-horizontal. This has been achieved either by building a long spectrometer, where gravity causes the neutrons to drop from their initially horizontal flight path but with dif-

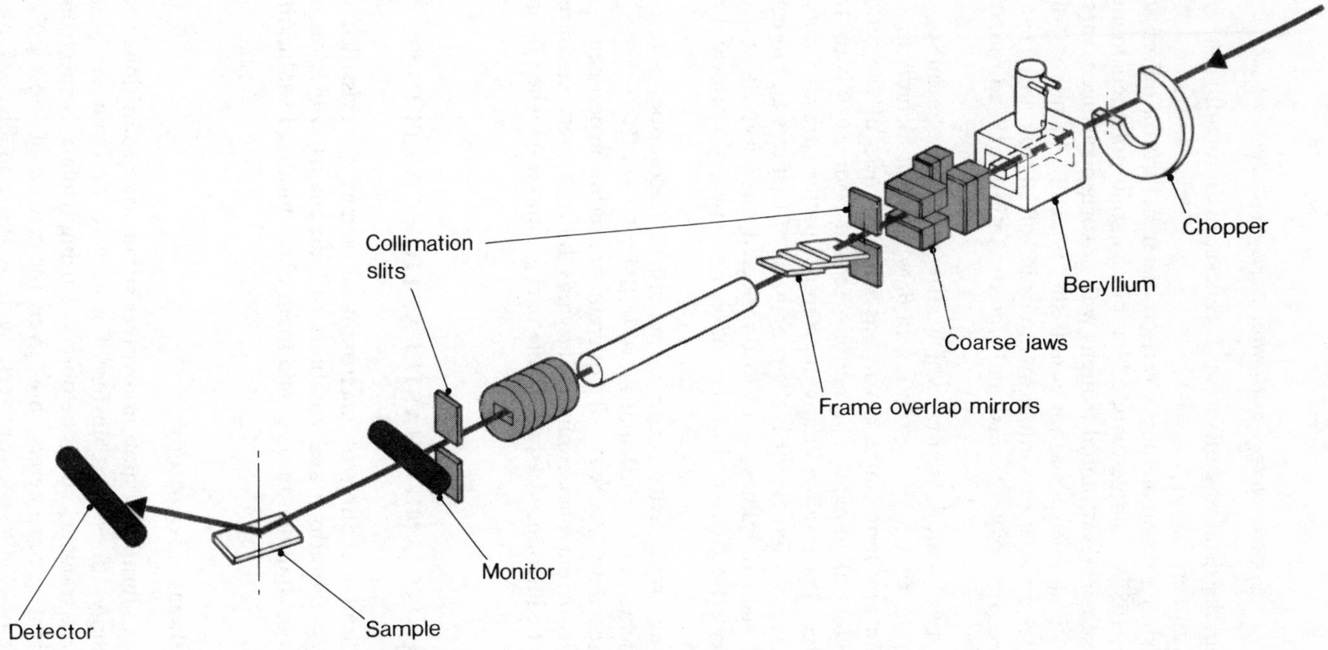

Fig. 3.7 Schematic diagram of the CRISP spectrometer at the ISIS neutron source.

ferent wavelengths dropping different distances, or by bending the beam, using a supermirror.

Figure 3.7 shows schematically a reflectometer built on a pulsed source (Penfold *et al.* 1987). Samples would be typically solid polymers loaded on optically flat substrates or liquid troughs with adsorbed surface layers. The incident beam is defined by a horizontal slit, typically 40 mm × 2–6 mm. This illuminates an area of sample of the order of 20–100 cm^2. Details of neutron reflectivity experiments and interpretation are considered in Chapter 10.

There are two major factors which affect the resolution of these experiments. The first is the resolution in θ, which will include any beam divergence and any long range deviations from flatness of the reflecting surface, including alignment. The major effect of poor $\Delta\theta/\theta$ is to 'round' the critical edge. The minima in any interference pattern are also smoothed out. A second resolution factor is the roughness of surfaces and interfaces. The effect of such roughness is to lose reflected intensity, via scattering, from the lower reflectivity end of the curve. Examples of both can be seen in Chapter 10.

In analysing reflectivity experiments these resolution effects are parameters included when calculating reflectivity from different models in order to fit the data. As with all coherent scattering experiments, there may be an incoherent background. This may be separately measured, or the level may be introduced as a variable background level when fitting the data.

3.4 INELASTIC AND QUASIELASTIC SCATTERING

The intramolecular motion of macromolecules may be classified under three headings: vibrations and rotations of side chains, vibration of the main chain in crystalline samples, and main-chain motion in solutions and melts.

3.4.1 Vibrational spectroscopy

Observation of vibrational motion involves stimulating transitions between quantized energy states via transfer of energy to or from the scattered neutrons. These transfers could be observed using triple-axis spectrometers or time-of-flight spectrometers, but given the nature of most polymeric samples, time-of-flight spectrometers, such as those in Fig. 3.2 are most usually employed.

The difficulties encountered in studying vibrational motion in macromolecules (side chain and main chain) arise from the samples themselves rather than from the nature of the motion. Samples of stereoirregular polymers form rubbers at higher temperatures and glasses at lower temperatures. Intramolecular potentials are highly anharmonic and give rise to broad inelastic features often difficult to identify. Stereo-regular polymers, on the other hand, rarely produce single-crystal textures, so that samples consist of small crystalline regions embedded in rubber or in glass (since the glass transition temperature is usually quite different from the crystallization temperature). Various preparative and spectroscopic techniques have been used to simplify the scattering so that it can be interpreted using the basic theory developed for small molecules. These will be discussed in detail later on (Chapter 9). Samples are usually hydrogenous polymers and are typically 'films' of thicknesses 0.25 mm in a beam of 2–3 cm diameter.

Treatment of the scattering data involves removal of background scattering, self-absorption, and other effects arising from sample geometry and detector efficiency factors. These corrections are undertaken using monitor counts and scattering from standard samples in an analogous way to the methods already described in detail for small angle scattering.

In a time-of-flight spectrometer such as that in Fig. 3.2, the number of neutrons arriving in each detector as a function of time gives directly the double differential cross section as a function of time-of-flight τ. This must be related then to the more usual cross section by

$$\frac{1}{\hbar}\frac{\partial^2\sigma}{\partial\Omega\,\partial\omega} = \frac{\partial^2\sigma}{\partial\Omega\,\partial E} = \frac{-\tau^3}{m}\frac{\partial^2\sigma}{\partial\Omega\,\partial\tau}. \tag{3.17}$$

Around the elastic peak the τ^3 term does not vary very fast, but across the inelastic spectrum this correction has an important effect on relative intensities of different parts of the spectrum. The energy resolution of such spectrometers depends both on the rotation rate of the choppers and/or crystals and the value of E_i chosen, but the best value of $\Delta E/E$ are typically 1 per cent. In terms of quasielastic scattering, this leads to a best resolution for 10 Å incident neutrons of around 20 μeV ($\omega \approx 3 \times 10^{10}$ Hz).

There is rarely any attempt made to remove instrumental resolution from the inelastic cross section finally obtained, but in cases where model scattering functions (for example the density of states, described in Chapter 9) are available these may be convoluted with the resolution function for comparison with the data.

3.4.2 Rotational motion

Rotational motion of side groups appears in the quasielastic scattering region. The rotational frequency is strongly temperature-dependent, and the required resolution may vary from better than 1 μeV up to several millielectronvolts. This then determines whether the back scattering or time-of-flight techniques are used. In both cases the hydrogenous samples will be films of around 0.25 mm thickness (calculated to give transmissions around 90 per cent) and 2–3 cm diameter. Background scattering and other corrections described above are similar for both techniques, but in this case instrumental resolution is important. It is usually observed via the incoherent elastic scattering from a slab of vanadium (about 2 mm thick). Generally no attempt is made to deconvolute this resolution function; rather, model scattering functions are convoluted with the observed resolution and fitted to the corrected experimental data. A recent discussion by Sivia *et al.* (1992) explores the possibilities of the deconvolution process in some detail. The data are usually described in terms of the incoherent scattering law $S_{\text{inc}}(\mathbf{q}, \omega)$ obtained from $(\partial^2\sigma/\partial\Omega\partial\omega)_{\text{inc}}$ via eqn (4.82).

3.4.3 The back-scattering spectrometer

The 10^{10} Hz resolution limit of most time-of-flight spectrometers arises mainly for reasons of flux limitation. It is normally possible to detect motion about one order of magnitude in energy less than the resolution of the spectrometer—thus down to around 10^9 Hz. However, the interesting range for polymer backbone motion often lies even lower in energy. In order to achieve better resolution the back-scattering spectrometer sacrifices the wide window in energy allowed on other spectrometers and concentrates the available flux in a narrow range observed with very high resolution.

The back-scattering technique uses a crystal monochromator set at a 90° Bragg angle, which gives an extremely sharp energy spectrum around 1 μeV or 10^9 Hz. Its basic construction is that of a triple axis spectrometer in which a mechanical Doppler drive shifts the neutron incident energies and scans small energy changes on scattering from the sample. A spectrometer in use at a reactor is shown schematically in Fig. 3.8. Neutrons arrive in a neutron guide from the reactor and are incident on a monochromator crystal. This is silicon, oriented so that the (1 1 1) planes Bragg reflect 6.2 Å neutrons at 180°. Because of this 180° reflection angle, θ, the wavelength spread (and hence resolution) which is proportional to $\cot(\theta/2)\,\mathrm{d}\theta$ is very narrow. This 6.2 Å beam travels back along its path to a graphite crystal and is there reflected at 45° through a chopper (the function of which will be explained later) to a sample. Neutrons scattered by the sample are incident on silicon crystal analysers oriented with the 111 planes again in back reflection for 6.2 Å neutrons. The analysers are on curved mountings focused on detectors just behind the sample position. If the sample scatters inelastically, the intensity counted will be diminished since some neutrons no longer reach the analysers with the necessary 6.2 Å wavelength. However, the monochromator crystal is mounted on a Doppler drive which imparts a small shift in energy to the 6.2 Å reflected neutrons. Only if they

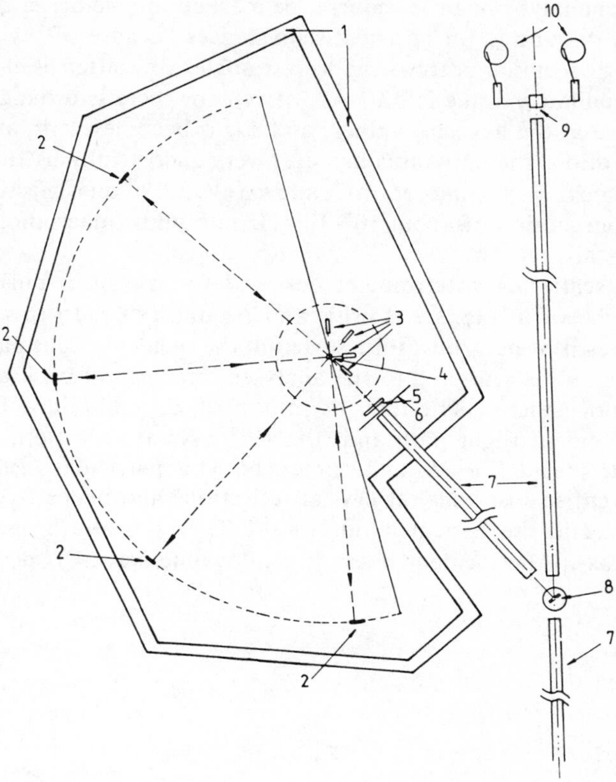

Fig. 3.8 Schematic diagram of the back scattering spectrometer, IN10 at the Institut Laue–Langevin in Grenoble. (ILL 1992.) (1) Shielding; (2) analyser crystals; (3) ^{3}He detectors; (4) sample; (5) monitor; (6) auxiliary chopper; (7) neutron guide; (8) graphite crystal; (9) monochromator crystal; (10) Doppler drive.

lose or gain in the sample the energy they have gained or lost at the Doppler drive, so that they again have exactly 6.2 Å wavelength, will the neutrons be reflected back at the analysers and reach the detector. (The auxiliary chopper coarsely pulses the neutrons and time-of-flight analysis allows discrimination against neutrons scattered directly into the detectors.) In this way, a small energy scan of about 30 μeV can be made with a very sharp (1 μeV) resolution. As can be seen, in this case resolution has been won at the expense of the width of the energy window.

Other limitations concern the q range and resolution. In order to increase the counting flux, the focusing analyser crystal banks are made relatively large, relaxing the angular resolution, and giving a $\Delta q/q$ of order 10 per cent. These analysers can, of course, be masked to give better q resolution but this is only done for special circumstances because of the severe flux penalties. For similar reasons the lowest angles of scatter used are limited and the smallest q value is 0.07 $Å^{-1}$. Even so, there is a danger of some interference at the lowest q values, and $\Delta q/q$ becomes very large. Within these limitations the apparatus has been very successfully used in observation of tunnelling splittings of order a few μeV and of quasielastic scattering arising from motion of about 10^9–10^{10} Hz in liquids, macromolecules, and liquid crystals.

A back-scattering spectrometer designed for use on a pulsed neutron source is shown in Fig. 3.9 (Carlile and Adams 1992). In this instrument a 'white' beam of neutrons, time sorted in the incident flight path, falls on the sample and is scattered to the analyser crystals and back at 175° into the detection system. The high definition of E_i is obtained by the long (36.5 m) incident flight path and, the 175° crystal reflection, slightly off back scattering, defines E_f quite precisely. This particular instrument has five different sets of analyser crystal reflections and hence five resolution functions, with the best at about 1 μeV. The relatively broad spread of wavelengths in the incident beam (the allowable spread is determined by

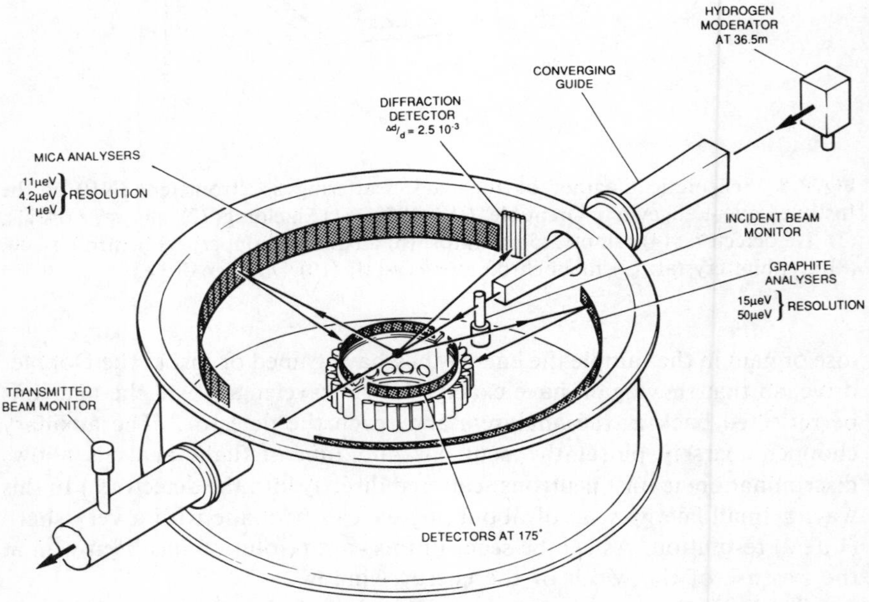

Fig. 3.9 Schematic diagram of the IRIS spectrometer at the pulse neutron source, ISIS.

frame overlap considerations) means that a wider energy scan can be achieved than for the reactor instrument. On this spectrometer an energy range of -0.4 to $+0.4$ meV with a resolution of 11 μeV is available while at 1 μeV resolution the range is still 100 μeV.

3.4.4 Main chain motion

The main chain diffusive motion of polymers is very slow compared to that of small molecules, at least at small values of q where motion of large sections is explored. Very high resolution spectrometers with resolution better than 1 μeV are required. In practice this has meant the back-scattering spectrometer for $0.2 < q < 1$ Å^{-1} and the spin-echo spectrometer for $q < 0.2$ Å^{-1}.

The solutions or molten samples are contained in aluminium cells for the higher q measurements, but since aluminium has a sizeable small-angle scattering signal, quartz or niobium containers have been used at low q for coherent scattering samples, particularly on the spin-echo spectrometer. Incoherently scattering melts are again required to be 0.25 mm thick, but the coherently scattering solution, where the solvent is the deuterated species, are of the order of 0.5 cm thick.

For the back-scattering spectrometer, instrumental resolution is again dealt with by convoluting suitable model functions with the observed resolution (obtained from vanadium scattering) and fitting to the corrected sample scattering. The spin-echo technique to be described in the next section, however, produces the Fourier transform of $S(q, \omega)$, $S(q, t)$ (Heidemann 1980).

3.4.5 The spin-echo spectrometer

This technique originally devised by Mezei (1972) (see Chapter 2, Section 2.8) uses the precession of the neutron spin in a magnetic guide field as a counter to measure very small changes in neutron velocity. A good introduction to the technique is given by Nicholson (1981). In order to explain the method, it is necessary to follow the path of neutrons through the apparatus, which is shown schematically in Fig. 3.10).

The spectrometer (Hayter 1978, 1980; Dagleish *et al.* 1980) has two identical arms on either side of the sample position, each consisting of a length of solenoid providing a magnetic guide field directed along the flight path, and a $\pi/2$ spin turn coil. The incoming neutrons are roughly monochromated using a velocity selector—the wavelength spread $\Delta\lambda/\lambda$ is about 10 per cent. This monochromator is, as will be seen, unnecessary for the *energy* resolution (Hayter 1980; Dagleish *et al.* 1980) but dominates the q-resolution. After the polarizer the neutron spins in the beam are aligned

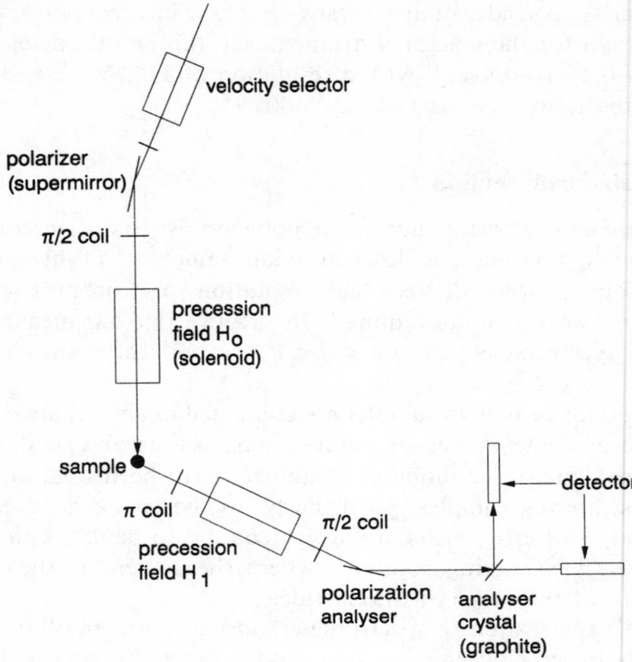

Fig. 3.10 Schematic diagram of the spin-echo spectrometer, IN11 at the Institut Laue–Langevin in Grenoble. (ILL 1992.)

along the flight path. The $\pi/2$ spin turn coil rotates the neutron spin direction which starts precessing about the guide field. The precession rate of each neutron is the same—given by the Larmor precession frequency ($= \omega_L = 2\mu_n B_0/h$ where μ_n is the magnetic moment and B_0 the guide field strength). However, the number of precessions will be governed by the length of time taken to traverse the guide field, i.e. the neutron velocity. A beam containing a wavelength spread loses its initial phase coherence as it travels through the guide field and in analogy with NMR the spins will have fanned out with respect to each other. In the π turn coil, located at the sample, the neutrons make a 180° precession *about a field perpendicular to the guide field* (H^1). This has the effect of reflecting the distribution in the H^1 axis. Those spins that were furthest behind in the fan of precession angles are now furthest ahead and vice versa. Thus, if the length of the second guide field is adjusted to be identical to the first, all the spins will once more be aligned at the second $\pi/2$ turn coil and 100 per cent polarization in the flight path direction will be observed on the analyser.

To observe quasielastic scattering, the number of precessions in each guide field is set equal for elastic events so that small changes in energy at the sample result in a neutron performing unequal numbers of precessions in the two fields thus producing a reduction in the polarization detected at the analyser.

The net polarization (P_z) is then given by

$$(P_z) = \int_0^\infty F(\lambda)\mathrm{d}\lambda \int_{-\infty}^{+\infty} P(\lambda, \delta\lambda) \cos \frac{2\pi N_0 \delta\lambda}{\lambda_0} \mathrm{d}(\delta\lambda) \tag{3.18}$$

where $F(\lambda)\mathrm{d}\lambda$ is the wavelength spread and $P(\lambda, \delta\lambda)$ is the probability that a neutron of wavelength λ will be scattered with a wavelength change $\delta\lambda$. Since $\delta\lambda$ can be transformed into an energy change $\Delta E = \hbar\omega$ this is, of course, just the scattering law $S(\mathbf{q}, \omega)$.

$$P(\lambda, \delta\lambda)\mathrm{d}(\delta\lambda) = S_{\mathrm{coh}}(\boldsymbol{q}, \omega)\mathrm{d}\omega \tag{3.19}$$

N_0 is the number of precessions made by a neutron of the mean wavelength λ_0. Note that because we follow the neutron spins it is $S_{\mathrm{coh}}(\boldsymbol{q}, \omega)$ which is measured. Now since

$$E = \frac{h^2}{2m}\frac{1}{\lambda^2} \tag{3.20}$$

$$\hbar\omega = \Delta E = \frac{h^2}{m}\frac{\delta\lambda}{\lambda^3} \quad \text{or} \quad \delta\lambda = \frac{m\lambda^3}{2\pi h}\omega$$

and

$$\langle P_z \rangle = \int_0^x F(\lambda)\mathrm{d}\lambda \int_{-\infty}^{\infty} S_{\mathrm{coh}}(\boldsymbol{q}, \omega) \cos\left\{\left(\frac{N_0 m\lambda^3}{h\lambda_0}\right)\omega\right\}\mathrm{d}\omega. \tag{3.21}$$

The expression $N_0 m\lambda^3/h\lambda_0$ has the dimensions of time, and can be designated $t(\lambda)$.

P_z then, is just the Fourier transform of $S_{\mathrm{coh}}(\boldsymbol{q}, \omega)$, i.e. the time correlation function $S_{\mathrm{coh}}(\boldsymbol{q}, t)$ is observed directly in spin-echo experiments. The experimental time scale $t(\lambda)$ is governed through N_0 by the strength of the magnetic guide field. For neutrons of 8 Å the time scale covers about two decades from 10^{-9} to 10^{-7} s. For study of polymer dynamics, direct observation of $S_{\mathrm{coh}}(\boldsymbol{q}, t)$ has considerable advantages as will appear below when analysis of the experimental data is considered.

As with the back-scattering spectrometer, the very high resolution (better than 0.1 μeV, 10^8 Hz) is won at the expense of q resolution. In this case, the angular definition of q is very good, since the beam has to be highly collimated within the guide fields. $\Delta\lambda/\lambda$ is, however, quite broad (of order 10 per cent) in order to increase flux and this imposes a 10 per cent limit on $\Delta q/q$. On the other hand, the tight collimation does allow use of small (of order 1°) scattering angles and the smallest q available is about 0.025 Å^{-1}.

One further important property of the spin-echo technique is its ability to distinguish coherent from incoherent scattering. Most incoherence arises from the neutron spin and since this machine explicitly follows the spin, it is possible to select only the coherently scattered neutrons. In cases where the coherent and incoherent scattering laws differ, this property is a great advantage. Also at low q the coherent scattering law has high intensity for macromolecular systems conveniently increasing the signal-to-noise ratio.

The data are corrected in the normal way for background (including cell or solvent scattering), self-absorption, detector efficiency, etc., but now the instrumental resolution can be directly removed. This is because the Fourier transform of a convolution is a simple product. The instrumental resolution is observed via the elastic scattering from a glassy polymer sample containing some labelled chains. It is then removed from the data by point-by-point division to leave the pure intermediate scattering function $S'_{\text{coh}}(\boldsymbol{q}, t)$ $(= S_{\text{coh}}(\boldsymbol{q}, t)/S_{\text{coh}}(\boldsymbol{q}, 0))$ of the sample. In principle of course, the $S(\boldsymbol{q}, \omega)$ data from the back-scattering spectrometer could be treated in the same way by Fourier transforming to $S'(\boldsymbol{q}, t)$, but experience shows that truncation errors generally lead to large uncertainty in the Fourier transform (Heidemann 1980).

We add just one final point to this chapter. No-one has yet constructed a spin-echo spectrometer for a pulsed source. Indeed at the time of going to press there are only two working spectrometers and one more under construction. This is not a technique that will be in common use for studying polymer motion. The results are, however, unique and it is for this reason that the *modus operandi* has been included in this book.

SUGGESTED FURTHER READING

Egelstaff (1965).
Willis (1973).
Windsor (1981).

4

Theoretical basis of scattering

4.1 INTRODUCTION

Some very simple ideas about the basis of neutron scattering theory have already been introduced in Chapter 1. It is now necessary to develop mathematically the formalism in which the angular and energy dependence of scattering from model systems can be calculated and then used to interpret experimental data. As was indicated, this formalism is based on two so-called structure factors: $S(q)$, the static structure factor and $S(q, \Delta E)$ the dynamic structure factor where q is the momentum transfer and ΔE the difference in energy between the incident neutron E_i and the scattered neutron E_f.

The strategy adopted in this chapter is to begin with the interactions between a neutron and a single nucleus at rest (without motion and not changing its quantum state) and thus to consider the basic physics of the neutron scattering length b and the neutron scattering cross section σ_s. Extension to an assembly of nuclei introduces the contrast between the bound atom and the free atom cross section and between coherent and incoherent scattering. In an experiment, measurements relate the fractions of neutrons scattered into a solid angle $d\Omega$ at a chosen scattering angle θ, and in a certain energy range dE to their initial intensity and energy. Formally, this relationship is expressed as a differential scattering cross section $\partial^2\sigma/\partial\Omega\partial E$. Theories must then relate the measured quantity $\partial^2\sigma/\partial\Omega\partial E$ to the positions and scattering cross sections of the scattering nuclei in the sample. If these positions are considered as a function of time, the dynamic properties of the sample are included and these will be studied in the second half of the chapter. If, as we shall do in the first half, the atoms are considered as immobile we restrict ourselves to the spatial arrangements only; in this case it is the structure alone that matters. The values of the neutron cross section of the nuclei composing the sample are in a certain way like the colour of an illustration and therefore irrelevant to the structure (and the dynamics) which are the focus of the experiments, though they are, of course, fundamental in producing a measured set of data. For this reason the differential cross sections are split up into the parts which are a function of the technique, the cross sections, and those which are properties of the sample, the structure factors, $S(q)$ and $S(q, E)$ (or

$S(q, \Delta E)$)). The aim of this chapter is to arrive at the definitions of these structure factors. Subsequent discussion then focuses on derivations of structure factors for different model scattering systems. In Chapters 5–8 we concentrate on the static scattering, giving structural information, in Chapter 9 we turn to the dynamics.

4.2 DEFINITION OF THE CROSS SECTION

Let us assume that we have a parallel beam of neutrons travelling in the direction defined by the vector $\boldsymbol{k}$, (Fig. 4.1). We define the intensity of the beam I_0 by the number of neutrons per unit area and per unit time, $I_0 = N/At$ where N is the number of neutrons arriving during the time t and A the area of the sample. This neutron beam of energy E_i is incident on a target which is small compared to the dimensions of the beam and which could be just one nucleus. Since the target is scattering neutrons it behaves like a new neutron source and will send neutrons in all the spatial directions with, in the general case, different energies. We define the scattering cross section σ_s by the total number of neutrons scattered per second divided by the intensity of the incident beam I_0.

If we consider a small solid angle $d\Omega$ (Fig. 4.1) and collect the neutrons scattered per second in it, their number will depend on their energy, on the orientation of $d\Omega$, and on I_0. This number will be called

$$I_0 \frac{\partial^2 \sigma_s}{\partial \Omega \partial E} d\Omega dE \tag{4.1}$$

the partial differential cross section. Integrating over all the final energies gives the differential cross section $\partial\sigma/\partial\Omega$ which corresponds to the total number of neutrons scattered per second into $d\Omega$, regardless of their energies.

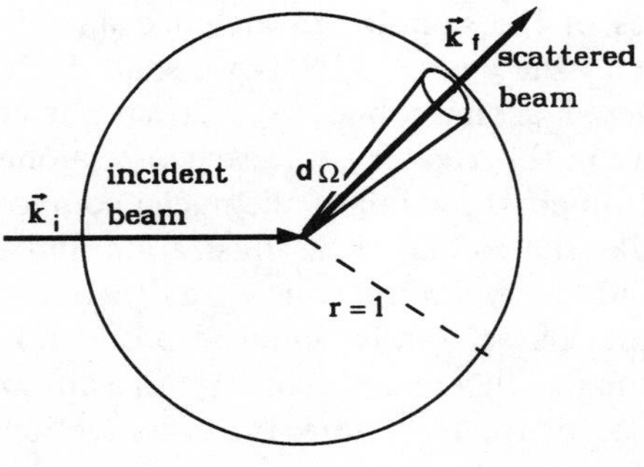

Fig. 4.1 The geometry of neutron scattering.

$$I_0 \mathrm{d}\Omega \int \frac{\partial^2 \sigma_s}{\partial \Omega \partial E} \mathrm{d}E = I_0 \frac{\partial \sigma}{\partial \Omega} \mathrm{d}\Omega. \tag{4.2}$$

Integrating over all directions gives, as it should, the product of the total scattering cross section σ_s by the incident intensity. One can also define the absorption cross section which measures the number of neutrons absorbed by the sample σ_a and the total cross section σ_t, writing

$$\sigma_t = \sigma_s + \sigma_a. \tag{4.3}$$

In the neutron-scattering experiments which will be discussed throughout this book one assumes that the effects of absorption have been corrected and that one measures only σ_s. The quantity σ has the dimension of area and is usually quoted in barns (10^{-28} m^2).

4.3 ELASTIC INTERACTION BETWEEN A NEUTRON AND A SINGLE NUCLEUS

Neutrons and nuclei are small particles for which wave–particle duality becomes important. A neutron behaves both as wave—it can show diffraction effects—and as a particle—it is detected at one point in space. While this duality does mean that neutrons are diffracted by the structural arrangements of nuclei within the sample and therefore forms the basis of the utility of neutron scattering as an experimental technique, it also means that the tools of quantum mechanics are required in order to describe the scattering processes. Fortunately, because the neutron energies are orders of magnitude less than those energies binding nuclei it is not necessary to embark on a nuclear physics course here and the details of neutron interactions with internal nuclear structure can be ignored.

As already explained, in quantum mechanics, a particle of velocity $\boldsymbol{v}$ is represented by a plane wave of wavelength λ related to its mass m and velocity v by the well-known de Broglie formula

$$\lambda = \frac{h}{mv}. \tag{4.4}$$

Let us assume that we have in the incident beam neutrons travelling in the direction of a vector $\boldsymbol{k}_i$. As we have seen in Chapter 1 the amplitude of this wave at a point $\boldsymbol{r}$ can be written, using complex notation as

$$\Psi(\boldsymbol{r}, t) = \exp(\mathrm{i}\omega t - \boldsymbol{r}\cdot\boldsymbol{k}_i) \tag{4.5}$$

where ω is the angular frequency and $2\pi/\lambda$ the quantity $|\boldsymbol{k}_i|$. If we replace λ by its value from eqn (4.4)

$$\boldsymbol{k}_i = \frac{2\pi}{h} m\boldsymbol{v} = \frac{m\boldsymbol{v}}{\hbar} \tag{4.6}$$

we see that $\boldsymbol{k}_i$ is the momentum of the neutrons, measured in $\hbar$ units. In the case of isotropic elastic scattering, a nucleus, struck by this wave will scatter, producing a spherical wave. This wave can be considered, far from its origin, as a plane wave with an amplitude decreasing as $1/|\boldsymbol{r}| = 1/r$ and can be expressed, for large distances as

$$\Psi'(\boldsymbol{r}) = \frac{\mathfrak{b}}{r}\exp i(\omega t - \boldsymbol{k}_f \cdot \boldsymbol{r}) \tag{4.7}$$

$$\Psi'(r) = \frac{\mathfrak{b}}{r}\exp i(\omega t - k_f r) \tag{4.7'}$$

since the vectors $\boldsymbol{k}_f$ and $\boldsymbol{r}$ have the same direction. The problem is now to determine the quantity $\mathfrak{b}$. Its square $\mathfrak{b}\mathfrak{b}^*$ describes the probability that an initial plane wave (with amplitude unity) will be scattered in the direction $\boldsymbol{r}$. In order to find the value of the quantity $\mathfrak{b}$ we need the time-independent Schrödinger equation

$$\left(\frac{-\hbar^2}{2m}\nabla^2 + V(\boldsymbol{r})\right)\Psi(\boldsymbol{r}) = E\Psi(\boldsymbol{r}) \tag{4.8}$$

where ∇^2 is the Laplacian operator. $V(\boldsymbol{r})$ is called the Fermi pseudopotential; it describes the interactions between the neutron and the nucleus. Since nuclear interactions are important only at very short distances (10^{-14} m) $V(\boldsymbol{r})$ can be replaced by the product of a delta function (see Appendix 1) and a constant b

$$V(\boldsymbol{r}) = -\frac{2\pi\hbar^2}{m}\, b\delta(\boldsymbol{r}). \tag{4.9}$$

We require a solution of eqn (4.8) which is stationary and satisfies the boundary conditions and, at large distances, is of the form

$$\Psi'(\boldsymbol{r}, t) = \exp(\mathrm{i}\omega t - \boldsymbol{r}\cdot\boldsymbol{k}_i) + \frac{\mathfrak{b}}{r}\exp i(\omega t - \boldsymbol{k}_f \cdot \boldsymbol{r}) \tag{4.10}$$

where the first term corresponds to the incident neutron beam. This solution is obtained by a perturbation method and the first approximation is called the first Born approximation.

Without going through the details of the calculation we give here only the result when the quantum state of the nucleus or its energy stays constant. Using this approximation the following solution is obtained

$$\Psi(\boldsymbol{r}) = \exp \mathrm{i}\boldsymbol{k}_i\cdot\boldsymbol{r} + \frac{\exp \mathrm{i}\boldsymbol{k}_i\cdot\boldsymbol{r}}{r}\frac{m}{2\pi\hbar^2}\iiint \exp -\mathrm{i}\boldsymbol{k}_f\cdot\boldsymbol{r}'\, V(\boldsymbol{r}')\, \exp \mathrm{i}\boldsymbol{k}_i\cdot\boldsymbol{r}'\, d\boldsymbol{r}' \tag{4.11}$$

(in this expression $\boldsymbol{r}'$ is just a dummy integration variable). We introduce the classical quantity $\boldsymbol{q} = \boldsymbol{k}_i - \boldsymbol{k}_f = 4\pi/\lambda \sin(\theta/2)$. Remember that $k_i = k_f$ since in an elastic collision the kinetic energy, and therefore the absolute value of the velocity does not change. Thus, comparing eqns (4.10) and (4.11) we obtain for $\mathscr{b}$

$$\mathscr{b} = \frac{m}{2\pi\hbar^2}\iiint \exp(-\,\mathrm{i}\boldsymbol{q}\cdot\boldsymbol{r}')V(\boldsymbol{r}')\mathrm{d}\boldsymbol{r}'. \tag{4.12}$$

This shows that $\mathscr{b}$ is the Fourier transform of the Fermi pseudopotential which has been approximated by a constant multiplied by a delta function (eqn (4.9)). Following Appendix 1, a delta function has a Fourier transform equal to unity and, after a simple identification of the terms we obtain

$$\mathscr{b} = -b \tag{4.13}$$

b is called the scattering length and can be positive, negative or complex. The sign convention is only important for problems where one evaluates the amplitude of the wave such as in reflection experiments (see Chapter 10). It does not have any effect in the case of scattering where one uses b^2 or $b_i b_j$. The convention chosen has the advantage of giving a positive sign for the majority of the b values usually tabulated. The case where b is complex corresponds to the capture of the neutron in a non-radiative process (e.g. absorption). In experimental practice one can correct for the effect of absorption (see Chapter 3). In this theoretical chapter we shall not discuss it further. The square of the amplitude corresponding to the scattered wave of eqn (4.5) is, at a distance r, given by: $\Psi\Psi^* = b^2/r^2 = \mathscr{b}\mathscr{b}^*/r^2$. Since $\Psi\Psi^*$ is the probability of having a neutron scattered at the point $\boldsymbol{r}$, the probability that it is scattered into the solid angle $\mathrm{d}\Omega$, which is the surface cut on the sphere (see Fig. 4.1) centred on the scattering nucleus having radius unity, is b^2.

From our earlier definition

$$\frac{\partial\sigma}{\partial\Omega} = b^2 \tag{4.14}$$

and the total cross section σ is

$$\sigma = \iint \frac{\partial\sigma}{\partial\Omega}\,\mathrm{d}\Omega = 4\pi b^2. \tag{4.15}$$

Tabulated values of b and σ (see Chapter 2) refer to bound-atom values. Clearly in most structures the atoms are bound into molecules, and the molecules are in a condensed state. In the case of free atoms, recoil is possible and the calculation of the momentum change has to be made in centre of mass coordinates. The free atom scattering length b_f is the value obtained from the bound atom by dividing by the reduced mass.

4.4 COHERENT AND INCOHERENT ELASTIC SCATTERING

4.4.1 Mixtures of isotopes

In the previous paragraph we have seen that the amplitude of the wave scattered by a nucleus can be written as b. If now we have a system made of N nuclei, each of them having a scattering length b_k, the generalization of eqn (4.7) gives

$$\Psi(\boldsymbol{r}) = \sum_k^N \frac{b_k}{r_k} \exp(\mathrm{i}\boldsymbol{q}\cdot\boldsymbol{r}_k) \tag{4.16}$$

where $\boldsymbol{r}_k$ defines the distance of the kth atom from the observer and b_k its scattering length. Since r_k is large it can be replaced by r, the distance between the target and the observer in the first term of eqn (4.16). The scattering intensity is obtained by taking the product $\Psi\Psi^*$ and averaging it using the symbol $\langle\,\rangle$. Generalizing eqn (4.14) leads to

$$\begin{aligned}\frac{\partial\sigma}{\partial\Omega} &= \left\langle \sum_k^N b_k \exp(\mathrm{i}\boldsymbol{q}\cdot\boldsymbol{r}_k) \sum_j^N b_j \exp(-\mathrm{i}\boldsymbol{q}\cdot\boldsymbol{r}_j)\right\rangle \\ &= \sum_{j,k}^N \langle b_j b_k \exp(\mathrm{i}\boldsymbol{q}\cdot(\boldsymbol{r}_k - \boldsymbol{r}_j))\rangle.\end{aligned} \tag{4.17}$$

We can divide the double sum into two parts: the terms for which $j = k$ and the terms for which $j \neq k$. This gives

$$\frac{\partial\sigma}{\partial\Omega} = \sum_{k=1}^N b_k^2 + \sum_{j\neq k}^N \langle b_j b_k \exp(\mathrm{i}\boldsymbol{q}\cdot(\boldsymbol{r}_k - \boldsymbol{r}_j))\rangle. \tag{4.18}$$

The first term is simply N times the average value of $\langle b^2\rangle$. In the second term b_j and b_k are independent since their values are not correlated with the positions of the atoms, and one can replace $b_j b_k$ by $\langle b_j\rangle\langle b_k\rangle = \langle b\rangle^2$. Equation (4.18) then becomes

$$\frac{\partial\sigma}{\partial\Omega} = N\langle b^2\rangle + \langle b\rangle^2 \sum_{j\neq k}^N \langle \exp(\mathrm{i}\boldsymbol{q}\cdot(\boldsymbol{r}_k - \boldsymbol{r}_j))\rangle. \tag{4.19}$$

Defining

$$\overline{\Delta b^2} = \langle b^2\rangle - \langle b\rangle^2 \tag{4.20}$$

where

$$\overline{\Delta b^2} = \langle (b - \bar{b})^2 \rangle = \langle b^2 - 2b\langle b \rangle + \langle b \rangle^2 \rangle = \langle b^2 \rangle - \langle b \rangle^2$$

we can write after adding and subtracting $\langle b^2 \rangle$

$$\frac{\partial \sigma}{\partial \Omega} = N\overline{\Delta b^2} + \langle b \rangle^2 \sum_{j,k}^{N} \langle \exp(i\mathbf{q} \cdot (\mathbf{r}_k - \mathbf{r}_j)) \rangle \tag{4.21}$$

where, in the double sum, j and k can have all the values between 1 and N. The first term in this equation depends only on the fluctuations in the value of the length b. This is the incoherent scattering discussed in Chapter 1, Section 8. From eqn (4.21) we define $\sigma_{\text{coh}} = 4\pi\langle b \rangle^2$ and $\sigma_{\text{inc}} = 4\pi\overline{\Delta b^2}$. The incoherent scattering in eqn (4.21) depends on nuclei with different scattering lengths occupying different positions. The usual source of such different scattering lengths would be different isotopes. Thus an isotopically pure sample would be expected to show no incoherent scattering.

4.4.2 The effect of nuclear spin

The effect we have just described is only rigorous if the nuclei carry no spin. But the neutron carries two possible spin values $+ 1/2$ or $- 1/2$ which, when they interact randomly with the nuclear spin state i, have two scattering lengths b^+ and b^-. For any spin quantum number L there are $2L + 1$ possible states arising from the relative orientation of the neutron–nuclear spin pairs so that for b^+ there are $2(i + 1/2) + 1 = 2(i + 1)$ states and for b^- there are $2(i - 1/2) + 1 = 2i$ states. The total number of states is, of course, $2i + 2 + 2i = 2(2i + 1)$. The probability of b^+ occurring is therefore $(i + 1)/(2i + 1)$ and of b^- is $i/(2i + 1)$ and hence

$$\langle b \rangle = \left[\frac{i + 1}{2i + 1}\right] b^+ + \left[\frac{i}{2i + 1}\right] b^- \tag{4.22}$$

and

$$\langle b^2 \rangle = \left[\frac{i + 1}{2i + 1}\right] b^{+2} + \left[\frac{i}{2i + 1}\right] b^{-2} \tag{4.23}$$

Now for hydrogen $i = 1/2$, $b^+ = 1 \times 10^{-12}$ cm and $b^- = -4.7 \times 10^{-12}$ cm are the experimental scattering lengths. In eqn (4.22) and (4.23) this leads to

$$\sigma_{\text{coh}} = 4\pi\langle b \rangle^2 = 1 \times 10^{-24}\,\text{cm}^2$$

$$\sigma_{\text{inc}} = 4\pi\overline{\Delta b^2} = 4\pi[\langle b^2 \rangle - \langle b \rangle^2] = 80 \times 10^{-24}\,\text{cm}^2.$$

Thus the coherent cross section of hydrogen is very large and, in fact, more than one order of magnitude larger than the incoherent cross section of other nuclei. It is considerably larger than all coherent cross sections as well and this has some very important consequences, as will be seen in later sections. The second term in eqn (4.19) depends on the position of all the scattering centres. This is the coherent scattering already mentioned in Chapter 1 and responsible for the angular dependence of the scattered intensity (Fig. 1.5). In the following sections we shall use for the coherent scattering length the letter b_c or b when there is no possible confusion. For the incoherent scattering we shall either use $\overline{\Delta b^2}$ if we want to remember the origin of this scattering or b_{inc}^2.

4.5 THE STATIC CORRELATION FUNCTION

Most of the discussions which we shall have in this book do refer to coherent scattering. It is therefore indispensable to discuss in more detail the meaning of the second term of eqn (4.21). In this general introduction we shall begin by considering a one component fluid, leaving the generalization to multicomponent systems for the following chapters. From what we have just seen the coherent scattering can be written as

$$\left(\frac{\partial \sigma}{\partial \Omega}\right)_{\text{coh}} = b^2 \left\langle \sum_{j=1}^{N} \exp(\mathrm{i}\boldsymbol{q}\cdot\boldsymbol{r}_j) \sum_{k=1}^{N} \exp(-\mathrm{i}\boldsymbol{q}\cdot\boldsymbol{r}_k) \right\rangle \tag{4.24}$$

where the sign $\langle\ \rangle$ indicates that we take an ensemble average over the value which is inside the brackets. One could also use a time average over one pair of molecules since, when the system is ergodic both averages coincide. In the preceding section b was the coherent scattering of a nucleus. Let us now consider small molecules. Since, in the low angle scattering approximation one can assume that all nuclei of the same molecule are situated at the same point $\boldsymbol{r}$, one can add all the scattering lengths of the nuclei of a single molecule and call b the coherent part of this sum: $b = \Sigma\, b_i$ extending the summation to all nuclei belonging to the same molecule. Transforming the product of sums into a simple double sum one obtains the equivalent formula

$$\left(\frac{\partial \sigma}{\partial \Omega}\right)_{\text{coh}} = b^2 \sum_{j=1}^{N} \sum_{k=1}^{N} \langle \exp(\mathrm{i}\boldsymbol{q}\cdot(\boldsymbol{r}_j - \boldsymbol{r}_k)) \rangle . \tag{4.25}$$

It is often useful to replace the discrete sum of eqns (4.24) or (4.25) with an integral. The method for doing so is explained in Appendix 1. Using the properties of the delta function we define what we shall call the local density of molecules by the expression

$$n(\boldsymbol{r}) = \sum_{j=1}^{N} \delta(\boldsymbol{r} - \boldsymbol{r}_j). \tag{4.26}$$

where *n(r)* is the density of molecules at any point $\boldsymbol{r}$. Since, as shown in Appendix 1

$$\int_V f(\boldsymbol{r})\delta(\boldsymbol{r}-\boldsymbol{r}_k)\mathrm{d}\boldsymbol{r} = f(\boldsymbol{r}_k). \tag{4.27}$$

it is evident that one can write

$$\begin{aligned}\sum_{k=1}^{N}\exp(-\mathrm{i}\boldsymbol{q}\cdot\boldsymbol{r}_k) &= \int_V \sum_{k=1}^{N}\exp(-\mathrm{i}\boldsymbol{q}\cdot\boldsymbol{r}_k)\delta(\boldsymbol{r}-\boldsymbol{r}_k)\mathrm{d}\boldsymbol{r}\\ &= \int_V n(\boldsymbol{r})\exp(-\mathrm{i}\boldsymbol{q}\cdot\boldsymbol{r})\mathrm{d}\boldsymbol{r}.\end{aligned} \tag{4.28}$$

This relationship allows us to write for the coherent scattered intensity

$$\left(\frac{\partial\sigma}{\partial\Omega}\right)_{\mathrm{coh}} = b^2\left\langle \int_V n(\boldsymbol{r})\exp(-\mathrm{i}\boldsymbol{q}\cdot\boldsymbol{r})\mathrm{d}\boldsymbol{r}\int_V n(\boldsymbol{r}')\exp(\mathrm{i}\boldsymbol{q}\cdot\boldsymbol{r}')\mathrm{d}\boldsymbol{r}'\right\rangle \tag{4.29}$$

or in the equivalent form

$$\left(\frac{\partial\sigma}{\partial\Omega}\right)_{\mathrm{coh}} = b^2\int_V\int_{V'}\exp(-\mathrm{i}\cdot\boldsymbol{q}(\boldsymbol{r}-\boldsymbol{r}'))\langle n(\boldsymbol{r})n(\boldsymbol{r}')\rangle\mathrm{d}\boldsymbol{r}\mathrm{d}\boldsymbol{r}' \tag{4.30}$$

In fact, in doing so, we have lost one term which is often neglected; we shall reintroduce it later.

In order to evaluate the integral in eqn (4.30) one has to take one point M at position $\boldsymbol{r}$, another M' at $\boldsymbol{r}'$ and evaluate the average value of the product $n(\boldsymbol{r})n(\boldsymbol{r}')$. If these two points are far apart it is evident that there is no correlation between the density fluctuations at M and M'. In this case the average value of the product is equal to the product of the averages

$$\text{if } |\boldsymbol{r}-\boldsymbol{r}'|\to\infty \text{ then } \langle n(\boldsymbol{r})n(\boldsymbol{r}')\rangle = \langle n(\boldsymbol{r})\rangle\langle n(\boldsymbol{r}')\rangle = \bar{n}^2 = n^2 \tag{4.31}$$

where $n=\bar{n}$ is the average density of scatterers in the medium. It is also evident that, in an isotropic medium, the correlations between density fluctuations at two points M and M' depend only on the distance between these points and neither on the orientation of the vector $\boldsymbol{MM}'$ nor on its position in the volume V. (One neglects surface effects.) This means that the quantity $\langle n(\boldsymbol{r})n(\boldsymbol{r}')\rangle$ is only a function of $|\boldsymbol{r}-\boldsymbol{r}'|$ and can be written as a function of the single variable $r''=|\boldsymbol{r}''|=|\boldsymbol{r}-\boldsymbol{r}'|$. One can therefore introduce the quantity $G(r'')$ by the relationship

$$G_0(r'') = \langle n(\boldsymbol{r})n(\boldsymbol{r}')\rangle = \langle n(0)n(\boldsymbol{r}'')\rangle \tag{4.32}$$

Introducing this definition in eqn (4.30) gives, using as variables $\boldsymbol{r}$ and $\boldsymbol{r}''$

$$\left(\frac{\partial\sigma}{\partial\Omega}\right)_{\text{coh}} = b^2 \int_V \int_{V''} \exp(-\mathrm{i}\boldsymbol{q}\cdot\boldsymbol{r}'')G_0(\boldsymbol{r}'')\mathrm{d}\boldsymbol{r}\mathrm{d}\boldsymbol{r}'' \tag{4.33}$$

after integration over $\boldsymbol{r}$

$$\left(\frac{\partial\sigma}{\partial\Omega}\right)_{\text{coh}} = b^2 \mathrm{V} \int_V \exp(-\boldsymbol{q}\cdot\boldsymbol{r}'')G_0(\boldsymbol{r}'')\mathrm{d}\boldsymbol{r}''. \tag{4.34}$$

If, instead of defining $G_0(\boldsymbol{r}'') = \langle n(0)n(\boldsymbol{r}' - \boldsymbol{r}\rangle$ we define another function by the relationship

$$G(\boldsymbol{r}'') = \langle \Delta n(0)\Delta n(\boldsymbol{r}' - \boldsymbol{r})\rangle \tag{4.35}$$

where $\Delta n(\boldsymbol{r})$ is the difference $n(\boldsymbol{r}) - n$, it is evident that

$$G_0(\boldsymbol{r}'') - n^2 = G(\boldsymbol{r}''). \tag{4.36}$$

In order to show this we write

$$G_0(\boldsymbol{r}) = \langle(\Delta n(\boldsymbol{r}) + n)(\Delta n(\boldsymbol{r}') + n)\rangle = \langle \Delta n(\boldsymbol{r})(\Delta n(\boldsymbol{r}'))\rangle$$
$$+ n\langle(\Delta n(\boldsymbol{r}) + \Delta n(\boldsymbol{r}'))\rangle + n^2\rangle \tag{4.37}$$

and since from its definition $\langle \Delta n(\boldsymbol{r})\rangle = 0$ the second term of the RHS is zero leading to the relationship (4.36). The scattered intensity is then

$$\left(\frac{\partial\sigma}{\partial\Omega}\right)_{\text{coh}} = b^2 \int_V \int_{V'} \exp(-\mathrm{i}\boldsymbol{q}\cdot\boldsymbol{r})\exp(\mathrm{i}\boldsymbol{q}\cdot\boldsymbol{r}')[G(\boldsymbol{r}' - \boldsymbol{r}) + n^2]\mathrm{d}\boldsymbol{r}\mathrm{d}\boldsymbol{r}'. \tag{4.38}$$

The term in n^2 does not contribute to any observable physical quantity (it contributes to scattering inside the incident beam and can therefore be neglected). Using the variables $\boldsymbol{r}$ and $\boldsymbol{r}'' = (\boldsymbol{r}' - \boldsymbol{r})$ allows us to write, after integration over $\boldsymbol{r}$

$$\left(\frac{\partial\sigma}{\partial\Omega}\right)_{\text{coh}} = Vb^2 \int_V \exp(-\mathrm{i}\boldsymbol{q}\cdot\boldsymbol{r})G(\boldsymbol{r})\mathrm{d}\boldsymbol{r} \tag{4.39}$$

which is identical to eqn (4.34).

We have said that, using continuous notation, we have lost a term which is usually negligible. In fact, we should really have written, instead of eqn (4.35)

$$G(\boldsymbol{r}) = \langle \Delta n(\boldsymbol{r})\Delta n(\boldsymbol{r}')\rangle + N\delta(\boldsymbol{r}). \tag{4.40}$$

This can be proved by the following argument. Going back to the delta function definition of the local density and replacing the variables $\boldsymbol{r}$ and $\boldsymbol{r}'$ by $\boldsymbol{r}'$ and $\boldsymbol{r}'' = \boldsymbol{r} - \boldsymbol{r}'$ we can write eqn (4.30) in the form

$$\left(\frac{\partial \sigma}{\partial \Omega}\right)_{\text{coh}} = b^2 \int_V \int_{V''} \exp(-\mathrm{i}\boldsymbol{q}\cdot\boldsymbol{r}'') \left(\sum_{j=1}^{N} \sum_{k=1}^{N} \langle \delta(\boldsymbol{r} + \boldsymbol{r}'' - \boldsymbol{r}_j)\delta(\boldsymbol{r} - \boldsymbol{r}_k) \rangle \right) \mathrm{d}\boldsymbol{r}\mathrm{d}\boldsymbol{r}''. \tag{4.41}$$

Now let us consider one of the N terms of the double sum for which $j = k$. For this term we can write

$$\int_V \int_{V''} \exp(-\mathrm{i}\boldsymbol{q}\cdot\boldsymbol{r}'')\delta(\boldsymbol{r}' + \boldsymbol{r}'' - \boldsymbol{r}_k)\delta(\boldsymbol{r}' - \boldsymbol{r}_k)\mathrm{d}\boldsymbol{r}'\,\mathrm{d}\boldsymbol{r}''. \tag{4.42}$$

Integrating first over $\boldsymbol{r}'$ we obtain

$$\int_{V''} \exp(-\mathrm{i}\boldsymbol{q}\cdot\boldsymbol{r}'')\delta(\boldsymbol{r}'')\mathrm{d}\boldsymbol{r}'' = 1 \tag{4.43}$$

which shows that since there are N terms similar to this one in the double sum the forgotten term is of the form $N\delta(\boldsymbol{r})$. Therefore in order to be complete we have to replace eqn (4.34) and (4.39) by

$$\left(\frac{\partial \sigma}{\partial \Omega}\right)_{\text{coh}} = b^2 \left[V \int_V \exp(-\mathrm{i}\boldsymbol{q}\cdot\boldsymbol{r})G(\boldsymbol{r})\mathrm{d}\boldsymbol{r} + N \right] \tag{4.44}$$

using for G the definition (4.32) or (4.35).

Quite often it is preferable to use the normalized function

$$\Gamma(\boldsymbol{r}) = \frac{1}{n^2}\langle \Delta n(\boldsymbol{r})\Delta n(\boldsymbol{r}')\rangle \tag{4.45}$$

obtaining for the scattering intensity

$$\left(\frac{\partial \sigma}{\partial \Omega}\right)_{\text{coh}} = b^2 N \left[1 + \frac{N}{V} \int_V \exp(\mathrm{i}\boldsymbol{q}\cdot\boldsymbol{r})\Gamma(\boldsymbol{r})\mathrm{d}\boldsymbol{r} \right]. \tag{4.46}$$

In Section 7.12 we shall define the function $g(\boldsymbol{r})$, called the radial distribution function. This function and $\Gamma(\boldsymbol{r})$ are simply related by

$$\Gamma(\boldsymbol{r}) = g(\boldsymbol{r}) - 1. \tag{4.47}$$

Equations (4.34) and (4.39) show that, in the case of a one-component system made of small molecules, the coherent scattering intensity per unit volume is equal to the Fourier transform of the correlation function $G_0(r)$ or $G(r)$ multiplied by the square of the coherent scattering length of the molecules. This result is important since its generalization to inelastic scattering leads to an important formula, (the Van Hove equation) which is the basis of the interpretation of many quasielastic results and will be discussed in Section 4.7.

Another interesting way of presenting the result in eqn (4.30) which can sometimes be useful is to introduce the Fourier transform of the local density which will be called $\hat{n}(\boldsymbol{q})$ and is related to $n(\boldsymbol{r})$ by

$$\hat{n}(\boldsymbol{q}) = \int_V n(\boldsymbol{r})\exp(-\mathrm{i}\boldsymbol{q}\cdot\boldsymbol{r})\mathrm{d}\boldsymbol{r}. \tag{4.48}$$

Then, from eqn (4.29):

$$\left(\frac{\partial\sigma}{\partial\Omega}\right)_{\mathrm{coh}} = b^2\langle\hat{n}(\boldsymbol{q})\hat{n}(-\boldsymbol{q})\rangle. \tag{4.49}$$

Making a Fourier transform of the density is effectively representing the system by a superposition of waves (here density waves) and the scattering in the $\boldsymbol{q}$ direction is just the Bragg diffraction corresponding to the waves of wavevector $\boldsymbol{q}$.

Equations (4.25), (4.39) or (4.49) may be used to calculate the elastic scattering arising from structure within samples. Through the Fourier transform, spatial correlations become variations in the scattering as a function of $\boldsymbol{q}$. Since $\boldsymbol{q}$ and $\boldsymbol{r}$ are Fourier pairs, it is immediately clear that large-scale structures will give scattering at small $\boldsymbol{q}$ values. This is why the interest of neutron scattering techniques for polymer samples centres heavily on small angle scattering. In the next few Chapters (5–8) we will develop the elastic scattering laws for polymeric systems and show how they can be used to interpret experimental data. For the rest of this chapter we turn to inelastic scattering and consider the dynamic scattering functions.

4.6 INELASTIC CROSS SECTION

The reader who looks more deeply into the theory of neutron scattering than the level that is presented in this book will find that we have greatly simplified the discussion which follows in order to bring out the main points in the physics. In particular, we have avoided the use of operators which the reader will find in most texts. We make no apology for these simplifications since we are not attempting to prove the theory, but merely to demonstrate where the main results arise. The formulae we give for the scattering laws are correct, even if the routes to them have only been sketched in.

To understand the form of the dynamic cross section one needs the formalism of quantum mechanics in order to describe correctly the interactions between a neutron and a nucleus. Before being struck by a neutron the nucleus is in one of a number of its possible energy states. During the collision it can change its state of energy either gaining or losing energy. This energy is obtained from the neutron and the energy of the scattered neutron will either be smaller (the atom gains energy) or larger (the atom loses energy). In more quantitative terms the energy of a neutron with momentum $\hbar \boldsymbol{k}_i$ is $E_i = \hbar\omega_i = \hbar^2 k_i^2/2m$. If an atom changes its energy from E_i to E_f, the energy of the neutron will change by the same quantity

$$\hbar\omega_i - \hbar\omega_f = E_i - E_f = \hbar\omega. \tag{4.50}$$

The momentum of the scattered neutron will then be given by

$$\hbar \boldsymbol{k}_i - \hbar \boldsymbol{k}_f = \hbar \boldsymbol{q} \tag{4.51}$$

where $\boldsymbol{q}$ is the momentum transfer to the nucleus. This implies that the energy of the scattered neutrons will have different values and one needs the calculation of $\partial^2\sigma/\partial\Omega\partial E$ in order to analyse the results. From now on we shall replace the energy E of the scattered neutron or more precisely its change of energy ΔE by the corresponding change in circular frequency ω, writing: $\Delta E = \hbar\omega$. For completeness an outline of the derivation of $\partial^2\sigma/\partial\Omega \mathrm{d}\omega$ will be given, but readers of an accepting frame of mind may jump forward to eqn (4.68) and understand it as an extension of eqn (4.39).

We have already defined the dynamic cross section as a function of q and E. We shall begin by considering a single nucleus, and define for this nucleus the dynamic scattering function $s(q, \omega)$

$$\frac{\partial^2\sigma}{\partial\Omega\partial\omega} = \hbar \frac{\partial^2\sigma}{\partial\Omega\partial E} = \frac{k_f}{k_i} b^2 s(q,\omega). \tag{4.52}$$

This equation contains, as a prefactor, the quantity k_f/k_i which was put equal to 1 in the elastic cross section calculation. This factor corrects for the fact that the flux of neutrons on the detector (which is what is measured) depends on the velocity of the neutrons. Its importance depends on the particular experimental set-up used, and it cannot be neglected except in quasielastic scattering where k_f and k_i are almost identical. This definition is also based on the fact that, regardless of the energy exchanges between the neutron and the nucleus, the scattering length is the same and can be factorized. Using the same method as in the case of elastic scattering in the Born approximation, we first consider the simple case of a nucleus going from the initial state α to the final β during the collision and look for the square of the scattering cross section $\mathfrak{f}^2$ corresponding to all the possible transitions. $s(q, \omega)$ can be defined, using the formalism of quantum mechanics, by a generalization of the treatment of Section 4.3 as a sum of terms

$$s(q,\omega) = \sum_{\alpha} p_{\alpha} \sum_{\beta} \{\langle\beta|\exp(\mathrm{i}\boldsymbol{q}\cdot\boldsymbol{r})|\alpha\rangle\}^2 \delta\left(\omega + \frac{E_{\alpha} - E_{\beta}}{\hbar}\right). \quad (4.53)$$

In this equation α designates the initial state of the nucleus and β its final state. In order to obtain the total value of the scattering cross section we have to add all the possible transitions after multiplication by a Boltzman factor p_{α} which is the probability of finding the nucleus in the initial α state

$$p_{\alpha} = \exp(-E_{\alpha}/k_{\mathrm{B}}T)/\sum_{\gamma}\exp(-E_{\gamma}/k_{\mathrm{B}}T). \quad (4.54)$$

The delta function is a simple way to explain that, in this transition, the energy of the scattered neutron is not the energy of the incident beam but satisfies energy conservation. We now use the definition of the delta function which can be written as the Fourier transform of the unit function and by the same token we introduce the time variable as the conjugate of ω. It has been shown in Appendix 1 that

$$\delta(\omega) = \frac{1}{2\pi}\int_{-\infty}^{\infty} \exp(\mathrm{i}\omega t)\mathrm{d}t \quad (4.55)$$

which, in this case can be written as

$$\delta\left(\omega + \frac{E_{a} - E_{\beta}}{\hbar}\right) = = \frac{1}{2\pi}\int_{-\infty}^{\infty} \exp\left[\mathrm{i}\left(\omega + \frac{E_{\alpha} - E_{\beta}}{\hbar}\right)t\right]\mathrm{d}t \quad (4.56)$$

using ω and t as conjugated variables. Now we replace the delta function in eqn (4.53) by its value in eqn (4.56)

$$s(q,\omega) = \frac{1}{2\pi}\sum_{\alpha} p_{\alpha}\sum_{\beta}\int_{-\infty}^{\infty} (\exp(\mathrm{i}\omega t)\mathrm{d}t)\{\langle\alpha|\exp(-\mathrm{i}\boldsymbol{q}\cdot\boldsymbol{r})|\beta\rangle\} \times$$

$$\langle\beta|\exp(-\mathrm{i}E_{\beta}t/\hbar)\exp(\mathrm{i}\boldsymbol{q}\cdot\boldsymbol{r})\exp(\mathrm{i}E_{\alpha}t/\hbar)|\alpha\rangle. \quad (4.57)$$

The Schrödinger equation can be written for the set of states α and β as

$$\mathcal{H}|\alpha\rangle = E_{\alpha}|\alpha\rangle \text{ and } \langle\beta|\mathcal{H}\rangle = \langle\beta|E_{\alpha}. \quad (4.58)$$

From this, one can show, expanding the exponential, that

$$\exp(\mathrm{i}E_{\alpha}t/\hbar)|\alpha\rangle = \exp(\mathrm{i}\mathcal{H}t/\hbar)|\alpha\rangle \quad (4.59)$$

and

$$\langle\beta|\exp(-\mathrm{i}E_{\beta},t/\hbar) = \langle\beta|\exp(-\mathrm{i}\mathcal{H}t/\hbar)\rangle \quad (4.60)$$

giving for the last matrix of eqn (4.57)

$$\langle\beta|\exp(-\mathrm{i}\mathcal{H}t/\hbar)\rangle\exp(\mathrm{i}\boldsymbol{q}\cdot\boldsymbol{r})\exp(\mathrm{i}\mathcal{H}t/\hbar)|\alpha\rangle. \qquad (4.61)$$

The rules of quantum mechanics, obtained using the Heisenberg representation, show that the time dependence of an operator O(τ) is given by

$$\mathrm{O}(t) = \exp(-\mathrm{i}\mathcal{H}t/\hbar)\{\mathrm{O}\}\exp(\mathrm{i}\mathcal{H}t/\hbar). \qquad (4.62)$$

Applying this to eqn (4.56) gives

$$s(q,\omega) = \frac{1}{2\pi}\sum_{\alpha}p_{\alpha}\sum_{\beta}\int_{-\infty}^{\infty}(\exp(\mathrm{i}\omega t)\mathrm{d}t)\cdot$$

$$\langle\alpha|\exp(-\mathrm{i}\boldsymbol{q}\cdot\boldsymbol{r}(0))|\beta\rangle\cdot\langle\beta|\exp(\mathrm{i}\boldsymbol{q}\cdot\boldsymbol{r}(t))|\alpha\rangle. \qquad (4.63)$$

From the classical results of quantum mechanics one can show that the two relationships

$$\sum_{\beta}|\beta\rangle\langle\beta| = 1 \qquad (4.64)$$

and

$$\sum_{\alpha}p_{\alpha}\langle\alpha|\mathrm{O}|\alpha\rangle = \langle\mathrm{O}\rangle \qquad (4.65)$$

where O is any operator, are always valid. This leads us to the final result

$$s(\boldsymbol{q},\omega) = \frac{1}{2\pi}\int_{\infty}^{\infty}\exp(\mathrm{i}\omega t)\mathrm{d}t\langle\exp(-\mathrm{i}\boldsymbol{q}\boldsymbol{r}(o))\exp(+\mathrm{i}\boldsymbol{q}\boldsymbol{r}(t))\rangle \qquad (4.66)$$

this expression has been obtained for one nucleus. It is easy to understand that, if we have many nuclei we have to modify it and take into account the interferences between nuclei, obtaining the final expression

$$\frac{\partial^2\sigma}{\partial\Omega\partial\omega} = \frac{k_\mathrm{f}}{k_\mathrm{i}}\frac{b^2}{2\pi}\sum_{ij}\int_{-\infty}^{\infty}\exp(\mathrm{i}\omega t)\ \langle\exp(-\mathrm{i}\boldsymbol{q}\cdot\boldsymbol{r}_\mathrm{i}(0))\exp(\mathrm{i}\boldsymbol{q}\cdot\boldsymbol{r}_\mathrm{j}(t))\rangle\mathrm{d}t. \qquad (4.67)$$

This formula was obtained by Van Hove (1954) and can be applied to classical motions as well as to quantum effects since it has been obtained by quantum mechanics but contains only classical quantities.

4.7 COHERENT AND INCOHERENT INELASTIC SCATTERING

We have already discussed for elastic scattering the case where all the nuclei do not have the same scattering length; it is interesting to examine what happens in the case of the inelastic scattering. Starting from eqn (4.67) in which one replaces b^2 by $b_i b_j$ and using exactly the same procedure as in Section 4.4 the generalization of eqn (4.21) becomes

$$\frac{\partial^2\sigma}{\partial\Omega\partial\omega} = \frac{k_f}{k_i}\frac{\Delta b^2}{2\pi}\sum_i\int_{-\infty}^{\infty}\exp(i\omega t)\langle\exp(-iq\cdot r_i(0))\exp(iq\cdot r_i(t))\rangle dt$$
$$+\frac{k_f}{k_i}\frac{b^2}{2\pi}\sum_{ij}\int_{-\infty}^{\infty}\exp(i\omega t)\langle\exp(-iq\cdot r_i(0))\exp(iq\cdot r_j(t))\rangle dt \quad (4.68)$$

The first term, which is called $\partial^2\sigma/\partial\Omega\partial\omega_{\text{inc}}$ (inc stands for incoherent) is no longer a constant as in elastic scattering. It gives information about the motions of the individual nuclei but nothing about their correlations since it is just the sum of the inelastic intensities scattered by each of the nuclei. The second term called $\partial^2\sigma/\partial\Omega\partial\omega_{\text{coh}}$ (coh stands for coherent) can, however, give information about the relative motion between nuclei i and j. It should be noted that this term includes pairs for which $i = j$ which means that coherent scattering contains a contribution from incoherent scattering. Another generalization which can be made here, as well as in the case of static scattering is to introduce the molecules and their scattering length. As long as we are working at q^{-1} values which are large compared to the dimensions of a monomer this is completely justified. This means that we will consider the scattering length b and the quantity $\sqrt{\Delta b^2}$ not as the coherent and incoherent scattering length of an atom but as the sum of the scattering lengths of the nuclei present in a small molecule or in a monomer repeating unit of a polymer since all the corresponding nuclei are scattering 'in phase'.

4.8 VAN HOVE EQUATION AND CORRELATION FUNCTION

4.8.1 The correlation function

Due to the similarity of the static and dynamic equations it is possible to generalize the spatial correlation function to time-dependent problems. It is sufficent to introduce the function $n(r, t)$ by a definition similar to eqn (4.26) writing

$$n(\boldsymbol{r}, t) = \sum_{j=1}^{N} \delta(\boldsymbol{r} - \boldsymbol{r}_j(t)). \tag{4.69}$$

where $n(\boldsymbol{r}, t)$ is the local density at time t. Using eqn (4.69) we write

$$\sum_{k=1}^{N} \exp(-i\boldsymbol{q}\cdot\boldsymbol{r}_k(t)) = \int_V \sum_{k=1}^{N} \exp(-i\boldsymbol{q}\cdot\boldsymbol{r}_k(t))\delta(\boldsymbol{r} - \boldsymbol{r}_k(t))\mathrm{d}\boldsymbol{r}$$

$$= \int_V n(\boldsymbol{r}, t)\exp(-i\boldsymbol{q}\cdot\boldsymbol{r})\mathrm{d}\boldsymbol{r}. \tag{4.70}$$

This relationship allows us to transform eqn (4.30) for coherent scattering into

$$\left(\frac{\partial^2\sigma}{\partial\Omega\partial\omega}\right)_{\mathrm{coh}} = \frac{k_\mathrm{f}}{k_\mathrm{i}}\frac{b^2}{2\pi}\int_{-\infty}^{+\infty} \exp(\mathrm{i}\omega t)\mathrm{d}t \iint_V \langle n(\boldsymbol{r}, 0)n(\boldsymbol{r}', t)\rangle$$

$$\exp \mathrm{i}\boldsymbol{q}\cdot(\boldsymbol{r}(0) - \boldsymbol{r}'(t))\mathrm{d}\boldsymbol{r}\mathrm{d}\boldsymbol{r}' \tag{4.71}$$

and to introduce, using as before the isotropy and the translational invariance of the system, the time-dependent Van Hove (1954) correlation function

$$G(\boldsymbol{r}, \boldsymbol{r}', t) = \langle n(\boldsymbol{r}, 0)n(\boldsymbol{r}', t)\rangle \tag{4.72}$$

this is the average value for two points, one at $\boldsymbol{r}$, the other at $\boldsymbol{r}'$ of the product of the densities at time 0 and t. Since in all isotropic systems the correlations depend only on $\boldsymbol{r}'' = |\boldsymbol{r}' - \boldsymbol{r}|$ we can write, after integration over $\boldsymbol{r}$

$$G(\boldsymbol{r}, t) = \int_V G(\boldsymbol{r}, \boldsymbol{r}'', t)\mathrm{d}\boldsymbol{r}'' \tag{4.72'}$$

$$\left(\frac{\partial^2\sigma}{\partial\Omega\partial\omega}\right)_{\mathrm{coh}} = \frac{k_\mathrm{f}}{k_\mathrm{i}} V \frac{b^2}{2\pi}\int_{-\infty}^{+\infty} \exp(\mathrm{i}\omega t)\int_V G(\boldsymbol{r}'', t) \exp(-i\boldsymbol{q}\cdot\boldsymbol{r}'')\mathrm{d}t\mathrm{d}\boldsymbol{r}''. \tag{4.73}$$

Let us now show that if one assumes that the nuclei are immobile one recovers the static expression. If $G(\boldsymbol{r}'', t)$ is independent of t, it reduces to the static correlation function $G(\boldsymbol{r}')$ and the integral over $\boldsymbol{r}$ is similar to eqn (4.36). Moreover $k_\mathrm{f} = k_\mathrm{i}$ and one obtains

$$\left(\frac{\partial^2\sigma}{\partial\Omega\partial\omega}\right)_{\mathrm{coh}} = Vb^2\frac{1}{2\pi}\int_{-\infty}^{+\infty} \exp(\mathrm{i}\omega t)\mathrm{d}t\left\{\int_V G(\boldsymbol{r}'')\exp(-i\boldsymbol{q}\cdot\boldsymbol{r}'')\mathrm{d}\boldsymbol{r}''\right\}. \tag{4.74}$$

The integration over t gives the delta function $\delta(\omega)$ (eqn (4.54)) and

$$\left(\frac{\partial^2\sigma}{\partial\Omega\partial\omega}\right)_{\text{coh}} = Vb^2\delta(\omega)\int_V G(\mathbf{r}'')\exp(-\mathrm{i}\mathbf{q}\cdot\mathbf{r}'')\mathrm{d}\mathbf{r}'' \qquad (4.75)$$

which shows that, as we already know, all the scattering is obtained for $\omega = 0$; there is no inelastic scattering.

4.8.2 The autocorrelation function

Remembering that, using the continuous notation, we forgot the case where $i = j$ we have to introduce the autocorrelation function G_s which corresponds both to the case of incoherent scattering and to the case where, in the coherent scattering $i = j$. Writing the first term in eqn (4.68), one obtains, after obvious simplification

$$\left(\frac{\partial^2\sigma}{\partial\Omega\partial\omega}\right)_{\text{inc}} = \frac{k_\mathrm{f}}{k_\mathrm{i}} N\frac{\overline{\Delta b^2}}{2\pi}\int_{-\infty}^{\infty}\exp(\mathrm{i}\omega t)\langle\exp\{\mathrm{i}\mathbf{q}\cdot(\mathbf{r}_1(0) - \mathbf{r}_1(t))\}\rangle\mathrm{d}t. \qquad (4.76)$$

We define the probability for one nucleus (or a molecule) starting at $\mathbf{r}$ at time zero to be at $\mathbf{r}'$ at time t and call it $G_s(\mathbf{r}'', t)$. Introducing this definition in eqn (4.80) gives (the probability is the same for all nuclei since they are identical)

$$\left(\frac{\partial^2\sigma}{\partial\Omega\partial\omega}\right)_{\text{inc}} = \frac{k_\mathrm{f}}{k_\mathrm{i}} N\frac{\overline{\Delta b^2}}{2\pi}\int_V\int_{-\infty}^{\infty}\exp(\mathrm{i}\omega t)\exp[\mathrm{i}\mathbf{q}\cdot\mathbf{r}'']G_\mathrm{s}(\mathbf{r}'', t)\mathrm{d}t\mathrm{d}\mathbf{r}''. \qquad (4.77)$$

As we have already pointed out this term appears also as an extra term in the expression for the coherent scattering. If we assume that the distances do not depend on time we see that $G_\mathrm{s}(\mathbf{r}'', 0)$, the autocorrelation function, is a delta function of $\mathbf{r}$ (the probability for the molecule to be at another place is zero): the integral over $\mathbf{r}''$ becomes unity and when we integrate over t we obtain the following function

$$\left(\frac{\partial^2\sigma}{\partial\Omega\partial\omega}\right)_{\text{inc}} = \frac{k_\mathrm{f}}{k_\mathrm{i}} N\frac{\overline{\Delta b^2}}{2\pi}\int_{-\infty}^{\infty}\exp(\mathrm{i}\omega t)\mathrm{d}t \qquad (4.78)$$

where the integral is a delta function in ω. We therefore find $N\Delta b^2$ for $\omega = 0$ as we should.

These formulae, once seen and understood, are very easy to remember: the only requirement is first to replace the static correlation function by a time-dependent correlation function and then to make the Fourier transform of the expression in order to replace the variable t by ω.

4.8.3 The static approximation

We have just seen that, when the nuclei are immobile the formulae in eqns (4.73) and (4.77) reduce to the formula obtained in the static approximation. The question one can ask is 'Can we, for a sample in which there are motions or changes in quantum states, recover from experiments done on spectrometers designed to measure the static scattering, information independent of the dynamics of the sample?' The question is not easy to answer since, if we integrate over all energies (or ω), we obtain

$$\int_{-\infty}^{+\infty}\left(\frac{\partial^2\sigma}{\partial\Omega\partial\omega}\right)_{\text{coh}}\mathrm{d}\omega = V\frac{b^2}{2\pi}\int_{\omega=-\infty}^{+\infty}\frac{k_\mathrm{f}}{k_\mathrm{i}}\int_V\int_{t=-\infty}^{+\infty}\exp(\mathrm{i}\omega t)G(\boldsymbol{r},t)\exp(-\mathrm{i}\boldsymbol{q}\cdot\boldsymbol{r}'')\mathrm{d}\omega\mathrm{d}t\mathrm{d}\boldsymbol{r}. \quad (4.79)$$

It is only when the function $G(r, t)$ is not sharply peaked around $t = 0$ (the motions are slow) that one can use the approximation which consists in neglecting the dependence on t of the function $G(r, t)$. Placzek (1954) derived a method for improving the static approximation, for the situation where $\hbar\omega$ is small. The correction appears as q-dependent additions to $\partial\sigma/\partial\Omega$ which are smaller for heavy scattering nuclei and for high incident energies. Fortunately these corrections are very small for $q \to 0$ so that in small-angle scattering the static approximation works well.

Some spectrometers give the result before integrating over t, i.e. the function called the intermediate scattering structure factor (which will be defined more precisely in the following paragraph)

$$S(q,t) = \int_V G(\boldsymbol{r},t)\exp(-\mathrm{i}\boldsymbol{q}\cdot\boldsymbol{r})\mathrm{d}\boldsymbol{r}. \quad (4.80)$$

For this case we see immediately that for $t = 0$ one obtains the static value.

4.8.4 Some remarks about notation

In the literature the relationships which we have established are used in many different ways, some of them described in Chapter 9. It is therefore useful to get acquainted with the language used in inelastic as well as in elastic studies. The profusion of terms comes from the fact that to go from $\partial^2\sigma/\partial\Omega\partial\omega$ to $G(r, t)$ or $G_\mathrm{s}(r, t)$ one has to use two Fourier transforms; one over $\boldsymbol{r}$ and the other over t. (We do not use r'' any more since it is an integration variable which does not appear in the final result.) Since it is possible to write two expressions for each Fourier transformation, either in the direct or the inverse form, the number of possible equations is large. On the other hand the symbols used to characterize these different quantities vary from one author to the other and this does not facilitate comprehension. This is the reason why, before beginning the experimental chapters some discussion about notation will probably help the reader. It is common practice to define the quantity $S_{\text{coh}}(q, \omega)$ as

$$S_{\text{coh}}(q, \omega) = \frac{1}{Nb_c^2} \frac{k_i}{k_f} \left(\frac{\partial^2 \sigma}{\partial \Omega \partial \omega} \right)_{\text{coh}} \tag{4.81}$$

and

$$S_{\text{inc}}(q, \omega) = \frac{1}{Nb_{\text{inc}}^2} \frac{k_i}{k_f} \left(\frac{\partial^2 \sigma}{\partial \Omega \partial \omega} \right)_{\text{inc}} . \tag{4.82}$$

This is the scattered intensity for a given q value divided by the number N of molecules in the sample, the square of the coherent scattering length and the ratio k_f/k_i. (The elimination of the factor $\hbar$ comes from the fact that we use as variable ω and no longer the energy ΔE.) This formulation has the advantage of excluding from the formulae parameters which are experimental quantities not immediately related to the structure of the sample and of leaving in the result only the structural properties of the system under study. This formalism is very convenient when dealing with a pure component but not when working with mixtures. In these cases one can use, like Flory (1965) and de Gennes (1970), the scattering divided by the total number of cells, a cell being defined by its volume, usually set equal to the volume of the solvent. This is quite convenient for an incompressible mixture of two constituents where there is only one parameter to characterize the scattering length, but problems arise for systems in which the different species have different scattering lengths (see Chapter 5). It will always be a sensible precaution to look for the exact definition of $S(q)$ when following authors who use this symbol.

We shall now use the definition of eqns (4.81) and (4.82) since we are considering only one component systems. Quite often, instead of $G(\boldsymbol{r}, t)$, the quantity $\Gamma(\boldsymbol{r}, t)$ is used. It is defined by a generalization of eqn (4.45) as

$$\Gamma(r, t) = \frac{1}{n^2} G(r, t). \tag{4.83}$$

It has the advantage of having a limiting value of unity, instead of n^2, when $\boldsymbol{r}$ or $t \to \infty$ and it is also dimensionless. With this notation, for instance, one can define $S_{\text{coh}}(q, \omega)$ using eqn (4.73) by

$$\begin{aligned} S_{\text{coh}}(q, \omega) &= \frac{V}{N} \frac{1}{2\pi} \int_{-\infty}^{+\infty} \exp(\mathrm{i}\omega t) \int_V G(r, t) \exp(-\mathrm{i}\boldsymbol{q} \cdot \boldsymbol{r}) \mathrm{d}t \mathrm{d}\boldsymbol{r} \\ &= \frac{n}{2\pi} \int_{-\infty}^{+\infty} \exp(\mathrm{i}\omega t) \int_V \Gamma(r, t) \exp(-\mathrm{i}\boldsymbol{q} \cdot \boldsymbol{r}) \mathrm{d}t \mathrm{d}\boldsymbol{r}. \end{aligned} \tag{4.84}$$

In all our formulae the quantities $G(r, t)$ and $G_s(r, t)$ are three-dimensional Fourier transformations over the variable $\boldsymbol{r}$. One can define therefore those quantities which will be called $S(q, t)$ (intermediate scattering), $\hat{G}(q, t)$ and $\hat{G}_s(q, t)$ by the relationships

$$S(q, t) = \hat{G}(q, t) = \iiint_V G(r, t) \exp(-\mathrm{i}\boldsymbol{q}\cdot\boldsymbol{r})\mathrm{d}\boldsymbol{r} \tag{4.85}$$

and

$$\hat{G}_s(q, t) = \iiint_V G_s(r, t) \exp(-\mathrm{i}\boldsymbol{q}\cdot\boldsymbol{r})\mathrm{d}\boldsymbol{r} \tag{4.86}$$

which means that we are taking the Fourier transform of G and G_s. With this notation we write for S_{inc}, using eqns (4.77), (4.82), and (4.86)

$$S_{\text{inc}}(q, \omega) = \frac{1}{2\pi} \int_{-\infty}^{\infty} \exp(\mathrm{i}\omega t)\, \hat{G}_s(q, t)\,\mathrm{d}t. \tag{4.87}$$

For the coherent part of the intensity one has to be more careful. As we already pointed out, the use of the function $G(r, t)$ leaves out all the pairs of identical points which would give a signal identical to the incoherent signal (there are no interferences). One has therefore to add this signal to the result of eqn (4.74), obtaining

$$\begin{aligned} S_{\text{coh}}(q, \omega) &= \frac{1}{2\pi} \int_{-\infty}^{\infty} \exp(\mathrm{i}\omega t)\, \hat{G}_s(q, t)\,\mathrm{d}t + \frac{1}{n}\frac{1}{2\pi} \int_{-\infty}^{+\infty} \exp(\mathrm{i}\omega t)\, \hat{G}(q, t)\,\mathrm{d}t \\ &= \frac{1}{2\pi} \int_{-\infty}^{\infty} \exp(\mathrm{i}\omega t) \left\{ \hat{G}_s(q, t) + \frac{1}{n}\, \hat{G}(q, t) \right\} \mathrm{d}t. \end{aligned} \tag{4.88}$$

Since these expressions are in the form of an inverse Fourier transform from the t variable to the ω variable there are no difficulties making the Fourier transformation which allows us to go back to t-space and we can write, without any calculation the definitions of $S_{\text{coh}}(q, t)$ and $S_{\text{inc}}(q, t)$

$$S_{\text{coh}}(q, t) = \hat{G}_s(q, t) + \frac{1}{n}\, \hat{G}(q, t) \tag{4.89}$$

$$S_{\text{inc}}(q, t) = \hat{G}_s(q, t). \tag{4.90}$$

A second Fourier transform would lead to $G(r, t)$ and $G_S(r, t)$ but due to the limitations of the experimental results this procedure is never used in practice.

4.9 SOME EXAMPLES

The general scattering laws developed in eqns (4.67) to (4.78) do not easily show the reader what the inelastic scattering from a sample will look like in an experiment. In order to make this connection clearer we shall go a little further with the inelastic scattering laws than we did with the elastic ones and consider some examples. By studying simple models of translational, vibrational, and rotational motion of individual molecules we shall demonstrate the behaviour of the scattering arising from these different motions. We consider an ensemble of identical molecules and rewrite the expression for the coherent intermediate structure factor, this time expressed as a discrete sum over all particles

$$S(q,t) = \sum_i \sum_j \langle \exp(-\mathrm{i}\boldsymbol{q}\cdot\boldsymbol{r}_i(0))\exp(-\mathrm{i}\boldsymbol{q}\cdot\boldsymbol{r}_j(t))\rangle. \qquad (4.91)$$

In order to separate translation, rotation, and vibration it is necessary to express the time-dependent position vector $\boldsymbol{r}$ in terms of three vectors, (Fig. 4.2) one for the centre of mass $\boldsymbol{c}_i$, one for rotations about the centre of mass $\boldsymbol{b}_i$ and a third one $\boldsymbol{u}_i$ characterizing the distance of the particle from its average position, thus writing

$$\boldsymbol{r}_i(t) = \boldsymbol{c}_i(t) + \boldsymbol{b}_i(t) + \boldsymbol{u}_i(t). \qquad (4.92)$$

With this notation eqn (4.91) becomes

$$S(q,t) = \sum_i \sum_j \langle \exp(-\mathrm{i}\boldsymbol{q}\cdot(\boldsymbol{c}_i(0) - \boldsymbol{c}_j(t)))\exp(-\mathrm{i}\boldsymbol{q}\cdot(\boldsymbol{b}_i(0) - \boldsymbol{b}_j(t)))$$
$$\exp(-\mathrm{i}\boldsymbol{q}\cdot(\boldsymbol{u}_i(0) - \boldsymbol{u}_j(t)))\rangle. \qquad (4.93)$$

Now we shall assume, as is frequently done in infrared spectroscopy, that these three vectors are independent. This is only an approximation in the case of rotation and vibration but quite sufficient for our actual purpose

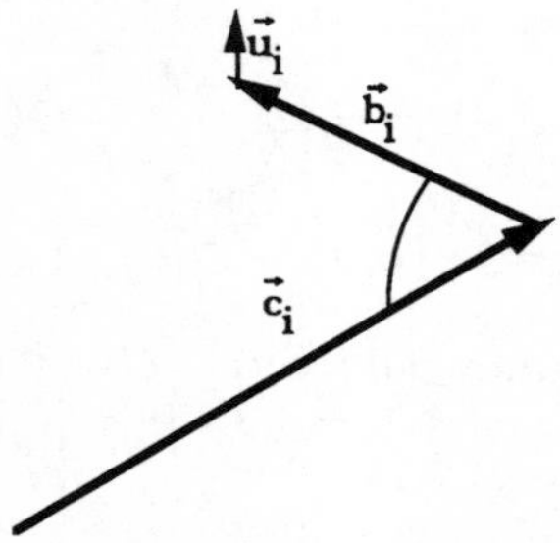

Fig. 4.2 The vectors characterizing translation, rotation, and vibration.

which is to introduce these effects qualitatively. This approximation allows us to replace the average value of the product by the product of three averages and to write

$$S(q,t) = \sum_i \sum_j \langle \exp - i\boldsymbol{q}\cdot(\boldsymbol{c}_i(0) - \boldsymbol{c}_j(t))\rangle$$
$$\otimes \langle \exp(-i\boldsymbol{q}\cdot(\boldsymbol{b}_i(0) - \boldsymbol{b}_j(t)))\rangle \otimes \langle \exp(-i\boldsymbol{q}\cdot(\boldsymbol{u}_i(0) - \boldsymbol{u}_j(t)))\rangle \tag{4.94}$$

where $S(q,t)$ is the double sum of the product of three quantities

$$S(qt) = \sum_i \sum_j s^{\mathrm{tr}}(qt) \otimes s^{\mathrm{rot}}(q,t) \otimes s^{\mathrm{vib}}(q,t) \tag{4.94$'$}$$

$s^{\mathrm{tr}}(q,t)$, $s^{\mathrm{rot}}(q,t)$, and $s^{\mathrm{vib}}(q,t)$ are respectively the contributions to the scattering of the translational, rotational, and vibrational motions of one pair of scattering centres. If the rotational or vibrational motions are absent, $\boldsymbol{b}_i(t)$ or $\boldsymbol{u}_i(t)$ is zero and the corresponding term is unity.

This equation, which we derived for coherent scattering, can also be applied to incoherent scattering if we make $i = j$ and reduce the double summation to a simple summation.

4.9.1 Translational motions—free diffusion

Incoherent scattering

We assume that the only motion of the particle is a translational motion which implies that $s^{\mathrm{rot}}(q,t) = s^{\mathrm{vib}}(q,t) = 1$ or that their effect is in a very different domain of time or frequency.

This could be the case for a gas of rigid particles or for large Brownian particles diffusing in a structureless solvent and observed at a very small angle in order to eliminate the effect of their form factor. If we look at the motion of such particles the path traced will be equivalent to a random walk and will obey the same rules. This path can be obtained starting from a point characterized by $\boldsymbol{c}_0$ and marking the position of the particle for times which are multiples of a defined time τ. It will show exactly the same shape and obey the same law as a chain of N segments. The average of the square of the distance covered by the particle during the time τ will be equivalent to $\overline{l^2}$ the mean square of the length of the statistical element of a chain of N units. We know (see Appendix 1) that the probability $w(\boldsymbol{c} - \boldsymbol{c}_0)$ for the end of the chain to be at a distance $\boldsymbol{c}$ from its origin $\boldsymbol{c}_0$ after N bonds is given by the Gaussian law (eqn (A1.5)) and that the only parameter is the value of the root mean square distance from the origin given by: $\overline{(c - c_0)^2} = N\overline{l^2}$. In a diffusion process, therefore, it suffices to replace N by t and to write

$$\overline{(c - c_0)^2} = 6Dt \tag{4.95}$$

where D is what is called the translational diffusion coefficient. If we consider now the first term of eqn (4.94) and restrict our interest to the incoherent scattering the average becomes an integral over the volume V of the probability function $w(\boldsymbol{c} = \boldsymbol{c}_0)$

$$s_{\text{inc}}^{\text{tr}}(q, t) = \int_V \exp(-\mathrm{i}\boldsymbol{q} \cdot (\boldsymbol{c}_i(0) - \boldsymbol{c}_i(t)))w(\boldsymbol{c}_i, t)\mathrm{d}\boldsymbol{c} \tag{4.96}$$

or, since we have assumed that $w(\boldsymbol{c})$ is Gaussian, we can write

$$s_{\text{inc}}^{\text{tr}}(q, t) = \int_V \exp(-\mathrm{i}\boldsymbol{q} \cdot (\boldsymbol{c}(0) - \boldsymbol{c}(t)))\left\{\frac{3}{2\pi\overline{(c - c_0)^2}}\right\}^{3/2}$$

$$\exp\left[\frac{-3(\boldsymbol{c} - \boldsymbol{c}_0)^2}{2\overline{(c - c_0)^2}}\right]\mathrm{d}(\boldsymbol{c} - \boldsymbol{c}_0). \tag{4.97}$$

The integration over $\boldsymbol{c} - \boldsymbol{c}_0$ is a classical problem since it is just the calculation of the Fourier transform of a Gaussian (eqn (A1.7)) and we obtain for one particle, using eqn (4.95)

$$s_{\text{inc}}^{\text{tr}}(q, t) = \exp(-q^2 Dt). \tag{4.98}$$

If we assume that we have a dilute solution all particles will obey the same law. This leads directly to the function $S_{\text{inc}}^{\text{tr}}(q, t)$, the intermediate incoherent translational scattering function

$$S_{\text{inc}}^{\text{tr}}(q, t) = N\exp(-q^2 Dt) \tag{4.99}$$

where N is the number of particles.

In order to obtain $S_{\text{inc}}^{\text{tr}}(q, \omega)$ we have to take the Fourier transform of the function $N\exp(-q^2Dt)$ with respect to the variable t. It is evident that the use of negative values of t has no physical meaning and therefore, without more elaborate justification, we replace t by $|t|$ and write $q^2 D = \Gamma$.

$$\frac{1}{2\pi}\int_{-\infty}^{\infty} \exp(\mathrm{i}\omega t)\exp(-\Gamma|t|)\mathrm{d}t = \text{Real part of }\left\{\frac{1}{\pi}\int_0^{\infty} \exp[(\mathrm{i}\omega - \Gamma t)]\mathrm{d}t\right\}$$

$$= \text{Real part of }\left\{\pi^{-1}\frac{1}{\Gamma - \mathrm{i}\omega}\right\}$$

$$= \pi^{-1}\frac{\Gamma}{\Gamma^2 + \omega^2} \tag{4.100}$$

and we obtain as a result (Berne and Pecora 1976)

$$S_{\text{inc}}^{\text{tr}}(q, \omega) = \frac{N}{\pi} \frac{\Gamma}{\Gamma^2 + \omega^2}. \tag{4.101}$$

It is easy to verify for this simple example that $S(q, t = 0)/N$ is unity, as one would expect, but $S(q, \omega = 0)$ has no simple meaning (it becomes a delta function when $q \to 0$). Using this example the general result that the static signal is obtained by integrating $S(q, \omega)$ over all values of ω can be simply verified. In order to do so we integrate $S(q, \omega)$ given by eqn (4.101) over all ω

$$\int_{-\infty}^{+\infty} \frac{1}{\pi} \frac{\Gamma}{\Gamma^2 + \omega^2} \mathrm{d}\omega = \frac{1}{\pi} \int_{-\infty}^{+\infty} \frac{1}{1 + \frac{\omega^2}{\Gamma^2}} \mathrm{d}\frac{\omega}{\Gamma} = \frac{2}{\pi}\left[\arctan\left(\frac{\omega}{\Gamma}\right)\right]_0^{\infty} = 1. \tag{4.102}$$

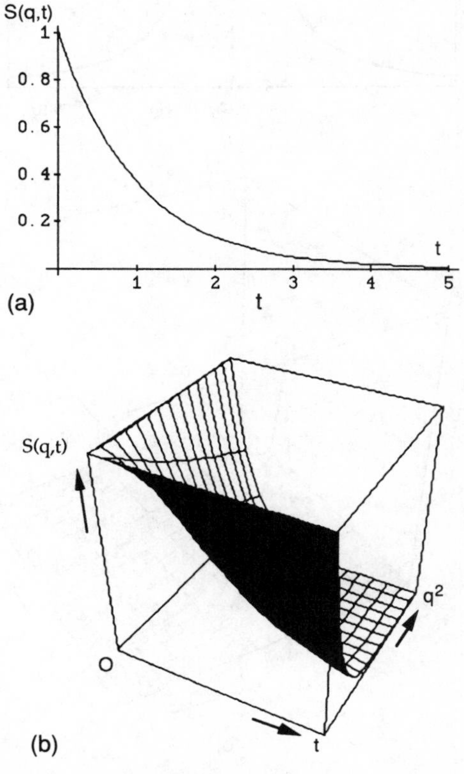

Fig. 4.3 (a) $S_{\text{inc}}^{\text{tr}}(q, t)$ at constant q, as function of t for a simple diffusive processs. The relaxation time, Γ^{-1}, is easy to obtain in a log(t) representation. (b) Three-dimensional plot of $S(q, t)$.

The formula we have just established gives an idea about the range of values of D which can be measured in one experiment since for such experiments to be possible Γt or $q^2 Dt$ should be of order unity.

With these results we now have two equations for the determination of the scattered intensity: $S^{\text{tr}}_{\text{inc}}(q, t)$ represented in Fig. 4.3 and its inverse Fourier transform $S^{\text{tr}}_{\text{inc}}(q, \omega)$ in Fig. 4.4. In the last case $\Gamma = \omega_{1/2}$ where

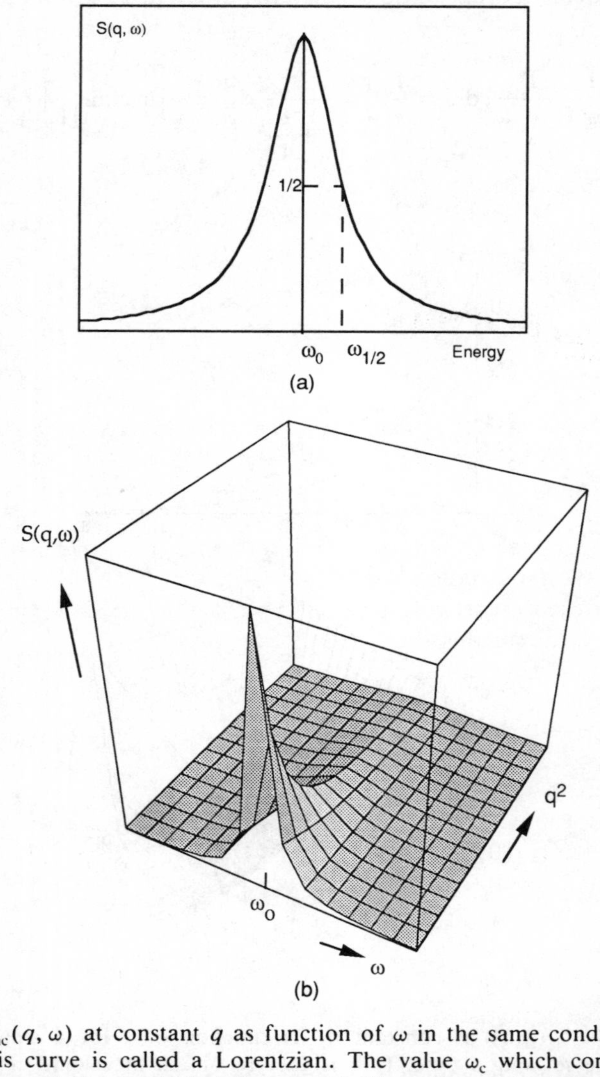

Fig. 4.4 $S^{\text{tr}}_{\text{inc}}(q, \omega)$ at constant q as function of ω in the same conditions as for Fig. 4.3. This curve is called a Lorentzian. The value ω_c which corresponds to $S(q, \omega_c) = S(q, 0)/2$ is such that $\Gamma = \omega_c$. (b) Three-dimensional plot of $S(q, \omega)$.

$\omega_{1/2}$ is the value of ω for which $S^{\text{tr}}_{\text{inc}}(q,\omega)$ has half its maximum value. In some types of experiments such as neutron spin-echo (see Chapter 3) or, in the field of light scattering (autocorrelation measurements) one obtains the quantity $S^{\text{tr}}_{\text{inc}}(q,t)$ directly. This means that a measurement as a function of t for a fixed value of q always gives a decay which, in the case of simple diffusive phenomena, is represented by the exponential $\exp(-\Gamma t)$ with $\Gamma = Dq^2$. In the case of spheres of radius R, in a solvent of viscosity η, the coefficient D is, given by the Stokes law $D = k_B T/6\pi\eta R$ and can be obtained from the incoherent scattering as long as qR is small compared to unity and the solution is dilute (see next paragraph). In the case of polymers the relationship between D and the characteristic length associated with the polymer radius of gyration (or end to end distance) is difficult to establish. The hydrodynamic radius R_H is introduced to replace R. This quantity is related to the diffusion coefficient by Stokes law and for linear chains with Gaussian statistics is related to the radius of gyration by the relationship: $R_G/R_H = 1.50$ (Joanny and Candau 1989). In a good solvent, the numerical value of the ratio is different. This technique is actually one of the best methods for measuring D and from it R_H.

Coherent scattering in dilute solutions

Up to this point we have been interested in the incoherent scattering by independent free particles, small compared to $1/q$. Is it possible to say something about the coherent scattering? In the next chapter it will be shown that the correlation function can be split into two terms: one term measuring the intramolecular scattering or the interferences between points belonging to the same molecule, the other, the intermolecular interferences, i.e. the interferences between points belonging to different molecules. If the solution is dilute this second term can be neglected and the scattering reduces to N times the scattering by one molecule. If one neglects the rotations one can say that each point k of this molecule ($0 \le k \le z$) which is at $\boldsymbol{c}_k(0)$ at $t = 0$ will be at $\boldsymbol{r}_k(t) = \boldsymbol{c}_k(0) + \Delta\boldsymbol{c}$ at time t, where $\Delta\boldsymbol{c}$ is the displacement of the molecule during this time. The sum $\Sigma_k \Sigma_l \exp[-\mathrm{i}\boldsymbol{q}\cdot(\boldsymbol{c}_k(0) - \boldsymbol{c}_l(t))]$ is therefore replaced by

$$\sum_k \sum_l \exp[-\mathrm{i}\boldsymbol{q}\cdot(\boldsymbol{c}_k(0) - \boldsymbol{c}_l(0) - \Delta\boldsymbol{c})]$$
$$= \exp(-\mathrm{i}\boldsymbol{q}\cdot\Delta\boldsymbol{c}) \sum_k \sum_l \exp[-\mathrm{i}\boldsymbol{q}\cdot(\boldsymbol{c}_k(0) - \boldsymbol{c}_l(0))]$$
$$= \exp(-\mathrm{i}\boldsymbol{q}\cdot\Delta\boldsymbol{c}) z^2 P(q) \tag{4.103}$$

where we have anticipated a result of Chapter 5, calling the double sum $z^2P(q)$.

The vector $\Delta\boldsymbol{c}$ can be considered as the displacement of the centre of mass during the time t. It is the only time-dependent parameter since we assume all the parts of the same molecule have the same displacement. The preceding development can be used and justifies the extension of eqn (4.99) to the intermediate coherent scattering function, giving now

$$S_{\mathrm{coh}}^{\mathrm{tr}} = Nz^2 \exp\left(-q^2 Dt\right) P(q). \tag{4.104}$$

This equation can be applied easily to dilute solutions of spheres since in this case the symmetry of the system suppresses the effect of the rotations (see Section 7.14). If the molecules are large and if we consider their rotations and their conformational changes as functions of time eqn (4.99) can be used only at the limit $q \to 0$. This can be demonstrated by noting that for $q \to 0$ the quantity $P(q) \to 1$. At small q therefore the molecules can be considered as points for which eqns (4.98) to (4.101) can be applied. Therefore, even in this case, one can determine the translational diffusion coefficient. In order to evaluate the combined effect of translation and rotation of rigid particles the calculations become more complex; they have been made by Maeda and Fujime (1984) for the case of rigid rods and show that it is possible to evaluate the effect of rotation in some special cases. In the case of flexible polymers where one has also to take into account the conformational changes this problem is difficult and will be discussed in Chapter 9.

4.9.2 Rotational motion

As a model for the rotational motion we use the following simple example: we assume that the particles can take two positions $\boldsymbol{r}_1$ and $\boldsymbol{r}_2$, jumping alternately through an energy barrier from $\boldsymbol{r}_1$ to $\boldsymbol{r}_2$ and from $\boldsymbol{r}_2$ to $\boldsymbol{r}_1$. We call $p(\boldsymbol{r}, t)$ the probability of finding the particle at time t in one of its possible positions. It is evident that, since the particle can only be in $\boldsymbol{r}_1$ or in $\boldsymbol{r}_2$, $p(\boldsymbol{r}_1, t) + p(\boldsymbol{r}_2, t) = 1$. Moreover this jumping process from $\boldsymbol{r}_1$ to $\boldsymbol{r}_2$ can be described by the set of rate equations:

$$\left.\begin{aligned} \frac{\mathrm{d}}{\mathrm{d}t}(p(\boldsymbol{r}_1, t)) &= -\frac{1}{\tau} p(\boldsymbol{r}_1, t) + \frac{1}{\tau} p(\boldsymbol{r}_2, t) \\ \frac{\mathrm{d}}{\mathrm{d}t}(p(\boldsymbol{r}_2, t)) &= -\frac{1}{\tau} p(\boldsymbol{r}_2, t) + \frac{1}{\tau} p(\boldsymbol{r}_1, t) \end{aligned}\right\} \tag{4.105}$$

assuming that the rate constant $1/\tau$ is the same for both types of jumps. Since $p(\boldsymbol{r}_1, t) + p(\boldsymbol{r}_2, t) = 1$, we can eliminate $p(\boldsymbol{r}_2, t)$ and write the following linear differential equation

$$\frac{\mathrm{d}}{\mathrm{d}t}(p(\boldsymbol{r}_1, t)) = \frac{1}{\tau}\{1 - 2p(\boldsymbol{r}_1, t)\}. \tag{4.106}$$

6

Form factors

6.1 INTRODUCTION

We have shown (Chapter 5) that, for an incompressible binary mixture where the volume of the scattering elements 0 and 1 are identical, the scattered intensity can be written as

$$\frac{\partial \sigma}{\partial \Omega} = I(q) = (b_1 - b_0)^2 S_{11}(q) \tag{6.1}$$

where 1 designates one of the two species present in the mixture (the polymer) and zero the other. Since the volume of the polymer is different from the volume of the solvent we assume that the polymer is made of z units having the same volume v_0 as the solvent (see Chapter 7 and Appendix 3). Therefore we use for b_1 the scattering length of what we call a 'monomer', i.e. an element of the molecule having the same volume v_0 as a solvent molecule. If we replace the scattering length b_1 of the total polymer by $b_1 v_0 / V_1$ where V_1 is the volume of the macromolecule we can split the scattered intensity into two terms, as already done in eqn (5.25), obtaining

$$I(q) = \left(b_1 \frac{v_0}{V_1} - b_0\right)^2 [Nz^2 P(q) + N^2 z^2 Q(q)]. \tag{6.2}$$

Using the definition $\bar{b} = b_1 \frac{v_0}{V_1} - b_0$ (eqn (A3.4)) we write

$$I(q) = \bar{b}^2 [Nz^2 P(q) + N^2 z^2 Q(q)] \tag{6.3}$$

or for $S(q)$

$$S(q) = Nz^2 P(q) + N^2 z^2 Q(q) \tag{6.3'}$$

where N is the number of molecules of species 1 in the scattering volume (which can be assumed to be unity) and z is the number of scattering units (of volume v_0 within each molecule. The term $P(q)$ comes from the intramolecular interferences and is characteristic of the size and the shape of the molecules of type 1. One of the major problems in neutron small angle scattering is to be able to separate and evaluate the contribution of $P(q)$ and $Q(q)$ to the scattered intensity. The first task is to evaluate the

A general solution is easy to find and in order to evaluate the constants we assume that at time 0 the particle is in $\boldsymbol{r}_1$ ($p(\boldsymbol{r}_1, 0) = 1$ and $p(\boldsymbol{r}_2, 0) = 0$) obtaining

$$p(\boldsymbol{r}_1) = \tfrac{1}{2}\left[1 + \exp\left(-\frac{2t}{\tau}\right)\right] \tag{4.107}$$

$$p(\boldsymbol{r}_2) = \tfrac{1}{2}\left[1 - \exp\left(-\frac{2t}{\tau}\right)\right]$$

which shows that for $t \to \infty$ at equilibrium, the probabilities of finding the particle in $\boldsymbol{r}_1$ or $\boldsymbol{r}_2$ are identical ($p = 1/2$).

In this simple example there are no correlations between the particles and the only thing one can obtain is the incoherent scattering characterized, for each particle, by the quantity $S_{\text{inc}}^{\text{rot}}(\boldsymbol{q}, t) = \langle \exp - i\boldsymbol{q}\cdot(\boldsymbol{r}(0) - \boldsymbol{r}(t))\rangle$. There are four possible states which are characterized by their probabilities

$$\left.\begin{aligned} p(\boldsymbol{r}_1, t; \boldsymbol{r}_1, 0) &= \tfrac{1}{2}\{1 + \exp(-2t/\tau)\} \\ p(\boldsymbol{r}_2, t; \boldsymbol{r}_1, 0) &= \tfrac{1}{2}\{1 - \exp(-2t/\tau)\} \end{aligned}\right\} \tag{4.108}$$

and by exchanging indices

$$\left.\begin{aligned} p(\boldsymbol{r}_2, t; \boldsymbol{r}_2, 0) &= \tfrac{1}{2}\{1 - \exp(-2t/\tau)\} \\ p(\boldsymbol{r}_1, t; \boldsymbol{r}_2, 0) &= \tfrac{1}{2}\{1 + \exp(-2t/\tau)\} \end{aligned}\right\} \tag{4.108$'$}$$

Multiplying the contribution of each state by its probability and adding them all gives the expression for $s_{\text{inc}}^{\text{rot}}(\boldsymbol{q}, t)$ corresponding to one jumping particle

$$\begin{aligned} s_{\text{inc}}^{\text{rot}}(\boldsymbol{q}, t) = {} & \tfrac{1}{2}\{1 + \exp(-2t/\tau)\} \\ & + \exp\{-i\boldsymbol{q}\cdot(\boldsymbol{r}_1(0) - \boldsymbol{r}_2(t))\}\tfrac{1}{2}\{1 - \exp(-2t/\tau)\} \\ & + \tfrac{1}{2}\{1 + \exp(-2t/\tau)\} \\ & + \exp - i\boldsymbol{q}\cdot(\boldsymbol{r}_2(0) - \boldsymbol{r}_1(t))\tfrac{1}{2}\{1 - \exp(-2t/\tau)\} \end{aligned} \tag{4.109}$$

or after simplification

$$s_{\text{inc}}^{\text{rot}}(\boldsymbol{q}, t) = A_0(\boldsymbol{q}) + A_1(\boldsymbol{q})\exp(-2t/\tau) \tag{4.110}$$

with

$$A_0(\boldsymbol{q}) = (1/2)\{1 + \cos\boldsymbol{q}\cdot(\boldsymbol{r}_2 - \boldsymbol{r}_1)\} \tag{4.110$'$}$$

and

$$A_1(\boldsymbol{q}) = (1/2)\{1 - \cos\boldsymbol{q}\cdot(\boldsymbol{r}_2 - \boldsymbol{r}_1)\}. \tag{4.110$''$}$$

Making a Fourier transform of eqn (4.110) over the variable t, as done in eqn (A1.58) and multiplying by N the number of particles (supposed to be identical and independent) gives the quantity $S_{\text{inc}}^{\text{rot}}(\boldsymbol{q}, \omega)$

$$S_{\text{inc}}^{\text{rot}}(\boldsymbol{q}, \omega) = N\left[A_0(\boldsymbol{q})\delta(\omega) + A_1(\boldsymbol{q})\frac{1}{\pi}\frac{2\tau}{4 + \omega^2\tau^2}\right]. \tag{4.111}$$

If the sample is isotropic, as for a liquid or a powder, one has to allow $\boldsymbol{q}$ to take all the equiprobable orientations obtaining

$$S_{\text{inc}}^{\text{rot}}(q, \omega) = \frac{N}{2}\left\{1 + \frac{\sin qd}{qd}\right\}\delta(\omega) + \frac{N}{\pi}\frac{\tau}{4 + \omega^2\tau^2}\left\{1 - \frac{\sin qd}{qd}\right\} \tag{4.112}$$

where d is the distance $|r_1 - r_2|$ (see eqn (A1.46)) and the quantity $\sin x/x$ is the zero order Bessel spherical function $J_0(x)$.

The scattered intensity is composed of two terms: the first is an elastic peak corresponding to the absence of motion, the second is a Lorentzian centred at $\omega = \omega_0$. The relative intensity depends on the value of q; as q tends to zero only the elastic peak remains (the particles behave like points and the delta function is broadened by the translational motion) but when q increases and qd becomes larger than 1 the second peak can be measured. In other words this effect, which is in addition to the translational motion, can only be measured for q values of the order of $1/d$. Fig. 4.5 where we

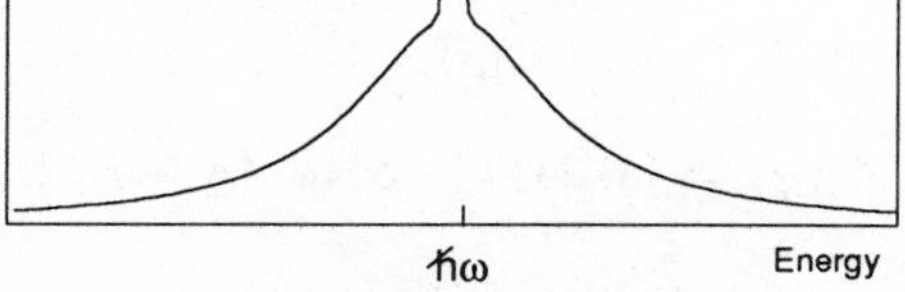

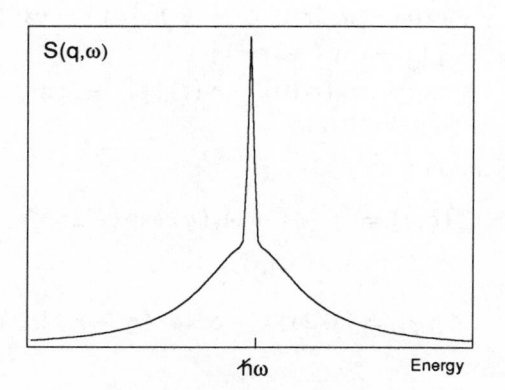

Fig. 4.5 Typical representation of the effect of rotations on the scattering intensity.

have replaced the delta function by a sharply peaked Gaussian which takes into account all the imperfections of the instrumentation (and the translational motion) shows qualitatively the shape of the function $S(q, \omega)$ for this model. It is evident that one can use more sophisticated models for the rotation of molecular groups with more than two possible positions and different probabilities of occupation but the qualitative aspect of the figure will stay unchanged.

4.9.3 Vibrational motion—one-phonon inelastic scattering

In this last section we will discuss the last term of eqn (4.94′), the effect of vibrations on the scattering phenomena. For this purpose we rewrite this last term

$$s^{\text{vib}}(q, t) = \langle \exp(-\mathrm{i}\boldsymbol{q}\cdot(\boldsymbol{u}_j(0) - \boldsymbol{u}_k(t)))\rangle. \tag{4.113}$$

In order to evaluate this average we make a series expansion of the exponential retaining only the first terms and taking the averages of each separately

$$\langle \exp(-\mathrm{i}\boldsymbol{q}\cdot(\boldsymbol{u}_i(0) - \boldsymbol{u}_j(t)))\rangle = 1 - \mathrm{i}\langle \boldsymbol{q}\cdot(\boldsymbol{u}_i(0) - \boldsymbol{u}_j(t))\rangle - \tfrac{1}{2}\langle\{\boldsymbol{q}\cdot(\boldsymbol{u}_i(0) - \boldsymbol{u}_j(t))\}^2\rangle \tag{4.114}$$

The term linear in $\boldsymbol{q}$ vanishes since the orientations of $\boldsymbol{u}$ and $\boldsymbol{q}$ are not correlated and for simplicity we shall keep only the q^2 term ignoring higher ones. This term can be written as

$$\tfrac{1}{2}\langle(\boldsymbol{q}\cdot\boldsymbol{u}_i(0))^2 + (\boldsymbol{q}\cdot\boldsymbol{u}_j(t))^2 - 2\boldsymbol{q}\cdot\boldsymbol{u}_i(0)\boldsymbol{q}\cdot\boldsymbol{u}_j(t)\rangle. \tag{4.115}$$

Since this is a sum we can split the average into three averages: the two first are identical and one obtains for s_{vib}

$$s^{\text{vib}}(q, t) = 1 - \langle(\boldsymbol{q}\cdot\boldsymbol{u}_i)^2\rangle + \langle \boldsymbol{q}\cdot\boldsymbol{u}_i(0)\boldsymbol{q}\cdot\boldsymbol{u}_j(t)\rangle. \tag{4.116}$$

Since we neglect higher terms we can also write

$$s^{\text{vib}}(q, t) = \exp(-\langle(\boldsymbol{q}\cdot\boldsymbol{u}_i)^2\rangle)\exp\langle \boldsymbol{q}\cdot\boldsymbol{u}_i(0)\boldsymbol{q}\cdot\boldsymbol{u}_j(t)\rangle = DW \otimes V_{\text{ib}} \tag{4.117}$$

where again we have split the problem into two parts, the first being the calculation of the purely elastic term $DW = \exp(-\langle(\boldsymbol{q}\cdot\boldsymbol{u}_i)^2\rangle$ leading to what is called the Debye–Waller effect and the second, the calculation of $V_{\text{ib}} = \exp\langle \boldsymbol{q}\cdot\boldsymbol{u}_i(0)\boldsymbol{q}\cdot\boldsymbol{u}_j(t)\rangle$ which is a true inelastic term.

The Debye–Waller effect

There was a question which puzzled the crystallographers at the beginning of the century. In a crystal the atoms are each at a well-defined equilibrium

position but, due to thermal motions, they are vibrating around this position, even at $T = 0$, since, from quantum mechanics there is, at this temperature, a residual quantum energy $k_B T/2$. This motion should modify the scattering function and the question was to find how it affects the static peaks in $S(q)$. This problem, although it was first studied for crystals, is more general since, even in an amorphous system, the molecules can vibrate around a more slowly varying equilibrium position. Study of the first term of eqn (4.117) immediately provides an answer.

If we assume that all the scattering particles have the same vibration amplitude, this first term can be moved outside the sum over i and j and the coherent static structure factor has to be multiplied by the quantity $\exp-\langle(\boldsymbol{q}\cdot\boldsymbol{u})^2\rangle = \exp(-q^2\langle u^2\rangle/3)$. (The factor 1/3 comes from the fact that we are working in three dimensions where $\overline{\cos^2\theta} = 1/3$.) By definition the quantity $\exp-\langle(\boldsymbol{q}\cdot\boldsymbol{u})^2\rangle$ is called $\exp(-2W)$, the Debye-Waller factor. The factor 2 comes from the fact that, in the general case $\bar{u}_i^2 \neq \bar{u}_j^2$ and one has to multiply each term of the double sum $\Sigma_{ij}\exp[-\mathrm{i}\boldsymbol{q}\cdot(\boldsymbol{r}_i - \boldsymbol{r}_j)]$ by the quantity: $b_i b_j \exp-(W_i + W_j)$.

The Debye-Waller coefficient shows that the effect of vibrations decreases the amplitude of the static peaks by a quantity increasing with q^2 and thus is only noticeable when $q^2\langle u^2\rangle$ is of the order of unity. Since, in the majority of the problems on polymers, one works at rather small q the Debye-Waller correction is usually completely negligible. In crystals where this factor affects the height of the Bragg diffraction peaks one can evaluate the order of magnitude of $\bar{u}^2$ and from it the effect of the factor W. If one replaces the oscillator by a classical harmonic oscillator in three-dimensional space its energy is $(3/2)k_B T$ and this is equal to $m\overline{v^2}/2$, where $\bar{v}^2$ is the average value of the square of the velocity. The velocity and the amplitude are related, for an harmonic oscillator, by: $\bar{u}^2 = \overline{v^2}/\bar{\omega}^2$, where $\bar{\omega}$ is the circular frequency of the oscillator. A simple calculation therefore gives

$$W = \frac{k_B T q^2}{2m\bar{\omega}^2}. \tag{4.118}$$

Vibrational inelastic scattering—phonon scattering

We have to analyse the contribution to the structure factor of the term

$$V_{ib} = \exp\langle\boldsymbol{q}\cdot\boldsymbol{u}_i(0)\boldsymbol{q}\cdot\boldsymbol{u}_j(t)\rangle \tag{4.119}$$

or, to a first order approximation

$$\{1 + \langle\boldsymbol{q}\cdot\boldsymbol{u}_i(0)\boldsymbol{q}\cdot\boldsymbol{u}_j(t)\rangle\}. \tag{4.119'}$$

If the scatterers are independent the two vectors $\boldsymbol{u}_i(0)$ and $\boldsymbol{u}_j(t)$ are not correlated, unless $i = j$, thus the second term of eqn (4.119′) is zero since

the average can be split into the product of two averages. In order to have a non-zero average we have to assume either that $i = j$ which leads to the incoherent term or that there are correlations between the orientations of the vectors $\boldsymbol{u}$ belonging to different molecules. We consider first the case of incoherent scattering.

If, in eqn (4.119′) we make $i = j$ we obtain the incoherent term

$$V_{ib_{\text{inc}}} = 1 + \langle \bar{q}\cdot\bar{u}_j(0)\bar{q}\cdot\bar{u}_j(t)\rangle = 1 + \langle q^2u^2(0)\rangle\cos\omega t \quad (4.120)$$

(ω is the circular frequency of the oscillator). In order to evaluate the quantity $S_{\text{inc}}(q, \omega)$ we have to make a Fourier transform of the variable t into ω. For this purpose we write

$$\cos\omega t = \tfrac{1}{2}\{\exp(\mathrm{i}\omega t) + \exp(-\mathrm{i}\omega t)\} \quad (4.121)$$

and we evaluate two integrals of the form

$$K = \frac{1}{2\pi}\int_{-\infty}^{+\infty} \frac{\exp(\pm\mathrm{i}\omega t)\exp(\mathrm{i}\omega_0 t)}{2}\,\mathrm{d}t \quad (4.122)$$

(we call ω_0 the observation frequency which is not the frequency ω of the oscillator). Using as variable $\omega' = \omega \pm \omega_0$ we obtain from the definition of the delta function (eqn (A1.9′))

$$K = \tfrac{1}{2}\,\delta(\omega \pm \omega_0).$$

This allows us to replace eqn (4.121) by its Fourier transform

$$\mathrm{FT}(V_{ib_{\text{inc}}}) = \delta(\omega_0) + \frac{\overline{q^2u^2}}{2}\{\delta(\omega + \omega_0) + \delta(\omega - \omega_0)\}. \quad (4.123)$$

If all the particles are identical the incoherent quasielastic scattering of N identical particles when one takes into account vibrations is given by substituting eqns (4.123) into (4.76)

$$\left(\frac{\partial^2\sigma}{\partial\Omega\partial\omega}\right)_{\text{inc}} = \frac{k_\text{f}}{k_\text{i}}\,\overline{\Delta b^2}\,N\exp(-2W)\left[\delta(\omega_0) + \frac{\overline{u^2q^2}}{2}\{\delta(\omega+\omega_0) + \delta(\omega-\omega_0)\}\right]. \quad (4.124)$$

The scattering intensity consists of three delta functions: the first corresponds to elastic scattering without change in the energy of the beam, the other two correspond either to the absorption or to the emission of a phonon from which the energy is either given to or subtracted from the incident neutron energy. Of course, as usual the delta functions are abstract symbols and all these peaks are broadened either by the experimental resolution or by other inelastic processes such as translation or rotation. Figure 4.6 gives a qualitative description of what is to be expected for the inelastic scattering if the vibrations are important.

Returning, now, to the coherent scattering, we have to reconsider

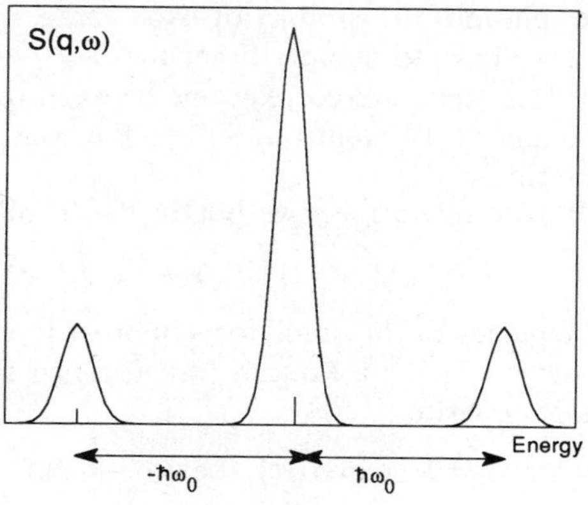

Fig. 4.6 Schematic diagram of the shape of the curve $S_{\mathrm{inc}}(q, \omega)$ as function of ω in a system where there is interaction between neutron and phonon.

eqn (4.119′) in the case where $i \neq j$ but there are correlations between the vibrational vector $\boldsymbol{u}_i$ of different particles. Such correlated vibrational motions are called phonons and are important in crystalline materials. In this case the centre of mass term $s_{\mathrm{tr}}(q, \omega)$ is no longer equal to unity and we have to consider, instead of eqn (4.124) the more general equation

$$\left(\frac{\partial^2 \sigma}{\partial \Omega \partial \omega}\right)_{\mathrm{coh}} = \frac{k_{\mathrm{f}}}{k_{\mathrm{i}}} \sum_i \sum_j b_i b_j \exp(-\mathrm{i}\boldsymbol{q}\cdot(\boldsymbol{c}_i - \boldsymbol{c}_j)) \exp - 2W \int_{-\infty}^{+\infty} \mathrm{d}t$$

$$\exp -\mathrm{i}\omega t [1 + \langle \boldsymbol{q}\cdot\boldsymbol{u}_j(0)\boldsymbol{q}\cdot\boldsymbol{u}_i(t)\rangle]. \quad (4.125)$$

As we have said, for the average of the quantity $\boldsymbol{q}\cdot\boldsymbol{u}_j(0)\boldsymbol{q}\cdot\boldsymbol{u}_i(t)$ to be non-zero the vectors $\boldsymbol{u}_j(0)$ and $\boldsymbol{u}_i(t)$ must be correlated. The scattering from the centre of mass positions of the nuclei in a crystal is a pattern of Bragg diffraction spots. This is the term $\exp(-\mathrm{i}\boldsymbol{q}\cdot(\boldsymbol{c}_i - \boldsymbol{c}_j))$. Each spot corresponds to $\boldsymbol{q} = \sigma$ where σ is a reciprocal lattice vector—i.e. a vector along a symmetry direction in the crystal, whose magnitude corresponds to $2\pi/a$ where a is the lattice plane spacing along that direction. When a phonon is passing, all the atoms are displaced from their rest positions and sequential planes are displaced by varying amounts corresponding to the phase of the propagating phonon. Some of the intensity of the Bragg spot is shifted in $\boldsymbol{q}$ by $\boldsymbol{\kappa}$ where $\boldsymbol{\kappa}$ is the wavevector of the phonon with the frequency $\omega(\boldsymbol{\kappa})$. Under these conditions the vectors $\boldsymbol{u}_i$ and $\boldsymbol{u}_j$ are correlated and the averages are non-zero. Without going into detail we quote the result: each of the inelastic functions in $\omega - \omega_0$ in eqn (4.124) becomes multiplied by a second δ function of the form

$$\delta(\boldsymbol{q} + \boldsymbol{\kappa} - \boldsymbol{\sigma}) \text{ or } \delta(\boldsymbol{q} - \boldsymbol{\kappa} - \boldsymbol{\sigma}). \tag{4.126}$$

There are very strong restrictions on where in $\boldsymbol{q}-\omega$ space intensity can be measured since two conservation conditions must be satisfied by the neutrons

$$\begin{aligned} \hbar\omega &= E_{\mathrm{i}} - E_{\mathrm{f}} = \hbar\omega_0(\boldsymbol{\kappa}) \\ \boldsymbol{q} &= \boldsymbol{k}_{\mathrm{i}} - \boldsymbol{k}_{\mathrm{f}} = \boldsymbol{\sigma} \pm \boldsymbol{\kappa}. \end{aligned} \tag{4.127}$$

For low molecular weight materials it is usual to prepare single crystals so that the direction of σ can be identified and $\omega(\boldsymbol{\kappa})$ vs $\boldsymbol{\kappa}$ values traced. For polymer materials, as we will see in Chapter 9, this is rarely possible.

SUGGESTED FURTHER READING

Bacon (1975).
Bée (1988)
Berne and Pecora (1976).
Collins (1989).
Hansen and McDonald (1986).
Lovesey (1984).
Springer (1972).
Squires (1978).

5

Labelling with deuterium—how, why, and when to use

5.1 INTRODUCTION

Anyone reading this chapter who has also been following the neutron scattering literature over the last 15 years cannot have failed to notice that deuterium labelling is at the centre of the small angle scattering technique as far as applications to polymer science are concerned. The reader may also have been puzzled by the question of how much labelling is required or desirable. In the literature there have been several swings which must give the impression that the neutron scatterers are very uncertain about the advantages and dangers of this tool. In the early papers very low concentrations of deuterated molecules were used. Moreover, extrapolations to zero concentration were applied in some cases. Then, in about 1980, levels around 50/50 of ordinary and deuterated species were reported. More recently, studies of the thermodynamic effects of mixing the two species have appeared and the fashion has switched back to lower levels of labelling. These changes have been caused by a deepening of the theoretical and experimental understanding of labelling and its consequences. Generally speaking, an historical approach to teaching science does not aid understanding (though it may well be amusing to follow the intuitive leaps that had to be made and discover some of the blind alleys of the subject). In this case it will be important to look carefully at both the physical interactions and their thermodynamical consequences in order to grasp the potential and the limitations of the technique.

The polymer scientists who first saw the potential of deuterium labelling for the study of molecular conformation in the bulk state were experienced in the use of light or X-ray scattering to study polymer solutions. In solution there are inevitably interactions between the solvent and the polymer molecules. These are favourable or unfavourable depending on the quality of the solvent, and lead to non-uniform distributions of molecules in the solution. The effects are usually described at low concentrations in terms of the so-called virial coefficient, A_2 (which will be discussed in Chapter 7). In a theta solvent A_2 is zero, and it had long been proposed that the bulk state would provide theta-conditions. It was understandable therefore

that the early experiments mimicked the light scattering technique and were run at low concentration to establish that A_2 was, indeed, zero. Why though, having established the Flory hypothesis of theta-conditions in the bulk did the experimenters persist in working at low concentration and thus considerably reducing their potential signal-to-noise ratio? The cynic might reply that deuterated molecules are very expensive and have to be paid for out of laboratory budgets. The neutrons which are usually even more expensive are paid for out of some distant central budget beyond the scientist's access.

The charitable answer lies in a very natural misconception about the scattering process which is prevalent among 'learners' today. If one imagines the labelled molecules as being 'painted' a different colour then in order to be 'seen' individually they would need to be well-separated spatially. More precisely, if there were a strong chance that in the mixture two deuterated chains could overlap each other, forming a dimer, then it seems intuitively obvious that the shape and dimensions recorded would be those of the dimer. In fact, as will be demonstrated shortly, since there is no special correlation between molecules that happen to overlap spatially, the accidental patterns from overlapping molecules all cancel and the only coherent signal in the scattering is that arising from the shape of any individual molecule. This fact was pointed out as soon as the theoreticians began to look more closely at what the polymer scientists were up to in their neutron scattering experiments (Cotton *et al.* 1974). The *caveat* was that there must be no special interactions in the system, i.e. the labelled and unlabelled molecules must behave identically. Deuteration was believed to be a relatively minor perturbation in chemical terms so, although there were a few small danger signs around, high labelling levels became *de rigueur* in bulk amorphous samples. It has been known for a long time (Strazielle and Benoît 1975), however, that the theta-temperature of polystyrene in cyclohexane is shifted about 5° up or down, depending on whether the polymer or the solvent is deuterated. Thermodynamically, therefore, deuteration is not so inert a process as one would wish since it introduces a change of 5/270 in the absolute temperature, i.e. a change of the order of 10^{-2} in the magnitude of the interactions.

The first signal that care should be taken in amorphous systems or solutions was when work began on semidilute and concentrated solutions. In the case of these experiments there was no real problem so long as the new theta-temperature was reported and used in describing the results. When studies began on crystalline polymers in the solid state, real problems arose. Polyethylene is the archetypal crystalline polymer just as polystyrene is the archetypal amorphous one. The first results on deuterated ordinary mixtures for this crystalline system, even at low labelling levels, gave ridiculously large values for molecular weights and dimensions (Schelten

et al. 1976). The cause was thought to lie in a different crystallization temperature for the two species. Thus during any but a fast crystallization process the deuterated molecules had a tendency to segregate and were no longer randomly distributed in the mixture. The difficulties posed by this effect considerably complicate the study of molecular conformation in crystalline samples. The approaches which can be used will be discussed in detail later.

Chemically dissimilar polymers were thought to mix rarely, though many more miscible or partially miscible systems are being discovered in recent studies. The difficulty lies in getting them into the mixed state since many show lower critical solution temperatures (LCST) (see Fig. A2.3, Appendix 2), that is they phase separate when the temperature is raised thus eliminating the normal way of reducing viscosity by going to temperatures well above the glass transition temperature T_g.

When neutron scattering studies on polymer blends began it soon became apparent that the cloud points or phase boundaries were often shifted by tens of degrees by deuterating one of the polymer components in a blend (Schelten *et al.* 1976). This should not have been surprising since in a polymer mixture the entropic component of the free energy of mixing becomes negligibly small so that other contributions such as the small thermodynamic changes due to deuteration may have disproportionately large effects. Fortunately (or unfortunately if one wanted to continue to use high levels of deuterum labelling with a quiet mind) these results prompted studies of the ultimately simple blend—polymers differing only in that one of them was deuterated. Perhaps it should not have come as a surprise that these mixtures behaved in the same way as blends of dissimilar polymers and that at very high molecular weights where entropy of mixing is small and enthalpy important, free energy curves and phase diagrams such as in Figs A2.1 and A2.2 of Appendix 2 were observed and phase separation detected. The details of these results and their consequences are left to the end of this chapter. The message can be summarized as follows:

1. The higher the molecular weight (the smaller the entropy of mixing per unit volume) the higher is the chance of phase separation effects occurring at high labelling concentrations.

2. In any system, the closer are the experimental conditions to a phase boundary (demixing in a blend or solution, or crystallization) the stronger will be the departure from ideality due to the effect of deuteration.

3. In an unknown system it is essential to look for the possible effects of deuteration by checking other physical properties as far as possible and, if in doubt, to keep the levels of deuteration as low as possible.

In the next few sections we will develop the mathematical basis of the deuteration labelling technique and try to give pictorial images of how it works.

5.2 SCATTERING LAWS FOR INCOMPRESSIBLE SYSTEMS

5.2.1 The basic scattering laws and the partial structure factors

In the static approximation we have shown that for a system containing only one type of scattering unit (eqn (4.25))

$$\frac{\partial \sigma}{\partial \Omega}(q) = b^2 \sum_{i=1}^{N} \sum_{j=1}^{N} \langle \exp[-\mathrm{i}\boldsymbol{q}\cdot(\boldsymbol{r}_i - \boldsymbol{r}_j)]\rangle \qquad (5.1)$$

which, introducing continuous notation and the density fluctuations

$$\Delta n(\boldsymbol{r}) = n(\boldsymbol{r}) - n = n(\boldsymbol{r}) - \frac{N}{V}$$

becomes

$$\frac{\partial \sigma}{\partial \Omega}(q) = b^2 \int_V \int_{V'} \exp[-\mathrm{i}\boldsymbol{q}\cdot(\boldsymbol{r}' - \boldsymbol{r})] \langle \Delta n(\boldsymbol{r}) \Delta n(\boldsymbol{r}') \rangle \mathrm{d}\boldsymbol{r}\mathrm{d}\boldsymbol{r}'. \qquad (5.2)$$

Equation (5.2) shows that the scattered intensity (which will be called $I(q)$ for simplicity) depends on density fluctuations in the system. If the molecules are uniformly distributed there will be no scattering beside the anomaly at $q = 0$. Only if there are regions where the density is above or below the mean value will scattering occur. The density fluctuations depend on the ease with which the system can be compressed and we will see in Chapter 7 that there is a formal relationship with the compressibility $(-\partial v/\partial p)_T$. Other factors come into play if the system contains more than one species. Let us consider a two species mixture with N_1 molecules of scattering lengths b_1 and N_2 of scattering length b_2. Collecting the terms belonging to each species and those of the cross terms together we obtain

$$I(q) = b_1^2 \sum_{i_1}^{N_1} \sum_{j_1}^{N_1} \langle \exp(-\mathrm{i}\boldsymbol{q}\cdot\boldsymbol{r}_{ij})\rangle + 2b_1 b_2 \sum_{i_1}^{N_1} \sum_{j_2}^{N_2} \langle \exp(-\mathrm{i}\boldsymbol{q}\cdot\boldsymbol{r}_{ij})\rangle$$
$$+ b_2^2 \sum_{i_2}^{N_2} \sum_{j_2}^{N_2} \langle \exp(-\mathrm{i}\boldsymbol{q}\cdot\boldsymbol{r}_{ij})\rangle \qquad (5.3)$$

or

$$I(q) = b_1^2 \iint_{V,V'} \langle n_1(\boldsymbol{r})n_1(\boldsymbol{r}')\rangle \exp[-\mathrm{i}\boldsymbol{q}\cdot(\boldsymbol{r}'-\boldsymbol{r})]\,\mathrm{d}\boldsymbol{r}\mathrm{d}\boldsymbol{r}'$$

$$+\, b_2^2 \iint_{V,V'} \langle n_2(\boldsymbol{r})n_2(\boldsymbol{r}')\rangle \exp[-\mathrm{i}\boldsymbol{q}\cdot(\boldsymbol{r}'-\boldsymbol{r})]\,\mathrm{d}\boldsymbol{r}\mathrm{d}\boldsymbol{r}'$$

$$+\, 2b_1 b_2 \iint_{V,V'} \langle n_1(\boldsymbol{r})n_2(\boldsymbol{r}')\rangle \exp[-\mathrm{i}\boldsymbol{q}\cdot(\boldsymbol{r}'-\boldsymbol{r})]\,\mathrm{d}\boldsymbol{r}\mathrm{d}\boldsymbol{r}' \quad (5.4)$$

where $n_i(\boldsymbol{r})$ is the local density of constituent i. Equation (5.4) reminds us that a different phenomenon arises in scattering from mixtures even if there is no fluctuation in overall density. If we take an element of volume $\mathrm{d}V$ large enough to contain a few of each species of molecule, the percentage of each within $\mathrm{d}V$ may vary with time. This is a local composition fluctuation, governed by the osmotic compressibility and hence the laws of thermodynamics and is discussed in Chapter 7. The two scattering processes are different and it is better not to mix them in the mathematical treatment. The easiest way is to assume that the two component system is incompressible or that the contribution from density fluctuations is very weak compared to concentration fluctuations and hence can be neglected. In fact this simplification is usually quite valid for most polymeric systems. From eqn (5.4) we define three partial structure factors, S_{11}, S_{22}, and S_{12} as

$$S_{ij} = \iint_{V,V'} \langle n_i(\boldsymbol{r})n_j(\boldsymbol{r}')\rangle \exp[-\mathrm{i}\boldsymbol{q}\cdot(\boldsymbol{r}'-\boldsymbol{r})]\,\mathrm{d}\boldsymbol{r}\mathrm{d}\boldsymbol{r}'. \quad (5.5)$$

We will now examine how we could simplify eqn (5.4) by finding relationships between these partial structure factors.

5.2.2 Relationship between the partial structure factors

To start simply, we will take a two component system where the two molecular volumes are identical and called v. The total volume will be

$$V = (N_1 + N_2)v \quad (5.6)$$

so that the volume fractions of the two components are respectively

$$\varphi_1 = \frac{N_1}{N_1 + N_2} \text{ and } \varphi_2 = \frac{N_2}{N_1 + N_2}. \quad (5.7)$$

At every point in space, since this is an incompressible system, we must have

$$n_1(\boldsymbol{r}) + n_2(\boldsymbol{r}) = n = \frac{N_1 + N_2}{V} = \frac{1}{v}. \quad (5.8)$$

Which also means that for variation away from the mean density we have

$$\Delta n_1(\boldsymbol{r}) + \Delta n_2(\boldsymbol{r}) = 0 \tag{5.9}$$

since it is assumed that v is independent of concentration. Multiplying by $\Delta n_1(\boldsymbol{r}')\exp(-\mathrm{i}\boldsymbol{q}\cdot(\boldsymbol{r}-\boldsymbol{r}'))$ and integrating over the whole scattering volume must still lead to zero, i.e.

$$\iint_{V,V'} \langle(\Delta n_1(\boldsymbol{r}) + \Delta n_2(\boldsymbol{r}))\Delta n_1(\boldsymbol{r}')\rangle \exp[-\mathrm{i}\boldsymbol{q}\cdot(\boldsymbol{r}-\boldsymbol{r}')]\mathrm{d}\boldsymbol{r}\,\mathrm{d}\boldsymbol{r}' = 0 \tag{5.10}$$

or:

$$\iint_{V,V'} \langle\Delta n_1(\boldsymbol{r})\Delta n_1(\boldsymbol{r}')\rangle \exp[-\mathrm{i}\boldsymbol{q}\cdot(\boldsymbol{r}-\boldsymbol{r}')]\mathrm{d}\boldsymbol{r}\,\mathrm{d}\boldsymbol{r}' + \iint_{V,V'} \langle\Delta n_2(\boldsymbol{r})\Delta n_1(\boldsymbol{r}')\rangle \exp[-\mathrm{i}\boldsymbol{q}\cdot(\boldsymbol{r}-\boldsymbol{r}')]\mathrm{d}\boldsymbol{r}\,\mathrm{d}\boldsymbol{r}' = 0. \tag{5.11}$$

Which can be written in terms of the partial structure factors as

$$S_{11}(q) + S_{12}(q) = 0. \tag{5.12}$$

We could also have multiplied by $\Delta n_2(\boldsymbol{r}')\exp\{-\mathrm{i}\boldsymbol{q}(\boldsymbol{r}-\boldsymbol{r}')\}$, obtaining the relationship

$$S_{12}(q) + S_{22}(q) = 0. \tag{5.13}$$

To begin with, we had for this system three structure factors, S_{11}, S_{22}, and S_{12}. Now we know from eqns (5.12) and (5.13) that (Daoud *et al.* 1975)

$$S_{11}(q) = S_{22}(q) = -S_{12}(q).$$

This means that eqns (5.3) can now be written as (Akcasu *et al.* 1980*b*)

$$I(q) = (b_1 - b_2)^2 S_{11}(q) = (b_1 - b_2)^2 S_{22}(q) = -(b_1 - b_2)^2 S_{12}(q). \tag{5.14}$$

It is a little surprising to see that the scattering of a two-component system seems to depend only on one of these components and to be independent of the other. We will attempt to give a pictorial view of what is going on later. Since this result is only valid for a system without density fluctuations the relationship (5.14) is often called 'the incompressibility hypothesis'.

5.2.3 Generalization to a mixture of more than two species

Let us assume that we have $p + 1$ different species i $(0 \leq i \leq p)$. Equation (5.3) is easily generalized to

$$I(q) = \sum_{i=0}^{p} b_i^2 S_{ii}(q) + 2 \sum_{i<j}^{p} b_i b_j S_{ij}(q). \tag{5.15}$$

Using exactly the same methods as before we introduce first the number density of segments for the species i: $n_i(\boldsymbol{r})$. As before, the segments have all the same volume v_0 and we obtain

$$\sum_{i=0}^{p} n_i(\boldsymbol{r}) = \frac{\Sigma N_i}{V} = n \tag{5.16}$$

and

$$\sum_{i=0}^{p} \Delta n_i(\boldsymbol{r}) = 0. \tag{5.17}$$

Multiplying eqn (5.17) by $\Delta n_0(\boldsymbol{r}) \exp - \mathrm{i}\boldsymbol{q}(\boldsymbol{r} - \boldsymbol{r}')$ and integrating over the whole scattering volume, as in eqn (5.10) gives

$$\sum_{i=0}^{p} S_{oi}(q) = 0. \tag{5.18}$$

Repeating the same operation with all the $\Delta n_i(\boldsymbol{r})$ gives p relationships obtained by replacing the index o in the preceding equation by any of the p indices

$$\sum_{i=0}^{p} S_{ki}(q) = 0. \tag{5.19}$$

(This formula can also be demonstrated in a simple fashion, see Section 5.3.2.) These results allow us, by elementary calculations, to eliminate all the terms with the index 0 and to generalize eqn (5.14) to

$$I(q) = \sum_{i=1}^{p} (b_i - b_0)^2 S_{ii}(q) + 2 \sum_{i<j} (b_i - b_0)(b_j - b_0) S_{ij}(q). \tag{5.20}$$

This removes all the structure factors involving the species zero since this species fills the vacancies left by the other species (Benoît *et al.* 1981).

This formula is quite general and is used in many circumstances. It is the basis of the interpretation of practically all the results obtained by small angle neutron scattering.

5.2.4 Decomposition of $S(q)$ into intra- and intermolecular interferences

Let us assume to begin with that we have only one species of molecules which can be supposed to be small enough to be considered as scattering points. In the static limit the scattered intensity is

$$I(q) = b^2 \sum_{i=i}^{N} \sum_{j=1}^{N} \langle \exp(-\mathrm{i}\boldsymbol{q}\cdot\boldsymbol{r}_{ij}) \rangle. \tag{5.21}$$

In this double summation there are N terms for which $i = j$. Since for these terms the exponential is equal to 1 it is possible to write

$$I(q) = b^2 \left[N + 2 \sum_{i<j}\sum \langle \exp(-\mathrm{i}\boldsymbol{q}\cdot\boldsymbol{r}_{ij}) \rangle \right]. \tag{5.22}$$

In the second term of the RHS of the preceding expression there are $N(N-1)$ terms which, as will be justified later, can be counted as N^2. If we consider one pair of points i and j and if we take the average over all possible relative positions of i and j the result will be independent of i and j. It is therefore possible to multiply by N^2 the value obtained for a simple term to obtain the correct result

$$I(q) = b^2 N [1 + N \langle \exp(-\mathrm{i}\boldsymbol{q}\cdot\boldsymbol{r}_{12}) \rangle]. \tag{5.23}$$

This equation shows that the scattered intensity depends on two terms: the first term corresponds to the scattering by individual molecules, the second corresponds to interferences between the waves scattered by different molecules. Without entering into details we shall generalize these results to molecules which cannot be considered as point scatterers.

Let us, remaining for a while with the problem of a one constituent system, rewrite the expression for $S(q)$ but, this time, taking into account the fact that the molecules are made of z segments; each unit, which can be called a 'monomer', has a scattering length b. There are N molecules, each with z segments. We define a scattering unit with two indices, one of which is characteristic of the chain p or q ranges from 1 to N. The second index is characteristic of the position of the monomer on the chain and will be called j or k and goes from 1 to z.

This gives for $S(q)$ the quadruple sum

$$S(q) = \sum_{p=1}^{N} \sum_{q=1}^{N} \sum_{j=1}^{z} \sum_{k=1}^{z} \langle \exp[-\mathrm{i}\boldsymbol{q}\cdot(\boldsymbol{r}_{p,j} - \boldsymbol{r}_{q,k})] \rangle. \tag{5.24}$$

Now, as before, we extract the terms for which $p = q$. There are N terms of this type which correspond to the interferences of two points on the same molecule and they are all identical. The terms which correspond to two different molecules are obtained for $q \neq p$ but since we take the average values they are all identical and their number is $N(N-1)$ which will be replaced by N^2. With these simplifications

$$S(q) = N \sum_{i_1=1}^{z} \sum_{j_1=1}^{z} \langle \exp(-\mathrm{i}\boldsymbol{q}\cdot\boldsymbol{r}_{i_1 j_1}) \rangle + N^2 \sum_{i_1=1}^{z} \sum_{j_2=1}^{z} \langle \exp(-\mathrm{i}\boldsymbol{q}\cdot\boldsymbol{r}_{i_1 j_2}) \rangle. \tag{5.25}$$

The first term depends only on the size and the shape of the macromolecule. The expression between the angled brackets in this term is formed of z^2 terms; if one divides it by z^2 one obtains what is usually called the form factor of the molecules. It is normalized to unity for $q = 0$. The tradition is to call it $P(q)$

$$P(q) = \frac{1}{z^2} \sum_{i_1=1}^{z} \sum_{j_1=1}^{z} \langle \exp(-i\boldsymbol{q}\cdot\boldsymbol{r}_{ij}) \rangle \,. \tag{5.26}$$

The second term also contains z^2 terms and we shall call it, after division by z^2 in order to have it equal to unity for $q = 0$, $Q(q)$

$$Q(q) = \frac{1}{z^2} \sum_{i_1=1}^{z} \sum_{j_2=1}^{z} \langle \exp(-i\boldsymbol{q}\cdot\boldsymbol{r}_{i_1 j_2}) \rangle. \tag{5.27}$$

We can now write

$$S(q) = Nz^2P(q) + N^2z^2Q(q) \tag{5.28}$$

remembering that the first term corresponds to intramolecular interferences and the second to intermolecular interferences. This equation shows that $P(q)$ and $Q(q)$ are essentially different: $P(q)$ is a dimensionless intensive property but $Q(q)$, on the other hand, is inversely proportional to N since the quantity $NQ(q)$ is dimensionless and intensive. Since N is always large this justifies the replacement of $N - 1$ by N.

5.3 APPLICATIONS OF THE GENERAL FORMULAE TO DEUTERATION

5.3.1 Two identical polymers, one deuterated the other not, in a melt or a glass

We assume that the two polymers are identical and differ only by coherent scattering length b_H and b_D and also that the volumes of the scattering units are identical. We assume also that the total number of molecules is N, xN of them being deuterated and $(1 - x)N$ being ordinary. The volume fractions are respectively x and $(1 - x)$. This allows use of the relationships between the partial structure factors developed earlier, and we can write

$$I(q) = (b_D - b_H)^2 S_{DD}(q) \tag{5.29}$$

or S_{HH} or $-S_{HD}$. Expressing these three quantities in terms of $P(q)$ and $Q(q)$ we obtain

$$\begin{aligned} xNz^2P(q) + x^2N^2z^2Q(q) &= (1 - x)Nz^2P(q) + (1 - x)^2N^2z^2Q(q) \\ &= -x(1 - x)N^2z^2Q(q). \end{aligned} \tag{5.30}$$

The last equation is obtained by observing that for the cross term S_{HD} two scattering points cannot belong to the same molecule and therefore there are no intramolecular interferences and only $(Nx)N(1-x)$ intermolecular interferences. In writing these expressions it has been assumed that for the two polymers one can use the same values of $P(q)$ and $Q(q)$. This is obvious for the Ps but needs more careful consideration for the Qs. Since physically the Qs are the interferences between two different molecules it is intuitively evident that they are identical between two deuterated molecules, two ordinary molecules and one deuterated and one ordinary molecule.

Equation (5.30) leads to relationships which allow us to evaluate Q as a function of P. Setting the difference between the first and the third terms in the equalities to zero, gives

$$Nz^2P(q) + N^2z^2Q(q) = 0$$

or

$$NQ(q) = -P(q).$$

Using this value of $Q(q)$ in any of the three relationships (5.30) gives

$$I(q) = (b_D - b_H)^2x(1-x)Nz^2P(q). \tag{5.31}$$

This result, first given by Cotton *et al.* (1974), Daoud *et al.* (1975) and Akcasu *et al.* (1980*b*), is simple and interesting for two reasons:

(a) The shape of the signal is independent of the composition x and depends only on the form factor of the molecules. As one might expect its intensity goes through a maximum for $x = 0.5$ and disappears for $x = 0$ or $x = 1$ since the hypothesis of incompressibility demands zero scattering for pure component systems.

(b) The signal, even though it is independent of the intermolecular interferences, allows us to determine them, since they are, neglecting the factor N, equal and of opposite sign to the form factor. This will be justified by a simple physical argument later on.

Once the contrast factor is removed at $q = 0$ the scattering from a polymer molecule just depends on the number of scattering nuclei, i.e. on the molecular weight and on the trivial factor $x(1-x)$. If $I(0)$ gives the correct value of the molecular weight for a deuterated-hydrogenous mixture then this is good evidence that the labelling technique is working correctly.

Earlier in this chapter we spoke about melts or glasses and this needs to be justified. Different experiments made above and below the glass transition temperature did not show any significant effect of the state of the

polymer on the scattering curves. The simple explanation of this result is to assume that, when the glass is formed, the molecules are frozen in the conformation they had above T_g and that the scattering corresponds to the scattering of a melt at approximatively the glass transition temperature.

5.3.2 The Babinet principle and some geometric justifications

The results which have just been presented are relatively simple and it is possible to justify them by geometrical considerations. Let us assume that we have a scattering system consisting of two substances, one of scattering length b_1 and the other b_2. For a given value of the scattering vector q, the scattered amplitude which is a complex number (in order not to forget the phase) will be called $A_1(q)$ and the intensity $A_1A_1{}^*$. Now let us assume that we take the 'negative' of the object, i.e. we replace the scattering length b_1 by b_2 and vice versa, obtaining for the same value of q the amplitude $A_2(q)$. Let us now superimpose the two scattering diagrams taking the phases into consideration; on the one hand the intensity should be zero because it corresponds to the scattering by a system of uniform density $(b_1 + b_2)$ but on the other it should be $\{A_1(q) + A_2(q)\}^2$. (See Fig. 5.1.)

This means that A_1 and A_2 are equal in magnitude but have a phase difference of π. In other words

$$A_1 = -A_2 \text{ and } A_1A_1^* = A_2A_2^* = -A_1A_2.$$

This is nothing else but eqn (5.14).

Now we shall try to explain eqn (5.31) geometrically. For this purpose one has to show that the quantity $Q(q) = -NP(q)$. In order to do so, we rewrite the general expression for $Q(q)$

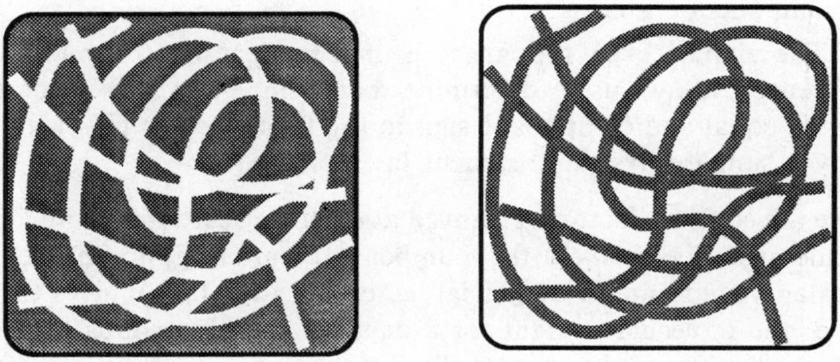

(a) (b)

Fig. 5.1 The Babinet principle. Diagram (a) is the negative of diagram (b), both figures have the same scattering pattern.

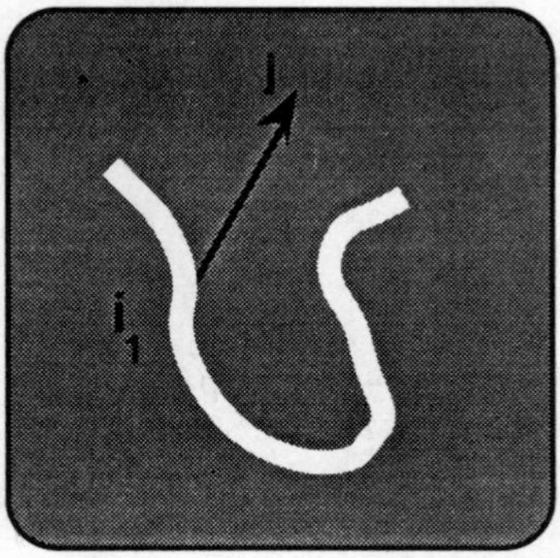

Fig. 5.2 The interpretation of the term $Q(q)$ in the scattering by a one-polymer material in the bulk.

$$Q(q) = \frac{1}{N^2 z^2} \sum_{p=1}^{N} \sum_{m=1}^{N} \sum_{i_p=1}^{z} \sum_{j_m=1}^{z} \langle \exp(-\mathrm{i}\boldsymbol{q}\cdot\boldsymbol{r}_{i_p j_m}) \rangle \quad (5.32)$$

with $p \neq m$. Since, on average the interference terms are identical we can replace the summation over p by a multiplication by N obtaining

$$Q(q) = \frac{1}{Nz^2} \sum_{i_1=1}^{z} \sum_{m=2}^{N} \sum_{j_m=1}^{z} \langle \exp(-\mathrm{i}\boldsymbol{q}\cdot\boldsymbol{r}_{i_1 j_m}) \rangle . \quad (5.33)$$

Now, considering this formula and before having made the first summation, we see that the last double sum is a summation over all the scattering centres of the sample except the points of molecule 1. This expression is the complex amplitude of the beam scattered by the whole sample (except molecule 1) with the phase origin in i_1 (Fig. 5.2).

Following the preceding demonstration, this sum can be substituted by its 'image', replacing the system by molecule 1 and changing the sign. This leads to

$$Q(q) = \frac{-1}{Nz^2} \sum_{i_1=1}^{z} \sum_{j_1=1}^{z} \langle \exp(-\mathrm{i}\boldsymbol{q}\cdot\boldsymbol{r}_{i_1 j_1}) \rangle = -\frac{1}{N} P(q). \quad (5.34)$$

5.3.3 The case of two different polymers

The application of the Babinet principle can easily be generalized to systems containing more than one constituent. Let us consider the case of a system made of two polymers of the same nature, one having a degree of polymerization z_D, the other z_H; the total number of polymer molecules is $N_D + N_H$. We take as before a point i on molecule 1 and we evaluate the total amplitude scattered by all molecules except 1. It will be

$$A(q) = \sum_{m=2}^{N_D} \sum_{j_m=1}^{z_D} \langle \exp(-iq \cdot r_{i_1 j_m}) \rangle + \sum_{p=1}^{N_H} \sum_{j_p=1}^{z_H} \langle \exp(-iq \cdot r_{i_1 j_p}) \rangle \tag{5.35}$$

From the Babinet principle this is equal to the amplitude scattered by molecule 1.

$$A_{i_1}(q) = -\sum_{j_1=1}^{z_D} \langle \exp(-iq \cdot r_{i_1 j_1}) \rangle \tag{5.36}$$

If we sum over all values of i_1 and multiply by N_D we obtain from eqn. (5.35)

$$N_D \sum_{i_1=1}^{z_D} \sum_{m=2}^{N_D} \sum_{j_m=1}^{z_D} \langle \exp(-iq \cdot r_{i_1 j_m}) \rangle + N_D \sum_{i_1=1}^{z_D} \sum_{p=1}^{N_H} \sum_{j_p=1}^{z_H} \langle \exp(-iq \cdot r_{i_1 j_p}) \rangle = -N_D z_d^2 P_d(q). \tag{5.37}$$

Looking carefully at the LHS of this equality shows that it is just $N_D^2 z_D^2 Q_{DD} + N_D N_H z_D z_H Q_{DH}$. We have therefore shown that

$$-N_D z_D^2 P_D(q) = N_D^2 z_D^2 Q_{DD}(q) + N_D N_H z_D z_H Q_{DH}(q). \tag{5.38}$$

This can be generalized to any number of constituents and is nothing but eqn (5.13) or its generalization (5.18).

5.3.4 Deuterated-hydrogenous mixture in a solvent

Let us take two polymers and dissolve them in a solvent which can be another polymer and has a scattering length b_0. We will take the precaution of having scattering units of the same size. Equation (5.20) gives

$$I(q) = (b_D - b_0)^2 S_{DD} + (b_H - b_0)^2 S_{HH} + 2(b_D - b_0)(b_H - b_0) S_{HD}. \tag{5.39}$$

Since both polymers are characterized by the same $P(q)$ and $Q(q)$ values

$$\begin{aligned} S_{DD} &= xNz^2P(q) + x^2N^2z^2Q(q) \\ S_{HH} &= (1-x)Nz^2P(q) + (1-x)^2N^2z^2Q(q) \\ S_{HD} &= -x(1-x)N^2z^2Q(q). \end{aligned} \tag{5.40}$$

A little arithmetic yields the following result

$$\begin{aligned} I(q) = {} & (b_D - b_H)^2 x(1-x) Nz^2 P(q) \\ & + (xb_D + (1-x)b_H - b_0)^2 [Nz^2P(q) + N^2z^2Q(q)]. \end{aligned} \tag{5.41}$$

This equation (Williams *et al.* 1979) is interesting because it offers to experimenters many possibilities: the average scattering contrast between the polymer and the solvent is evidently

$$\langle b \rangle = x b_D + (1 - x) b_H - b_0.$$

The term having this coefficient represents the scattering by a system where all polymers have the same average scattering length. This scattering arises from concentration fluctuations of the polymer. To this scattering one has to add the scattering represented by the first term; it is due to the fluctuations, at constant concentration of polymer, of the composition in the deuterated and normal polymer; this contribution is identical to the scattering of the mixture of polymers in the melt we have just been studying.

This suggests the following experiment: if one adjusts the scattering length b_0 of the solvent by using instead of a pure liquid, a mixture of deuterated and ordinary solvent such that

$$x b_D + (1 - x) b_H - b_0 = 0 \tag{5.42}$$

the average polymer–solvent contrast and hence the corresponding scattering becomes zero and one is left with the following scattering intensity

$$I(q) = (b_D - b_H)^2 x(1 - x) N z^2 P(q). \tag{5.43}$$

The experiment thus yields directly the form factor of the polymer molecules in solution even at high concentration of polymer. It is also possible to change the composition x, keeping the total concentration in polymer constant. By comparison of the different scattering intensities measured as functions of x, one can determine both functions $P(q)$ and $Q(q)$ (see Chapter 8).

5.3.5 Mixture of deuterated and ordinary polymer in any system

We assume that we have a system made of any number of constituents containing among them xN deuterated species and $(1 - x)N$ ordinary polymers. Using the same procedure it can be shown that

$I(q) = (b_D - b_H)^2 x(1 - x) N z^2 P(Q)$ + the signal one would get if the N molecules of polymer were identical and had the coherent scattering length $\langle b \rangle$

$$\langle b \rangle = x b_D + (1 - x) b_H.$$

In other words, to obtain the form factor of a polymer in a given system one can perform the two following experiments. First, a polymer is prepared where all the monomers are partially labelled in order for each monomer to have the scattering length $\langle b \rangle$ and the scattered intensity is

measured (one cannot use a statistical copolymer of D and H monomers since they are randomly distributed in one molecule of copolymer and this could give an extra scattering). Second, this polymer is replaced by a mixture of a deuterated and an ordinary polymer with a volume fraction x of the deuterated species and the scattered intensity is measured. Subtracting the intensity of the first experiment from the intensity measured in the second experiment gives $x(1-x)Nz^2P(Q)$, i.e. the form factor of the polymer (Benoît *et al.* 1984). These experiments require that both samples have rigorously the same structure and up to the present have not been attempted in practice.

5.4 THE SYMMETRIES OF $Q(q)$

5.4.1 Linear polymers

It is useful to look a little more closely at the quantity $Q(q)$. It is the average value of $\exp(-\mathrm{i}\boldsymbol{q}\cdot\boldsymbol{r}_{ij}) = s(i,j)$ taken over pairs of scattering centres on different molecules. To show what this means a matrix can be drawn, as shown in Fig. 5.3, with $z \times z$ elements corresponding to the z repeat units in each molecule. In each element of the matrix is written the appropriate value of $\langle\exp(-\mathrm{i}\boldsymbol{q}\cdot\boldsymbol{r}_{ij})\rangle$ called *sij*. $Q(q)$ is just the sum of these z^2 terms divided by z^2.

For linear polymers the matrix obeys several symmetry rules. One obvious one is the symmetry on either side of the main diagonals. A closer look indicates also that

s1,1	s2,1	... s(z−1)/2,1	s(z+1)/2,1	... sz−1,1	sz,1
s1,2	s2,2	... s(z−1)/2,2	s(z+1)/2,2	... sz−1,2	sz,2
...	...	... ...	...	... ...	...
s1,(z−1)/2	s2,(z−1)/2	... s(z−1)/2,(z−1)2	s(z+1)/2,(z−1)2	... sz−1,(z−1)/2	sz,(z−1)/2
s1,(z+1)/2	s2,(z+1)/2	s(z−1)/2,(z+1)2	s(z+1)/2,(z+1)2	sz−1,(z+1)/2	sz,(z+1)/2
...	...	... ...	...	... ...	...
s1,(z−1)	s2,(z−1)	... s(z−1)/2,(z−1)	s(z+1)/2,(z−1)	... sz−1,(z−1)	sz,(z−1)
s1,z	s2,z	... s(z−1)/2,z	s(z+1)/2,z	... sz−1,z	sz,z

Fig. 5.3 Diagram showing the symmetries of the matrix $Q(q)$.

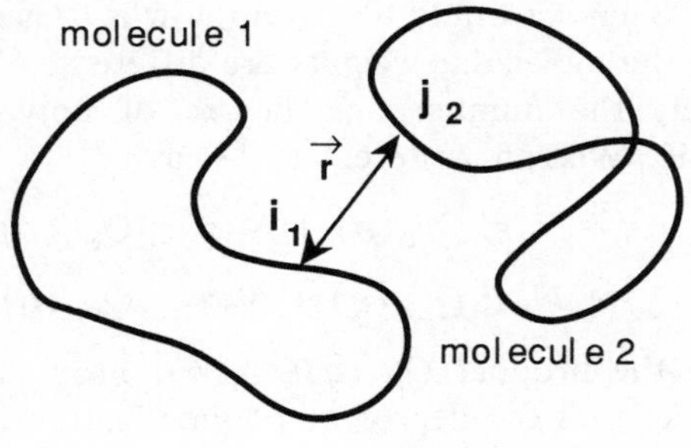

Fig. 5.4 Interactions between two ring-polymers.

$$\langle \exp(-\mathrm{i}\boldsymbol{q}\cdot\boldsymbol{r}_{ij})\rangle = \langle \exp(-\mathrm{i}\boldsymbol{q}\cdot\boldsymbol{r}_{i,z-j})\rangle \tag{5.44}$$

and

$$\langle \exp(-\mathrm{i}\boldsymbol{q}\cdot\boldsymbol{r}_{ij})\rangle = \langle \exp(-\mathrm{i}\boldsymbol{q}\cdot\boldsymbol{r}_{z-i,j})\rangle$$

there are two further symmetry axes—the vertical and horizontal lines passing through the centre as in the Fig. 5.3 where it has been supposed that z is odd. This means that the term $Q(q)$ can be evaluated taking the average value either of the whole table or of one-quarter of it.

5.4.2 The case of a ring molecule

If the molecules form rings there is another interesting property for the intermolecular interference term $Q(q)$. Since every point on the ring plays the same role, all the $\langle \exp - \mathrm{i}\boldsymbol{q}\boldsymbol{r}_{ij}\rangle$ terms are identical and $Q(q)$ is just the quantity $\langle \exp - \mathrm{i}\boldsymbol{q}\boldsymbol{r}_{ij}\rangle$ which can be evaluated for any pair of points so long as they are on different molecules (see Fig. 5.4). This allows us to evaluate the quantity $\langle \exp - \mathrm{i}\boldsymbol{q}\boldsymbol{r}_{ij}\rangle$ in a system made only of rings and one obtains:

$$\langle \exp(-\mathrm{i}\boldsymbol{q}\boldsymbol{r}_{ij})\rangle = -\frac{P(q)}{N} \tag{5.45}$$

regardless of the position of i and j ($P(q)$ is the form factor of a ring (see eqn 6.89)).

5.5 APPROXIMATE METHODS FOR THE EVALUATION OF THE SCATTERED INTENSITY

5.5.1 A mixture of two polymers of different molecular dimensions, one deuterated the other not

In Section 5.3.1 we have been able to evaluate the scattered intensity from two polymers of the same molecular weight, differing only by their coherent

scattering length. What we would like to do now is to generalize this result to the case where the molecular weights are different. N_D and z_D, N_H and z_H are respectively the number and degree of polymerization of the polymers D and H. We can write eqn (5.13) as

$$z_D^2 P_D(q) + N_D z_D^2 Q_{DD}(q) + N_H z_D z_H Q_{DH}(q) = 0$$
$$z_H^2 P_H(q) + N_H z_H^2 Q_{HH}(q) + N_D z_D z_H Q_{DH}(q) = 0. \tag{5.46}$$

Now let us write the product $Q_{DD}(q)Q_{HH}(q)$. From its definition, and introducing the fact that the degree of polymerization z_D and z_H are different; we have

$$N_D^2 z_D^2 Q_{DD}(q) N_H^2 z_H^2 Q_{HH}(q) = \sum_{p=1}^{N_D} \sum_{q=1}^{N_D} \sum_{j=1}^{z_D} \sum_{k=1}^{z_D} \langle \exp[-\mathrm{i}\boldsymbol{q}\cdot(\boldsymbol{r}_{jp} - \boldsymbol{r}_{kq})] \rangle \sum_{m=1}^{N_H} \sum_{n=1}^{N_H} \sum_{j=1}^{z_H} \sum_{k=1}^{z_H} \langle \exp[-\mathrm{i}\boldsymbol{q}\cdot(\boldsymbol{r}_{jm} - \boldsymbol{r}_{kn})] \rangle \tag{5.47}$$

with the condition that $p \neq q$ and $m \neq n$. We can transform this product of two sums into a single multiple sum of the quantity

$$\langle \exp[-\mathrm{i}\boldsymbol{q}\cdot(\boldsymbol{r}_{jp} - \boldsymbol{r}_{kq})] \rangle \langle \exp[-\mathrm{i}\boldsymbol{q}\cdot(\boldsymbol{r}_{jm} - \boldsymbol{r}_{kn})] \rangle. \tag{5.48}$$

We now exchange terms in the exponential functions and make the hypothesis that this exchange, which does not modify the instantaneous value, (Benoît *et al.* 1991) will not modify the average value, writing

$$\langle \exp[-\mathrm{i}\boldsymbol{q}\cdot(\boldsymbol{r}_{jpD} - \boldsymbol{r}_{kqD})] \rangle \langle \exp[-\mathrm{i}\boldsymbol{q}\cdot(\boldsymbol{r}_{jmH} - \boldsymbol{r}_{knH})] \rangle =$$
$$\langle \exp[-\mathrm{i}\boldsymbol{q}\cdot(\boldsymbol{r}_{jpD} - \boldsymbol{r}_{knH})] \rangle \langle \exp[-\mathrm{i}\boldsymbol{q}\cdot(\boldsymbol{r}_{jmH} - \boldsymbol{r}_{kqD})] \rangle. \tag{5.49}$$

When we sum over this last quantity we obtain $[N_D N_H z_D z_H Q_{DH}(q)]^2$. Our final result is therefore

$$Q_{DD}(q) Q_{HH}(q) = [Q_{DH}(q)]^2. \tag{5.50}$$

Now let us define $S(q)$ by the relationship

$$S(q) = \frac{I(q)}{(b_D - b_H)^2}.$$

From the result of the incompressibility hypothesis (eqn (5.14)) we can write

$$S(q) = N_D z_D^2 P_D(q) + N_D^2 z_D^2 Q_{DD}(q) = N_H z_H^2 P_H(q) + N_H^2 z_H^2 Q_{HH}(q)$$
$$= -N_H N_D Q_{DH}(q).$$

Evaluating Q_{DD}, Q_{HH}, and Q_{HD} from the preceding relationships and putting these values thus obtained in eqn (5.50) gives

$$S(q) = \frac{N_D N_H z_D^2 z_H^2 P_D(q) P_H(q)}{N_D z_D^2 P_D(q) + N_H z_H^2 P_H(q)} \tag{5.51}$$

or, taking the inverse

$$\frac{(b_D - b_H)^2}{I(q)} = \frac{1}{N_H z_H^2 P_H(q)} + \frac{1}{N_D z_D^2 P_D(q)}. \tag{5.52}$$

This is the well-known formula given by de Gennes (1970) following a method called the 'random phase approximation'. If $N_D = N_H$ one recovers eqn (5.31), established for two identical molecules. We have to remember that this is not the general case; it has been assumed that the interactions between the polymers did not depend on their nature and other methods exist which allow us to take these interactions into account. Our procedure justifies in some respect the name 'random phase approximation' since going from eqn (5.48) to eqn (5.49) we modify the way we take the averages of the phases represented by these exponentials.

5.5.2 The case of a symmetrical block copolymer in the bulk

In the preceding section it has been shown that it is possible, without heavy mathematic formalism, to obtain equations allowing the calculation of the intensity scattered by mixtures of polymers (even when the polymers do not have the same degree of polymerization) at any composition. In this section we want to extend what we have already learned to the case of copolymers, considering first the simplest copolymer which is a linear chain of z monomers $z/2$ being normal and the other $z/2$ being deuterated.

At first glance this does not appear very different from a 50/50 mixture of deuterated and hydrogenous molecules for which

$$I(q) = (b_D - b_H)^2 S_{DD}(q).$$

Since the system is, as always, assumed to be incompressible, we have

$$S_{DD} = S_{HH} = -S_{DH}.$$

The expressions for S_{DD} and S_{HH} developed previously in terms of $P(q)$ and $Q(q)$ still hold but S_{DH} now contains both $P(q)$ and $Q(q)$ since it is possible to find cross terms on the same molecule.

P_{DD} and P_{HH} are the form factors of the deuterated and hydrogenous sections which must be the form factor of a molecule of length $z/2$. P_{DH} on the other hand is an interference term for units D and H on the same molecule. If the two parts of the molecule are not identical then the total form factor $P_T(q)$ must be given by

$$z^2 P_T = z_D^2 P_{DD} + z_H^2 P_{HH} + 2 z_D z_H P_{DH} \tag{5.53}$$

where z_D and z_H are the numbers of monomers of parts D and H (with $z = z_D + z_H$). Now for the 50/50 diblock $z_D = z_H = z/2$ so that (since $P_{DD} = P_{HH}$)

$$2P_T = P_{DD} + P_{DH} \tag{5.54}$$

and, since $Q_{DD} = Q_{HH}$, we obtain the two relationships

$$S_{DD} = N\left(\frac{z}{2}\right)^2 P_{DD} + N^2\left(\frac{z}{2}\right)^2 Q_{DD} \tag{5.55}$$

and

$$S_{DH} = N\left(\frac{z}{2}\right)^2 P_{DH} + N^2\left(\frac{z}{2}\right)^2 Q_{DH}.$$

We now look at the relationship between Q_{DD} and Q_{DH}. In the matrix form (Fig. 5.3) these are the first and second (or third) quadrants and, given the symmetry we remarked on, these are equal, and equal to a quarter of the value for the whole polymer.

However all the $Q(q)$ are normalized by the total number of elements in the table. $Q(q)$ is the average value of a term in the matrix. Since the four quadrants are identical the average values are the same taken for any of the quadrants or for the total matrix. It follows that

$$Q_{DD} = Q_{DH} = Q_T.$$

Remembering also (eqn (5.34)) that

$$NQ(q) = -P(q)$$

we find that

$$S_{DD}(q) = N\left(\frac{z}{2}\right)^2 P_{DD} - N^2\left(\frac{z}{2}\right)^2 \frac{P_T}{N} \tag{5.56}$$

$$= N\left(\frac{z}{2}\right)^2 [P_{DD} - P_T]$$

$$= \tfrac{1}{2} N\left(\frac{z}{2}\right)^2 [P_{DD} - P_{DH}]$$

or

$$I(q) = (b_D - b_H)^2 N\left(\frac{z}{2}\right)^2 (P_{DD} - P_T). \tag{5.57}$$

This formula (Leibler 1980, Leibler and Benoît 1981) is rather surprising and leads to a very special scattering curve (shown in Fig. 5.5) observed for the first time by Boué *et al.* (1978).

It starts from zero at $q = 0$, increases to a maximum of $I/N(b_D - b_H)^2 = 0.17$ for $qR = 1.9$ (where R is the radius of gyration of one branch). Then the curve returns to zero when q tends to infinity. One can explain this result by the following argument: at $q = 0$, since the system is incompressible, one measures the composition fluctuations in a large volume (since it is for distances larger than $1/q$). In a copolymer there are, on this scale, no concentration changes since the composition of the system cannot be different from the composition of one molecule. Thus the scattering is zero. When q increases the distances at which one observes the fluctuations decrease and it is expected that the largest fluctuations will occur for distances of the order of the radius of gyration of one branch. When q increases still further one observes the normal decrease of the scattering intensity, since, when one looks on a small scale, the scattering curve should not be very different from the curve of a mixture of homopolymers of degree of polymerization $z/2$. The peak has therefore nothing to do with order. It is just the result of the fact that the scattered intensity is zero at zero angle and finite elsewhere. Before closing this paragraph we would like to remind the reader that the shape of the copolymer has never been introduced specifically in our discussion. This means that, as long as the copolymer is symmetric $P_a(q) = P_b(q)$ and 50/50 in composition the formulae above can be used regardless of the structure of the copolymer. If, in an extreme case, one considers an alternating copolymer, eqn (5.57) is still

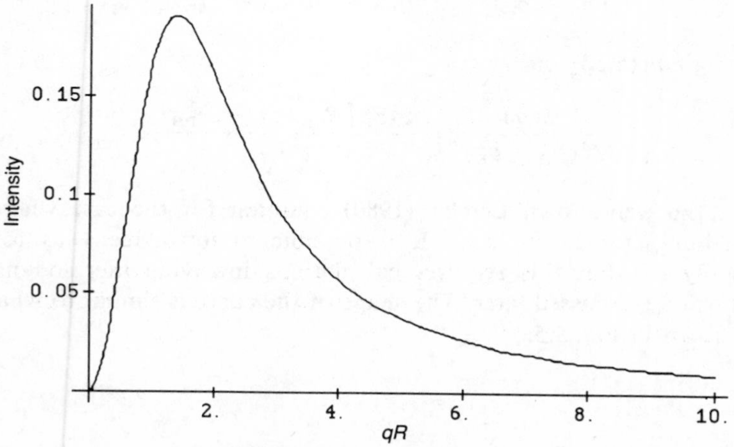

Fig. 5.5 The intensity scattered by a symmetrical Gaussian copolymer as function of qR (R being the radius of gyration of one branch).

valid but, evidently the signal is very small and the peak, which depends on qR will be situated at large q since R, the radius of gyration of one branch, is small (Benoît and Hadziioannou 1988).

5.5.3 Copolymers of any composition and structure in bulk

The formula which we have just demonstrated is rigorous and it would be interesting, as we have done for homopolymers, to extend it to unsymmetrical copolymers using the approximation (eqn (5.50))

$$Q_{\mathrm{DH}}^2 = Q_{\mathrm{D}} Q_{\mathrm{H}}. \tag{5.58}$$

This is very easy to do. Calling z_{D} the number of monomers of the deuterated block and z_{H} the corresponding quantity for the block H, we introduce the quantity P_{DH} which is, in the total form factor of one molecule, the term characterizing the interferences between the H and D blocks (see Section 6.3.6 for a more precise definition) and we write

$$\begin{aligned} I(q) &= (b_{\mathrm{D}} - b_{\mathrm{H}})^2 [N z_{\mathrm{D}}^2 P_{\mathrm{DD}} + N^2 Z_{\mathrm{D}}^2 Q_{\mathrm{DD}}] \\ I(q) &= (b_{\mathrm{D}} - b_{\mathrm{H}})^2 [N z_{\mathrm{H}}^2 P_{\mathrm{HH}} + N^2 z_{\mathrm{H}}^2 Q_{\mathrm{HH}}] \\ I(q) &= (b_{\mathrm{D}} - b_{\mathrm{H}})^2 [N z_{\mathrm{D}}^2 z_{\mathrm{H}} P_{\mathrm{DH}} + N^2 z_{\mathrm{D}} z_{\mathrm{H}} Q_{\mathrm{DH}}]. \end{aligned} \tag{5.59}$$

Evaluating Q_{DD}, Q_{HH}, and Q_{DH} as functions of $I(q)$, P_{DD}, P_{HH}, and P_{DH} and putting these values in eqn (5.58) one obtains

$$\frac{I(q)}{N(b_{\mathrm{D}} - b_{\mathrm{H}})^2} = \frac{z_{\mathrm{D}}^2 z_{\mathrm{H}}^2 [P_{\mathrm{DD}} P_{\mathrm{HH}} - P_{\mathrm{DH}}^2]}{z_{\mathrm{D}}^2 P_{\mathrm{DD}} + z_{\mathrm{H}}^2 P_{\mathrm{HH}} + 2 z_{\mathrm{D}} z_{\mathrm{H}} P_{\mathrm{DH}}} \tag{5.60}$$

or, using eqn (5.53)

$$\frac{I(q)}{N(b_{\mathrm{D}} - b_{\mathrm{H}})^2} = \frac{z_{\mathrm{D}}^2 z_{\mathrm{H}}^2 [P_{\mathrm{DD}} P_{\mathrm{HH}} - P_{\mathrm{DH}}^2]}{z^2 P_{\mathrm{T}}}. \tag{5.61}$$

This is the well-known Leibler (1980) equation for the case where the interaction parameter χ is 0. It is possible to introduce χ as for the homopolymers but this requires calculations involving thermodynamics which will be discussed later. The shape of the curve is similar to what has been drawn in Fig. 5.5.

5.6 DEUTERATION EFFECTS ON THERMODYNAMICS

5.6.1 Observation and explanation

Having developed the principles of deuterium labelling assuming that the normal and the deuterated molecules are identical we must return to the question of whether this assumption is true in practice. The potential effect of deuteration first became the subject of severe worries in the SANS studies of amorphous systems because of the large shifts observed in the miscibility limits of polymer blends when one component is deuterated. The most dramatic effect is seen in very high molecular weight blends of two polymers differing only in the fact that one is the deuterated analogue of the other. Bates *et al.* (1985) investigated blends of deutero with normal polybutadiene and deutero with normal polystyrene and showed that in extreme conditions the D–H pairs may become immiscible.

As described in Appendix 2, there are both entropic and enthalpic contributions to the Gibbs free energy of mixing of binary mixtures, g_c, where g_c is defined per repeat unit or per reference cell volume. For small molecules the entropic term is dominated by the configurational contribution and is favourable for mixing (i.e. g_c is negative). The enthalpic contribution is often unfavourable to mixing and (assuming no volume changes on mixing), it is the relative magnitude of the two terms which decides if the system will mix or not. The possible variation of g_c with ϕ (volume fraction of one of the constituents) is shown in Fig. A2.1 of Appendix 2. If the system is partially miscible, with a phase diagram as in Fig. A2.2, then the spinodal (the absolute miscibility limit) is defined by

$$\frac{1}{k_B T}\frac{\partial^2 g_c}{\partial \varphi^2} = \frac{1}{z_1\varphi_1} + \frac{1}{z_2\varphi_2} - 2\chi = 0 \tag{5.62}$$

z_i and φ_i are the degree of polymerization and volume fraction of component i respectively and χ is the Flory–Huggins interaction parameter which in the majority of the cases and specially for mixtures of ordinary with deuterated polymers is positive. The system is not miscible when the LHS of eqn (5.62) is negative. For such partial miscibility in isotopic mixtures, where χ is very small, very high degrees of polymerization are necessary in order to increase the importance of the unfavourable enthalpic term. This is exactly what Bates *et al.* (1986) observed. By increasing z they eventually achieved a situation where the samples became cloudy over a period of time indicating that a two phase structure was developing. For polybutadiene z was about 4×10^3 and for polystyrene about 10^4. In each case the composition was about 0.5. For higher molecular weight samples, the miscibility limits occur at higher temperatures (eqn (5.62)). Since it is always necessary

to observe phase separation above the glass transition temperature, the much higher value of T_g for polystyrene means that much higher values of the molecular weight are required before phase separation can be observed. The implication of the above discussion is that deuteration adds a mixing term which is unfavourable. The simple F-H theory does not allow for volume changes on mixing. However, it is well known that isotopic substitutions change the molar volume of most systems. For typical organic molecules $\Delta v/v$ is of the order of 10^{-2} to 10^{-3}. It is the deuterated species which has the larger volume despite having the smaller zero point motion. This effect arises because the larger zero point oscillation in C–H bonds leads to a larger attractive potential energy between them as observed in vibrational spectroscopy. Buckingham and Hentschel (1980) considered the miscibility of isotopic species by conceptually splitting the free energy of mixing into two terms—a compression or expansion of the molecules until they occupied uniform molar volumes and then the normal F–H mixing contributions. They showed that the interaction term is effectively zero so that eqn (5.62) becomes a balance between the entropy and the volume compression term. They also showed that the critical temperature below which phase separation might occur is given by

$$T_c = \frac{vz\left[\dfrac{\Delta v}{v}\right]^2}{4\beta R} \tag{5.63}$$

β is the average isothermal compressibility $(-1/v)(\mathrm{d}v/\mathrm{d}p)$, v the average molar volume of the monomer repeat units in the mixture, R the gas constant and Δv the difference in molar volumes of the deuterated and the ordinary species $v_D - v_H$. This expression for T_c occurs for completely matched molecular weights of the two species and in this case the diagram in Fig. A2.2 is symmetrical so that φ_c is 0.5. Generally, however, degrees of polymerization cannot be matched properly. For unmatched mixtures, T_c is reduced by a term $(1 - 3\varphi^2/16)$ and φ_c shifts towards the lower fractions of the larger molecules by an amount $3\varphi/8$.

Buckingham and Hentschel (1980) showed that for typical values of $\Delta v/v$, T_c for low molecular weight species is inaccessible (2×10^{-2} K for benzene for example). However, because $\Delta v/v$ refers to a monomer, and is multiplied by z in eqn (5.63), for polymers T_c may well become accessible.

For 1,4-polybutadiene, Bates *et al.* (1986) report that β is $5.3 \times 10^{-10}\,\mathrm{Pa}^{-1}$, $\Delta v/v$ is 41×10^{-3} and $v = 10^{-2}\,\mathrm{cm}^3$ so that T_c for the symmetrical case would occur at 396 K for $z = 5.2 \times 10^3$. Thus, their observed phase separation of isotopic mixtures of 1,4-polybutadiene with $z = 4 \times 10^3$ is in good agreement with the Buckingham and Hentschel (1980) predictions.

5.6.2 Consequences of the isotope effect—when is deuteration safe?

Introduction of an interaction parameter

It is now clear why the use of high levels of labelling in order to increase signal intensity plunged the neutron scatterers into difficulties. It has always been the aim of those using isotopic labelling to match closely the molecular weights of the deuterated and hydrogenous species. Thus φ_c is about 0.5 and effects of phase separation may be observable at the experimental temperatures. By working with labelling levels at the 0.05 level the mixture is placed at the extreme limits of the phase diagram; the temperature for phase separation is dramatically lowered, and often lies below the glass transition. For the matched case the condition for phase separation to occur is given by Buckingham and Hentschel (1980) as

$$T \leq \frac{vz}{2\beta R}\left(\frac{\Delta v}{v}\right)^2 \left\{\frac{\varphi_H - \varphi_D}{\ln\varphi_H - \ln\varphi_D}\right\}. \tag{5.64}$$

For the 1,4 polybutadiene case with $\varphi_D = 0.05$, z as before 5.2×10^3, T must be less than 181 K. Thus at typical experimental working temperatures such a mixture is well away from the phase boundary and observable effects should be very small. The message is clear: lower degrees of polymerization, higher working temperatures and low labelling levels all minimize effects of isotopic differences.

What, however, might be the consequences of staying close to phase separation conditions? The scattering from binary blends will be considered in subsequent chapters, but to complete the present discussion, the scattering law for such a system can be quoted. At $q = 0$ the scattering per repeat unit is given by (see eqn (7.63) or (A3.26) for precise definitions)

$$i(0) = \left(\frac{I(0)}{N_T}\right) = (b_D - b_H)^2 \frac{k_B T}{\dfrac{\partial^2 g_c}{\partial \varphi^2}} \tag{5.65}$$

As already discussed, for these isotopic mixtures the χ contribution to $\partial^2 g_c/\partial\varphi^2$ in eqn (5.62) is given purely by the volume effects. When these are zero, and for the matched conditions, $I(0)$ per repeat unit in eqn (5.62) is simply determined by the molecular weight just as in eqn (5.31) (with $P(q) = 1$ and dividing by Nz).

Thus the importance of the isotopic effects should be judged in terms of deviations from the expected molecular weight obtained from the forward scattered intensity of an isotopic mixture. Experience shows that in practice and under the conditions outlined of reasonable z, high T, and low φ such effects are rarely observed.

Isotopic exchanges

All the discussion in the preceding paragraph was based on the fact that no isotopic exchange between deuterium and hydrogen took place in the mixture between the deuterated and the ordinary polymers. However, it is possible that, if the deuterium atom is in a position where its reactivity is enhanced one can have an exchange reaction between a deuterium and a hydrogen atom. Let us take the example of a mixture of a polymer and its deuterated counterpart. When the situation is favourable for this reaction we will have after a certain time a mixture of two partially deuterated polymers, i.e. a mixture of copolymers of D and H monomers and it is evident that, when the reaction proceeds, the signal will decrease and eventually disappear. Fortunately this effect, although frequent for hydrosoluble biological polymers, has rarely been observed in classical polymers. Detailed experimental results and the way of interpreting them can be found in the literature. The possible existence of this effect has to be considered since it could happen under some circumstances, specially when there are some catalyst residues in the sample. Forgetting this possibility might result in data which cannot be interpreted. One case, that of transesterification, is described in Chapter 8.

SUGGESTED FURTHER READING

Cotton (1974).
Cotton (1991).
de Gennes (1979).
Flory (1953).

form factor $P(q)$. In some cases it is easy. For instance, we have already shown that if we make a mixture of identical polymers, one being deuterated the other normal, we obtain $P(q)$ immediately (see eqn (5.31)). Another way of obtaining $P(q)$ will be discussed in Chapter 8 and here we shall only give a short summary. We first introduce the quantity (independent of the model) which is defined as $b_v = b_1/V_1 - b_0/v_0$, i.e. the difference between the scattering per unit volume between the polymer and the solvent. With this notation one writes for the scattering per unit volume $I(q)/V$

$$I(q)/V = b_v^2 \varphi V_1 [P(q) + NQ(q)] \tag{6.4}$$

where V_1 is the volume of the polymer molecule and φ the volume fraction occupied by the polymer. Alternatively

$$\frac{I(q)}{\varphi V} = b_v^2 V_1 [P(q) + \varphi \frac{V}{V_1} Q(q)] = b_v^2 [V_1 P(q) + \varphi Q'(q)] \tag{6.4'}$$

with $Q'(q) = \frac{V}{V_1} Q(q)$.

This is a linear function of φ, or of the concentration if one uses different units and contains two terms. The first is independent of the concentration and the second proportional to it. This means that, by extrapolation to infinite dilution, only the first term remains. In other words, by extrapolating the quantity $I(q)/V\varphi$ to infinite dilution of the solute, one obtains the quantity $b_v^2 V_1 P(q)$ and, after renormalization to unity of the zero angle scattered intensity, the form factor $P(q)$. Playing with contrast factors leads to other ways of obtaining $P(q)$ (see Chapter 5); some examples will be discussed in Chapter 8. What we intend to do in this chapter is to explore the information one can obtain once $P(q)$ has been experimentally determined. The only available method is to work with models. Starting from a model the form factor $P(q)$ is evaluated and compared to the experimental data. The discussion will be divided into four sections. In the first the behaviour of $P(q)$ at small values of q will be discussed, in the second typical examples of form factors will be given, in the third some rules concerning the behaviour at large q values will be discussed and in the final section the effect of polydispersity will be introduced.

6.2 THE BEHAVIOUR AT SMALL q VALUES

6.2.1 Series expansion of $P(q)$

The form factor has been defined as

$$P(q) = \frac{1}{z^2} \sum_{i=1}^{z} \sum_{j=1}^{z} \langle \exp(-\mathrm{i}\boldsymbol{q}\cdot\boldsymbol{r}_{ij}) \rangle \tag{6.5}$$

where $\boldsymbol{r}_{ij}$ is the vector joining two scattering points of a molecule and the sum extends to its z^2 scattering points. The sign $\langle\rangle$ indicates that an average value has to be taken over all orientations and conformations. (This average can be taken either over the sum or over each term of the sum as written in eqn (6.5).) If one is interested in the value of $P(q)$ in the small q range it is helpful to expand $P(q)$ in increasing powers of q. Before doing so, however, we shall average the quantity under the summation sign over all orientations. This is easy to do; the probability for a given orientation in polar coordinates is given by $(1/4\pi)\sin\theta\mathrm{d}\theta\mathrm{d}\phi$ where θ and ϕ are the classical polar angles. (See Fig. 6.1.)

The orientational average of a function $f(\theta,\phi)$ is given by

$$\langle f(\theta,\phi) \rangle = \frac{1}{4\pi} \int_{\phi=0}^{2\pi} \int_{\theta=0}^{\pi} f(\theta,\phi)\sin\theta\,\mathrm{d}\theta\,\mathrm{d}\phi \tag{6.6}$$

applying this formula to $\exp(-\mathrm{i}\boldsymbol{q}\cdot\boldsymbol{r})$ (see Appendix 1) gives

$$\frac{\sin q r_{ij}}{q r_{ij}}. \tag{6.7}$$

The expansion of the formula

$$P(q) = \frac{1}{z^2} \sum_{i=1}^{z} \sum_{j=1}^{z} \left\langle \frac{\sin q r_{ij}}{q r_{ij}} \right\rangle \tag{6.8}$$

where the average has to be taken only over the distances r_{ij} is now straightforward

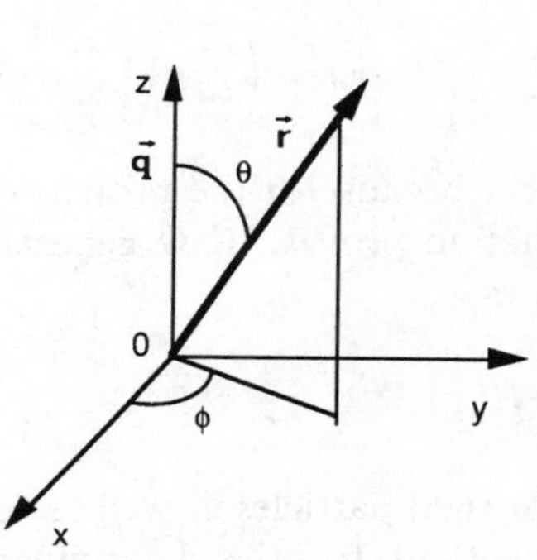

Fig. 6.1 Definition of the coordinate system.

$$P(q) = 1 - \frac{q^2}{6z^2} \sum_{i=1}^{z} \sum_{j=1}^{z} \langle r_{ij}^2 \rangle + \frac{q^4}{5!\,z^2} \sum_{i=1}^{z} \sum_{j=1}^{z} \langle r_{ij}^4 \rangle -$$

$$\frac{q^6}{7!\,z^2} \sum_{i=1}^{z} \sum_{j=1}^{z} \langle r_{ij}^6 \rangle + \ldots . \tag{6.9}$$

6.2.2 The radius of gyration

One of the most important properties of small angle neutron scattering is that the second term of this expansion has a very simple geometrical meaning, independent of the scattering properties of the sample; it gives the radius of gyration (Guinier 1956). The square of the radius of gyration of a particle made of z identical elements, around its centre of gravity, is given by

$$\overline{R^2} = \frac{1}{z} \sum_{i=1}^{z} \langle r_i^2 \rangle \tag{6.10}$$

where $\boldsymbol{r}_i$ is the vector joining the scattering point i to the centre of gravity. This quantity is also given by the formula

$$\overline{R^2} = \frac{1}{2z^2} \sum_{i=1}^{z} \sum_{j=1}^{z} \langle r_{ij}^2 \rangle . \tag{6.11}$$

In order to show this result let us write $\boldsymbol{r}_{ij} = \boldsymbol{r}_i - \boldsymbol{r}_j$ and introduce this quantity in expression (6.11); we obtain

$$\overline{R^2} = \frac{1}{2z^2} \sum_{i=1}^{z} \sum_{j=1}^{z} (\langle r_i^2 \rangle + \langle r_j^2 \rangle - 2\langle \boldsymbol{r}_i \cdot \boldsymbol{r}_j \rangle) . \tag{6.12}$$

The two first terms are identical and equal to $z\Sigma\langle r_i^2 \rangle$ and the last term is equal to zero since

$$\sum_{i=1} \sum_{j=1} \langle \boldsymbol{r}_i \cdot \boldsymbol{r}_j \rangle = \left\langle \sum_i \boldsymbol{r}_i \right\rangle \left\langle \sum_j \boldsymbol{r}_j \right\rangle = 0. \tag{6.13}$$

This last result is evident because of the properties of the centre of mass.

The preceding calculation proves, quite generally, that the form factor can always be written as

$$P(q) = 1 - \frac{1}{3} q^2 \overline{R^2} + \ldots . \tag{6.14}$$

This relation is valid for rigid particles as well as for flexible molecules and gives a very elegant method for the determination of the size of the molecules. The initial slope of the curve $I(q)$ as function of q^2 is equal to

$(\overline{R^2})/3$. In fact this is a comparison of the size of the molecule to the factor q which depends on the wavelength and on the direction of observation. For practical purposes $q^2\overline{R^2}$ has to be neither too small, since the deviation from 1 would be difficult to detect, nor too large, because the influence of the remaining terms in the expansion can no longer be neglected. From an experimental point of view slow neutrons are perfect for molecules having a radius of gyration between a few to a few hundred angstroms. The upper and lower limits are difficult to establish rigorously since they depend on the contrast, on the wavelength of the neutrons, and on the quality of the instrument.

6.2.3 The radius of gyration for various geometrical shapes

This is a purely geometric calculation and we give only the results

$$\overline{R^2} = \frac{3}{5} r^2 \tag{6.15a}$$

for a sphere of radius r

$$\overline{R^2} = \frac{1}{5}(a^2 + b^2 + c^2) \tag{6.15b}$$

for an ellipsoid of half axes a,b,c; it reduces to a sphere if $a = b = c$

$$\overline{R^2} = \frac{L^2}{12} + \frac{b^2}{2} \tag{6.15c}$$

for a rod of length L and transverse radius b

$$\overline{R^2} = \frac{L^2}{12} \tag{6.15d}$$

for a rod of length L and negligible diameter

$$\overline{R^2} = \frac{b^2}{2} \tag{6.15e}$$

for a thin disc of radius b.

6.2.4 The radius of gyration for a Gaussian chain

Since macromolecular chains are of particular interest, the value of the radius of gyration for Gaussian chains will be evaluated. We call freely jointed chain or Gaussian a chain made of z independent units attached sequentially and without correlation between the orientation of any pair of segments. (See Fig. 6.2.) It is shown in Appendix 1 that the average of the square of the end to end distance $\overline{L^2}$ is given by the relation

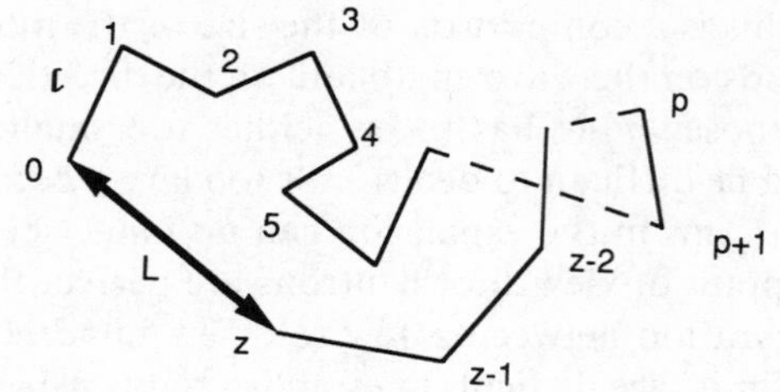

Fig. 6.2 Schematic diagram of a freely jointed chain starting at O and finishing at z.

$$\overline{L^2} = z\overline{l^2} \tag{6.16}$$

where $\overline{l^2}$ is the average of the square of the length of any of the elementary steps. This is true for any chain length

$$\langle r_{ij}^2 \rangle = |\,i - j\,|\,\overline{l^2} \tag{6.16$'$}$$

where the sign | . . . | indicates that one has to take the absolute value of $i - j$. Since r_{ij}^2 is always positive. Instead of replacing the summation in eqn (6.11) by an integration, assuming that z is large, we shall make the rigorous calculation, valid for any z using the following method. All the values of $\langle r_{ij}^2 \rangle / \overline{l^2}$ are placed in a square matrix having $z + 1$ rows and columns since the chain has z segments (see Fig. 6.3). We consider as scattering

j \ i	0	1	...	p	...	z–1	z
0	0	1	...	p–	...	z–1	z
1	1	0	...	p–1	...	z–2	z–1
...	...	...	...	...	...	...	...
p	p	p–1	...	0	...	z–p–1	z–p
...	...	...	...	...	...	...	...
z–1	z–1	z–2	...	z–p	...	0	1
z	z	z–1	...	z–p	...	1	0

Fig. 6.3 A matrix for evaluating $\overline{R^2}$.

units the ends of the segments. One sees immediately that the major diagonal is a symmetry axis. Grouping the elements on lines parallel to the direction of this diagonal gives

$$\sum_i \sum_j \frac{\langle r_{ij}^2 \rangle}{l^2} = \sum_i \sum_j |i-j| = 2[z + (z-1)2 + (z-2)3 + \ldots + (z-p+1)p + \ldots + 2(z-1) + z] = 2S. \tag{6.17}$$

The evaluation of this sum is straightforward

$$S = z \sum_{p=1}^{z} (z+1-p)p = (z+1)\sum_{p=1}^{z} p - \sum_{p=1}^{z} p^2 \tag{6.18}$$

and, consulting tables of formulae, one finds

$$S = \frac{z(z+1)(z+2)}{6}. \tag{6.19}$$

Remembering that our chain is made of $z+1$ elements and applying eqn (6.11) we obtain for its radius of gyration (Debye 1946)

$$\overline{R^2} = l^2 \frac{z}{6} \frac{z+2}{z+1}. \tag{6.20}$$

As soon as z is large enough the second fraction in eqn (6.20) is unity and

$$\overline{R^2} = \frac{1}{6} \overline{L^2} \tag{6.21}$$

where $\overline{L^2}$ is the average square of the end to end distance. This calculation, made for a simple example demonstrates a general property which we shall use later: if one writes a double sum or a double integral of the form

$$\sum_{i=0}^{N} \sum_{j=0}^{N} f(|i-j|)$$

it is possible to transform it into a simple sum or integral simply by writing: $|i-j| = p$. This gives the equality, valid only for N large

$$\sum_{i=0}^{N} \sum_{j=0}^{N} f(|i-j|) = 2\sum_{p=0}^{N} (N-p) f(p). \tag{6.22}$$

The relation has been proved for a simple example using our matrix demonstration but can be applied to any function $f(|i-j|)$.

6.2.5 The ring polymer

Physical chemists are interested in the properties of ring polymers, i.e. polymers having both ends linked together in order to make a loop without end. We shall first calculate the distance between two arbitrary units in a loop in order to be able to evaluate its radius of gyration. Let us assume that we have a ring, made of z elements; the initial element is arbitrary and they are numbered from 0 to $z-1$ (see Fig. 6.4). We should like to know the probability to go from 0 to p on the loop. This requires that we arrive at the point p either via the left side or via the right along a chain of p or of $z-p$ segments respectively. The left side is a random walk of p segments and the probability of arriving at r will be

$$w_p(r) = \left(\frac{3}{2\pi p\overline{l^2}}\right)^{\frac{3}{2}} \exp - \frac{3r^2}{2p\overline{l^2}}. \tag{6.23}$$

By the right path it is

$$w_{z-p}(r) = \left(\frac{3}{2\pi(z-p)\overline{l^2}}\right)^{\frac{3}{2}} \exp - \frac{3r^2}{2(z-p)\overline{l^2}}. \tag{6.24}$$

Since we have to be on the ring the total probability to be at a distance r, $w_{0p}(r)$ is the product of these two probabilities and, writing for the moment the normalization constant as K, we obtain

$$w_{0p}(r) = w_p(r)w_{z-p}(r) = K \exp\left[-\frac{3r^2}{2\overline{l^2}}\left(\frac{1}{p}+\frac{1}{z-p}\right)\right]. \tag{6.25}$$

This expression shows that the probability is still Gaussian and, if we put it in the form $\exp - (3r^2/2\overline{L^2_{z,p}})$ we obtain, calling $\overline{L^2_{z,p}}$ the root mean square distance between 0 and p

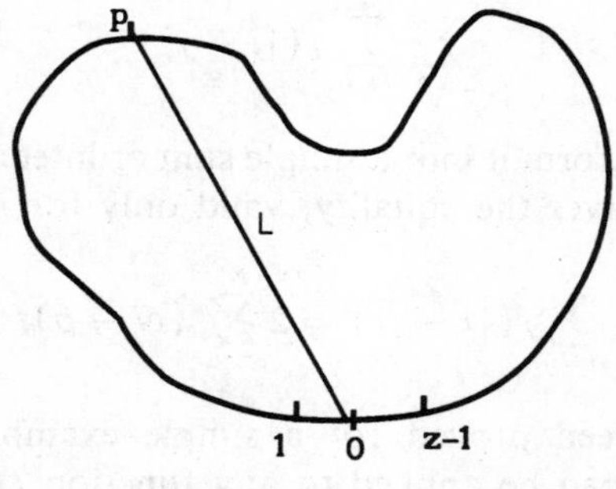

Fig. 6.4 Schematic diagram of a ring molecule.

$$\frac{1}{\overline{L_{z,p}^2}} = \frac{1}{\overline{l^2}}\left(\frac{1}{p} + \frac{1}{z-p}\right) \tag{6.26}$$

or:

$$\overline{L_{p,z}^2} = \overline{r_{i,\,i+p}^2} = \overline{l^2}p\left(1 - \frac{p}{z}\right). \tag{6.27}$$

Applying eqn (6.11) to this result one obtains, for the radius of gyration of a ring

$$\overline{R^2} = \frac{z\overline{l^2}}{12}. \tag{6.28}$$

It is half the value obtained for a linear chain having the same number of segments.

6.2.6 The case of copolymers

We have seen (Chapter 5) that in the case of a copolymer, made of two different kinds of monomers A and B, one has to introduce three form factors S_{AA}, S_{BB}, and S_{AB}. These form factors are associated with three radii of gyration $\overline{R_{AA}^2}$, $\overline{R_{BB}^2}$, and $\overline{R_{AB}^2}$. It is important to understand the physical meaning of these quantities. If one looks at the part A of the copolymer the meaning of $\overline{R_{AA}^2}$ is clear and eqn (6.11) can be applied. The same is true for the part B. The question is less clear for the cross term but, by analogy, we define the 'cross radius of gyration' by a similar expression. We have therefore the three following expressions

$$\overline{R_A^2} = \frac{1}{2z_A^2}\sum_{i=0}^{z}\sum_{j=0}^{z}\overline{r_{i_A j_A}^2} \tag{6.29}$$

$$\overline{R_B^2} = \frac{1}{2z_B^2}\sum_{i=0}^{z}\sum_{j=0}^{z}\overline{r_{i_B j_B}^2} \tag{6.30}$$

$$\overline{R_{AB}^2} = \frac{1}{2z_A z_B}\sum_{i=0}^{z}\sum_{j=0}^{z}\overline{r_{i_A j_B}^2}. \tag{6.31}$$

In order to clarify the physical meaning of $\overline{R_{AB}^2}$ we write

$$\boldsymbol{r}_{i_A j_B} = \boldsymbol{A}_i\boldsymbol{G}_A + \boldsymbol{G}_A\boldsymbol{G}_B + \boldsymbol{G}_B\boldsymbol{B}_j \tag{6.32}$$

replacing the direct route between the scattering points A_i and B_j by a detour through G_A and G_B, where G_A and G_B are the centre of mass of the

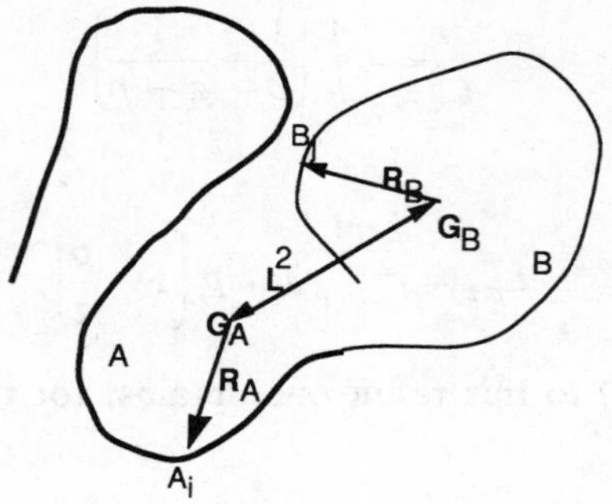

Fig. 6.5 Schematic diagram of a diblock copolymer and geometrical meaning of the quantity L.

two sequences. (See Fig. 6.5.) Squaring this last expression and making the double summation gives, for $\bar{R}^2_{AB}$, the following expression where all the cross terms disappear (Benoît and Wippler 1960)

$$2\overline{R^2_{AB}} = \bar{R}^2_A + \bar{R}^2_B + \overline{G_A G^2_B}. \tag{6.33}$$

From the definitions of the form factor of the whole polymer we have

$$(z_A + z_B)^2 P_T = z^2_A P_A + z^2_B P_B + 2 z_A z_B P^2_{AB} \tag{6.34}$$

and for its radius of gyration

$$(z_A + z_B)^2 \overline{R^2_T} = z^2_A \overline{R^2_A} + z^2_B \overline{R^2_B} + 2 z_A z_B \overline{R^2_{AB}}. \tag{6.35}$$

Introducing the value of $\overline{R^2_{AB}}$ obtained from eqn (6.33) into the definition of $\overline{R^2_T}$, calling x the fraction of monomer A in the molecule $x = z_A/(z_A + z_B)$ and $\overline{L^2} = \overline{G_A G^2_B}$ the mean square distance between the centres of mass of the part A and the part B of the molecule, we arrive at the formula

$$\overline{R^2_T} = x\overline{R^2_A} + (1 - x)\overline{R^2_B} + x(1 - x)\overline{L^2}. \tag{6.36}$$

This formula is quite general and valid for any type of molecule made of two parts; it is useful in the discussion of the scattering from copolymers when, by changing the contrast of the solvent, one looks for the changes in the scattering envelope as will be shown later.

As a simple example, we shall determine the value of $\overline{L^2}$ for a linear copolymer made of two monomers of the same size. For this purpose we replace in eqn (6.36) $\overline{R_T^2}$, $\overline{R_A^2}$, $\overline{R_B^2}$ and x by their values as function of z_A, z_B, and $\overline{l^2}$ obtaining

$$\frac{z_A + z_B}{6}\overline{l^2} = \frac{z_A^2 \overline{l^2}}{6(z_A + z_B)} + \frac{z_B^2 \overline{l^2}}{6(z_A + z_B)} + \frac{z_A z_B}{(z_A + z_B)^2}\overline{L^2}. \tag{6.37}$$

After simplification this gives

$$\overline{L^2} = 2(\overline{R_A^2} + \overline{R_B^2}) = 2\overline{R_T^2}. \tag{6.38}$$

6.3 THE COMPLETE FORM FACTOR

6.3.1 The mathematical methods

In the first part of this chapter we have discussed the form factor for small q values. In this section we shall try to obtain a complete analytical expression valid for any value of q. Various methods are possible. The most classical starts from eqn (6.5)

$$P(q) = \frac{1}{z^2}\sum_{i=1}^{z}\sum_{j=1}^{z}\langle \exp - \mathrm{i}\boldsymbol{q}\cdot\boldsymbol{r}_{ij}\rangle. \tag{6.39}$$

Since the molecules are randomly oriented one can average over all orientations obtaining

$$P(q) = \frac{1}{z^2}\sum_{i=1}^{z}\sum_{j=1}^{z}\left\langle\frac{\sin q r_{ij}}{q r_{ij}}\right\rangle. \tag{6.40}$$

This formula is very simple since it contains only distances and no imaginary numbers. Instead of making the summation over all pairs of scattering points one can count the number $n(r)$ of pairs being at the distance r and write

$$P(q) = \frac{1}{\int n(r)\mathrm{d}r}\int n(r)\left(\frac{\sin qr}{qr}\right)\mathrm{d}r. \tag{6.41}$$

This expression is used mainly for continuous bodies for which the sums are transformed into integrals. The function $n(r)$ has to be compared with the radial distribution function $g(r)$ introduced in Chapter 4; its definition is similar but in the present case we are dealing with one molecule and not a large number of small molecules. If the molecule has a centre of symmetry it is more convenient to use the following approach: first, the amplitude of the scattered wave $A(q)$ is calculated for a given orientation of the molecule using the centre of symmetry as the origin for the phases and for the numbering of the scattering elements. After normalization to unity a $q = 0$, we call $A(\boldsymbol{q})$, $A_0(\boldsymbol{q})$

$$A_0(\boldsymbol{q}) = \frac{1}{z} \sum_{i=-z/2}^{+z/2} \exp(-\mathrm{i}\boldsymbol{q}\cdot\boldsymbol{r}_i). \tag{6.42}$$

The symmetry implies that the phase difference between the total scattered wave and the wave scattered by the centre of symmetry is zero; this makes $A(\boldsymbol{q})$ real since one can always associate the points $\boldsymbol{r}_i$ and $-\boldsymbol{r}_i$

$$\exp(-\mathrm{i}\boldsymbol{q}\cdot\boldsymbol{r}_i) + \exp(+\mathrm{i}\boldsymbol{q}\cdot\boldsymbol{r}_i) = 2\cos(\boldsymbol{q}\cdot\boldsymbol{r}_i). \tag{6.43}$$

The scattered intensity for a given orientation is

$$A_0^2(\boldsymbol{q}) = \frac{1}{z^2}\left[\sum_{-z/2}^{+z/2} \cos(\boldsymbol{q}\cdot\boldsymbol{r}_i)\right]^2 \tag{6.44}$$

and the form factor $P(q)$ is obtained after averaging over all orientations

$$P(q) = \frac{1}{z^2}\left\langle\left[\sum \cos(\boldsymbol{q}\cdot\boldsymbol{r}_i)\right]^2\right\rangle_{\text{orientations}} \tag{6.45}$$

As an example of the use of this method we shall calculate the form factor of a sphere.

6.3.2 The sphere

This case is extremely simple since the scattering intensity does not depend on the orientation of the sphere. One evaluates the amplitude scattered by the sphere (eqn (6.42)) transformed from discrete to continuous notation and normalized to unity for $q = 0$

$$A_0(\boldsymbol{q}) \approx \frac{1}{V} \iiint_V \exp(-\mathrm{i}\boldsymbol{q}\cdot\boldsymbol{r}) r^2 \sin\theta \, \mathrm{d}\theta \, \mathrm{d}\varphi \, \mathrm{d}r, \tag{6.46}$$

the factor V comes from the fact that our normalization condition $A^2(0) = 1$ has to be satisfied. Using the direction of $\boldsymbol{q}$ as the z axis ($\boldsymbol{qr} = qr\cos\theta$) and as new variable $u = \cos\theta$, one obtains after integration over φ

$$A_0(q) \approx \frac{2\pi}{V} \int_{u=-1}^{+1} \int_{r=0}^{R} \exp(-\mathrm{i}qru) r^2 \mathrm{d}u \mathrm{d}r. \tag{6.47}$$

Integrating first over u we recover the well-known result

$$A_0(q) \approx \frac{3}{R^3} \int_0^R \frac{\sin qr}{qr} r^2 \mathrm{d}r. \tag{6.48}$$

The integration over r is made by parts leading (with $v = qR$) to the expression

$$A_0(q) = \frac{3}{v^3}(\sin v - v\cos v). \tag{6.49}$$

As expected this expression depends only on R and the form factor (Rayleigh 1914) is

$$P(q) = [A_0(q)]^2 = \frac{9}{v^6}(\sin v - v\cos v)^2. \tag{6.50}$$

In some texts this formula is given in an equivalent form as function of the Bessel function

$$J_{\frac{3}{2}}(v) = \sqrt{\frac{2}{\pi v}}\,\frac{\sin v - v\cos v}{v}\,.$$

Expanding around q (or v) = 0 we obtain

$$P(q) = 1 - \frac{v^2}{5} + \frac{3v^4}{175} - \frac{4v^6}{4725} + \frac{2v^8}{72765} + \dots. \tag{6.51}$$

If one writes $P(q)$ as $1 - (q^2R^2/3)$ the second term gives the correct value for the radius of gyration of a sphere (eqn (6.15a)). Figure 6.6(a) shows $P(q)$ as a function of $v = qR$.

Equation (6.50) shows that the scattering intensity becomes zero for all the values of v (except zero) satisfying the equation

$$v = \text{tangent}(v)$$

or practically: $v = (2n + 1)\pi/2$ (n being a positive integer) and goes through a maximum between two consecutive zeros. The main maximum at $q = 0$ is much more important than the secondary maxima (Fig. 6.6(b)). It is almost two orders of magnitude larger but it is quite possible to detect several maxima from well-defined systems. In order to see these minima and maxima more clearly Fig. 6.6(c) shows the logarithm of the form factor as function of $v = q^2R^2$.

6.3.3 Other objects with spherical symmetry (for example, shells)

The method we have just used can be applied to many other systems having spherical symmetry. If, instead of having an uniform density, the sphere has a density $\rho(r)$ of scattering elements which depends on the distance r to the centre, one can immediately write for the scattering amplitude, by generalization of eqn (6.48)

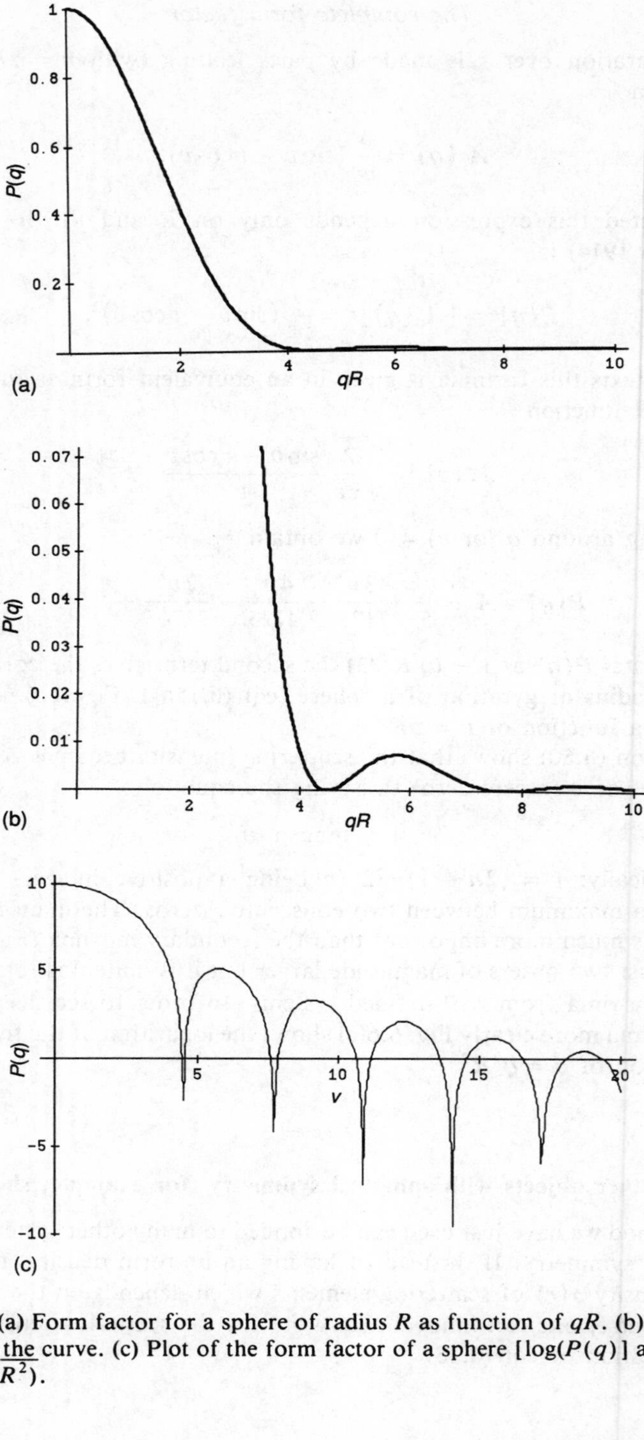

Fig. 6.6 (a) Form factor for a sphere of radius R as function of qR. (b) Enlarging the tail of the curve. (c) Plot of the form factor of a sphere [$\log(P(q)$] as function of $v = q^2R^2$).

$$A_0(q) = \frac{1}{C}\int n(r)\,\frac{\sin qr}{qr}\,r^2\mathrm{d}r. \tag{6.52}$$

The constant C is defined by the normalization condition $A(0) = 1$

$$C = \int n(r) r^2 \mathrm{d}r \tag{6.53}$$

and one obtains for the form factor

$$P(q) = \left[\frac{1}{C}\int n(r)\left(\frac{\sin qr}{qr}\right) r^2\mathrm{d}r\right]^2. \tag{6.54}$$

As an example let us assume that we want to evaluate the form factor of a hollow sphere where there are only scattering points between the radius R_{int} and the external radius R_{ext} (see Fig. 6.7). It suffices to take n = constant for $R_{\text{int}} < r < R_{\text{ext}}$ and $n = 0$ everywhere else, obtaining

$$A_0(q) = \frac{1}{C'}\int_{R_{\text{int}}}^{R_{\text{ext}}} \frac{\sin qr}{qr}\,r^2\mathrm{d}r = \frac{3}{q^3R_{\text{int}}^3}\,(\sin qR_{\text{int}} - qR_{\text{int}}\cos qR_{\text{int}})$$

$$- \frac{3}{q^3R_{\text{ext}}^3}\,(\sin qR_{\text{ext}} - qR_{\text{ext}}\cos qR_{\text{ext}}). \tag{6.55}$$

The normalization constant C' is obtained by writing that for $q = 0$, $A_0(q)$ should be equal to 1. C' is equal to the volume of the shell divided by 4π

$$C' = \int_{R_{\text{int}}}^{R_{\text{ext}}} r^2\mathrm{d}r = \frac{1}{3}\,(R_{\text{ext}}^3 - R_{\text{int}}^3). \tag{6.56}$$

The form factor is evidently $(A_0(q))^2$. If the difference $R_{\text{ext}} - R_{\text{int}}$ is small compared to q

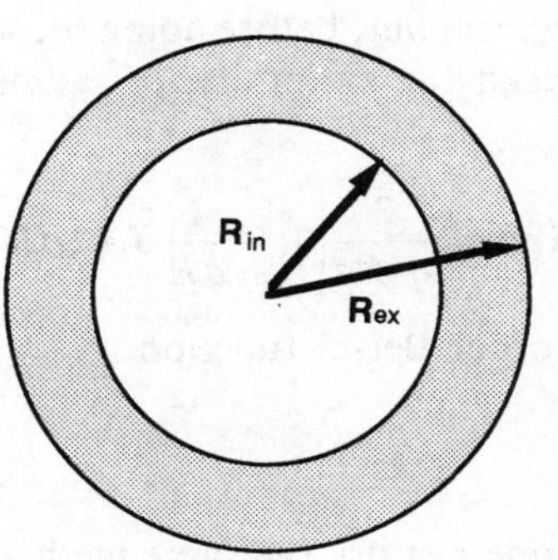

Fig. 6.7 Model for an empty sphere.

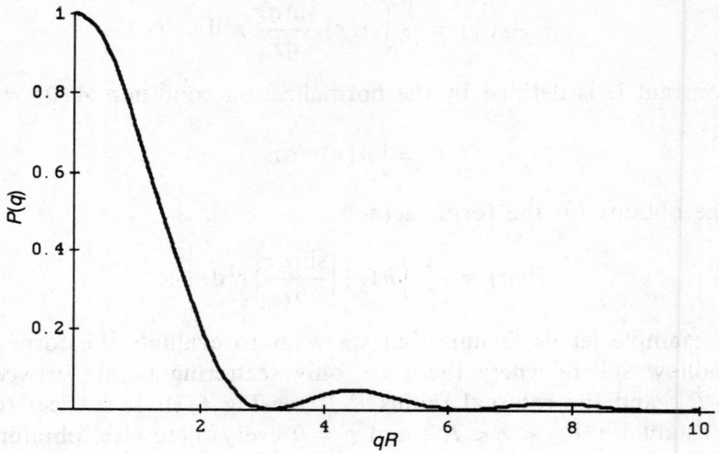

Fig. 6.8 Form factor of a hollow spherical shell of negligible thickness.

$$P(q) = \left[\frac{\sin qR}{qR}\right]^2 \tag{6.57}$$

with $R \approx R_{\mathrm{int}} \approx R_{\mathrm{ext}}$. (See Fig. 6.8.)

6.3.4 Other simple shapes – discs and rods

The disc

Until this point we have discussed the most simple case, the sphere. We could go on and discuss many other geometrical objects which can be used as models for molecular structure. This is purely a mathematical exercise and we shall discuss in detail only the linear molecules which are the most common model for polymers but, before doing so, we give here for the sake of completeness, the result for a thin disc of radius R (Kratky and Porod 1949*a*) (see Fig. 6.9).

$$P(q) = \frac{2}{q^2R^2}\left[1 - \frac{1}{qR} J_1(2qR)\right] \tag{6.58}$$

where $J_1(x)$ is a first order Bessel function.

The rod

For a rod of length L one can use the same method and taking the centre of the rod as origin calculate the scattering amplitude. Using the classi-

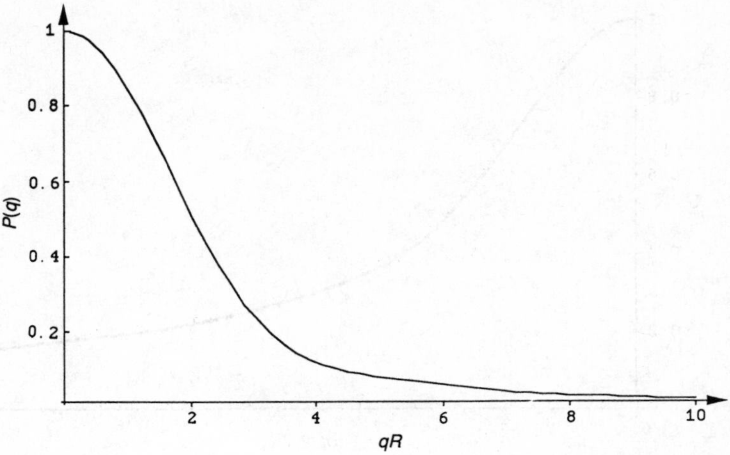

Fig. 6.9 Form factor for a thin disc of radius R as a function of qR.

cal coordinate system with the z axis in the direction of the vector $\boldsymbol{q}$ ($\boldsymbol{q}\cdot\boldsymbol{r} = qr\cos\theta$) we immediately obtain

$$A_0(q) = \frac{1}{L}\int_{-\frac{L}{2}}^{\frac{L}{2}} \exp(-\,\mathrm{i}qr\cos\theta)\,\mathrm{d}r \tag{6.59}$$

and, after performing the integration

$$A_0(q) = \frac{2}{qL\cos\theta}\sin\left(\frac{qL\cos\theta}{2}\right). \tag{6.60}$$

Now we take the square of this expression and integrate over all orientation

$$P(q) = \frac{1}{2}\int_{\theta=0}^{\pi} [A_0(q)]^2 \sin\theta\,\mathrm{d}\theta.$$

After integration by parts (Neugebauer 1943)

$$P(q) = \frac{2}{qL}S_i(qL) - \frac{\sin^2\frac{qL}{2}}{\left(\frac{qL}{2}\right)^2} \tag{6.61}$$

where $S_i(x)$ is the sine integral function

$$S_i(x) = \int_0^x \frac{\sin u}{u}\,\mathrm{d}u. \tag{6.62}$$

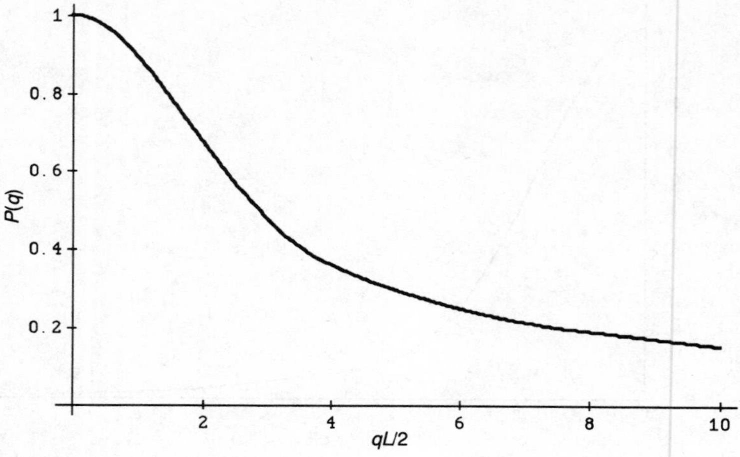

Fig. 6.10 $P(q)$ for a rod of length L as function of $qL/2$.

Figure 6.10 shows $P(q)$ for rods of length L as function of $qL/2$.

Expanding eqn (6.61) as a function of $x = qL/2$, around q or $x = 0$, we obtain

$$P(q) = 1 - \frac{1}{9}x^2 + \frac{2}{225}x^4 - \frac{6}{2205}x^6 + \frac{2}{127\,575}x^8 \ldots \tag{6.63}$$

Remembering from Section 6.2 that the x^2 term is $1/3\; q^2\overline{R^2}$ we recover for the radius of gyration of a rod, $\overline{L^2}/12$.

6.3.5 Gaussian chains

The simplest method for evaluating the form factor of Gaussian chains is to start from the definition of the form factor given by eqn (6.5)

$$P(q) = \frac{1}{z^2}\sum_{i=0}^{z}\sum_{j=0}^{z}\langle \exp(-\mathrm{i}\boldsymbol{q}\cdot\boldsymbol{r}_{ij})\rangle \tag{6.64}$$

where the average has to be taken not only over the orientations but also over the distances since in a Gaussian chain these distances depend on the conformation of the chain. The first step is to take the average value of one term and then to sum these values in order to obtain $P(q)$. From the definition of an average value we can write

$$\langle \exp(-\mathrm{i}\boldsymbol{q}\cdot\boldsymbol{r}_{ij})\rangle = \iiint w(r_{ij})\exp(-\mathrm{i}\boldsymbol{q}\cdot\boldsymbol{r}_{ij})\mathrm{d}^3\boldsymbol{r}_{ij} \tag{6.65}$$

where $w(r_{ij})$ is the probability for the segments i and j to be at the distance r_{ij}. This expression is nothing but the Fourier transform of $w(r_{ij})$. If the chain is made of z elements without any orientation correlations we have shown in Appendix 1 that one can calculate $w(r_{ij})$ knowing the probability of the elementary step $w_0(r_{i,i+1})$. More precisely if $\omega_0(q)$ is the Fourier transform of w_0, the Fourier transform of $w(r_{ij})$, $\omega_{ij}(q)$, is given by

$$\omega_{ij}(q) = \omega_0(q)^{|i-j|} \tag{6.66}$$

$P(q)$ then has the form

$$P(q) = \frac{1}{z^2} \sum_{i=0}^{z} \sum_{j=0}^{z} \omega_0(q)^{|i-j|}. \tag{6.67}$$

In these expressions q is a number and not a vector since it has been demonstrated (see Appendix 1) that if the probability w depends only on the distances and not on the orientations, w and ω depend only on r and q. One could replace the summation by an integration but, in order to have an expression valid even for small z, we shall perform the exact calculation. We write, as has been done for the radius of gyration, the ωs as a square matrix, $z^2P(q)$ being the sum of all the terms of this matrix (see Fig. 6.11). The index i runs from 1 to z along the rows and the index j down the columns. $\omega_0(q)^{|i-j|}$ is written as ω^p. The main diagonal of this matrix is a symmetry axis. Since on this diagonal $i = j$, we have $\omega^0 = 1$ and the

1	ω	ω^2	ω^3	…	ω^p	…	ω^z
ω	1	ω	ω^2	…	ω^{p-1}	…	ω^{z-1}
ω^2	ω	1	ω	…	…	…	ω^{z-2}
ω^3	ω^2	ω	1	…	…		ω^{z-3}
…	…	…	…	…	…	…	…
ω^p	ω^{p-1}	ω^{p-2}	…	…	…	…	ω^{z-p}
…	…	…	…	…	…	…	…
ω^z	ω^{z-1}	ω^{z-2}	…	…	…	ω	1

Fig. 6.11 Matrix explaining the calculation of the form factor of a Gaussian chain.

sum of these terms is z. Let us call $2T$ the sum of all the other terms. T can be evaluated by grouping the terms of one-half of the matrix by lines. We write therefore

$$z^2P(q) = z + 2T$$

$$T = \omega[(1+\omega+\omega^2+\omega^3+\ldots\ldots+\omega^{z-2})+(1+\omega+\omega^2+\omega^3+\ldots+\omega^{z-3}) + (1+\omega+\omega^2+\ldots+\omega^{z-4})+\ldots+(1+\omega)+(1)]. \tag{6.68}$$

Each line is a geometrical series which can be easily summed

$$T = \omega \sum_{1}^{z-1} \frac{1-\omega^p}{1-\omega}. \tag{6.69}$$

This simple sum can also be evaluated giving

$$P(q) = \frac{1}{z^2}\left\{z\frac{1+\omega}{1-\omega} - 2\omega\frac{1-\omega^z}{(1-\omega)^2}\right\}. \tag{6.70}$$

This formula allows $P(q)$ to be evaluated for any value of z; it is rigorous, even for $z = 1$. When z is large one can simplify $P(q)$, replacing $\omega_0(q)$ by the two first terms of its expansion

$$\omega_0(q) = 1 - \frac{q^2\overline{l^2}}{6} \tag{6.71}$$

which is equivalent to the assumption that $\omega_0(q)$ (or $w_0(r)$) obeys a Gaussian law

$$\omega_0(q) = \exp\left(-\frac{q^2\overline{l^2}}{6}\right). \tag{6.72}$$

With this approximation, neglecting the higher order terms in eqn (6.70) and assuming also that $q^2\overline{l^2}/6$ *is small* compared to unity gives the well known Debye (1946) formula

$$P(q) = \frac{2}{x^2}[x - 1 + \exp(-x)] \tag{6.73}$$

with $x = \dfrac{q^2 z\overline{l^2}}{6} = q^2\overline{R^2}$.

An easier method for obtaining this result is to start from eqn (6.67) and to use the Gaussian approximation for $\omega_0(q)$, obtaining

$$P(q) = \frac{1}{z^2}\sum_{i=0}^{z}\sum_{j=0}^{z}\exp\left(-|i-j|\frac{q^2\overline{l^2}}{6}\right). \tag{6.74}$$

We can transform the double sum into a simple sum using eqn (6.22)

$$P(q) = \frac{2}{z^2} \sum_{p=0}^{z} (z - p) \exp\left(-p \frac{q^2 \overline{l^2}}{6}\right). \tag{6.75}$$

(Neglecting the term z^{-1} which is negligible when z is large.) Transforming the sum into an integral and using the variable $u = p/z$ one obtains

$$P(q) = 2 \int_0^1 (1 - u) \exp\left(-\frac{uzq^2 \overline{l^2}}{6}\right) \mathrm{d}u \tag{6.76}$$

and by integration by parts eqn (6.73) is recovered.

6.3.6 Chains of different architecture

In this section we would like to show that it is relatively easy to extend these calculations of form factors to chains made of Gaussian segments but having different geometries. For this purpose we will study two cases. The first one will be the case of what is called by the polymer community 'stars'. These polymers, in the simplest case, are made of m identical chains joined together by a universal joint at one of their extremities. This model is idealized since it is impossible if m is large, to achieve junctions which have a negligible volume and allow for all the relative orientations of the chains. Nevertheless stars are the object of intense study mainly for their rheological and thermodynamic properties and it is important to be able to interpret their form factors. We shall also discuss the case of the 'ring polymers'. These polymers are made of one linear chain where both extremities are chemically attached. They make chains without end. They are also of rheological interest because it is difficult to apply to them the concept of reptation (de Gennes 1971).

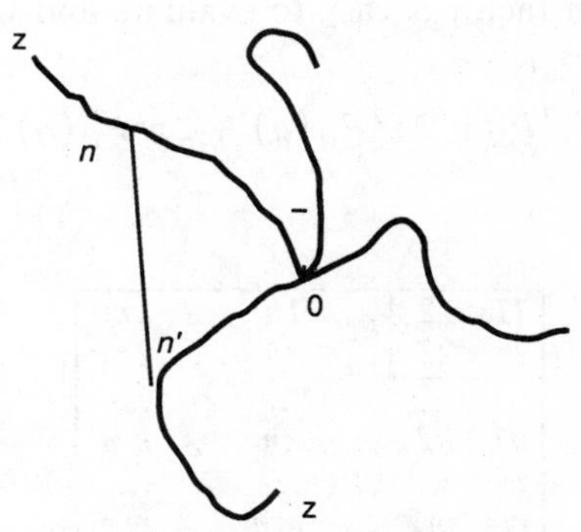

Fig. 6.12 Model for a star polymer.

The star polymers

Let us assume that we have a 'star' polymer made of m linear chains attached together at one end (see Fig. 6.12). Each chain has z segments; the total degree of polymerization is $Z = mz$. One scattering element is characterized by the double index i_n; n designates the chain and i the position on the chain. The i's $(0 \leq i \leq z-1)$ are arranged in order to have the index zero at the junction point. We consider the quadruple sum

$$\sum_{n=1}^{m} \sum_{p=1}^{m} \sum_{i_n}^{z} \sum_{j_p}^{z} \exp\left[-(i_n + j_p)\frac{q^2\overline{l^2}}{6}\right] \tag{6.77}$$

(if $n = p$, one has to replace $i_n + j_n$ by $i_n - j_n$ in the exponential). This sum corresponds to the scattering by a pair of points having a Gaussian probability and a mean square distance $(i_n + j_p)\overline{l^2}$ or $(i_n - j_n)\overline{l^2}$ and can be split, by summation over i and j into m^2 units each having z^2 terms. One has m units for which $n = p$ called nn and $m(m-1)$ units for which $n \neq p$ called np. All the units called nn are identical; moreover they are equal to z^2 times the form factor of a linear Gaussian chain of z segments which will be called $z^2 P_{11}(q)$. The units np are also equal; they correspond to the interferences between two points on different chains and they will be called $z^2 P_{12}(q)$. Grouping all these terms in a square matrix (Fig. 6.13) gives

$$Z^2 P^*(q) = m^2 z^2 P^*(q) = mz^2 P_{11}(q) + m(m-1)z^2 P_{12}(q) \tag{6.78}$$

or

$$m^2 P^*(q) = mP_{11}(q) + m(m-1)P_{12}(q).$$

In order to evaluate the term $P_{12}(q)$ consider the four first cells of the matrix. They give the form factor of a two-branch star, i.e. a linear polymer of length $2z$. Its form factor is easy to evaluate and will be called $P_{22}(q)$. It follows that

$$4z^2 P_{22} = 2z^2 P_{11}(q) + 2z^2 P_{12}(q). \tag{6.79}$$

11	12	...	1n	...	1m
21	22	...	2n	...	2m
...	...	...	...	...	...
n1	n2	...	nn	...	nm
...	...	...	...	...	...
m1	m2	...	mn	...	mm

Fig. 6.13 Table explaining the calculation of the form factor of a star.

Eliminating $P_{12}(q)$ between eqns (6.78) and (6.79) leads to the result

$$P^*(q) = 2\left(1 - \frac{1}{m}\right)P_{22}(q) - \left(1 - \frac{2}{m}\right)P_{11}(q). \tag{6.80}$$

This relationship shows that the form factor of one star can be expressed by knowing the structure factor of one branch and of the linear polymer made of two branches.

Remark: if one wants to evaluate $P^*(q)$ for more complicated cases one cannot say that a two-branch star is identical to a double size linear polymer (consider for instance the case of a star with rod-like arms). Instead of using this short cut to evaluate $P_{12}(q)$ one can evaluate it directly (Benoît 1953). Keeping in mind that the root mean square distance between i and j on different branches is, in the case of a universal joint at the junction point

$$\overline{r_{ij}^2} = \overline{r_{i0}^2} + \overline{r^2}_{0j} = \overline{l^2}(i_n + j_n) \tag{6.81}$$

one obtains immediately

$$P_{12}(q) = \frac{1}{z^2}\sum_{n}^{z}\sum_{n'}^{z}\exp\left[-\frac{1}{6}q^2\overline{l^2}(n+n')\right]. \tag{6.82}$$

Replacing the sums by two integrations leads to

$$P_{12}(q) = \left[\frac{1 - \exp -\left(\frac{1}{6}q^2\overline{l^2}z\right)}{\frac{1}{6}q^2\overline{l^2}z}\right]. \tag{6.83}$$

We can easily verify that this expression is equal to $2P_{2z} - P_{1z}$. This method can also be used to evaluate the cross form factor $P_{AB}(q)$ in the case of Gaussian copolymers. Expanding $P^*(q)$ as function of q in the small q range, following eqn (6.14), allows us to evaluate the radius of gyration. Applying this method to eqn (6.80) gives

$$\overline{R^2} = 2\left(1 - \frac{1}{m}\right)\overline{R^2_{2z}} + \left(\frac{2}{m} - 1\right)\overline{R^2_z} \tag{6.84}$$

or, since we have assumed that the chain is Gaussian

$$\overline{R^2} = \left(3 - \frac{2}{m}\right)\frac{\overline{l^2}}{6}z. \tag{6.85}$$

For $m = 1$ and 2 we recover the classical results. Also, as in the case of $P^*(q)$ when m becomes large these expressions reach, as we could have guessed, a limiting value independent of m.

Ring macromolecules

In a ring macromolecule the distances between two points i and j are Gaussian and we can rewrite eqn (6.74)

$$P(q) = \frac{1}{z^2}\sum_{i=0}^{z-1}\sum_{j=0}^{z-1}\exp\left(-\overline{L_{ij}^2}\frac{q^2}{6}\right) \tag{6.86}$$

(we have z scattering points numbered from 0 to $z-1$) where $\overline{L_{ij}^2}$ is the root mean square distance between the scattering points i and j. If we draw the same square as in the study of linear molecules we realize that all lines contain exactly the same sum of terms; we can therefore write

$$P(q) = \frac{1}{z}\sum_{p=0}^{z-1}\exp-\left(\frac{q^2\overline{L_p^2}}{6}\right). \tag{6.87}$$

From eqn (6.27) we can replace $\overline{L_p^2}$ by its value: $\overline{l^2}p(1-p/z)$. Writing $\lambda = (q^2\overline{l^2}z/6)$, transforming the sum into an integral and using as variable $u = p/z$ we obtain the simple expression

$$P(q) = \int_0^1 \exp\{-\lambda u(1-u)\}\mathrm{d}u. \tag{6.88}$$

This expression cannot be integrated to give classical functions. It is usually transformed to a new integral by completing the square of the expression in the exponential and writing $v = \sqrt{\lambda}\,(u - 1/2)$ (Cassasa 1965)

$$P(q) = \frac{2}{\sqrt{\lambda}}\exp-\frac{\lambda}{4}\int_0^{\frac{\sqrt{\lambda}}{2}}\exp(v^2)\,\mathrm{d}v. \tag{6.89}$$

Figure 6.14 shows the result of the numerical integration.

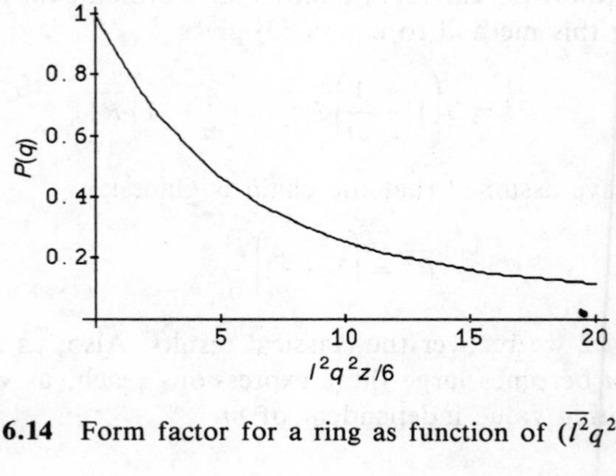

Fig. 6.14 Form factor for a ring as function of $(\overline{l^2}q^2Z/6)$.

6.4 THE INTERMEDIATE AND HIGH *q* RANGE

6.4.1 Qualitative interpretation of the different *q* domains

Up to now we have discussed the methods for the calculation of the radius of gyration and the form factor $P(q)$ but we have not considered how it might be possible to compare experimental data to the curves evaluated for models. At first sight this seems to be very difficult since, forgetting the relatively small oscillations which are very frequently erased by polydispersity, all these curves look very similar. In Fig. 6.15 we have collected the curves calculated for rods, Gaussian chains, and spheres having the same radius of gyration; they look very similar except for the fact that, at large angles, they decrease more or less rapidly. It is this behaviour, at large q that we would like to discuss but, before doing so, it is interesting to see qualitatively what one can expect from neutron scattering in different q domains.

It is evident that, since the scattering intensity as a function of q is the Fourier transform of the pair distribution of scattering centres in the sample, one can say that $\mathbf{q}$ space and $\mathbf{r}$ space are conjugated, or in other words that small $\mathbf{q}$ corresponds to large values of $\mathbf{r}$ and vice versa. Looking at a scattering diagram as function of q is just like looking at it with a magnifying glass of changing power. When the power increases the field decreases and is of the order of q^{-1}. In this discussion we shall assume that we are studying a dilute solution or a system in which the scattering

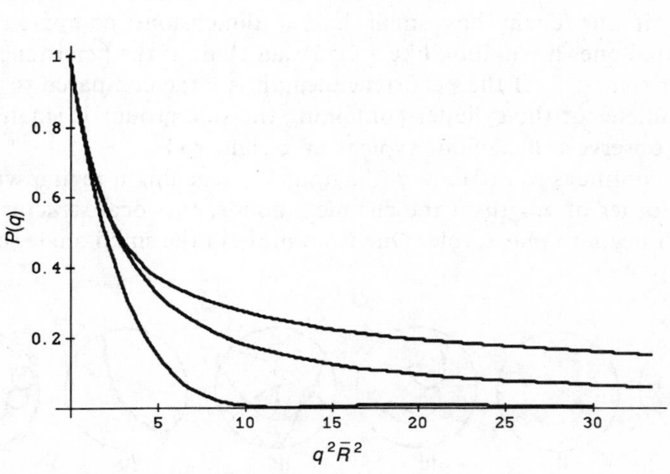

Fig. 6.15 $P(q)$ for, in decreasing order, a rod, a Gaussian chain, and a sphere as function of $q^2\overline{R^2}$.

diagram is completely described by the form factor; we shall generalize these results to any system later.

Working at low q is equivalent to using a low power magnifying glass. In this situation each molecule is practically a point and the only thing one can do is to count the number of points which is equivalent to a measure of the molecular weight. If one wants to be more precise one can say that neutron scattering gives $\overline{\Delta N^2}$ which, for an ideal solution, is equal to N. To summarize the situation: at $q = 0$ one measures only thermodynamic aspects of the solution and one does not obtain any information about its structure. (Fig. 6.16 I.)

Now we increase q in order to have q^{-1} of the order of the radius of gyration. What one sees is approximately represented in diagram II. One does not see the details of the shape and the structure of the molecules but only their dimensions. In this domain, which is called the Guinier domain, one measures the radius of gyration.

If we increase q still further we reach either domain III or III′ depending on the concentration: diagram III corresponds to dilute solution, diagram III′ to moderately concentrated solution (called semidilute by de Gennes (1979)). From diagram III it is clear that one sees only a part of a molecule; the scattering does not depend on molecular weight and one obtains information about the statistics of the chain and its persistence length. If the solution is concentrated (diagram III′) one sees parts belonging to different chains and one obtains information about the length of the chain between two contact points with other chains; this length has been called the correlation length and will be discussed in Chapter 8.

Increasing q will leave only one part of one chain in the field (diagram IV) and if the chain has small lateral dimensions compared to the longitudinal ones it will look like a Gaussian chain if the persistence length is smaller than q^{-1}. If the persistence length is large compared to q^{-1} and if the diameter of the cylinder containing the side groups is small enough one will observe a behaviour typical of a rigid rod.

If one continues to increase q (diagram V), reaching a region where q^{-1} is of the order of length of the chemical bonds, the local structure of the chain will begin to play a role. One is no more in the small angle scattering

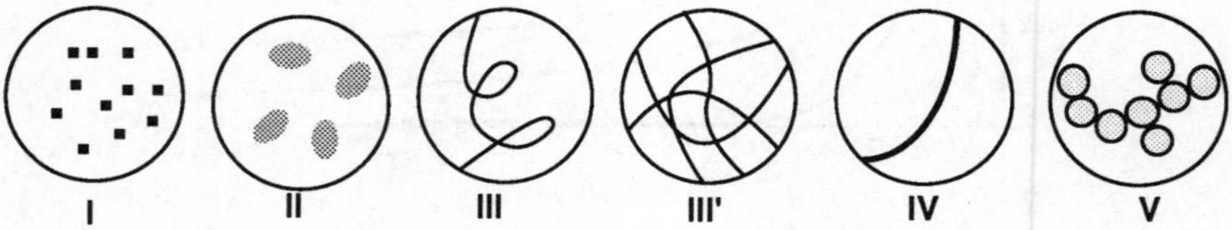

Fig. 6.16 Schematic representation of the different q domains.

range and other methods have to be used, for example, the diffuse scattering technique mentioned in Chapter 3.

From this analysis some conclusions can be deduced:

1. In what is called the intermediate range (diagram III or III′ and IV) the scattering intensity should not depend on the molecular weight. The length of the chain is detected only by the effects of chain ends. The probability of observing one of them in the field of observation decreases when q increases. For large q it is a rare event (specially for large molecular weights) which does not affect the results. If this is true, neutron scattering will give, in this range, information which is independent of the chain length and polydispersity and depends only on chain statistics.

2. We have assumed for the sake of simplicity (except in diagram III′) that we had a dilute solution. If the concentration increases the analysis of domains I and II becomes difficult since the intermolecular interactions are at distances of the order of the size of the molecules. One cannot determine the radius of gyration. In domain IV the concentration has no effect since with such a large magnification one will very rarely have two different molecules in the field of observation. One can conclude from this remark that for q larger than $1/R$ and smaller than $1/l$ the scattering depends only on the chain statistics and not on the concentration.

3. For many reasons these limits, which have been arbitrarily fixed at R, the radius of gyration and l, the statistical element of the chains, are not precise. One should, to be correct, speak about 'crossover' regions where the dominant factor influencing the scattering changes continuously. To summarize one can say that for a single linear chain there are three distinct domains: (1) the zero angle and the Guinier domain, where one can measure the molecular weight and the radius of gyration (R_g) at least for dilute solutions; (2) the intermediate domain $1/R_g < q < 1/l$ where the scattering depends only on the chain statistics and sometimes a third domain $1/l < q$ where, if the monomer is thin enough, the chain behaves like a rigid rod. This last point is important for rigid polymers like cellulosic derivatives and polymers able to give liquid crystal like phases and the determination of this last crossover has been used to characterize what is called the persistence length of the polymer.

4. Finally it is evident that the effect of polydispersity decreases when q increases and disappears completely at large q since, as we have said the results are independent on the length of the chains.

All these considerations are useful when one encounters difficulties in the interpretation of data from a complicated system. But even if they suggest some interpretation they are not sufficient for extracting from experiments as much information as possible. We shall therefore, in the following sections, discuss more quantitatively the shape of the scattering curves at high q values. We shall begin, before using the expressions which have already been established, by a demonstration using what is called scaling arguments. This demonstration is perhaps not absolutely necessary but, since this type of argument is used quite often in polymer science it is important to be familiar with it.

6.4.2 The use of scaling arguments for the determination of exponents at high q values

If we neglect the interactions which, as it has just been shown, have no influence at high q values we can write

$$S(q) = Nz^2 P(qR_g) \tag{6.90}$$

or

$$= V\varphi z P(qR_g)$$

where V is the volume of the sample and φ the volume fraction occupied by the scattering units of degree of polymerization z. We write the form factor as $P(qR_g)$ since it is a dimensionless quantity and we have seen in all the previous equation that q is associated with a characteristic dimension of the model. We assume now that, at large q values, $P(qR_g)$ has an asymptotic expansion beginning with a leading term of the form $(qR_g)^{-\alpha}$. The quantity we want to determine is the parameter α which can have any value. It will be shown now that, if we assume that $I(q)$ does not depend on z we can obtain the value of α. For this purpose we first write the relationship between R and z (or dimensions and mass).

for a Gaussian chain	$R_g \approx z^{\frac{1}{2}}$	
for a chain with excluded volume (Flory)	$R_g \approx z^{0.6}$	(6.91)
for a rod	$R_g \approx L = z$	
or, quite generally	$R_g \approx z^a$.	

Replacing R_g in eqn (6.90) by its expression as function of z gives

$$S(q) \approx V\varphi z(qz^a)^{-\alpha} = V\varphi q^{-\alpha} z^{(1-a\alpha)}. \tag{6.92}$$

In order to have $S(q)$ independent of z, the power at which z is taken in eqn (6.92) must be zero giving

$$\alpha = \frac{1}{a}.$$

In other words $S(q)$ should vary as $q^{\frac{-1}{a}}$.

This leads to, for the rod $$S(q) \approx \frac{1}{q}$$

for the Gaussian chain $$S(q) \approx \frac{1}{q^2}$$

and for the chain with excluded volume $$S(q) \approx \frac{1}{q^{1.66}}. \qquad (6.93)$$

This demonstration gives the correct result for these linear objects but has to be modified for two-dimensional objects such as lamellae with small and constant thickness. In this case we can first say that $S(q)$ is proportional to the total scattering surface $S_T \approx Nz$ and then express everything as function of the surface S of the objects for which we evaluate $P(qR)$ as $P(qS^{\frac{1}{2}})$. We first replace after a simple calculation, Nz^2 in eqn (6.90) by $Nv^2 \approx S_T S^2$ obtaining

$$S(q) \approx S_T S P(qS^{\frac{1}{2}}) \approx S_T S (qS^{\frac{1}{2}})^{-\alpha} \qquad (6.94)$$

which, after using the hypothesis that $S(q)$ is proportional to the total surface or to the concentration (since the thickness is constant) leads to the exponent $\alpha = 2$. This coincidence between the exponent of the Gaussian chains and the thin lamella is surprising and has been explained by des Cloizeaux and Jannink (1987) who showed that a Gaussian chain is a two-dimensional object.

The last case we shall discuss is the case of three-dimensional objects like spheres. If we return to our image of the magnifying glass we realize that, in order to have a signal, the surface of separation between the two media has to be in the field of observation. Inside or outside the scattering object there is no scattering. This means that the scattering should be proportional to the total surface of the sample. Since on the one hand $R_g \approx S^{\frac{1}{2}}$ and $Nz^2 \approx NV^2 \approx NS^3$ and the total surface S_T of the sample to which the intensity is proportional is, on the other hand, proportional to NS we obtain for the scattering intensity

$$S(q) \approx S_T S^2 P(qS^{\frac{1}{2}}) \approx S_T S^2 (qS^{\frac{1}{2}})^{-\alpha}. \qquad (6.95)$$

Applying the condition that the intensity is proportional to S_T and does not depend on S, gives $\alpha = 4$. This law ($I(q) \sim q^{-4}$)* is quite general and has been called the Porod law. Porod (1951) has given a rigorous demonstration of this law with the exact value of the front factor.

*One can illustrate this law by the following consideration: clouds are usually colourless. They scatter light following the Rayleigh law which predicts that the scattering is proportional to λ^{-4} (the sky is blue) times the form factor $P(qr)$ of the water droplets. In order to have no colour the intensity must be independent of λ: the power law for $P(qr)$ has to scale like q^{-4} since q is proportional to λ^{-1}.

6.4.3 The case of objects with rough surfaces

All the models which have been considered so far were geometrical figures. This hypothesis may be sufficient at large-scale distances but this is not the case at high q were details of the molecular structure have to be taken into account. Even if, in low angle scattering, one does not examine the fine structure of the samples it is useful to see how the behaviour at large q values is affected by the departure of the physical object from the geometrical model. We shall make an approximate treatment for the rod, the Gaussian chain, and the lamella and will say only a few words in the case of three-dimensional objects.

The case of the rod

In all the preceding treatments the rods were supposed to have no lateral dimensions. What happens when it is impossible to neglect their transverse dimensions? In order to study this problem we shall assume that the dimensionless axis of the rod is uniformly covered with spheres of radius a. Each of these spheres scatters independently of its orientation with an amplitude given by eqn (6.48) or (6.52), in phase with the amplitude scattered by its centre. It is therefore possible in the calculations made previously to say that the scattering amplitude of a point of the rod is multiplied by the amplitude scattered by the sphere A. This leads to the result: $P_{\text{dec}}(q) = A^2(q)P(q)$ calling $P(q)$ the form factor of the thin rod, P_{dec} the form factor of the decorated rod, and A^2 the form factor of the spheres covering the rod. If the transverse radius of gyration, R_{T}, of these spheres is small one can expand $A^2(q)$ around $q = 0$ assuming that, even if q is large enough to justify the use of the asymptotic form of the form factor of the rod (eqn (6.110)) the quantity $q\rho$ remains small. This leads to the following result:

$$P_{\text{dec}}(qL) = \left(\frac{\pi}{qL} - \frac{2}{(qL)^2}\right)\left(1 - \frac{q^2 R_{\text{T}}^2}{3}\right) \tag{6.96}$$

(This formula is only valid for q such that $qR_{\text{T}} < 1$ and $qL > 1$.) Remark: using the trick of placing spheres with centres at each point of the rod the radial density of the rod is not constant and decreases when the transverse distance increases. One could correct for this effect using spheres with variable density.

The case of chains and lamellae

A similar result can be obtained for statistical chains if one assumes, as in hydrodynamics, that all the mass of the chain is concentrated in spheres located at the junction point of the segments. The same type of argument can also be used for lamellae. If one decorates the lamellae, as we have done for the rod with contiguous spheres one obtains a symmetrical profile of scattering density and the same type of formula can be used.

The case of three-dimensional objects

The simplest way to describe these objects is to assume that one deals with spheres. Since at large angles, the scattering is a local property, the formulae obtained by this method can be generalized easily to the general case, assuming that the radius of curvature of the scattering surface is large compared with the thickness of the rough layer. Using eqn (6.54) it is easy to show that the intensity scattered by spheres with a scattering density $n(r)$ which depends only on the distance r from their centre r is expressed by

$$I(q) = N\left(\frac{b_1}{v_1} - \frac{b_0}{v_0}\right)^2 \left[\int n(r)\,\frac{\sin qr}{qr}\,4\pi r^2 \mathrm{d}r\right]^2. \qquad (6.97)$$

The quantity $b_1/v_1 - b_0/v_0$ is the difference between the scattering length per unit volume of the spherical object and its surrounding and can be called b_v (eqn (A3.5)). It is often preferable to use the scattering length density $\rho_b(r)$ which is equal to $b_v n(r)$ the product of the scattering length per unit volume by the density of scattering elements inside the sphere (eqn (A3.33)) thus rewriting eqn (6.97) as

$$I(q) = N\left[\int \rho_b(r)\,\frac{\sin qr}{qr}\,4\pi r^2 \mathrm{d}r\right]^2. \qquad (6.97')$$

We assume now that the sphere has a radius R, a contrast factor density ρ_{b_1}, and is covered with a layer of uniform density of thickness z and contrast factor ρ_{b_2}. We assume also, in order to apply these formulae to the case of spheres coated with a layer of polymer that these scattering length densities are constant. The constant density of the layer is introduced here for simplicity but this formula can be extended easily to any density profile. Dividing the integral in two parts we obtain

$$I(q) = N\left\{\left[\int_0^R \rho_{b_1}\frac{\sin qr}{qr}\,4\pi r^2 \mathrm{d}r\right]^2 + \left[\int_R^{R+z} \rho_{b_2}\frac{\sin qr}{qr}\,4\pi r^2 \mathrm{d}r\right]^2 + 2\int_0^{R_i} \rho_{b_1}\frac{\sin qr}{qr}\,4\pi r^2 \mathrm{d}r \int_R^{R+z} \rho_{b_2}\frac{\sin qr}{qr}\,4\pi r^2 \mathrm{d}r\right\} \qquad (6.98)$$

where N is the number of spheres per unit volume. In this equation the first term represents the structure factor of a sphere of radius R the second term is the scattering by a lamella of thickness z; it is an integral between R and

$R+z$ which can be simplified when z is small and R is large. The third term is a cross term represented by the product of the amplitudes scattered by the sphere and the spherical lamella. These calculations are easy and after replacing the functions by their values at large q (using the first term of eqn 6.122), we obtain

$$\frac{1}{N} I(q) = \rho_{b_1}^2 V^2 \frac{9}{2q^4R^4} + \rho_{b_2}^2 \frac{8\pi^2 R^2}{q^2} z^2 + 2\rho_{b_1}\rho_{b_2} V \frac{6\pi z}{q^4 R^2}. \qquad (6.99)$$

Since everything should be expressed as function of the surface S of the spheres, we write for the scattering intensity divided by the volume fraction occupied by the spheres

$$\frac{1}{NV} I(q) = \rho_{b_1}^2 \frac{2\pi}{q^4} \frac{S}{V} + \rho_{b_2}^2 \frac{2\pi}{q^2} z^2 \frac{S}{V} + 2\rho b_1 \rho_{b_2} z \frac{2\pi}{3q^4} \left(\frac{S}{V}\right)^2. \qquad (6.100)$$

This expression is similar to an expression derived by Auvray and Cotton (1987). It as been used to study the adsorption of polymers on silica beads.

6.4.4 Scattering by fractals

During recent years the notion of 'fractal' has been introduced quite successfully in physical chemistry (Mandelbrot 1982) and has been used to interpret the scattering results of neutrons, X-rays or light. This book would be incomplete if we did not say a few words on this aspect of the interpretation of the data. Let us look at a geographical map of England and examine the coastline. Now take another map at a much smaller scale and look also at the coastline. The outline is different but it has a very similar look. It is impossible to tell anything about the scale of the map: one says that the figure is 'self-similar' or better 'statistically self-similar'. This means that regardless of its scale it looks the same. Of course this self-similarity does not extend from zero to infinity. It is impossible to define it for dimensions smaller than a few metres at one limit and larger than the earth at the other. On consideration it is evident that many curves have this kind of property and are not the regular curves one uses in geometry. Another interesting feature is the following. Assume that we want to measure the distance apart along the coast line of two harbours situated at a distance L apart in a straight line. For this purpose we take a sphere in three dimensions (or a disc in two dimensions) of diameter L; one needs only one disc to join both points. Now we take discs of diameter L/δ and repeat the measurement putting the disc side by side along the coastline. We shall need N discs and it is evident that N is larger than L/δ. The smaller the disc the larger will be the discrepancy between N and L/δ. It is assumed that, because of the self-similarity, the relation between N, L, and δ takes the form $N=(L/\delta)^D$.

One can also take a point 0 on the fractal curve and draw a sphere (or a disc) of centre 0 and radius R and measure the mass $M(R)$ of the curve which is inside the sphere. If we change the radius R the mass will change according to the relation

$$M(R) \approx R^D.$$

It can be shown that for statistically self-similar curves both definitions lead to an identical parameter D.

From the last definition it is clear that we can predict the exponent in the scattering laws. In eqn (6.92) we have written (replacing z by the mass of the particle and the volume fraction by the concentration)

$$S(q) \approx VcM(qM^a)^{-\alpha} \approx cq^{-\alpha}M^{(1-a\alpha)} \tag{6.101}$$

where V is the scattering volume and where we assume that the relationship between M and R is of the form $R = M^a$. In order to obtain a result independent of the mass the exponent α must satisfy the relationship

$$a = \frac{1}{D} \text{ and } I(q) \approx q^{-D}. \tag{6.102}$$

This law allows us to recover all the results obtained previously except the result for three-dimensional objects. Before studying this case one has to remark that the Gaussian chain has a fractal dimension of two, and the chain with excluded volume of 1.66 (which is smaller than 2 but larger than the value 1) corresponds to rods. This has been discussed some years ago by des Cloizeaux and Jannink (1987).

6.4.5 Surface fractals

One can also consider surfaces with fractal dimensions or on which one has deposited material with fractal properties. In this case one will also introduce a fractal dimension writing $S \approx R^{Ds}$. For a smooth surface Ds is evidently two and can never be larger than three, the dimension of a compact object. Applying the relationship (6.90) which claims that the scattering is of the form

$$S(q) \approx NM^2(qR)^{-a} \approx NSf(q) \tag{6.103}$$

and is proportional, at large q values to the total surface $S_T = NS$, and replacing M by R^3 and S by R^{Ds} gives, since the exponent of R has to be the same on both sides of the last equality

$$a = 6 - Ds \text{ and } I(q) \approx q^{-6+Ds}. \tag{6.104}$$

If the surface is smooth, $a = 4$, the value already obtained (the Porod law). A value less than 3 is impossible.

Great care is needed if fractal dimensions are to be obtained from scattering results. It is, of course very simple to plot log $(I(q))$ vs log (q) and to measure the slope obtained in this type of diagram. But, in order to be meaningful the domain of q where a straight line is observed has to be large and cover at least one decade otherwise the interpretation in terms of fractal exponents could be meaningless. Here are a few fractal exponents collected by Schaefer (1987)

Rigid rod	1
Linear Gaussian chain	2
Chain with excluded volume	5/3
Gaussian chain randomly branched	16/7
Swollen branched chain	2
Smooth two-dimensional objects	2
Three-dimensional objects with smooth surfaces	4
fractal surfaces	3 to 4

6.5 PRACTICAL METHODS FOR CHARACTERIZATION OF THE SCATTERING CURVES

It is clear that, for the models we have presented, it is impossible to interpret unambiguously the experimental curves since, at high q values, all scattering curves decrease to zero; the intensity is small and the power laws are difficult to establish. This has suggested to the experimentalist different methods of plotting the results; we now discuss these methods and study systematically how it is possible to interpret the data. This analysis will concern not only the high q end of the curve but also its low q beginning and review the different possible approaches. Most approaches aim to turn the experimental curve into a straight line for comparison with models in order to obtain parameters from the slope and the intercept.

6.5.1 The plots at low angles

If one neglects the interactions by working at very low concentration and if one plots log $[I(q)]$ as function of q^2 one can write using eqn (6.14)

$$\log I(q) = \log I(0) - q^2 \frac{\overline{R^2}}{3} + \dots . \tag{6.105}$$

The initial part is therefore a straight line and its slope allows the determination of the radius of gyration. This is called the Guinier plot (Guinier and

Fournet 1955) and is used extensively for the interpretation of small angle X-ray measurements. One of the difficulties of this method is that the range of q values for which one obtains a linear plot is limited. For the curve to be linear in a large range of q values the scattered intensity should be represented by a Gaussian function $I(q) = I(0)\exp(-q^2\overline{R^2}/3)$ and, as we have seen, this approximation is not satisfied for all the models we have studied.

Another method consists in using what has been called the Zimm plot (Zimm 1948*b*) which has the advantage of being useful for both the beginning and the end of the curve: one plots $1/S(q)$ as function of q^2. The initial part becomes

$$\frac{Nz}{S(q)} = \frac{1}{z}(1 + q^2\frac{\overline{R^2}}{3} + \ldots \tag{6.106}$$

which becomes if one uses $I(q)$

$$\frac{Nz\bar{b}^2}{I(q)} = \frac{1}{z}(1 + q^2\frac{\overline{R^2}}{3} + \ldots .$$

The initial slope of the curve is also $\overline{R^2}/3$. The two methods look equivalent but, due to the limitation of the experimental precision, they could lead to different results. In the second method one has to know the extrapolated value at $q = 0$ with precision in order to calculate the slope. In the Guinier method this is not necessary. If both methods do not give the same result one has to be very cautious and revise the experiments.

6.5.2 The plots at large angles

When the power law which should characterize the scattering of a given sample is known it is usual to plot $I^{-1}(q)$ as function of q^a, where a is the exponent defined in eqns (6.91–6.94), in order to obtain a straight line. If the points at high values of q are on a straight line one has verified the hypothesis and can proceed further, looking, for instance, at the intercept or at the different characteristics of the scattering curve. For Gaussian chains, which are characterized by a value of $a = 2$, clearly the Zimm (1948) representation has the advantage of giving useful results for high and small values of q. The Zimm plot is also used for fluids in the vicinity of their critical point since the Ornstein–Zernike (1914) theory gives (see Section 7.3):

$$S(q) = \frac{A}{1 + q^2\xi^2} \tag{6.107}$$

which is also an asymptotic law in q^{-2}. If a straight line is not obtained one has to use an alternative stratagem and plot $\log I(q)$ as function of log (q). If the plot is linear over a sufficiently large range of q, the characteristic exponent can again be extracted from the slope. A third method, developed specially for the large q range, is called the Kratky plot: $I(q)q^a$, where a is the exponent expected from the sample under investigation, is plotted against q. If the plot has a horizontal asymptote the value of the exponent a is confirmed. This method is more precise than the method consisting in plotting $\log I(q)$ against log (q) and measuring the slope.

6.5.3 The Gaussian chain

The curve corresponding to the Gaussian chain eqn (6.73) is shown in Fig. 6.11 in the Zimm representation ($(I(q))^{-1}$ vs q^2). As expected from our previous discussion it shows a linear asymptote. In order to obtain the equation of this asymptote we take the value of eqn (6.73), neglecting the exponential which is small at large x

$$S(q) = 2Nz^2\left[\frac{1}{x} - \frac{1}{x^2}\right] \tag{6.108}$$

with $x = q^2\overline{l^2}z/6$. Taking the inverse and using eqn (6.4) gives

$$\frac{\bar{b}^2\varphi V}{I(q)} = \frac{1}{V_1}\left[\frac{x}{2} + \frac{1}{2}\right]. \tag{6.108'}$$

If we use the molecular weight M_1 (in daltons) instead of the volume of the polymer, the contrast factor per unit volume $b_v = \bar{b}/v_0$ and the specific volume of the polymer v_s this formula becomes

$$\frac{b_v{}^2 c V}{I(q)} = \frac{N_A}{v_s^2}\left[\frac{1}{2M} + \frac{q^2\overline{l^2}}{12m}\right] \tag{6.109}$$

where m is equal to M/z.

The straight line obtained when we plot $I^{-1}(q)$ as function of q^2 (Fig. 6.17) has a slope $(\overline{l^2}/12m)$ which does not depend on the parameter z since $\overline{l^2}/12\,m = \overline{R^2}/M$ and which is universal in the sense that this value is obtained regardless of the geometrical structure and length of the Gaussian chain. These factors have an influence on the second term which is the intercept of the asymptote. The only point we would like to make is that this intercept $(1/2M)$ allows the determination of the molecular weight. When the polydispersity is discussed it will be shown that, in the case of a polydisperse system, $I(0)$ gives the weight average and the intercept of the high q asymptote, the number average molecular weight.

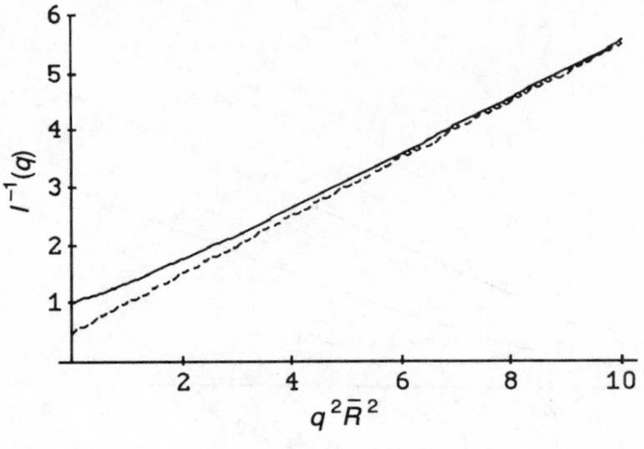

Fig. 6.17 Scattered intensity for a Gaussian chain in Zimm's representation; $I^{-1}(q)$ as function of $q^2\bar{R}^2$. The dotted line is the asymptote.

6.5.4 Star molecules

It suffices to expand in eqn (6.78), $P_1(q)$ and $P_2(q)$ keeping only the terms in q^{-2} and q^{-4} and substituting $x = q^2\overline{l^2}z/6$, where z is the degree of polymerization of one branch, to obtain

$$mP(q) \approx (2 - m)\left(\frac{2}{x} - \frac{2}{x^2}\right) + 2(m - 1)\left(\frac{1}{x} - \frac{1}{2x^2}\right). \qquad (6.110)$$

This gives us, for the asymptote, by analogy with eqn (6.109)

$$\frac{b_v^2 c V}{I(q)} = \frac{N_A}{v_s^2}\left[\frac{3 - m}{4M_0} + \frac{q^2 l^2}{12\mu}\right] \qquad (6.111)$$

where M_0 is the molecular weight of one branch.

Following eqn (6.108) the slope of the asymptote is $\overline{l^2}/12\mu$ as for linear chains. The intercept, however, depends on the number of branches m. It is positive for $m < 3$ and negative for $m > 3$. Figure (6.18) shows these results for a two-branch star, a three-branch star (the intercept is zero), and a 10-branch star. μ is the mass of a monomer (M_0/z).

6.5.5 Ring polymers

From what we know we should find similar results for rings. The calculations are more difficult, since the expression for $P(q)$ (eqn (6.89)) cannot be integrated; we shall not undertake the derivation here and we give only the result: at large values of q the form factor of a ring can be expanded as

$$[P(q)]^{-1} = \frac{x}{2} - 1. \qquad (6.112)$$

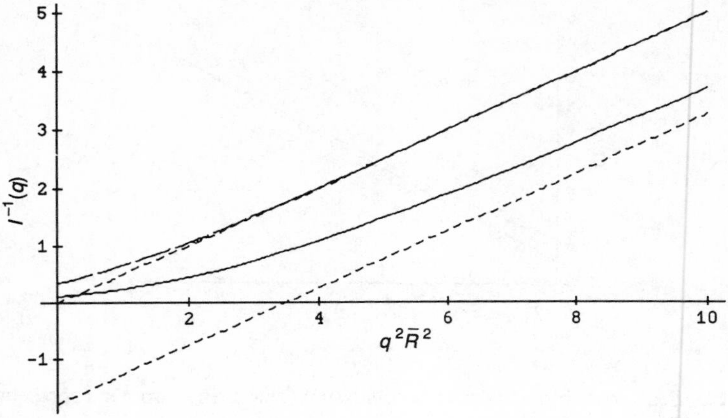

Fig. 6.18 Inverse of the scattered intensity for a three-branch star (dashed line) and a 10-branch star (solid line). The straight dotted lines are the corresponding asymptotes (eqn. (6.111)).

The intercept is negative (see Fig. 6.19) and curiously the curve crosses its asymptote. This means that terms of higher order are necessary for a complete description of the curves.

To complete this discussion of Gaussian chains we show, in Fig. 6.20, the Kratky diagram ($q^2 I(q)$ vs q) for the type of chains which have been

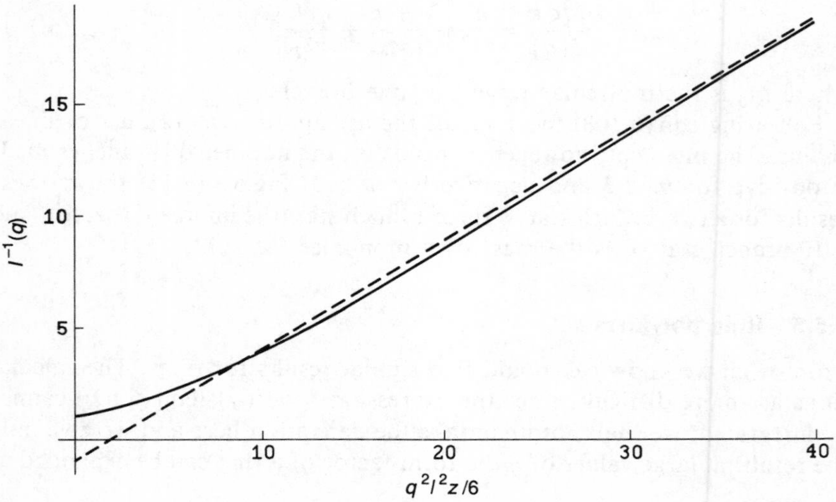

Fig. 6.19 $I^{-1}(q)$ for a ring molecule as function of $(q^2l^2/z/6)$; the straight dotted line is the asymptote.

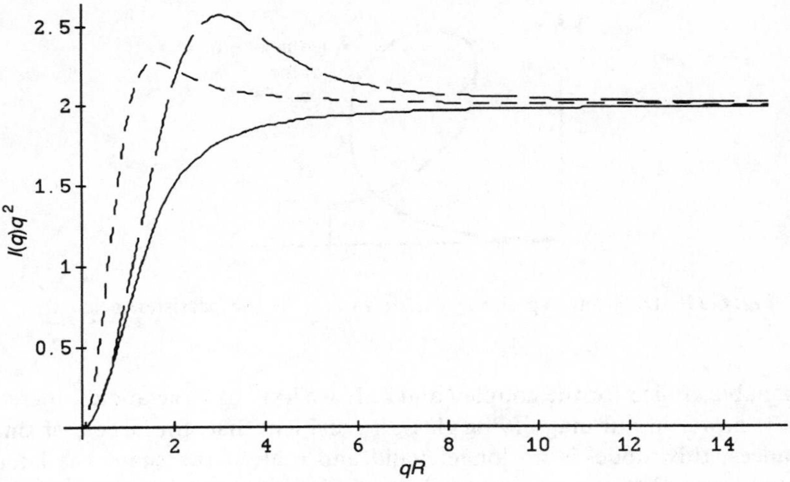

Fig. 6.20 Kratky plot of $I(q)\ q^2$ vs qR for a linear chain l(—), a ring (— — —), and a four-branch star (---).

studied (linear, stars, and rings) as functions of $(q^2\overline{l^2}/6/z$, where z is either the degree of polymerization, or the length of a branch in the case of stars. All the curves have the same asymptote but the linear chain always increases while the ring and the star go through a maximum around $qR = 1$ and decrease slowly afterwards. This can be explained qualitatively by saying that rings and stars have a more compact structure than the linear chain. The number of pairs at a distance of the order of R is larger than in a linear chain. Since the scattering in a given range of q is proportional to the number of pairs at a distance R such that $qR = 1$ the intensity in this range will be higher and thus will go through a maximum.

6.5.6 Determination of the persistence length

As we have seen, as long as chain molecules are of sufficient molecular weight and the values of q are sufficiently low the scattering diagram can be interpreted safely by the Debye form factor (eqn (6.73)) or its modifications taking into account the architecture of the chains (branches or rings). We have said that the chains are fractal objects with dimension 2 but, as we know, there is, for each object, a lower limit of the range of validity of the scaling laws. We can put this another way: when we say that the radius of gyration of a Gaussian chain is $\overline{R^2} = l^2z/6$ the choice of l and z is arbitrary. If z is divided by x then l must be multiplied by $\sqrt{x}$ and none of the experiments we have described can help to select a

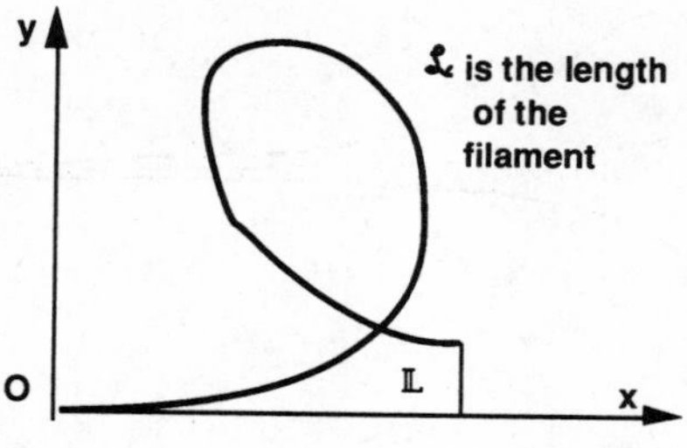

Fig. 6.21 Diagram explaining the definition of the persistence length.

reasonable choice for the couple l and z. If we look at large angles, increasing the power of our magnifying glass, it is evident that, for large q or small distances, this model is no longer valid and that, if the chain has lateral dimensions which can be neglected, it will be more similar to a rod than to a coil. In order to describe such chains theoreticians have introduce the concept of rigid chains: the Kratky and Porod (1949*a*, *b*) worm-like chain which uses as parameters the length of the chain $\mathcal{L}$ and what is called the persistence length $\mathbb{L}$. This persistence length can be described qualitatively as in Fig. 6.21

We start from the origin following the direction Ox and plot one conformation of the chain. The projection of its end is on the axis Ox at a distance $\mathbb{L}$ from the origin and we take the average value of $\mathbb{L}$ for all the conformations of the chain subject to the condition that the initial slope is zero. $\mathbb{L}$ will depend on the contour length (the length of the filament) $\mathcal{L}$ and increases with it but reaches a maximum value $\mathbb{L}^*$ when $\mathcal{L}$ goes to infinity. Roughly speaking one can say that if $\mathcal{L} < \mathbb{L}^*$ chain behaves as a rod and if $\mathcal{L} > \mathbb{L}^*$ as a Gaussian chain. One can express the radius of gyration of such a chain as function of $\mathcal{L}$ and $x = \mathcal{L}/\mathbb{L}^*$ by the relation (Benoît and Doty 1953)

$$\frac{\overline{R^2}}{\mathbb{L}^{*2}} = \frac{x}{3} - 1 + \frac{2}{x} - \frac{2}{x^2}\,[1 - \exp(-x)]. \tag{6.113}$$

When $\mathcal{L} > \mathbb{L}^*(x \gg 1)$ this simplifies to

$$\overline{R^2} = \frac{1}{6}\,2\mathcal{L}\,\mathbb{L}^* \tag{6.114}$$

and when $\mathcal{L} < \mathbb{L}^*(x \ll 1)$ to

$$\overline{R^2} = \frac{\mathcal{L}^2}{12}. \tag{6.115}$$

In the first case, we recover the Gaussian behaviour $\overline{R^2} \sim \mathcal{L}$ and in the second the behaviour of a rod $\overline{R^2} \sim \mathcal{L}^2$. It is evident that there is a crossover in the curve: scattered intensity $I(q)$ as a function of q, in the q range where the transition between the asymptotic behaviour of a chain and of a rod occurs. In the first case $I(q)$ is proportional to q^{-2} but, when we look at distances smaller than $\mathbb{L}^*$, $I(q)$ is proportional to q^{-1} (see Fig. 6.22).

This crossover has been shown by des Cloizeaux (1973) to occur between $q\mathbb{L}^* = 0.25$ and 1.5 but its position is too imprecise to be used quantitatively. Moreover the model assumes no lateral dimensions of the chain. If the persistence length is of the order of magnitude of the diameter it is impossible to see this crossover and one has to use computer calculations to interpret $P(q)$.

One important problem for the experimenter is to define precisely the statistical element which describes a polymer molecule. The first question one can ask is what can we learn from experiments on these molecules. If one looks carefully at all we have said it is clear that only two experimental quantities are available: the molecular weight and the radius of gyration. Even the measurement of the slope of the asymptote at large angles in the Zimm representation which is supposed to give $\overline{l^2}$ gives $\overline{R^2}/M$, i.e. the ratio of two already known quantities. From the molecular weight one can determine the number of monomer units and from it the length of the completely extended chain. We usually call this the contour length, $\mathcal{L}$ which is z time the length of a monomer. Kuhn (1934), in order to determine a value of l which is not model-dependent, wrote the two relationships

$$\overline{L^2} = 6\overline{R^2} = l^2 z \text{ and } \mathcal{L} = zl \tag{6.116}$$

and called the value of l which satisfies them the statistical element. This is the most commonly used method. It simply requires that the sum of the

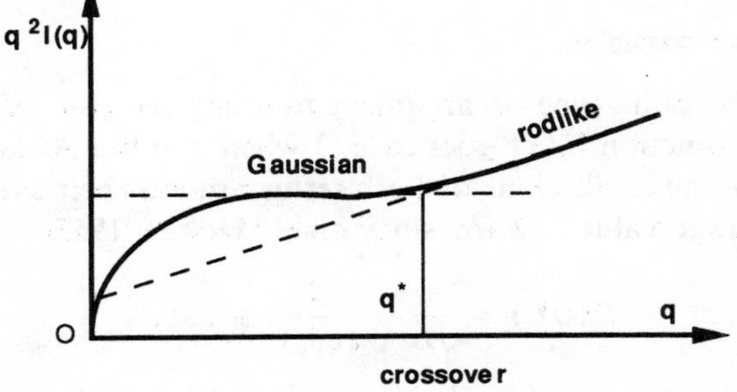

Fig. 6.22 Kratky plot for a Gaussian chain becoming rod-like at small distances.

length of the statistical elements building the chain is equal to the length of the completely extended molecular chain.

A second method uses the worm-like chain model. Looking at eqn (6.114) it is clear that, for long chains, the end to end distance is given by $\overline{L^2} = 2\mathbb{L}^*\mathcal{L}$. Since we have assumed that one can determine $\mathcal{L}$ it is evident that one can obtain the persistence length $\mathbb{L}^*$ and characterize the rigidity of the chain by this quantity which, in fact, is half the Kuhn statistical element.

Quite often in this book we define a 'monomer' as a part of a polymer having the same volume as the solvent. This is done in order to facilitate the calculations. One could, at first sight, assume that this choice obliges us to define the statistical element as the length of this unit. This is not correct, since if one looks carefully at all the developments one sees that the calculation of the form factors is completely independent of the choice we make for the statistical element and the number of monomeric units in the chain.

Another method for defining the rigidity of the chain is to use for z the number of C–C bonds in the skeleton and to define l by the relation $l^2 z = \overline{L^2} = 6\overline{R^2}$. If the C–C bonds were universal joints one should obtain 1.54 Å. If the chain is a free rotating chain one should find

$$\overline{L^2} = zl^2 \frac{1 + \cos\theta}{1 - \cos\theta}$$

where θ is the supplement of the valence angle. In fact, the experiment values of $\overline{L^2}$ are usually larger than this value (called L_f^2) because of the short range interactions between neighbouring segments and in the *Polymer Handbook* (Bandrupt and Immergut 1975) the rigidity is characterized by the ratio $\sigma^2 = \overline{L^2}/L_f^2$ which, for the majority of the vynilic polymer is of the order of two. This is only valid in theta solvents since it is only in these conditions that the Gaussian statistics can be used.

6.5.7 Rod-like particles

We can use the same type of argument to study the case of the rod. In eqn (6.61) the function $S_i(x)$ goes to $\pi/2$ when $x \to \infty$. At large q values one can ignore the oscillations of both terms around their average values. Using the average value 1/2 for $\sin^2 x$ gives (Holtzer 1955)

$$P(qL) = \frac{\pi}{qL} - \frac{2}{(qL)^2} + \dots \quad (6.117)$$

We use now the quantity $b_\mathrm{v} = b_1/v_1 - b_0/v_0$ which is the contrast factor per unit volume and write for the scattering intensity per unit volume

$$\frac{b_v{}^2 c V}{I(q)} = \frac{N_A}{v_s^2}\left[\frac{qL}{\pi M} + \frac{2}{\pi^2 M}\right]. \tag{6.118}$$

This result is similar to the result of eqn (6.109). At large q, $I^{-1}(q)$ is a linear function of q and its slope (in the representation $I^{-1}(q)$ as a function of q) depends only on the mass per unit length (M/L) and is independent of the length of the rod. The second term provides a method for measuring the molecular weight which has rarely been used. If, on the rod there are some accidents, like a change of direction, a universal joint or a branching point, only the second term will be modified.

Many chains become rigid Porod–Kratky chains when q^{-1} is large and $I^{-1}(q)$ becomes proportional to q. If the lateral dimensions are small z compared to the average radius of curvature of the chain, the portion of chain seen in the magnifying glass of Section 6.4 looks like a rigid rod and $I^{-1}(q)$ is proportional to q^{-1}. In the Kratky plot one will therefore see a change of behaviour and a crossover point which allows the measurement of the persistence length.

6.5.8 The case of two-dimensional objects

Using our scaling arguments we have been able to show that two-dimensional objects scatter at high q values as $1/q^2$ but it would be interesting to know the coefficient which has to be used since we know that this law is, as for one-dimensional objects, universal. The simplest way is to take one example of such an object and to determine this coefficient for this object. It has been shown (eqn (6.57)) that for an infinitely thin sphere $P(q)$ has (if one forgets the oscillations), the form

$$P(q) = \left[\frac{\sin qR}{qR}\right]^2 \approx \frac{1}{2q^2R^2}. \tag{6.119}$$

We recover, as expected, the q^{-2} law. If we now write the scattered intensity using the mass of a shell M and its surface S we obtain

$$\frac{q^2 I(q)}{cV} = b_v^2 v_s^2 \frac{M}{2R^2} = b_v^2 v_s^2 2\pi \frac{M}{S} \tag{6.120}$$

and we see that the scattered intensity depends only on M/S which is defined as the mass per unit surface. One should not forget that, if q^{-1} becomes larger than the film thickness, the scatterer cannot be considered as a two-dimensional object and this approximation is no longer valid.

6.5.9 The case of three-dimensional objects

We shall utilize the same method using the sphere as an example. For spheres we have shown in eqn (6.50) and in Appendix 3 that $I(q)$ can be written as

$$I(q) = b_v^2 N V_1^2 \frac{9}{(qR)^6} (\sin qR - qR\cos qR)^2 \tag{6.121}$$

where V_1 is the volume of the spheres and R their radius. Evaluating the square and replacing $\cos^2 qR$ and $\sin^2 qR$ by 1/2 (their average values) and $\sin qR \cos qR = 1/2 \sin 2qR$ by 0 gives

$$I(q) = b_v^2 N V_1^2 \frac{9}{2} \left\{ \frac{1}{q^4 R^4} + \frac{1}{q^6 R^6} \right\}. \tag{6.122}$$

Expressing V_1^2/R^4 as function of the surface S of a sphere one obtains for the first term

$$q^4 I(q) = b_v^2 2\pi N S = b_v^2 2\pi S_T \tag{6.123}$$

where $S_T = NS$, is the total surface of the spheres present in the sample. This verifies the scaling argument we gave in the preceding paragraph which was based on the fact that, at high q values, the scattering is proportional to the total surface of the sample. Moreover it gives the numerical coefficient of the q^{-4} Porod (1951) law.

6.6 THE EFFECT OF POLYDISPERSITY

Until this point we have considered the scattering by one molecule or an ensemble of N identical and independent molecules. The effect of the interactions will be studied in the next chapter but, since it is very rare to have a system made of rigorously identical molecules or particles, it is important to consider the effect of heterogeneity of dimensions.

6.6.1 Some definitions for polydisperse macromolecular systems

We will assume that all molecules are built from the same monomer and differ only in their degree of polymerization. Using discrete notation we introduce the probability $\nu_i = N_i/\Sigma N_i$ of having a molecule of degree of polymerization z_i; using continuous notation we introduces the probability $w(z)\,dz$ for a degree of polymerization between z and $z + dz$. This allows us to introduce different averages of the degree of polymerization z, averages which are related to the different moments of the molecular weight distribution. The number average degree of polymerization is defined by

$$\langle z\rangle_{\mathrm{n}} = \frac{\sum N_i z_i}{\sum N_i} = \sum \nu_i z_i = \int w(z) z \,\mathrm{d}z. \tag{6.124}$$

The weight average degree of polymerization is defined as

$$\langle z\rangle_{\mathrm{w}} = \frac{\sum N_i z_i^2}{\sum N_i z_i} = \frac{\sum \nu_i z_i^2}{\sum \nu_i z_i} = \frac{\int w(z) z^2 \,\mathrm{d}\nu}{\int w(z) z \,\mathrm{d}z}. \tag{6.125}$$

This is, in fact, the ratio of the second moment to the first moment. We introduce also what has been called traditionally the 'z average' degree of polymerization defined as the ratio of the third moment to the second

$$\langle z\rangle_z = \frac{\sum N_i z_i^3}{\sum N_i z_i^2} = \frac{\sum \nu_i z_i^3}{\sum \nu_i z_i^2} = \frac{\int w(z) z^3 \,\mathrm{d}z}{\int w(z) z^2 \,\mathrm{d}z}. \tag{6.126}$$

Sometimes it is preferable to use concentration rather than the number of molecules to characterize the composition of the system; since $c_i \approx n_i z_i$ we can rewrite the averages as

$$\langle z\rangle_{\mathrm{n}} = \frac{\sum c_i}{\sum \frac{c_i}{z_i}} \quad \langle z\rangle_{\mathrm{w}} = \frac{\sum c_i z_i}{\sum c_i} \quad \langle z\rangle_z = \frac{\sum c_i z_i^2}{\sum c_i z_i}. \tag{6.127}$$

This list can be extended but these moments are the classical ones and are the most used in the literature. It can be shown that always

$$\langle z\rangle_{\mathrm{n}} \le \langle z\rangle_{\mathrm{w}} \le \langle z\rangle_z \tag{6.128}$$

while the terms are equal only for monodisperse systems.

6.6.2 An example of the effect of polydispersity

Tradition, arising from the experimental possibilities, characterizes the heterogeneity of a sample by the ratio $r = \langle z\rangle_{\mathrm{w}}/\langle z\rangle_{\mathrm{n}}$. The more polydisperse the system the larger this number. Samples with $r = 1.05$ or less are, in fact, possible to prepare. In classical radical polymerization corresponding to industrial product, r is of the order of 1.2 to 2. For Ziegler–Natta polymerization of ethylenic monomers r can reach values of 5 or even more. Each experimental technique which allows the determination of the molecular weight gives, for a polydisperse sample, a specific type of average. It is easy to show that osmotic pressure or any other method of measurement of the colligative properties of solutions gives the number average molecular weight $\langle z \rangle_n$. The scattering technique measures the weight average molecular weight. As we have seen, by neutron scattering at zero angle and without interactions, one obtains

$$I(0) = \bar{b}^2 N z^2 \approx cM. \tag{6.129}$$

For a polydisperse system

$$I(0) = \bar{b}^2 \sum N_i z_i^2 \approx c \frac{\sum N_i z_i^2}{\sum N_i z_i} = c\langle z \rangle_{\mathrm{w}} \tag{6.130}$$

since $c \approx \sum N_i z_i$. It is important, in order to study the effect of polydispersity on the scattering results to be able to characterize the polydispersity by a molecular weight distribution. If one works with continuous notation the problem is to know which distribution, or more precisely which function, $w(z)$, one has to use. For practical reasons we use quite often what is called the Schultz–Zimm distribution (Schultz 1939; Zimm 1948*b*) which is defined as

$$w(u)\mathrm{d}u = \frac{k^k}{\Gamma(k)} u^{k-1} \exp(-ku)\mathrm{d}u \tag{6.131}$$

where $\Gamma(k)$ is the gamma function, k a parameter which characterizes the width of the distribution and u the ratio $z/\langle z \rangle_{\mathrm{n}}$. It is easy to verify that this function is normalized to unity and that

$$\int_0^\infty w(u) z \mathrm{d}u = \langle z \rangle_{\mathrm{n}}. \tag{6.132}$$

Moreover

$$\langle z \rangle_{\mathrm{w}} = \langle z \rangle_{\mathrm{n}} \frac{\int_0^\infty w(u) z^2 \mathrm{d}u}{\int_0^\infty w(u) z \mathrm{d}u} = \langle z \rangle_{\mathrm{n}} \frac{k+1}{k} \tag{6.133}$$

showing that the parameter r used for characterizing the polydispersity is equal to $1 + 1/k$. The function $w(c)$ (which corresponds to the kind of curves one obtains from size exclusion chromatography) is shown in Fig. 6.23.

We should point out that for $k = 1$ the function $w(u)$ takes the form e^{-u}; it is what is called the most probable distribution and corresponds to

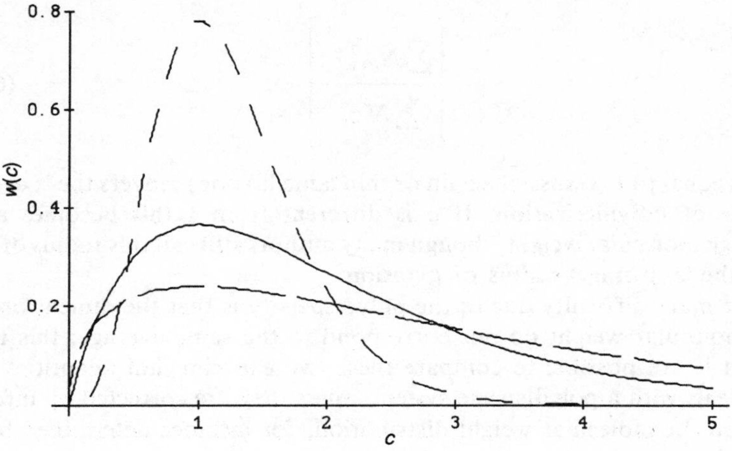

Fig. 6.23 Polydispersity for the 'Zimm–Schultz' distribution for different values of k: (——) $k = 1$, $\langle z\rangle_w/\langle z\rangle_n = 2$; (----) $k = 4$, $\langle z\rangle_w/\langle z\rangle_n = 1.25$; (——) $k = 0.5$, $\langle z\rangle_w/\langle z\rangle_n = 3$.

radical polymerization ending by disproportionation and also to random degradation of a linear polymer of large molecular weight; the numbers $\langle z_n\rangle$, $\langle z_w\rangle$, $\langle z_z\rangle$ are thus in the ratio 1 : 2 : 3.

6.6.3 The radius of gyration in polydisperse systems

These definitions being established it is easy to write the expression for the radius of gyration in a polydisperse system. Starting from eqn (6.14) and writing the scattered intensity by adding the intensity scattered by each of the constituents one obtains

$$I(q) \approx \sum N_i z_1^2 - \frac{q^2}{3} \sum N_i z_1^2 \overline{R_i^2} + \ldots . \tag{6.134}$$

which defines the apparent radius of gyration as

$$\overline{R^2} = \frac{\sum N_i z_1^2 \overline{R_i^2}}{\sum N_i z_1^2} . \tag{6.135}$$

Let us now assume that the radius of gyration obeys a power law

$$\overline{R^2} = c^2 z^a \tag{6.136}$$

the radius of gyration one measures is the radius of gyration of a hypothetical molecule having as degree of polymerization

$$\left[\frac{\sum N_i z_1^{2+a}}{\sum N_i z_1^2}\right]^{\frac{1}{a}}. \tag{6.137}$$

If a is equal to 1 (Gaussian chain or thin lamella) one recovers the 'z average' degree of polymerization. If a is different from 1 this becomes a new average molecular weight, though many authors still call this radius of gyration the 'z average' radius of gyration.

The main difficulty due to the polydispersity is that the dimensions and the molecular weight do not correspond to the same average; this means that it is not possible to compare these two experimental quantities when one deals with a polydisperse system unless they are corrected by information on the molecular weight distribution, for instance determined by size exclusion chromatography.

6.6.4 The effect of polydispersity at high q values

Since we have said that polydispersity does not affect the intermediate domain it is important to see what happens for a polydisperse system in this range of q values. We take as an example, the case of Gaussian chains. For a polydisperse system we write the Debye equation as

$$I(q) = \bar{b}^2 \sum N_i z_i^2 P_i(qR_i) \tag{6.138}$$

with

$$P_i(qR_i) = \frac{2}{\lambda^2 z_i^2}[\lambda z_i - 1 + \exp(-\lambda z_i)]$$

and

$$\lambda = \frac{q^2 \overline{l^2}}{6}.$$

Using the asymptotic form of the form factor (eqn (6.103)) we obtain, after separation of the two terms

$$\frac{I(q)}{\bar{b}^2 \varphi} = \frac{1}{\sum N_i z_i}\left[\frac{2}{\lambda}\sum N_i z_i - \frac{2}{\lambda^2}\sum N_i\right] \tag{6.139}$$

where φ is the total volume fraction occupied by the polymer, or

$$\frac{\bar{b}^2 \varphi}{I(q)} = \frac{\lambda}{2} + \frac{1}{2}\frac{\sum N_i}{\sum N_i z_i} = \frac{\lambda}{2} + \frac{1}{2\langle z \rangle_n}. \tag{6.140}$$

Not only do we recover the result that the slope of the asymptote is independent of z but additional information is also obtained: the intercept of the asymptote with the y-axis in a Zimm representation gives the reciprocal of the number average molecular weight. This makes it possible, if the chains are linear, to measure directly on one sample not only the weight average molecular weight but also its number average value if data are available over a wide enough q range.

In the preceding section we have seen that this intercept depends also on the geometry of the polymer (presence of branches or ring structure); it is therefore an interesting parameter which should be measured more often. In order to summarize this discussion we can calculate the expression for the form factor for a polydiperse Gaussian chain obeying the Zimm–Schultz distribution (eqn (6.127)). In Fig. 6.24 the inverse scattering curve for polymers having different polydispersity are compared. The calculation is straightforward and could be made easily by the reader. We give its result here. Instead of giving the structure factor we give the quantity $\sum N_i z_i^2 - P_i(q)$ evaluated using eqn (6.127) and the quantity $\lambda = q^2\overline{l^2}/6$

$$I(q) = N\bar{b}^2 \frac{2}{\lambda^2}\left[\lambda z_n - 1 + \left(1 + \frac{\lambda z_n}{k}\right)^{-k}\right] \qquad (6.141)$$

with

$$\frac{z_w}{z_n} = 1 + \frac{1}{k}\,.$$

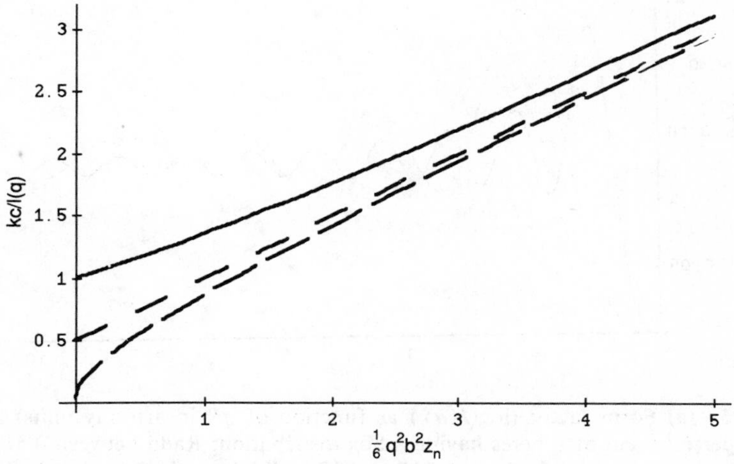

Fig. 6.24 Inverse scattered intensity from a monodisperse system of chains (——) and a polydisperse system having the same $\langle z_n \rangle$ but different polydispersities: (---) $z_w/z_n = 2$; (———) $z_w/z_n = 50$.

The curves are in fact, except for the low q range, rather insensitive to polydispersity, which has to be taken into consideration only when making very precise measurements. We can also note that the curves are not far from straight lines over the whole q range; this justifies the use of the Zimm representation by experiments, since the only curve one can parametrize easily is the straight line.

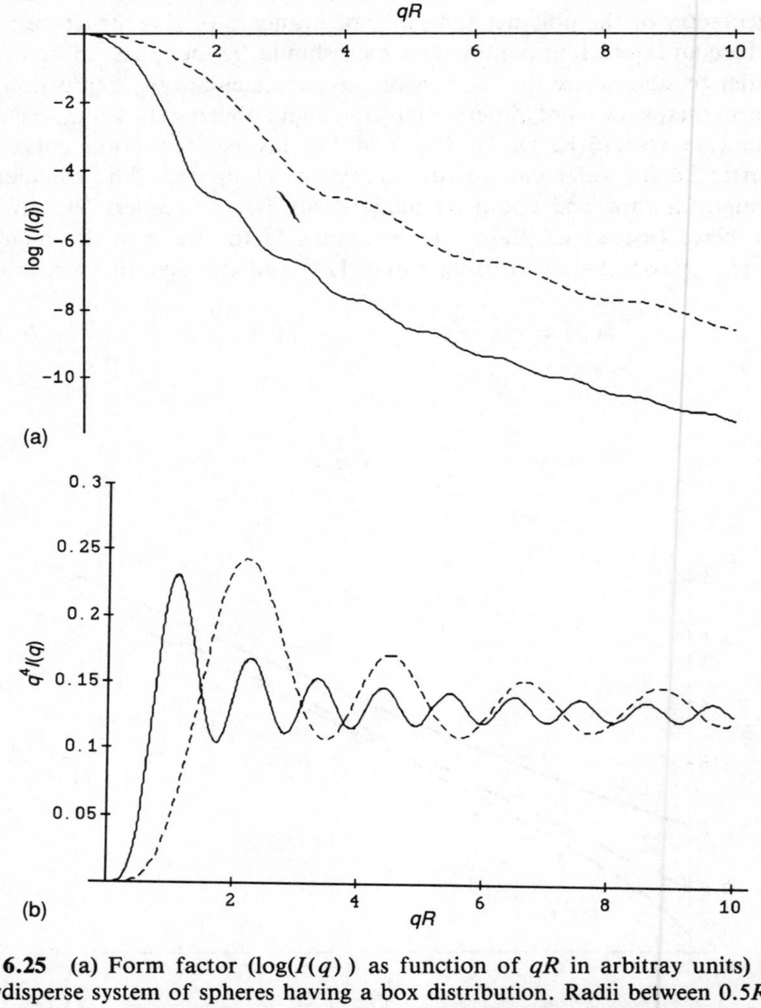

Fig. 6.25 (a) Form factor (log($I(q)$) as function of qR in arbitray units) for a polydisperse system of spheres having a box distribution. Radii between $0.5R$ and $1.5R$ (Dotted line). Radii between $0.33R$ and $3R$ (solid line). (b) Same result plotted as $q^4 I(q)$ vs qR (in arbitrary units). Porod regime appears faster for the more polydisperse polymer.

6.6.5 Polydisperse spheres

In order to show a case where polydispersity is important, let us consider the case of a suspension of spheres. For a monodisperse system of spheres we have evaluated the form factor (eqn (6.50)). This gives for the scattering intensity

$$I(q) = b_v^2 N V^2 \left[3\frac{\sin qR - qR\cos qr}{(qR)^3} \right]^2. \tag{6.142}$$

Since the volume replaces the degree of polymerization in continuous notation one has to utilize the coherent contrast factor per unit volume b_v and not, as we have usually done, the contrast factor per scattering centre $\bar{b}$. The evaluation of the scattering is made replacing N by either Nv_i or $Nw(R)\mathrm{d}R$ and making the summation or the integration. It is not worthwhile repeating these calculations and we give only two examples which are illustrated by Fig. 6.25(a) and (b). Comparing these figures to the figures for a monodisperse system (Fig. 6.6 (a–c)) it is clear that the oscillations disappear quickly for polydisperse systems.

SUGGESTED FURTHER READING

Guinier and Fournet (1955).
Glatter and Kratky (1982).
Lindner and Zemb (1991).

7

Interacting systems

Part 1 Zero angle scattering

7.1 DENSITY AND CONCENTRATION FLUCTUATIONS–ONE OR TWO COMPONENTS

7.1.1 Introduction

Up to this point we have discussed the scattering in general terms and shown that the expression for coherent scattering at any angle from a system containing only one species of molecules was given by the quantity

$$I(q) = b^2 \sum_i^N \sum_j^N \langle \exp(-\mathrm{i}\boldsymbol{q}\cdot\boldsymbol{r}_{ij}) \rangle \tag{7.1}$$

where b is the coherent scattering length of the N molecules and each pair is separated by the vector $\boldsymbol{r}_{ij}$. As explained in the Chapter 4, this can also be written as

$$I(q) = b^2 \iint_V \langle n(\boldsymbol{r}) n(\boldsymbol{r}') \rangle \exp[-\mathrm{i}\boldsymbol{q}\cdot(\boldsymbol{r}-\boldsymbol{r}')]\,\mathrm{d}\boldsymbol{r}\,\mathrm{d}\boldsymbol{r}' \tag{7.2}$$

where n is the local density of scattering centres: $\bar{n} = N/V$. In order to eliminate the anomaly at $q = 0$ (see Chapter 4) it is better to write

$$I(q) = b^2 \iint_V \langle \Delta n(\boldsymbol{r}) \Delta n(\boldsymbol{r}') \rangle \exp[-\mathrm{i}\boldsymbol{q}\cdot(\boldsymbol{r}-\boldsymbol{r}')]\,\mathrm{d}\boldsymbol{r}\,\mathrm{d}\boldsymbol{r}' \tag{7.3}$$

with $\Delta n(\boldsymbol{r}) = n(\boldsymbol{r}) - \bar{n}$ and $\Delta n(\boldsymbol{r}') = n(\boldsymbol{r}') - \bar{n}$.

The first section of this chapter will be devoted to the discussion of the scattering at zero angle, i.e. the quantity obtained when $q = 0$ in the preceding equation

$$I(q=0) = b^2 \int_V \int_{V'} \langle \Delta n(\boldsymbol{r}) \Delta n(\boldsymbol{r}') \rangle\,\mathrm{d}\boldsymbol{r}\,\mathrm{d}\boldsymbol{r}'. \tag{7.4}$$

Let us assume that we want to define the quantity $\overline{\Delta N^2}$ which is the average of the square of the fluctuations in the volume V. We write

$$\overline{\Delta N^2} = \left\langle \left[\int_V \Delta n(\mathbf{r})\,\mathrm{d}\mathbf{r} \right]^2 \right\rangle = \left\langle \int_V \Delta n(\mathbf{r})\,\mathrm{d}\mathbf{r} \int_{V'} \Delta n(\mathbf{r}')\,\mathrm{d}\mathbf{r}' \right\rangle$$

$$= \int_V \int_{V'} \langle \Delta n(\mathbf{r})\,\Delta n(\mathbf{r}') \rangle\,\mathrm{d}\mathbf{r}\,\mathrm{d}\mathbf{r}' \qquad (7.5)$$

which is exactly the integral of the preceding equation. This leads us to the conclusion that the scattered intensity at zero angle is proportional to the mean square value of the fluctuations of the scattering centres in the scattering volume (Einstein 1910).

$$I(q = 0) = b^2\,\overline{\Delta N^2}. \qquad (7.6)$$

This expression is very simple since the neutron scattering problem at $q = 0$ is reduced to the problem of the calculations of the fluctuations in a system in equilibrium. In the first part of this chapter we shall first discuss the case of a one-component system, and extend the method to the case of an incompressible mixture of two constituents. In the second section we shall use a more general method to treat the problem of a multicomponent incompressible system. This requires some knowledge of fundamental thermodynamics which can be found in Appendix 2. In the last section we shall say a few words about multicomponent compressible systems which form the general case.

7.1.2 A one-component system

In order to show how $\overline{\Delta N^2}$ is related to thermodynamic variables let us make the following mental experiment (Cabannes 1929). In a system supposed to scatter neutrons we introduce a cylinder, enclosing N molecules with a piston, as schematized in Fig. 7.1. Instead of looking at the fluctuations of the number of molecules inside the cylinder we consider the

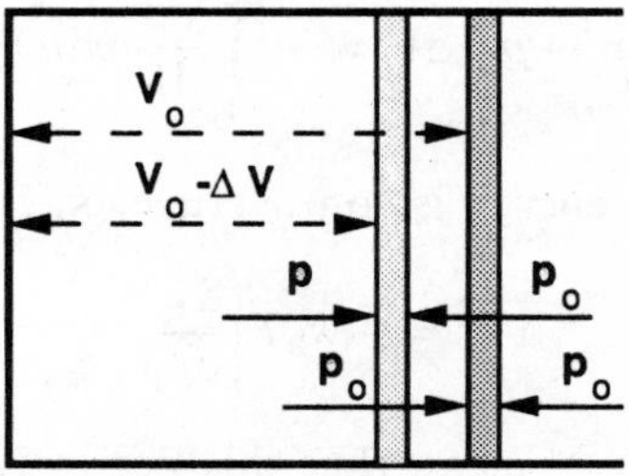

Fig. 7.1 Explanation of the density fluctuations in a one-component system using a cylinder and a piston.

fluctuations of volume at constant number of molecules. Since the number of molecules per unit volume $n = N/V$ is, to a first approximation, constant

$$\frac{\Delta N}{N} = \frac{\Delta V}{V}$$

or

$$\overline{\Delta V^2} = \frac{\overline{\Delta N^2}}{n^2}. \tag{7.7}$$

We can therefore evaluate $\overline{\Delta V^2}$ and, from it, recover $\overline{\Delta N^2}$ and the scattered intensity. Under the influence of the Brownian motion, since the pressure on the right face of the piston is p_0 and on the left side p, the position of the piston will fluctuate and the volume v inside it will change continuously giving rise to the fluctuations we want to determine. This piston has one degree of freedom and, applying the theorem of equipartition of energy, its energy is $k_B T/2$ where k_B is the Boltzmann constant and T the temperature. The problem now is to evaluate the value of the fluctuations of volume which correspond to this energy. In order to push the piston from its original position ($v = v_0$) to the position where $v = v_0 - \Delta v$, work w must supplied

$$w = -\int_{V_0}^{V_0 - \Delta V} (p - p_0)\,\mathrm{d}V. \tag{7.8}$$

Replacing the instantaneous value of the volume V by the quantity $v = V_0 - V$, the instantaneous value of p is given by

$$p = p_0 + \frac{\mathrm{d}p}{\mathrm{d}V}\mathrm{d}V \tag{7.9}$$

where $(\mathrm{d}p/\mathrm{d}V)^{-1}$ is proportional to the isothermal compressibility. Replacing $p - p_0$ in the integral by its value in eqn (7.9) we obtain

$$w = -\int_{V_0}^{V_0 - \Delta V} (p - p_0)\,\mathrm{d}V = -\int_0^{\Delta V} \frac{\mathrm{d}p}{\mathrm{d}V} v\mathrm{d}v = -\frac{\mathrm{d}p}{\mathrm{d}V}\tfrac{1}{2}\Delta V^2. \tag{7.10}$$

Remembering that this energy is, on average, $k_B T/2$ we obtain

$$\overline{\Delta V^2} = -k_B T\left(\frac{\mathrm{d}V}{\mathrm{d}p}\right)_{T,N} \tag{7.11}$$

where the subscripts T, N mean that this derivative is taken at constant temperature and constant number of molecules. The result of eqn (7.10) is obvious if one thinks about the analogy between a spring and our device.

In the case of a spring one would get for the energy $1/2k_{\mathrm{el}}\overline{(\Delta x^2/2)}$ where k_{el} is the elastic constant of the spring. Introducing now the isothermal coefficient of compressibility

$$\beta = -\frac{1}{V}\left(\frac{\mathrm{d}V}{\mathrm{d}p}\right)_T \tag{7.12}$$

we obtain, from eqns (7.6) and (7.11)

$$I(q=0) = b^2\frac{N^2}{V}k_{\mathrm{B}}T\beta = b^2 Nnk_{\mathrm{B}}T\beta. \tag{7.13}$$

This shows clearly the relationship between scattering at zero angle and the compressibility of the system.

7.1.3 Incompressible solutions

It is easy to use the same method for an incompressible solution. We have only to replace the piston by a semipermeable membrane transparent to the solvent and opaque to the solute. (See Fig. 7.2.) We want to describe the fluctuations of the number of solvent molecules in a given volume, writing eqn (7.6) but, this time N will be called N_0 and is the number of molecules of solvent in the volume V; $\bar{b}$ is not the coherent scattering length of the solute but the difference between its scattering length and the scattering length of the same volume of solvent. The pressure acting on the system is what is called the osmotic pressure π which is defined by the relationship

$$\mu_0 - \mu_0^0 = -\pi v_0 \tag{7.14}$$

where μ_0 is the chemical potential of the solvent in the solution and μ_0^0

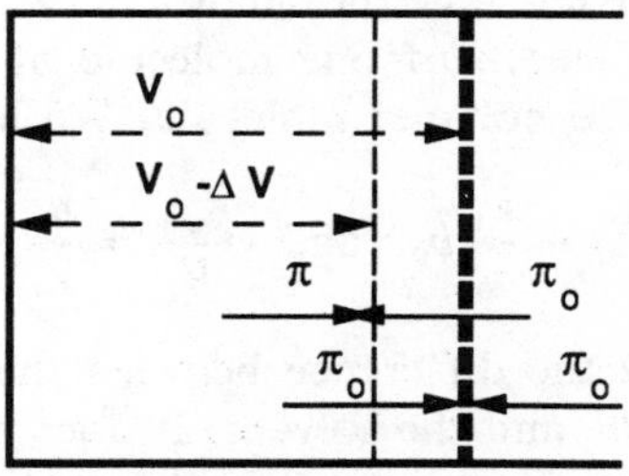

Fig. 7.2 The same diagram but this time for a solution with a semipermeable piston.

the chemical potential of the pure solvent which has v_0 as partial molar volume.

Replacing the pressure in eqn (7.11) by the osmotic pressure we arrive at the following equation

$$\overline{\Delta V^2} = -k_B T \left(\frac{dV}{d\pi} \right)_{T,N} = -\frac{k_B T}{\left(\frac{d\pi}{dV} \right)_{T,N}}. \tag{7.15}$$

In order to obtain the result in more practical units we introduce the volume fraction

$$\varphi = \frac{N_1 V_1}{V}$$

where V_1 is the volume of one molecule of solute and V the volume of the solution. Differentiating eqn (7.15) gives

$$\frac{\Delta N_1}{N_1} = \frac{\Delta V}{V}. \tag{7.16}$$

This allows us to write the scattering intensity as function of $\overline{\Delta V^2}$ (Putzeys and Brosteaux 1935; Debye 1944) as:

$$I(q=0) = b^2_{\text{app}} \frac{k_B T \varphi V}{\left(\frac{\partial \pi}{\partial \varphi} \right)_T V_1^2}. \tag{7.17}$$

Applying this formula to experimental results requires some transformations. Since the use of intensive variables is always more convenient, we shall introduce the scattered intensity per unit volume $I(0)/V$, dividing the total intensity by the volume. The coherent scattering length b_{app} which has been used represents the difference between the coherent scattering length $\mathscr{b}_1$ of the dissolved molecule and the coherent scattering length of the same volume of solvent, i.e. the quantity: $\mathscr{b}_1 - V_1 b_0/v_0$, where b_0 is the coherent scattering length of one molecule of solvent of volume v_0 and V_1 the volume of the polymer molecule. We have therefore

$$b_{\text{app}} = \mathscr{b}_1 - \frac{V_1}{v_0} b_0 \quad \text{and} \quad \frac{b_{\text{app}}}{V_1} = \frac{\mathscr{b}_1}{V_1} - \frac{b_0}{v_0}. \tag{7.18}$$

The quantity b_{app}/V_1 is the difference between the scattering length per unit volume of the solute and the solvent. It does not depend on the size of molecules and will be defined as b_v in Appendix 3. This allows us to write eqn (7.17) as:

$$\frac{1}{V} I(q=0) = b_{\mathrm{v}}^2 \frac{k_{\mathrm{B}} T \varphi}{\dfrac{\partial \pi}{\partial \varphi}}. \tag{7.19}$$

Instead of volume fraction, we can use concentration as variable. The calculations are very similar and we give only the result

$$\frac{1}{V} I(q=0) = \{(v_1^{\mathrm{s}})^2 b_{\mathrm{v}}^2\}\, k_{\mathrm{B}} T \frac{c}{\dfrac{\partial \pi}{\partial c}} \tag{7.19'}$$

where c is the concentration (mass of polymer per unit volume) and v_1^{s} its specific volume. This type of unit is less convenient for neutron scattering and is used mainly by people working in light scattering since, in that case, the experimental parameter $\mathrm{d}n/\mathrm{d}c$ corresponds to the term in the curly brackets in eqn (7.19′).

7.1.4 The example of a gas or a dilute solution

The first use of this formula has been its application to dilute gases and solutions of polymers. If the gas follows the perfect gas equation $pV = Nk_{\mathrm{B}}T$ then

$$\beta = -\frac{1}{V}\left(\frac{\partial V}{\partial p}\right)_T = p^{-1}. \tag{7.20}$$

Putting this value in eqn (7.13) gives

$$I(q=0) = Nb^2$$

which is the simple result one expects from independent molecules. Knowing the equation of state, and from it the compressibility, allows the scattering intensity to be evaluated for a real gas. Reciprocally, one can use scattering phenomena to determine the compressibility and from it the properties of the system. This technique works only for systems which are in thermodynamic equilibrium and cannot be applied to systems out of equilibrium like glasses or multiphase systems. We know from the theory of polymer solutions (Flory 1953) that we can expand the osmotic pressure as a function of the volume fraction, writing

$$\frac{\pi}{c} = k_{\mathrm{B}} T \left(\frac{1}{M_1} + A_2 c + A_3 c^2 + A_4 c^3 + \dots\right). \tag{7.21}$$

This is called a virial expansion and it is theoretically possible, from scattering data, to determine all the different coefficients. Writing the inverse of the scattered intensity per unit volume gives (see Appendix 3)

$$\frac{kc}{I(q=0)} = \frac{1}{M_1} + 2A_2c + 3A_3c^2 + 4A_4c^3 \tag{7.22}$$

k being proportional to b^2_{app}. The first virial coefficient is the molecular weight and this technique is one of the most reliable techniques for the determination of molecular weights. It can be used with neutrons but, for this purpose, the data must be calibrated absolutely, which is not a simple procedure. It is also necessary to extrapolate the experimental data to zero angle, which for some solid samples can be a problem. The formula which we have established is valid at any concentration. The scattered intensity depends on what has been called the osmotic compressibility $(\partial\pi/\partial\varphi)_T$, a quantity which plays an important role in the theory of semidilute solutions and gels.

7.2 FLUCTUATIONS IN MULTICOMPONENT SYSTEMS

7.2.1 Introduction

Up to this point we have considered systems, made of identical molecules (except for the case of a solution) and we would like to generalize the formula (7.6)

$$I(q=0) = b^2\,\overline{\Delta N^2}$$

to systems containing p different species of molecules. This has been done first by Kirkwood and Goldberg (1950), Stockmayer (1950), and presented in a more compact form by des Cloizeaux and Jannink (1980). Each species is characterized by its number of molecules $N_k (1 \leqslant k \leqslant p)$ and has a coherent scattering length b_k. Splitting the double summation of eqn (7.1) into p^2 double sums gives

$$I(q=0) = \sum_{k=1}^{p} b_k^2 \sum_i^{N_k} \sum_j^{N_k} \langle \exp(-\mathrm{i}\boldsymbol{q}\cdot\boldsymbol{r}_{i_k j_k}) \rangle$$
$$+2\sum_{k<g}^{p} \cdot \sum^{p} b_k b_g \sum_{i=1}^{N_k} \sum_{j=1}^{N_g} \langle \exp - (\mathrm{i}\boldsymbol{q}\cdot\boldsymbol{r}_{i_k j_g}) \rangle . \tag{7.23}$$

In order to simplify this equation we introduce the partial structure factors which have already been defined in Chapter 5

$$I(q=0) = \sum_{k=1}^{p} b_k^2 S_{kk}(q=0) + 2b_k b_g \sum_{k<g}^{p} \sum^{p} S_{gk}(q=0). \tag{7.24}$$

It is evident that, following eqn (7.6) the $S_{gk}(q = 0)$ are defined as

$$S_{kk}(0) = \sum_{i}^{N_k} \sum_{j}^{N_k} \overline{\Delta N_{i_k} \Delta N_{j_k}} \tag{7.25}$$

and

$$S_{gk}(0) = \sum_{i}^{N_k} \sum_{j}^{N_k} \overline{\Delta N_{i_g} \Delta N_{j_k}} \,. \tag{7.26}$$

This procedure can be summarized as follows: using the quantity

$$\mathcal{B} = \sum_{k=1}^{p} b_k N_k \tag{7.27}$$

the scattering intensity takes the form

$$I(q = 0) = \overline{\Delta \mathcal{B}^2} \tag{7.28}$$

which will be used later in this chapter. Since we shall soon need the formalism of vector analysis it is worth while introducing it now. The S_{gk} can be written as a square matrix of p rows and columns which will be called $[S(0)]$

$$[S(0)] = \begin{pmatrix} S_{11}(0) & S_{12}(0) & \dots & S_{1k}(0) & \dots & S_{1p}(0) \\ S_{21}(0) & S_{22}(0) & \dots & S_{2k}(0) & \dots & S_{2p}(0) \\ \dots & \dots & \dots & \dots & \dots & \dots \\ S_{k1}(0) & S_{k2}(0) & \dots & S_{kk}(0) & \dots & S_{kp}(0) \\ \dots & \dots & \dots & \dots & \dots & \dots \\ S_{p1}(0) & S_{p2}(0) & \dots & S_{pk}(0) & \dots & S_{pp}(0) \end{pmatrix} . \tag{7.29}$$

This matrix is symmetrical. One can also define the column vector $[\mathscr{b}]$ having the b_k as elements and the row vector $[\mathscr{b}]^T$, the transpose of $[\mathscr{b}]$ having also the b_k as elements. With this notation and using the definition of the matrix products, eqn (7.24) becomes

$$I(q = 0) = [\mathscr{b}]\,[S(0)]\,[\mathscr{b}]^T . \tag{7.30}$$

7.2.2 Evaluation of $\overline{\Delta N_i \Delta N_j}$

This evaluation is a pure problem of thermodynamics and the detailed calculation can be found in Appendix 2. We just give the result

$$\overline{N_i^2} - (\bar{N}_i)^2 = \overline{\Delta N_i^2} = k_B T \left(\frac{\partial \bar{N}_i}{\partial \mu_i}\right)_{V,T} \tag{7.31}$$

and

$$\overline{\Delta N_i \Delta N_j} = k_B T \left(\frac{\partial \bar{N}_i}{\partial \mu_j}\right)_{V,T,N_k} = k_B T \left(\frac{\partial \bar{N}_j}{\partial \mu_i}\right)_{V,T,N_k} . \tag{7.32}$$

The thermodynamic quantities which have been introduced are the derivatives of the number of molecules with respect to the chemical potentials μ at constant volume and temperature.

The problem is now theoretically solved since it suffices to write

$$S_{ij}(0) = k_B T \left(\frac{\partial \bar{N}_j}{\partial \mu_i}\right)_{V,T,N_k} \tag{7.33}$$

and to use eqn (7.24) to describe the scattering at zero angle. In fact the situation is less simple because, when we examine terms like $(\partial \bar{N}_j / \partial \mu_i)$ we realize that they do not correspond to measurable physical quantities. We have to proceed via two steps to arrive at a useful formula

$$(1) \text{ replace } \left(\frac{\partial \bar{N}_j}{\partial \mu_i}\right)_{V,T} \text{ by } \left(\frac{\partial \mu_i}{\partial N_j}\right)_{V,T}$$

and

$$(2) \text{ replace } \left(\frac{\partial \mu_i}{\partial N_j}\right)_{V,T} \text{ by } \left(\frac{\partial \mu_i}{\partial N_j}\right)_{p,T} .$$

These calculations are detailed in the appendix and we quote here only the results.

If we write the p by p matrix $[M]$ of the $\partial \bar{N}_i / \partial \mu_j$ (as we have done for the S_{ij}) we can show that the matrix inverse of $[M]$, $[M]^{-1}$, is the matrix of the $(\partial \mu_i / \partial \bar{N}_i)_{V,T}$ (the inverse matrix is such that $[M].[M]^{-1} = [1]$). From this result, and using eqn (7.33), we arrive at the simple formula

$$I(q=0) = k_B T [\mathscr{b}] \left[\left(\frac{\partial \bar{N}_j}{\partial \mu_i}\right)_{V,T} \right] [\mathscr{b}]^{\mathrm{T}} . \tag{7.34}$$

The second step of our procedure is now to transform $(\partial N_j/\partial\mu_i)_{V,T}$ into $(\partial\bar{N}_j/\partial\mu_i)_{p,T}$. This is also done in Appendix 2 (eqn A2.17) where it is shown that

$$\left(\frac{\partial\mu_i}{\partial N_k}\right)_{V,T} = \left(\frac{\partial\mu_i}{\partial N_k}\right)_{p,T} + \frac{v_i v_k}{\beta V} \tag{7.35}$$

where v_i and v_k are the partial molar volumes of the constituents i and k and β the compressibility.

7.2.3 Application to a one-component system

In the case of a one-component system, clearly

$$\left(\frac{\partial\bar{N}}{\partial\mu}\right)_{V,T} = \frac{1}{\left(\dfrac{\partial\mu}{\partial\bar{N}}\right)_{V,T}} \tag{7.36}$$

and, using eqn (7.35),

$$\left(\frac{\partial\mu}{\partial N}\right)_{V,T} = \left(\frac{\partial\mu}{\partial N}\right)_{p,T} + \frac{v^2}{\beta V}. \tag{7.37}$$

Since we have a one-component system $(\partial\mu/\partial N)_{p,T} = 0$ (μ being an intensive variable depends only on p and T). On the other hand since $v = V/N$ we obtain

$$I(q=0) = b^2\frac{N^2 k_B T}{V}\beta \tag{7.38}$$

which is identical to eqn (7.13).

7.2.4 Application to a two-component system

We have shown that we have in matrix form (Benoit, 1991)

$$[S_{ij}(q=0)] = k_B T\left[\left(\frac{\partial\bar{N}_i}{\partial\mu_j}\right)_{V,T}\right].$$

This allows us to write for $1/k_B T[S]^{-1}$ (after using eqn (7.35) to transform the derivatives at constant V into derivatives at constant p)

$$[S(0)]^{-1}/k_{\mathrm{B}}T = \begin{pmatrix} \left(\frac{\partial\mu_0}{\partial N_0}\right)_p + \frac{v_0^2}{\beta V} & \left(\frac{\partial\mu_0}{\partial N_1}\right)_p + \frac{v_1 v_0}{\beta V} \\ \left(\frac{\partial\mu_0}{\partial N_1}\right)_p + \frac{v_1 v_0}{\beta V} & \left(\frac{\partial\mu_1}{\partial N_1}\right)_p + \frac{v_1^2}{\beta V} \end{pmatrix} \tag{7.39}$$

and we have to evaluate $[S]$, the inverse of $[S]^{-1}$. This procedure begins by the calculation of its determinant which is done using the Gibbs-Duhem relationship (eqn A2.10) at constant p and T

$$N_0 \frac{\partial\mu_0}{\partial N_i} + N_1 \frac{\partial\mu_1}{\partial N_i} = 0 \quad (\text{for } i = 0 \quad \text{or } 1) \tag{7.40}$$

which enables us to express all the partial derivatives as functions of one of them. This leads to

$$\mathrm{Det} = -\left(\frac{\partial\mu_0}{\partial N_1}\right)_p \frac{V}{\beta N_0 N_1}\,. \tag{7.41}$$

The terms of the inverse of a matrix are the cofactors divided by the determinant. These cofactors are also easy to determine, giving

$$\begin{aligned} \left(\frac{\partial N_0}{\partial\mu_0}\right)_V &= \frac{\overline{\Delta N_0^2}}{k_{\mathrm{B}}T} = \frac{\beta N_0^2}{V} - \frac{N_0 N_1 v_1^2}{V^2}\left(\frac{\partial\mu_0}{\partial N_1}\right)^{-1} \\ \left(\frac{\partial N_0}{\partial\mu_1}\right)_V &= \frac{\overline{\Delta N_0 \Delta N_1}}{k_{\mathrm{B}}T} = \frac{\beta N_0 N_1}{V} + \frac{N_0 N_1 v_0 v_1}{V^2}\left(\frac{\partial\mu_0}{\partial N_1}\right)^{-1} \\ \left(\frac{\partial N_1}{\partial\mu_1}\right)_V &= \frac{\overline{\Delta N_1^2}}{k_{\mathrm{B}}T} = = \frac{\beta N_1^2}{V} - \frac{N_0 N_1 v_0^2}{V^2}\left(\frac{\partial\mu_0}{\partial N_1}\right)^{-1}. \end{aligned} \tag{7.42}$$

In order to evaluate the scattered intensity we write eqn (7.24) for two components

$$I(q = 0) = b_0^2 S_{00} + b_1^2 S_{11} + 2b_0 b_1 S_{01}\,. \tag{7.43}$$

Replacing the S_{ij} in terms of measurable quantities (eqn 7.42) we obtain

$$I(q = 0) = (b_0 N_0 + b_1 N_1)^2 \frac{k_{\mathrm{B}}T\beta}{V} - \frac{N_0 N_1 (b_1 v_0 - b_0 v_1)^2}{V^2} k_{\mathrm{B}}T \left(\frac{\partial\mu_0}{\partial N_1}\right)^{-1}. \tag{7.44}$$

This is the general equation for a two-component system. If $N_1 = 0$ we recover eqn (7.13) valid for a one-component system. If $\beta = 0$, the system is incompressible and, after some algebra we recover eqn (7.19) for an incompressible mixture of two constituents.

If we had written the scattering intensity in the form of eqn (7.43) we would have obtained three terms S_{00}, S_{11}, and S_{01}. By a correct association of the terms we have succeeded in decoupling density and composition fluctuations, thus obtaining only two terms. This decoupling shows that the term for the density fluctuations has a contrast factor.

$$\langle b \rangle = \frac{N_0 b_0 + N_1 b_1}{N_0 + N_1} \tag{7.45}$$

which is the average scattering length of the medium. The second term, which characterizes the composition fluctuations, has a contrast factor proportional to $b_v = (b_1/v_1 - b_0/v_0)$ and leads to a formula identical to eqn (7.19), evaluated in the case of an incompressibility hypothesis. *This means that the scattering due to the compressibility and the scattering due to the composition fluctuations are additive, at least for a two-component system.* It has been generalized by des Cloizeaux and Jannink (1980) to any number of components.

7.3 A GENERAL THEORY FOR ZERO ANGLE SCATTERING

7.3.1 The exchange chemical potential

The method we have used to describe the scattering is straightforward but it presents some difficulties due to the fact that, when the system is incompressible the determinant of S_{ij} is zero which makes the calculations impossible. Moreover this calculation leads to $p(p + 1)/2$ partial structure factors $S(0)$ and we have already seen in Chapter 5 that only $p(p - 1)/2$ of these coefficients are necessary to describe the scattering by an incompressible system. It is therefore useful to use another method for the description of these systems; this can be done easily by using what are called the exchange chemical potentials (Benoît 1991).

In Appendix 2 we describe the exchange chemical potential in detail. The idea, which has been already used in Chapter 5, consists in characterizing the solute species not by their chemical potential per molecule but by units having the same size as the solvent or the reference molecule and using volume fraction to depict the state of the mixture. We summarize here the notation: z_i is the number of segments of the molecule i defined as

$$z_i = \frac{v_i}{v_0} \tag{7.46}$$

where v_i and v_0 are the volumes of the species i and the reference molecule respectively. (This reference molecule will always have the index 0.) The total volume of the solution is

$$V = N_0 v_0 + \sum_{i=1}^{p} N_i v_i = \left(N_0 + \sum N_1 z_i\right) v_0 = \left(N_0 + \sum N_i'\right) v_0 \qquad (7.47)$$

where N_i' is the number of segments of species i in the solution ($N_i' = N_i z_i$) and φ_i its volume fraction

$$\varphi_i = \frac{N_1'}{N_0 + \sum N_i'} = \frac{z_i N_i}{N_0 + \sum z_i N_i}, \; \varphi_0 = \frac{N_0}{N_0 + \sum z_i N_i}. \qquad (7.48)$$

As a model one can imagine, as in the Flory–Huggins theory, a lattice with cells having the size of a molecule of solvent, each cell being occupied either by one molecule of solvent or one segment of polymer. The polymer is divided into z units each of which has the size of a molecule of solvent. The Gibbs free energy, which is

$$G = N_0 \mu_0 + \sum_{i=1}^{p} N_i \mu_i$$

can be written as

$$G = N_0 \mu_0 + \sum N_i' \frac{\mu_i}{z_i}. \qquad (7.49)$$

Let $\mu_i / z_i = \mu_i'$ and define the free energy g_c per unit cell, dividing G by the number of cells. This is also the free energy per volume of a solvent molecule and obeys the relationship

$$g_c = \frac{G}{N_0 + \sum z_i N_i} = \mu_0 \left(1 - \sum \varphi_i\right) + \sum \mu_i' \varphi_i = \mu_0 + \sum (\mu_i' - \mu_0) \varphi_i. \qquad (7.50)$$

Defining the exchange potential or $\bar{\mu}_i$ as the quantity $(\mu_i' - \mu_0)$ and N_T the sum $N_0 + \Sigma N_i z_i$

$$G = N_T g_c = N_T \left[\mu_0 + \sum \bar{\mu}_i \varphi_i\right] = N_T \mu_0 + \sum N_i' \bar{\mu}_i. \qquad (7.51)$$

Many of the classical thermodynamic relationships which are established for μ_0 and μ_i can be generalized to μ_0 and $\bar{\mu}_i$ (see Appendix 2). The most interesting for the present applications are the following

$$\frac{\partial g_c}{\partial \varphi_i} = \bar{\mu}_i, \; \left(\frac{\partial \bar{\mu}_i}{\partial p}\right)_{v, \varphi_i, z_i} = 0, \; \left(\frac{\partial \bar{\mu}_i}{\partial \varphi_j}\right)_{p,T} = \left(\frac{\partial \bar{\mu}_i}{\partial \varphi_j}\right)_{V,T} \qquad (7.52)$$

and the Gibbs–Duhem relationship

$$\mathrm{d}\mu_0 + \sum_{i=1}^{p} \varphi_i \mathrm{d}\bar{\mu}_i = p\mathrm{d}v - S\mathrm{d}T. \qquad (7.53)$$

The difficulty lies in the definition of the parameter z. It is practically always assumed to be a constant. In fact it could depend on temperature and pressure if the expansion coefficient and the compressibility of the constituents are different. z is a thermodynamic property, since it can be defined by the relationship

$$z_i = \frac{\left(\frac{\partial \mu_i}{\partial p}\right)_{T,N}}{\left(\frac{\partial \mu_0}{\partial p}\right)_{T,N}} \tag{7.54}$$

where the differentiations are made at constant temperature and number of molecules and could depend on the thermodynamic parameters.

7.3.2 The scattering equation

Starting from eqn (7.27) and (7.28) with these new notations, we rewrite the expression for $\mathcal{B}$ as

$$\mathcal{B} = = N_{\mathrm{T}} b_0 + \sum_{i=1}^{p} N_i' \left(\frac{\mathscr{b}_i}{z_i} - b_0\right) = N_{\mathrm{T}} \left[b_0 + \sum_{1}^{p} \left(\frac{\mathscr{b}_i}{z_i} - b_0\right) \varphi_i \right]. \tag{7.55}$$

Now we shall introduce what we call the contrast factor defined as

$$\bar{b}_i = \frac{\mathscr{b}_i}{z_i} - b_0. \tag{7.56}$$

This is the difference between the coherent scattering length of a unit of the polymer and the solvent occupying the same volume and has already been used. With this notation the scattering intensity becomes

$$I(0) = \overline{\Delta\mathcal{B}^2} = b_0^2 \Delta \overline{N_{\mathrm{T}}^2} + 2N_{\mathrm{T}} b_0 \sum_i \bar{b}_i \overline{\Delta N_{\mathrm{T}} \Delta\varphi_i} + N_{\mathrm{T}}^2 \sum_i \sum_k \bar{b}_i \bar{b}_k \overline{\Delta\varphi_i \Delta\varphi_k} \tag{7.57}$$

or, using the second eqn (7.55),

$$\overline{\Delta\mathcal{B}^2} = \langle b \rangle^2 \Delta \overline{N_{\mathrm{T}}^2} + 2N_{\mathrm{T}} \sum_i \langle b \rangle b_i \overline{\Delta N_{\mathrm{T}} \Delta\varphi_i} + N_{\mathrm{T}}^2 \sum_i \sum_k \bar{b}_i \bar{b}_k \overline{\Delta\varphi_i \Delta\varphi_k} \tag{7.58}$$

where $\langle b \rangle$ is the average scattering length of the sample, i.e. the quantity

$$\langle b \rangle = b_0 + \sum_{1}^{p} \left(\frac{\mathscr{b}_i}{z_i} - b_0\right) \varphi_i = \sum_{0}^{p} \frac{\mathscr{b}_i}{z_i} \varphi_i. \tag{7.59}$$

If the system is incompressible and the z_i do not depend on pressure N_T does not fluctuate and the only term left in eqn (7.44) is the term due to composition fluctuations. If the system is compressible we have to consider the three terms but, we know that the cross term, which depends on correlations between density and composition fluctuation, is equal to zero (des Cloizeaux and Jannink 1980). Since the compressibility term can be evaluated as in the case of a one component system our only task is the evaluation of the quantities: $\overline{\Delta\varphi_i \Delta\varphi_k}$. This is done following exactly the same procedure as for the $\overline{\Delta N_i \Delta N_k}$. One writes, using eqn (7.49), the free energy as function of N_T and the φ_is and finds, after a straightforward calculation

$$(\overline{\Delta\varphi_i \Delta\varphi_k})_{V,N_T} = \frac{k_B T}{N_T}\left(\frac{\partial\varphi_i}{\partial\bar{\mu}_k}\right)_{N_T,\mu_m \neq i} = \frac{k_B T}{N_T}\left(\frac{\partial\varphi_k}{\partial\bar{\mu}_i}\right)_{N_T,\mu_m \neq k} \tag{7.60}$$

We have now, as we did before, to invert the matrix $[\partial\varphi_i/\partial\mu_k]$, obtaining

$$\left[\frac{\partial\varphi_i}{\partial\bar{\mu}_k}\right] = \left[\frac{\partial\bar{\mu}_i}{\partial\varphi_k}\right]^{-1}. \tag{7.61}$$

This last quantity is also equal to the matrix of the second derivatives of g_c. Consequently we arrive at the well-known result

$$[S] = N_T k_B T\left[\frac{\partial^2 g_c}{\partial\varphi_i \partial\varphi_j}\right]^{-1} \tag{7.62}$$

Using the intensity scattered by the volume of a molecule of solvent, $s(q) = S(q)/N_T$, as new volume variable, we can write for eqn (7.62) the expression

$$[s(0)] = k_B T\left[\frac{\partial^2 g_c}{\partial\varphi_i \partial\varphi_j}\right]^{-1} \tag{7.62'}$$

Since we have assumed that the system was made of $p+1$ constituents (the solvent plus the p solutes) the matrix is p by p, one order less than in the general case and we write

$$I(q=0) = N_T k_B T[\bar{b}]\left[\frac{\partial^2 g_c}{\partial\varphi_i \partial\varphi_j}\right]^{-1}[\bar{b}]^T \tag{7.63}$$

where, as already explained, $[\bar{b}]$ is the column vector of the $\bar{b}_i$ defined in eqn (7.56). It should be noted that we did not specify whether the differentiations were to be made at constant v or constant p. This is due to the fact that, from eqn (7.52) these quantities are identical.

7.3.3 Application to polymer solutions

This last equation shows that the determination of the intensity scattered at zero angle requires knowledge of the Gibbs free energy of the system or the function g_c. In order to obtain useful results one has to introduce a model for this free energy. Here we shall use the Flory–Huggins theory (Flory 1942; Huggins 1942) which is currently adopted, as a first approximation, by all the workers in the polymer field. From Appendix 2 (eqn (A2.55)) we write

$$g_c = k_B T \left[\varphi_0 \log \varphi_0 + \sum_{i=1}^{p} \frac{\varphi_i}{z_i} \log \varphi_i + \sum_{i<j} \varphi_i \varphi_j \chi_{ij} + \varphi_0 \sum_{i=1}^{p} \varphi_i \chi_{0,i} \right]. \tag{7.64}$$

Since the solvent is ignored as a constituent in scattering equations we have to replace φ_0, volume fraction of the solvent by $1 - \Sigma_{i=1}^{p} \varphi_i$. The quantities χ_{ij} are what are called the Flory interaction parameters and are supposed to depend only on temperature in the original Flory theory. The evaluation of the $\partial^2 g_c / \partial\varphi_i \partial\varphi_j$ is easy giving

$$\frac{1}{k_B T} \frac{\partial^2 g}{\partial\varphi_i \partial\varphi_k} = \frac{1}{\varphi_0} + \chi_{ik} - \chi_{i0} - \chi_{k0} = v_{ik}$$

$$\frac{1}{k_B T} \frac{\partial^2 g}{\partial\varphi_i^2} = \frac{1}{z_i \varphi_i} + \frac{1}{\varphi_0} - 2\chi_{i0} = \frac{1}{z_i \varphi_i} + v_{ii}. \tag{7.65}$$

The right-hand side of these equations defines new quantities: the excluded volume parameters v_{ij}, frequently used in polymer physical chemistry and discussed in more details in Section 7.2

$$v_{ij} = \frac{1}{\varphi_0} + \chi_{ik} - \chi_{i0} - \chi_{k0} \tag{7.66}$$

with the convention $\chi_{ii} = 0$. These excluded volume parameters are not constant for χ_{ik} constant but are useful parameters and sometimes much more convenient to handle than the χs. The matrix $1/k_B T[\partial^2 g / \partial\varphi_i \partial\varphi_k]$ now takes the form

$$\begin{pmatrix} \frac{1}{z_1 \varphi_1} + v_{11} & v_{12} & \cdots & v_{1p} \\ v_{21} & \frac{1}{z_2 \varphi_2} + v_{22} & \cdots & v_{2p} \\ \cdots & \cdots & \cdots & \cdots \\ v_{p1} & v_{2p} & \cdots & \frac{1}{z_p \varphi_p} + v_{pp} \end{pmatrix} \tag{7.67}$$

with $v_{ij} = v_{ji}$, and can, for simplicity, be written as the sum of two matrices: the first is a diagonal matrix having the $1/z_i\varphi_i$ on the first diagonal and zeros every where else and will be called $[x]^{-1}$. This gives for the inverse matrix $[x]$ a matrix having $z_i\varphi_i$ on the first diagonal and zero's anywhere else; the second is the matrix of the v_{ij} and will be called $[v]$. Clearly

$$\frac{1}{k_B T}\left[\frac{\partial^2 g}{\partial\varphi_i\partial\varphi_k}\right] = [x]^{-1} + [v] \tag{7.68}$$

which leads to the relationship

$$[s]^{-1} = [x]^{-1} + [v] \tag{7.69}$$

defining the matrix $[s]$ as giving the scattering intensity by a volume equal to the volume of a molecule of solvent (this comes from the use of g_c rather than $N_T g_c$).

7.3.4 Divergence of the intensity at zero angle

All the expressions we have encountered for the intensity scattered at zero angle have a denominator and we have to say a few words about what happens when this denominator goes to zero. The equations tell us that the intensity should become infinite. It is evident that a physical quantity like scattering intensity, cannot become infinite but is important to understand the physics of this phenomenon. We shall consider the simplest case, the case of a one-component system, a van der Waals gas. From eqn (7.13), divergence can occur only when the compressibility β (i.e. $-1/V\ \partial v/\partial p_T$) becomes infinite. Using the classical p, V diagram and plotting the isothermal curves, one sees that, in order to have $\beta^{-1} = 0$ the tangent to the isotherm must be horizontal. This can happen only in two specific cases: either at the critical point where, as it is known, the compressibility becomes infinite or at temperatures below the critical points where the isotherms have the following shape (see Fig. 7.3). In fact, below the critical point the real isotherms are made of the curve a, the straight line AB where there are two coexisting phases (liquid and vapour) followed by the curve b. It seems therefore that $\beta^{-1} = 0$ can never occur. The curves AA′ and B′B correspond to metastable equilibria and we know that, if nucleation is avoided, the system can be obtained in a metastable state represented by one point on the curves AA′ and BB′. This is impossible for the portion of the curve between A′ and B′ since on this curve dp/dV is positive and the scattering intensity is negative. This portion is therefore absolutely forbidden. A′ and B′ are the borders between possible states and thermodynamically impos-

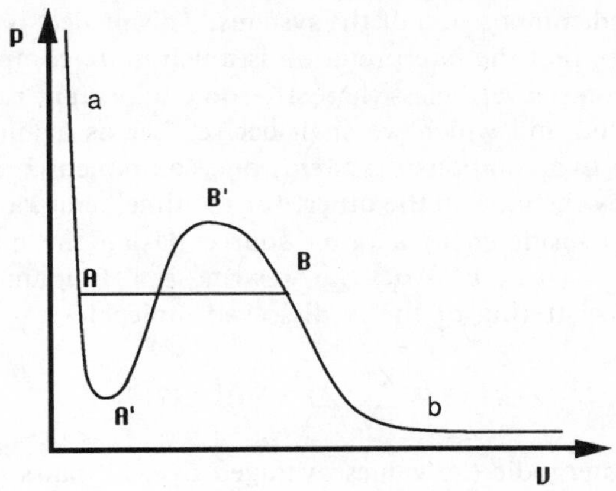

Fig. 7.3 Typical isotherm for van der Waals gas.

sible states: they are also the points for which the intensity diverges. We can therefore say that when the intensity diverges the system is, either at the critical point, or at one of the points A′ or B′, the limit between the metastable and absolutely unstable states for the system. The locus of these points is called the spinodal; the critical point is also evidently on this locus which obeys the equation $I^{-1}(0) = 0$. This definition can be generalized to multicomponent systems (see Appendix 2) and, from eqn (7.62), we define the spinodal (which can be a curve, a surface or an hypersurface) by the relationship

$$\mathrm{Det}\left|\frac{\partial^2 g}{\partial\varphi_i\partial\varphi_j}\right| = 0. \tag{7.70}$$

Scattering is a very elegant method of finding the spinodal which, of course, could be obtained by pure thermodynamical arguments.

Part 2 Finite angle scattering

7.4 DILUTE SOLUTION SCATTERING

7.4.1 The radial distribution function

In the preceding section we have been discussing extensively the law governing the scattering at zero angle and have shown that the results depend essentially on the thermodynamic quantities characterizing the system. In this section we proceed to the interpretation of the total scattering curve at any angle and composition of the systems. This project is very ambitious and we shall see that the interpretation is much more complex than in the preceding section. Nevertheless there are some important rules which have to be understood and which we shall derive. Let us again start from an incompressible two component system, one component being the solvent characterized by the index 0 the other, for the time being, a molecule small enough to be considered as a point source. Using the notation already introduced ($\bar{b} = (b_1 v_0/v_1 - b_0)$ we rewrite the fundamental equation expressing the scattering of the N dissolved molecules

$$I(q) = \bar{b}^2 \sum_{i=0}^{N} \sum_{j=0}^{N} \langle \exp(-\mathrm{i}\boldsymbol{q}\cdot\boldsymbol{r}_{ij}) \rangle. \tag{7.71}$$

If the system is ergodic the values averaged over all pairs ij are identical; we now consider only one pair and multiply by the number of pairs $i \neq j$ which is $N(N-1)$ and will be replaced by N^2 (which for large N is equivalent). We thus write

$$I(q) = \bar{b}^2 N[1 + N\langle \exp(-\mathrm{i}\boldsymbol{q}\cdot\boldsymbol{r}_{12}) \rangle]. \tag{7.72}$$

In this chapter we shall assume that *the scattering volume is equal to unity* so that N is the number of molecules per unit volume. The quantity $I(q)/\bar{b}^2$ will be called $S(q)$. It is the structure factor of the polymer for a scattering volume equal to unity

$$S(q) = N[1 + N\langle \exp(-\mathrm{i}\boldsymbol{q}\cdot\boldsymbol{r}_{12}) \rangle]. \tag{7.73}$$

The problem of the evaluation of the effect of the interaction is therefore the problem of the calculation of the last term of the RHS of the preceding equation. Let us call $w(\boldsymbol{r}_1, \boldsymbol{r}_2)\, \mathrm{d}\boldsymbol{r}_1 \mathrm{d}\boldsymbol{r}_2$, the probability of having the molecule 1 in $\mathrm{d}\boldsymbol{r}_1$ and the molecule 2 in $\mathrm{d}\boldsymbol{r}_2$. This allows us to write

$$S(q) = N + N^2 \int_{V_1} \int_{V_2} \exp(-\mathrm{i}\boldsymbol{q}\cdot\boldsymbol{r}_{12})\, w(\boldsymbol{r}_1, \boldsymbol{r}_2)\, \mathrm{d}\boldsymbol{r}_1 \mathrm{d}\boldsymbol{r}_2. \tag{7.74}$$

We can also say that this probability is the product of the probability $w_1(\boldsymbol{r}_1)$ of having the molecule 1 at $\boldsymbol{r}_1$ with the conditional probability $w(\boldsymbol{r}_2|\boldsymbol{r}_1)$ of having the molecule 2 at $\boldsymbol{r}_2$ when the molecule 1 is at $\boldsymbol{r}_1$. This last probability describes the influence of the molecule 1 at $\boldsymbol{r}_1$ on the presence of the molecule 2 at $\boldsymbol{r}_2$. This influence clearly depends only on the relative position of molecules 1 and 2 and not on the position of the pair in the scattering volume V. Moreover, in an isotropic medium, this probability does not depend on the orientation of the vector $\boldsymbol{r}$ joining the points 1 and 2 but only on its magnitude. This allows us to write

$$w_1(\boldsymbol{r}_1)\, \mathrm{d}\boldsymbol{r}_1 = \frac{\mathrm{d}V}{V} = \mathrm{d}V, \quad \text{since from our hypothesis } V = 1$$

$$w(\boldsymbol{r}_2|\boldsymbol{r}_1) = w(|\boldsymbol{r}_2 - \boldsymbol{r}_1|) = w(\boldsymbol{r}). \tag{7.75}$$

Using the variables $\boldsymbol{r}_1$ and $\boldsymbol{r} = \boldsymbol{r}_2 - \boldsymbol{r}_1$ and integrating over V and over the orientations of $\boldsymbol{r}$ gives

$$S(q) = N + N^2 \int_V w(\boldsymbol{r}) \exp(-\mathrm{i}\boldsymbol{q}\cdot\boldsymbol{r})\,\mathrm{d}\boldsymbol{r}. \tag{7.76}$$

Now we replace $w(\boldsymbol{r})$ by $g(r)$, defining the function $g(r)$ as the radial distribution function. It is the factor by which the probability $\mathrm{d}V$ (or $\mathrm{d}\boldsymbol{r}$) of finding a molecule at at a given position is multiplied when there is another molecule situated at a distance r from it. With this definition and after averaging over all orientations (see Appendix 1) we obtain

$$S(q) = N + N^2 \int_V g(r) \frac{\sin qr}{qr} 4\pi r^2 \mathrm{d}r. \tag{7.77}$$

In order to simplify the preceding equation we use the simple trick of replacing $g(r)$ by $1 - (1 - g(r))$ which allows us to write

$$S(q) = N\left[1 - N\int_V (1 - g(r)) \frac{\sin qr}{qr} 4\pi r^2 \mathrm{d}r\right]. \tag{7.78}$$

The integral $\int_v \sin qr/(qr) 4\pi r^2 \mathrm{d}r$ is identical to the integral of $\exp(-\mathrm{i}\boldsymbol{q}\cdot\boldsymbol{r})$ over $\boldsymbol{r}$ and it is known that this latter integral is a delta function peaked at $q = 0$ which does not contribute to the scattering.

7.4.2 Properties of the function *g(r)*

As soon as the distance r becomes larger than the distances of interaction between two molecule (a few tens of angstroms), the presence of a molecule at this distance does not affect the second molecule. This means that $g(r)$ goes to 1 when r goes to infinity and $g(r) - 1$ goes to zero; the upper limit of the integral in eqn (7.78) can be replaced by infinity rather than V since, for distances larger than the interaction distance, the value of the integral is zero.

If, for instance, the molecules are hard spheres and if we assume that they are scattering like points, it is impossible to have two centres at a distance smaller than D, the diameter of these spheres. If there are no other interactions then $g(r) = 0$ for $r < D$ and 1 for $r > D$ (see Fig. 7.4). More generally, if there are repulsions, $g(r)$ is less than 1 (the probability of finding a molecule is smaller than the average) and if there are attractions, $g(r) > 1$. In a dilute solution the occurrence of contacts of higher order (when for instance more than two molecules are interacting together) can be neglected and $g(r)$ is replaced by the Boltzmann factor,

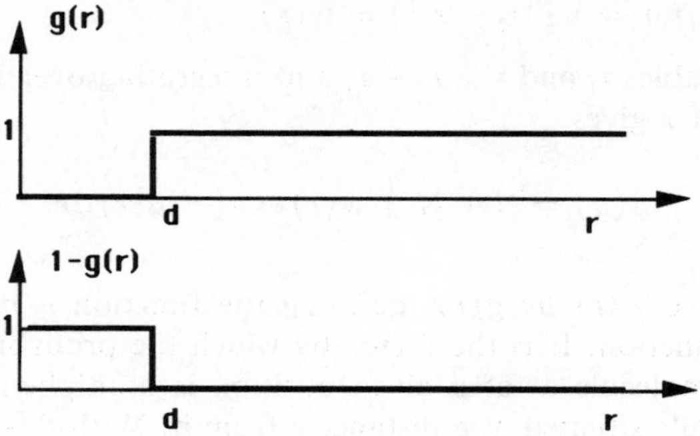

Fig. 7.4 The function $g(r)$ for hard spheres of diameter D without other interactions.

$\exp(-U(r)/k_B T)$. This gives the probability that a molecule will be at a distance r from the first molecule, assuming that the potential energy of the pair is $U(r)$

$$S(q) = N\left[1 - N\int_V \left(1 - \exp - \frac{U(r)}{k_B T}\right) \frac{\sin qr}{qr} 4\pi r^2 \mathrm{d}r\right]. \tag{7.79}$$

The integral on the right-hand side of this equation is a useful quantity which plays an important role in the theory of polymer solutions; it will be called $\mathrm{v}(q)$ the excluded volume and defined by

$$\mathrm{v}(q) = \int_0^\infty \left(1 - \exp - \frac{U(r)}{k_B T}\right) \frac{\sin qr}{qr} 4\pi r^2 \mathrm{d}r. \tag{7.80}$$

Since the interactions are usually short range (of the order of the volume of the solvent or of the molecules) the q dependence of $\mathrm{v}(q)$ which appears in eqn (7.79) can be neglected, except in the vicinity of the critical point and we shall use the quantity v defined as

$$\mathrm{v} = \int_0^\infty \left(1 - \exp\left[-\frac{U(r)}{k_B T}\right]\right) 4\pi r^2 \mathrm{d}r. \tag{7.81}$$

The ratio of this quantity to the volume of the monomer (or of the solvent), $v = \mathrm{v}/v_0$, is a dimensionless quantity, called the excluded volume parameter which is frequently used. It is perhaps worth recalling that in the statistical theory of liquids the quantity $\mathrm{v}(q)$, as defined by eqn (7.80) is called $-c(q)$

In Fig. 7.5 we have plotted a typical form of $U(r)$, which corresponds to van der Waals interactions, and the corresponding quantity $1 - g(r) = 1 - \exp[-U(r)/k_B T]$. From eqn (7.81) v is the integral of the lower

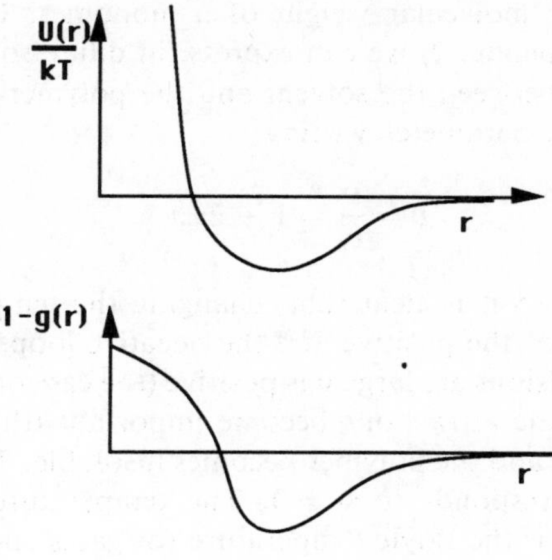

Fig. 7.5 An example of an interaction potential between two molecules and the corresponding 1-$g(r)$ function.

curve multiplied by $4\pi r^2$. The repulsive part of the interactions corresponds to a positive contribution to $\mathfrak{v}$ and a negative contribution arises from attractions.

7.4.3 Relationship with the second virial coefficient

In Section 7.1 we showed (eqn (7.22)) that the scattered intensity can be written as

$$I(0) = kcM[1 - 2A_2cM + \ldots]. \tag{7.82}$$

Comparing this equation and eqn (7.79) at $q = 0$, since the quantity in the square bracket is dimensionless, we have

$$2A_2cM = N\mathfrak{v} \tag{7.83}$$

or

$$A_2 = \frac{N_A\mathfrak{v}}{2M^2} \tag{7.84}$$

where N_A is the Avogadro constant and M the molecular weight of the solute. For a polymer, if one assumes that the contacts concern only one pair of monomers there are z^2 possibilities of contact; $\mathfrak{v}$ is replaced by $z^2\mathfrak{v}$ giving

$$A_2 = \frac{N_A\mathfrak{v}}{2m^2} \tag{7.85}$$

where m is now the molecular weight of a monomer. Using the Flory-Huggins theory (Appendix 2) we can express, in dilute solution, the interaction parameter χ between the solvent and the polymer as a function of the excluded volume parameter writing

$$v = \frac{\mathrm{v}}{v_0} = 1 - 2\chi. \tag{7.86}$$

Going back to Fig. 7.5 it is clear that, changing the temperature changes the respective areas of the positive and the negative loops of the $1 - g(r)$ function. If the repulsions are large v is positive (the case of a good solvent), if, on the contrary the attractions become important (the case of a poor solvent) v is negative and the polymer becomes insoluble. The limit between these two cases corresponds to $\mathrm{v} = 0$. The temperature for which this condition is fulfilled is the Boyle temperature for gases and is called, in the field of polymers, the Flory θ temperature.

7.4.4 Geometrical interpretation of the excluded volume parameter

Let us write eqn (7.72) in terms of $S(q)$

$$S(q) = N + N^2\langle \exp(-\mathrm{i}\boldsymbol{q}\cdot\boldsymbol{r}_{12})\rangle. \tag{7.87}$$

Qualitatively speaking point 2 cannot be within the excluded volume because of the presence of molecule 1 (see Fig. 7.4). More precisely the probability of being excluded is $(1 - g(r))4\pi r^2\mathrm{d}r$. If we apply the Babinet principle, as already done (see Section 5.3.2), instead of evaluating the average in eqn (7.87) we can evaluate the amplitude of the signal scattered by the forbidden region and put a minus sign in front of it. This gives

$$N^2\langle \exp(-\mathrm{i}\boldsymbol{q}\cdot\boldsymbol{r}_{12})\rangle = N^2\int_V (1 - g(r))\exp(-\mathrm{i}\boldsymbol{q}\cdot\boldsymbol{r})\,4\pi r^2\mathrm{d}r = N^2\mathrm{v}. \tag{7.88}$$

This allows us to establish the following rule (which has already been used in Chapter 5): evaluate the amplitude scattered by the forbidden región around one molecule, taking its centre as the phase reference and subtract this intensity from the total intensity.

7.5 THE ZIMM FORMULA (SINGLE CONTACT APPROXIMATION)

It is now straightforward to generalize this treatment to the case of a dilute polymer solution. Let us consider two polymer molecules of degree of polymerization z. We split, as has been done many times before, the intermolecular from the intramolecular part of the scattering law, writing

$$S(q) = Nz^2P(q) + N^2 \sum_{i_1=1}^{z} \sum_{j_2=1}^{z} \langle \exp(-\mathrm{i}\boldsymbol{q}\cdot\boldsymbol{r}_{i_1 j_2}) \rangle. \tag{7.89}$$

Let us consider a point i_1. We have to evaluate all the forbidden configurations where the chain 1 has contacts with the chain 2. Following what has just been said, we use what has been called the single contact approximation: we neglect the cases where there is more than one point of contact between the chains as in Fig. 7.6(b) and consider only the scattering from configurations such as those in Fig. 7.6(a). Since the configuration depicted in Fig. 7.6(a) is forbidden we evaluate its scattering and subtract it from the intramolecular scattering. We write

$$\boldsymbol{i}_1\boldsymbol{i}_2 = \boldsymbol{i}_1\boldsymbol{j}_1 + \boldsymbol{j}_1\boldsymbol{j}_2 + \boldsymbol{j}_2\boldsymbol{i}_2 \tag{7.90}$$

and take the exponential of this argument multiplied by $-\mathrm{i}\boldsymbol{q}$. Since the three vectors $\boldsymbol{i}_1\boldsymbol{j}_1$, $\boldsymbol{j}_1\boldsymbol{j}_2$, and $\boldsymbol{j}_2\boldsymbol{i}_2$ are independent it is easy to show that

$$\langle \exp(-\mathrm{i}\boldsymbol{q}\cdot\boldsymbol{i}_1\boldsymbol{i}_2) \rangle = \langle \exp(-\mathrm{i}\boldsymbol{q}\cdot\boldsymbol{i}_1\boldsymbol{j}_1) \rangle \langle \exp(-\mathrm{i}\boldsymbol{q}\cdot\boldsymbol{j}_1\boldsymbol{j}_2) \rangle \langle \exp(-\mathrm{i}\boldsymbol{q}\cdot\boldsymbol{j}_2\boldsymbol{i}_2) \rangle. \tag{7.91}$$

We can therefore integrate the three terms separately to obtain the scattered amplitude. The integrated value of the middle term of the RHS is nothing but the excluded volume $-\boldsymbol{v}$ which is independent of j_1 and j_2 and can be factorized. The two other terms are identical and after summation over the four variables eqn (7.89) becomes

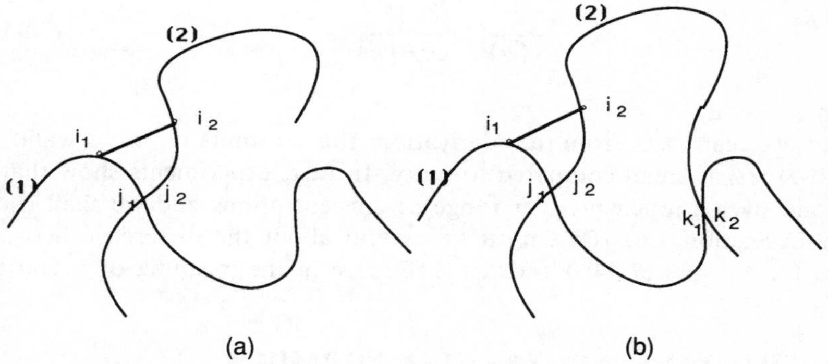

Fig. 7.6 Typical forbidden configuration in the single contact approximation. Single contacts (a). The multiple contacts (b) are not taken into consideration in the calculation.

$$N^2 \sum_{i_1=1}^{z} \sum_{j_2=1}^{z} \langle \exp(-\mathrm{i}\boldsymbol{q}\cdot\boldsymbol{r}_{i_1 i_2}) \rangle = -N^2 \mathfrak{v} \left[\sum_{i_1} \sum_{j_1} \int w(r_{i_1 j_1}) \exp(-\mathrm{i}\boldsymbol{q}\cdot\boldsymbol{r}_{i_1 j_1}) \mathrm{d}\boldsymbol{r}_{i_1 j_1} \right]^2. \tag{7.92}$$

Since i_1 and j_1 are anywhere on the chain we have to sum over all their values (from 1 to z), recovering the form factor of the polymer $z^2P(q)$. Thus

$$N^2 \sum_{i_1=1}^{z} \sum_{j_2=1}^{z} \langle \exp - \mathrm{i}\boldsymbol{q}\cdot\boldsymbol{r}_{i_1 i_2} \rangle = -\mathfrak{v}\,[Nz^2P(q)]^2 \tag{7.93}$$

and for the structure factor $S(q)$ and the scattered intensity we obtain

$$\frac{I(q)}{\bar{b}^2} = S(q) = Nz^2P(q) - \mathfrak{v}\,[Nz^2P(q)]^2. \tag{7.94}$$

Assuming that the second term is smaller than the first one, this expression can be inverted, leading, as a first approximation (Zimm 1948*a*) to

$$\frac{1}{S(q)} = \frac{1}{Nz^2P(q)} + \mathfrak{v}. \tag{7.94$'$}$$

We can express this in different notation. First we introduce the structure factor $s(q) = S(q)/N_{\mathrm{T}}$ which corresponds to the scattering intensity from a volume equal to that of a molecule of solvent with $N_{\mathrm{T}} = V/v_0$ where V is the volume of the solution (which we have chosen as unity) and v_0 the volume of one molecule of solvent. Secondly we define the concentration of polymer by its volume fraction $\varphi = N_z/N_{\mathrm{T}}$. This allows us to write

$$\frac{1}{s(q)} = \frac{1}{\varphi z P(q)} + v \tag{7.94$''$}$$

with $v = \mathfrak{v}/v_0$.

As we can see from its derivation this formula is only valid if $\varphi z P(q)\mathfrak{v}/v_0$ is small compared to unity. In fact, experiments show that it is valid over a much broader range of concentrations and we shall show why in Section 7.62. (We must be careful about the difference between eqns (7.94′) and (7.94″) and the difference in the meaning of $\mathfrak{v}$ and v.)

7.6 THE ORNSTEIN–ZERNIKE FORMULA

7.6.1 The case of small molecules

In the previous paragraph we have considered dilute solutions and shown that it is possible to describe these by subtracting from the intensity scattered by the N molecules a term taking into consideration the possibility of two molecules interacting. It is possible to generalize this procedure and to consider the configurations of the medium which are also forbidden. As a first step we shall take into account a linear chain of small molecules and as a second a more general procedure (Stanley 1971), obtaining a result which, as shown in statistical mechanics books (Hansen and McDonald 1986) is rigorous. Subsequently we shall apply this procedure to polymers and show that it can also be used for copolymers and for mixtures of different polymers.

In Section 7.9 we have subtracted the intensity scattered by a pair of molecules but we could also consider larger aggregates and subtract terms coming from situations like the one shown in Fig. 7.7. The scattering molecules 1 and 2 are in contact with a third molecule p; there is excluded volume between 1 and p and 2 and p and this configuration is also forbidden. In order to take the indirect contacts $1 - p$, $2 - p$ into account we write

$$\langle \exp(-iq \cdot r_{12}) \rangle = \langle \exp(-iq \cdot r_{1p}) \rangle \langle \exp(-iq \cdot r_{p2}) \rangle \tag{7.95}$$

and after averaging over all positions we obtain for one molecule p

$$\int_V \langle \exp(-iq \cdot r_{12}) \mathrm{d}r \rangle = \int_V \langle \exp(-iq \cdot r_{1p}) \mathrm{d}r \rangle \int_V \langle \exp(-iq \cdot r_{p2}) \rangle \mathrm{d}r. \tag{7.96}$$

Since the molecule p can be any of the $N - 2$ molecules other than 1 or 2 we have to multiply by $N - 2 \approx N$. The integrals on the right-hand side of eqn (7.96) are both equal to $-\mathfrak{v}$. After adding the self term N and the single contact term we obtain

$$S(q) = N - N^2 \mathfrak{v} + N^3 \mathfrak{v}^2. \tag{7.97}$$

If instead of considering only interactions between two and three molecules we consider all the possible chains of contacts with 2,3, ... p molecules (see Fig. 7.7(b)) the scattered intensity will be given by

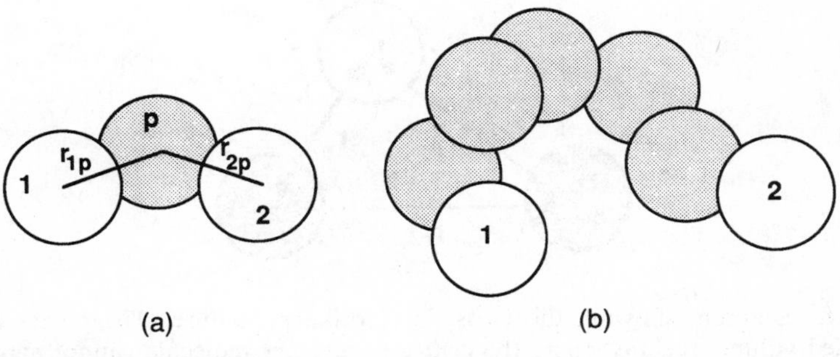

Fig. 7.7 Forbidden configurations for the interferences between particles 1 and 2. (a) One intermediate molecule. (b) Many intermediate molecules.

$$S(q) = N[1 - N\upsilon + N^2\upsilon^2 - N^3\upsilon^3 + N^4\upsilon^4 \ldots]. \tag{7.98}$$

This geometric series can be summed immediately, giving

$$S(q) = \frac{N}{1 + \upsilon N} \tag{7.99}$$

or taking the inverse

$$S^{-1}(q) = \frac{1}{N} + \upsilon. \tag{7.99$'$}$$

The method we have used is interesting since it allows us to understand the physics but it is not rigorous since it is not clear why we have selected this series of forbidden configurations and have left out many others. A more general treatment will be presented now; it is due to Ornstein and Zernike (1914). We consider the direct interaction υ and a quantity h which we define as the total interaction obtained by taking into account all the interactions between two particles or molecules in the system. We assume that we know this quantity and define it precisely as the interaction part of the scattering (see eqn (7.74)) writing

$$S(q) - N = N^2 h. \tag{7.100}$$

Let us consider the diagram in Fig. 7.8. We assume that we know $\hat{h}(r)$, the inverse Fourier transform of $h(q)$ (it is in fact the function $g(r) - 1$) and following Ornstein and Zernike (1914) we write, without giving the demonstration (see, for instance, Hansen and McDonald 1986) the following integral equation

$$\hat{h}(r_{12}) = -\hat{\upsilon}(r_{12}) - N\iiint \hat{\upsilon}(r_{1p})\,\hat{h}(r_{p2})\,\mathrm{d}r_{1p} = -\hat{\upsilon}(r_{12}) - N\iiint \hat{\upsilon}(r_{1p})\,\hat{h}(r_{12} - r_{1p})\,\mathrm{d}r_{1p} \tag{7.101}$$

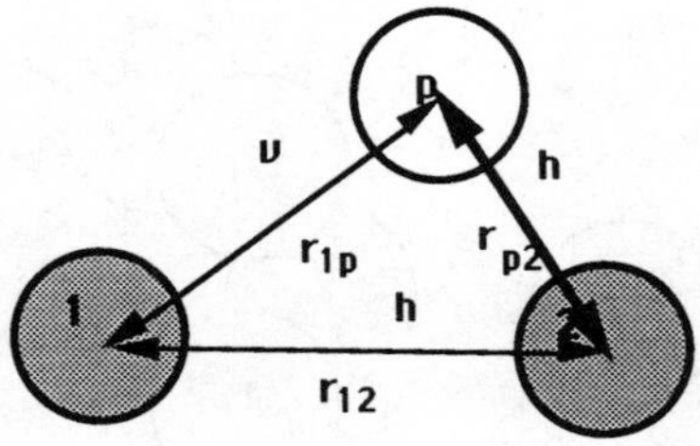

Fig. 7.8 Diagram showing the Ornstein–Zernike procedure. The circles are the excluded volume regions where the centre of another molecule cannot enter. The grey molecules are the scattering molecules.

$\hat{\upsilon}(r_{1p})$ is the inverse Fourier transform of $\upsilon(q)$. This means that, in order to evaluate the total interaction h going from 1 to 2, one can first go from 1 to p with the direct interaction $-\upsilon(r)$ and afterwards from p to 2 with the total interaction $\hat{h}(r)$.

Using the Fourier transform of eqn (7.101) and applying the convolution theorem (see Appendix 1) gives

$$h(q) = -\upsilon(q) - Nh(q)\upsilon(q) \tag{7.102}$$

or

$$h(q) = \frac{-\upsilon(q)}{1 + N\upsilon(q)} \tag{7.102$'$}$$

we see that this is exactly the same result as eqn (7.99)* since writing $S(q)$ from eqn (7.100) gives

$$S(q) = N + N^2h = N - \frac{N^2\upsilon}{1 + N\upsilon} = \frac{N}{1 + N\upsilon} \tag{7.103}$$

and

$$\frac{1}{S(q)} = \frac{1}{N} + \upsilon(q). \tag{7.103$'$}$$

Equation (7.102) is rigorous in the sense that this relationship between $h(q)$, the total interaction, and $\upsilon(q)$, the pair interaction, is always satisfied. Unfortunately a rigorous calculation of either $\upsilon(q)$ or $h(q)$ is difficult; it is the object of the statistical mechanics of simple fluids which attempts to relate the thermodynamic properties to the energy of interaction between molecules. In the context of our discussion of neutron scattering problems by polymeric systems we shall not discuss the problem of the calculation of υ or h. We consider them as concentration-dependent parameters, and use for the most part the expression given by the Flory–Huggins theory.

7.6.2 Application of the O–Z method to macromolecules

Since the method of accounting for an infinite series of interactions has been so successful it is tempting to use it in the case of macromolecules (Yerukhimovich 1979; Benoît and Benmouna 1984). In order to do so we shall transform the diagram in Fig. 7.6 and replace it by the diagram in Fig. 7.9. The two chains are replaced by two vertical lines and the 'contacts'

* One could as well interchange the order of the operations going first from 1 to p with the total interaction h and from p to 2 with the direct interaction υ in the last term of the LHS of eqn (7.101).

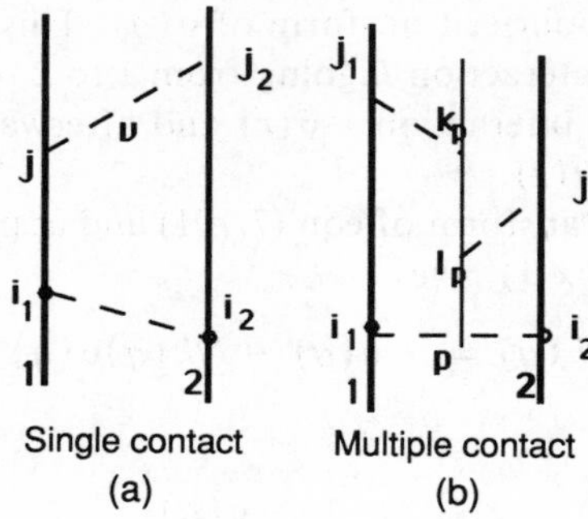

Fig. 7.9 The vertical lines represent the macromolecules which do not have to be linear; the dashed lines represent contacts between two molecules.

by dotted lines joining the interacting groups. The evaluation of the forbidden intensity for the two scattering points i_1 and i_2 can be made using the following procedure: we replace a dotted line joining two arbitrary points (j_1 and j_2) on different chains which are represented by vertical lines, by the excluded volume v, and replace the chains by the quantity $Nz^2P(q)$.

In order to evaluate the term which corresponds to the interactions between i_1 and i_2 we travel on the diagram from i_1 to i_2; each time we join two points on the same polymer molecule we introduce the factor $x = Nz^2P(q)$; when we go from one polymer to the other we multiply by the factor $-\mathrm{v}$. Looking at Fig. 7.9(a) we evaluate directly the Zimm expression, writing

$$S(q) = x - \mathrm{v}x^2 \tag{7.104}$$

recovering eqn (7.92).

If we take into account the case of Fig. 7.9(b) we obtain

$$S(q) = [x - x^2\mathrm{v} + x^3\mathrm{v}^2]. \tag{7.105}$$

The generalization to an infinite series of terms is obvious, giving

$$\begin{aligned} S(q) &= x[1 - x\mathrm{v} + (x\mathrm{v})^2 - (x\mathrm{v})^3 + \ldots \\ &= x\left[\frac{1}{1 + \mathrm{v}x}\right] \end{aligned} \tag{7.106}$$

and

$$S^{-1}(q) = \frac{1}{x} + \mathrm{v} = \frac{1}{Nz^2P(q)} + \mathrm{v}. \tag{7.106'}$$

This is the Zimm formula (1948*a*) in its inverse form and it justifies the approximation used by the experimenters and discussed in Chapter 8. It is obvious that we could have obtained this formula using the Ornstein–Zernike procedure which relates υ and h for molecules small compared to λ. In this case we introduce the quantity $h(q)$ defined by the relationship

$$S(q) = x + h(q)x^2 \tag{7.107}$$

and replace the Ornstein–Zernike relationship eqn (7.102) by

$$h(q) = -\upsilon - \upsilon x h(q). \tag{7.108}$$

The factor x is due to the fact that we have introduced an arbitrary chain in our calculation (see Fig. 7.10(a)). In fact there are three chains but the two scattering chains are already taken into account by the factor x^2 in eqn (7.106). We can therefore write

$$h(q) + \frac{-\upsilon}{1 + \upsilon x}. \tag{7.108$'$}$$

The quantity $S(q)$ becomes

$$S(q) = x - x^2 \frac{\upsilon}{1 + \upsilon x} = \frac{x}{1 + \upsilon x} \tag{7.109}$$

or

$$S^{-1}(q) = \frac{1}{Nz^2 P(q)} + \upsilon. \tag{7.109$'$}$$

This is eqn (6.72) but this time obtained by a more rigorous route (Daoud *et al.* 1975).

7.6.3 Relationship with thermodynamics

Studying the intensity scattered at zero angle we have shown (eqn (7.69) specialized to the case of one polymer) that one has

$$\frac{1}{s(0)} = \frac{1}{\varphi z} + v \tag{7.110}$$

where $s(q) = S(q)/N_T$ and φ is the volume fraction occupied by the polymer. Using the same notation we obtain here

$$\frac{1}{s(q)} = \frac{1}{\varphi z P(q)} + v. \tag{7.111}$$

The formulae (7.109) and (7.110) are identical, except for the presence of the form factor $P(q)$ of the molecules. This leads us to believe that, in order to write a formula valid at any angle, one has first to evaluate in terms of volume fractions the formula describing the scattering at zero angle and afterwards to replace the molecular weight z_i of any species by the product

$z_i P_i(q)$ where $P_i(q)$ is the form factor of the species i at the concentration and in the environment of the solution. Evidently this result is limited to the case of linear molecules and cannot be extrapolated to two- or three-dimensional objects. Even in the case of linear molecules it is not rigorous since we did not take into account the possibility of multiple contacts between two chains. Anyway it gives a simple method for describing the scattered intensity which, until now, has been quite successful. This is the reason why we shall generalize it it now for two examples: for a mixture of homopolymers and for copolymers.

7.7 MIXTURE OF TWO POLYMERS

7.7.1 In the presence of solvent

Since we are dealing with a three-component system, assumed to be incompressible, we have three partial structure factors and we have to write

$$I(q) = \bar{b}_1^2 S_{11} + \bar{b}_2^2 S_{22} + 2\bar{b}_1 \bar{b}_2 S_{12} \tag{7.112}$$

where the $\bar{b}_i$ are the differences between the coherent scattering length of a monomer unit and a molecule of solvent and where we assume that the scattering volume is unity.

Following eqn (7.105) we define the h_{ij} by the relationships

$$\left.\begin{aligned} S_{11}(q) &= x_1 + x_1^2 h_{11}(q) \\ S_{22}(q) &= x_2 + x_2^2 h_{22}(q) \\ S_{12}(q) &= x_1 x_2 h_{12}(q) \end{aligned}\right\} \tag{7.113}$$

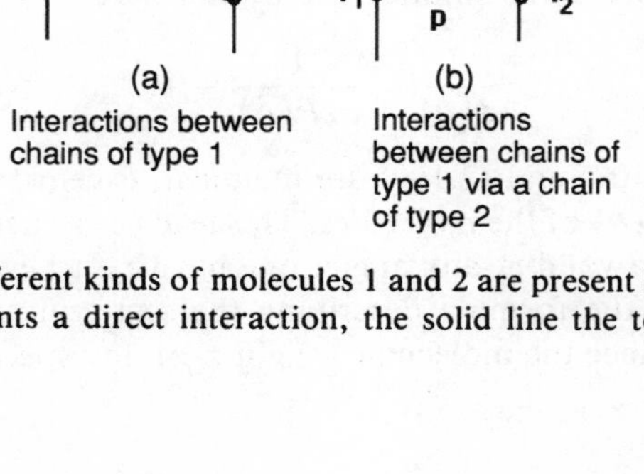

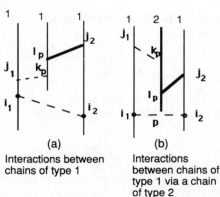

Fig. 7.10 Two different kinds of molecules 1 and 2 are present in the medium. The dashed line represents a direct interaction, the solid line the total interaction.

where $x_i = N_i z_i^2 P_i(q)$ and h_{ij} is, following Ornstein–Zernike, the total interaction between molecules i and j.

Let us now evaluate the quantity $h_{11}x_1^2$. For this we examine Fig. 7.10 where the two possible cases are shown. In Fig. 7.10(a), we add, as we did before, to the direct interaction $-v_{11}x_1^2$, the indirect term $-v_{11}x_1^3 h_{11}(q)$. In the second case we have to consider the interaction between two chains of type 1 via a chain of type 2. The excluded volume corresponding to this interaction will be called v_{12} and the total interaction h_{12}. This term, following the method discussed earlier, will be written as $-x_2 x_1^2 v_{12} h_{12}(q)$. Adding these two possibilities as well as the direct contact, one obtains for the extension of the O–Z formula to the case of a two-component system in a solvent

$$x_1^2 h_{11}(q) = -x_1^2 v_{11} - v_{11} x_1^3 h_{11}(q) - v_{12} x_1^2 x_2 h_{12}(q) \tag{7.114}$$

or dividing by x_1^2

$$h_{11}(q) = -v_{11} - v_{11} x_1 h_{11}(q) - v_{12} x_2 h_{12}(q). \tag{7.115}$$

The calculation of $h_{22}(q)$ is similar and, exchanging the indices 1 and 2, we obtain

$$h_{22}(q) = -v_{22} - v_{22} x_2 h_{22}(q) - v_{12} x_1 h_{12}(q). \tag{7.115'}$$

In order to evaluate the cross term h_{12} we need the help of Fig. 7.11 since the total interaction can be formed using as intermediate molecule either a molecule of type 1 or of type 2. In Fig. 7.11(a) the indirect interaction is $-v_{11} x_1^2 x_2 h_{12}$ and in Fig. 7.11(b), $-v_{12} x_2^2 x_1 h_{12}(q)$. Adding all the terms and dividing by $x_1 x_2$ gives

$$h_{12}(q) = -v_{12} - v_{11} x_1 h_{12}(q) - v_{12} x_2 h_{22}(q) \tag{7.116}$$

and by exchanging 1 and 2

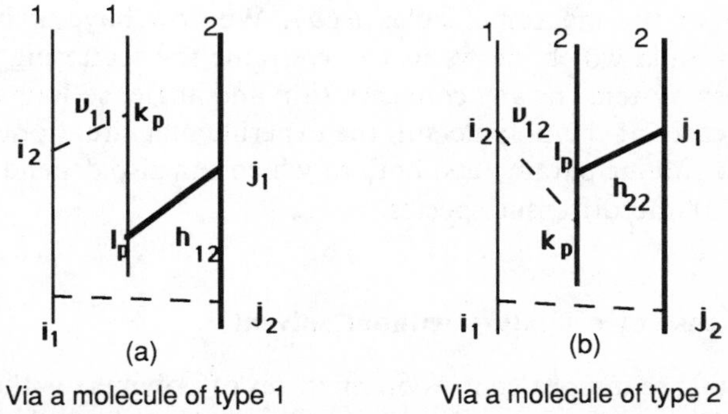

Via a molecule of type 1 Via a molecule of type 2

Fig. 7.11 Interactions between molecules 1 and 2.

$$h_{12}(q) = -\upsilon_{12} - \upsilon_{22}x_2h_{12}(q) - \upsilon_{12}x_1h_{11}(q). \qquad (7.116')$$

Since we have four relationships connecting the h_{ij} we can evaluate the $S_{ij}(q)$ as functions of the υ_{ij} and the x_i. Due to the properties of the functions S_{ij} it is clear that $S_{ij} = S_{ji}$ which means that, in fact, three relationships would be sufficient. Looking carefully at the eqns (7.114)–(7.116) it is clear that there is a matrix equation, of the same form as the matrix of eqn (7.108)

$$[h] = -[\upsilon]([1] + [x][\upsilon])^{-1} \qquad (7.117)$$

where [1] is the unit matrix having 1 on its diagonal and zero everywhere else. $[h]$ and $[\upsilon]$ are the matrices built from the h_{ij} and the υ_{ij} respectively and $[x]$ a diagonal matrix with only x_i on its first diagonal. This equation is a generalization of eqn (7.107′) for multicomponent systems and, using eqn (7.112), gives for the matrix $[S(q)]$

$$[h][x] = [S][x]^{-1} - [1] \qquad (7.118)$$

or

$$[S(q)]^{-1} = [[x(q)]^{-1} + [\upsilon]]. \qquad (7.119)$$

Using now the scattering intensity per cell, where the cell has the volume of a solvent molecule leads to

$$[s(q)]^{-1} = [[x(q)]^{-1} + [v]] \qquad (7.120)$$

where $x(q) = x(q)/N_{\mathrm{T}} = \varphi_i z_i P_i(q)$ and $v = \upsilon/v_0$.

It is easy to show that, if we increase the number of species present in the system, this matrix equation can be generalized without any difficulty. Comparing this equation with eqn (7.69) we see that, in order to extend the results obtained at zero angle, it is sufficient to replace the diagonal matrix $[x]$ by the matrix $[x(q)]$ introducing the q dependence through the form factor of the macromolecules $P(q)$. We now have at our disposal a general formula which allows us to determine the scattering intensity of any polymeric system for any concentration and angle, so long as we know the form factors of the polymers in the experimental situation studied and the excluded volume parameters, both of which can also depend on the concentrations of the different species.

7.7.2 The case of a mixture without solvent

In many cases the interest centres on mixtures of polymers without solvent. In the problem we have discussed so far the solvent was always present. Now we would like to extend the discussion to provide relationships valid for mixtures. In fact, unless one uses the method of de Gennes called the 'random phase approximation', it is not possible to obtain these formulae directly. There are, fortunately, two indirect methods which can be used easily. The first one consists of using the fact that the formulae are valid at any concentration of polymers and solvent: the concentration of the solvent is decreased until it reaches zero. This can be done easily but it has the disadvantage of obliging us to start the calculation with $(P + 1)$ components when the problem requires only P components.

The second method is based on the following argument. In eqn (7.69) which is the zero angle value of eqn (7.121) the solvent could also be a polymer by an obvious change in the equation of the free energy of mixing, as is done in Appendix 2 (eqn (A2.61)). The equation is identical except for the fact that the volume fraction of the solvent φ_0 is replaced by the quantity $z_0\varphi_0 P_0(q)$ where z_0 is the ratio of the volume of the polymer to an arbitrary volume (the same for all the species present in the system) and $P_0(q)$ the form factor of the polymer considered as a solvent. In order to take this route we use the Flory–Huggins theory and replace the v_{ij} by their new definition

$$v_{ik} = \frac{1}{\varphi_0 z_0 P_0(q)} + \chi_{ik} - \chi_{i0} - \chi_{k0} \tag{7.121}$$

where $P_0(q)$ is the form factor of the polymer used as a solvent.

As an example of this procedure let us consider the case of a polymer in a solution for which the Zimm equation can be written as

$$\frac{1}{s(q)} = \frac{1}{\varphi z P(q)} + \frac{1}{\varphi_0} - 2\chi. \tag{7.122}$$

This will give for a solution of polymer 1 in polymer 2

$$\frac{1}{s(q)} = \frac{1}{\varphi_1 z_1 P_1(q)} + \frac{1}{\varphi_0 z_0 P_0(q)} - 2\chi. \tag{7.123}$$

This formula is symmetrical with respect to both polymers and could also be obtained starting from eqn (7.11) by elimination of the solvent. In Chapter 5 we obtained the same result by a simple approximation (see eqn (5.52)), except for the fact that the parameter χ was missing. Equation (7.123) justifies that result and extends it to the more general case where thermodynamic interactions are taken into account.*

* It is also possible to obtain the term 2χ starting from eqn (5.52) by the following argument: at $q = 0$, $s^{-1}(0)$ is the second derivative of g_c. One can generalize this result by saying that, at any q, $s^{-1}(q)$ is the Fourier transform of the second derivative with respect to the volume fraction of the polymer of the functional $g_c(\varphi, r)$ which takes into account the composition fluctuations. From eqn (5.52) one evaluates the function $g_c(\varphi, r)$ without interactions. One adds the term $\chi\varphi(1 - \varphi)$ to this function and takes its second derivative with respect to φ assuming χ independent of r. This leads just to the term -2χ which is added to $s^{-1}(q)$.

7.7.3 The case of copolymers

Discussing the case of homopolymers we found a clear relationship between the $s_{ij}(q)$ and the second derivatives of the chemical potentials. This is no longer true in the case of co-polymers; they need a special treatment. In order to understand this let us consider an incompressible diblock copolymer in bulk. From a thermodynamic point of view it is a one-component system which, being incompressible, does not scatter neutrons. This is correct at zero angle (only if the system is monodisperse in composition) but we have seen that, at $q \neq 0$, $I(q)$ is different from zero. From an optical point of view, since the scattering lengths of the two species of monomer are different they will scatter differently depending on the scattering length of the solvent. One has therefore to introduce different partial structure factors $S_{ij}(q)$ which exist even if the two types of monomers are on the same molecule. This justifies a special treatment of copolymers. In fact this treatment is just an extension of the preceding arguments. We do not intend to discuss it in detail and in order to show the methods utilized we will discuss only the second virial coefficient in a dilute solution of copolymer and give only the general formula as an extension of eqn (7.121).

Since we have already discussed the case of copolymers several times we shall be brief. We assume that we have in solution N molecules of copolymer each one made of z_a monomers of type a and z_b monomers of type b. The degree of polymerization of this copolymer is $z = z_a + z_b$. It will be convenient to introduce the volume fraction of part a of the polymer, $u = z_a/z$ and of part b, $v = z_b/z$, $(u + v = 1)$. The monomers of type a have a constrast factor $\bar{b}_a$ with the solvent, the monomer b, $\bar{b}_b$.

As has already been discussed we have to introduce three form factors: the form factor of the monomers of type a, $P_a(q)$, the form factor of the monomers of type b, $P_b(q)$ and the cross form factor $P_{ab}(q)$. These quantities depend on the structure of the polymer and can be evaluated using the methods described in Chapter 6[†]. We also introduce, as in the case of homopolymers, the three excluded volumes v_{aa}, v_{bb}, and v_{ab} corresponding to contacts between two monomers a, two monomers b and one monomer a and one monomer b, respectively.

7.7.4 Generalization of the Zimm equation

We recall that in this case, following eqn (7.5), the scattering intensity has to be written as

[†] For instance if the copolymer is a statistical copolymer with short sequences one has evidently $P_{aa} = P_{bb} = P_{ab}$.

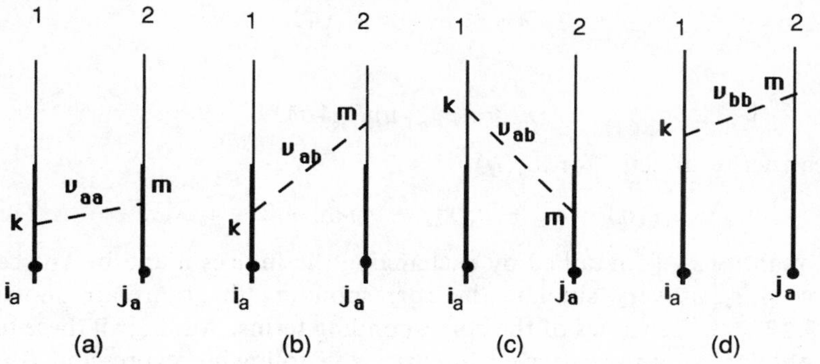

Fig. 7.12 The heavy lines represent part a of the copolymer. (a) Contacts between two a monomers, (b) and (c) contacts between a and b, (d) corresponds to contacts between two b monomers.

$$I(q) = \bar{b}_a^2 S_{aa}(q) + \bar{b}_b^2 S_{bb}(q) + 2\,\bar{b}_a \bar{b}_b S_{ab}(q) \tag{7.124}$$

The calculation of S_{aa} follows the method used in the case of homopolymers: we subtract from the intensity scattered by the part (a) of the molecules $(Nz_a^2 P_a(q))$ the forbidden configurations considering only single contacts. But if we look at diagrams similar to Fig. 7.9(a) we realize immediately that we have to take into account the four possibilities of Fig. 7.12. Diagram (a) corresponds to a contact between a monomers and its contribution to the scattering is, as in the case of homopolymers:

$$-N^2 z_a^4 P_a^2(q)\, v_{aa}.$$

In the diagram b (or c) there is one monomer (a) interacting with one monomer (b) the excluded volume contribution is v_{ab}. When going from i_a to j_a (case b) one goes first from i_a to k which gives $N_a z_a^2 P_a(q)$ then from k to m $(-v_{ab})$ and finally from m to j_a. In this last step we go from one b point of the copolymer to an a point and the corresponding form factor is clearly $P_{ab}(q)$. Summarizing, we have for the four cases

$$\text{(a)}\ -N^2 z_a^4 P_a^2(q)\, v_{aa}, \qquad \text{(b)}\ -N^2 z_a^3 z_b P_a(q) P_{ab}(q) v_{ab},$$

$$\text{(c)}\ -N^2 z_a^3 z_b P_a(q) P_{ab}(q) v_{ab}, \quad \text{(d)}\ -N^2 z_a^2 z_b^2 P_{ab}^2(q) v_{ab}.$$

Now, using the quantities $s(q)$ (corresponding to the scattering by the unit cell), instead of $S(q)$ and therefore the quantities x_{aa}, x_{bb}, and x_{ab} defined as

$$x_{aa} = \frac{N_a z_a^2 P_a(q)}{N_T} = \frac{N_a z_T^2 u^2 P_a(q)}{N_T} = \varphi_T z_T u^2 P_a(q) \tag{7.125}$$

where φ_T and z_T are the volume fraction and the total molecular weight of the copolymer

$$x_{bb} = \varphi_T z_T v^2 P_b(q) \qquad (7.126)$$

and

$$x_{ab} = \varphi_T z_T uvP_{ab}(q)$$

we can write directly for $s_{aa}(q)$

$$s_{aa}(q) = x_{aa} - v_{aa}x_{aa}^2 - v_{bb}x_{ab}^2 - 2v_{ab}x_{aa}x_{ab}. \qquad (7.127)$$

The quantity s_{bb} is obtained by exchanging the indices a and b. The calculation of s_{ab} is very similar; the corresponding diagrams are shown in Fig. 7.13 with the values of the corresponding terms. Adding all these terms one obtains for the scattering intensity the following expression (Benoît *et al.* 1985)

$$\frac{I(q)}{N_T} = \bar{b}_a^2[x_{aa} - v_{aa}x_{aa}^2 - v_{bb}x_{ab}^2 - 2v_{ab}x_{aa}x_{ab}] + \bar{b}_b^2[x_{bb} - v_{bb}x_{bb}^2 - v_{aa}x_{ab}^2 - 2v_{ab}x_{bb}x_{ab}] + 2\bar{b}_a\bar{b}_b[x_{ab} - v_{aa}x_{aa}x_{ab} - v_{bb}x_{bb}x_{ab} - (x_{aa}x_{bb} + x_{ab}^2)v_{ab}]. \qquad (7.128)$$

This expression is rather complex and shows that the behaviour of copolymers in dilute solution is very different from the behaviour of homopolymers and depends drastically on the solvent. For more details of this subject it is worth referring to the literature.

7.8 GENERAL EQUATION

From what has been discussed already the treatment of the copolymers is essentially the same as the treatment of the homopolymers but is a little

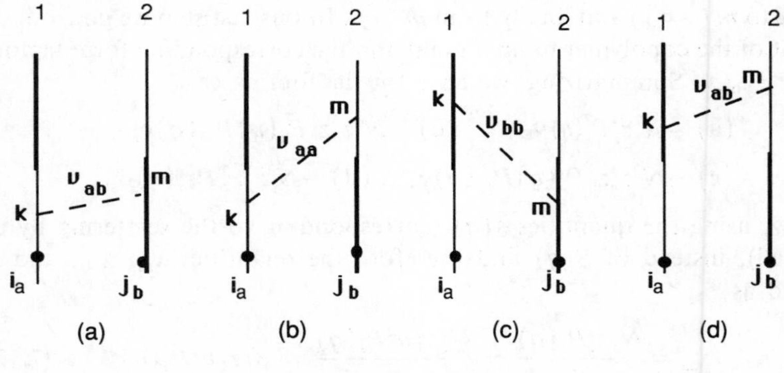

$-x_{aa}x_{bb}v_{ab}$ $\quad -x_{aa}x_{ab}v_{aa}$ $\quad -x_{ab}x_{bb}v_{bb}$ $\quad -x_{ab}^2v_{ab}$

Fig. 7.13 Evaluation of the term $S_{ab}(q)$ for a block copolymer.

more complex since we have to take into account more diagrams. The problem is the same when we apply the Ornstein–Zernike formalism. We shall not derive the corresponding calculations. Anyone interested can consult the literature (Benoît *et al.* 1990) and we give here only the results for a copolymer made of blocks of species a and b.

The matrix formulation of the problem yields

$$\begin{pmatrix} s_{aa}(q) & s_{ab}(q) \\ s_{ab}(q) & s_{bb}(q) \end{pmatrix} \begin{pmatrix} 1 + v_{aa}x_{aa} + v_{ab}x_{ab} & v_{ab}x_{bb} + v_{aa}x_{ab} \\ v_{ab}x_{aa} + v_{bb}x_{ab} & 1 + v_{bb}x_{bb} + v_{ab}x_{ab} \end{pmatrix} = \begin{pmatrix} x_{aa} & x_{ab} \\ x_{ab} & x_{bb} \end{pmatrix}. \tag{7.129}$$

The second matrix on the LHS can be written as

$$[1] + \begin{pmatrix} x_{aa} & x_{ab} \\ x_{ab} & x_{bb} \end{pmatrix} \begin{pmatrix} v_{aa} & v_{ab} \\ v_{ab} & v_{bb} \end{pmatrix} = [1] + [x]\,[v] \tag{7.130}$$

and, after elementary transformation eqn (7.130) becomes

$$[s(q)]^{-1} = [x(q)]^{-1} + [v]. \tag{7.131}$$

This result is especially simple. In order to include the case of copolymers it is sufficient to replace the diagonal matrix of the x_i in eqn (7.121) by a matrix containing also the non-diagonal terms used in the definition of the copolymers. It is clear that this matrix equation is valid for any copolymer regardless of its structure, the number and the relative position of the sequences or the number of species of monomers. As a simple example we give the final formula for a copolymer composed of two monomers a and b which was discussed earlier. Resolving eqn (7.131) gives

$$i(q) = \frac{\bar{b}_a^2 x_a + \bar{b}_b^2 x_b + 2\bar{b}_a\bar{b}_b x_{ab} + (x_a x_b - x_{ab}^2)(\bar{b}_a^2 v_b + \bar{b}_b^2 v_a - 2\,\bar{b}_a\bar{b}_b v_{ab})}{1 + v_a x_a + v_b x_b + 2v_{ab}x_{ab} + (v_a v_b - v_{ab}^2)(x_a x_b - x_{ab}^2)}. \tag{7.132}$$

If we eliminate the solvent by letting its volume fraction go to zero using for the v_{ij} the expressions given in eqn (7.65) we obtain

$$\frac{1}{i(q)} = (b_a - b_b)^2 \frac{x_a + x_b + 2x_{ab}}{x_a x_b - x_{ab}^2} - 2\chi_{ab} \tag{7.133}$$

which is again the equation obtained in eqn (5.61) but this time taking into account the interactions between the two different species of monomers.

7.9 THE CASE OF POLYDISPERSE SYSTEMS

Before completing the discussion of these results it is useful to discuss rapidly the case of polydisperse systems. Going back to the case of monodisperse systems we note, looking at all the results which have been obtained, that the polymers of one species enter the formulae, only via the quantity $Nz^2P(q)$ or if one uses the quantities $s(q)$ via $\varphi\, zP(q)$. Nothing would be changed if, instead of N identical molecules of degree of polymerization z and structure factor $P(q)$ we considered a polydisperse system made of different molecules, each of them being characterized by the number of individual molecules N_α of degree of polymerization z_α and structure factor $P_\alpha(q)$. The only difference would be that the quantity $x_i = N_i z_i^2 P_i(q)$ is replaced by

$$\langle x_i(q)\rangle = \sum_\alpha N_{i\alpha} z_{i\alpha}^2 P_{i\alpha}(q) \tag{7.134}$$

or, when evaluating $x_i(q)$, by

$$\langle x_i(q)\rangle = \sum_\alpha N_{i\alpha}\varphi_{i\alpha} z_{i\alpha} P_{i\alpha}(q). \tag{7.135}$$

In the case of copolymers the definitions are identical except for the cross term which, considering the general case where the copolymers can have more than two different monomers, has to be written as

$$\langle x_{\mathrm{ab}}(q)\rangle = \frac{1}{N_{\mathrm{T}}}\sum_i \sum_j N_{i\mathrm{a}j\mathrm{b}} z_{i\mathrm{a}} z_{j\mathrm{b}} P_{i\mathrm{a}j\mathrm{b}}(q) \tag{7.136}$$

where $N_{i\mathrm{a}j\mathrm{b}}$ is the number of molecules made with both monomers (a) and (b). These formulae can be applied to any polydisperse system containing homo- and copolymers as long as the scattering length of all the monomers is independent of z.

Copolymers are no different from homopolymers since we just have to add a term in x_{ab} in the matrix $[x]$ in order to take into account the fact that both monomers belong to the same molecule. As a simple example let us consider the case of the pure copolymer a, b and suppose that it is made of two very special copolymers, the first one made of only A monomers ($u = 1, v = 0$) and the second one of B monomers ($u = 0, v = 1$). For each of them $x_{\mathrm{ab}} = 0$ and x_{a} and x_{b} correspond to the first and the second monomer respectively. For this system eqn (7.133) immediately becomes eqn (7.124) valid for a mixture of homopolymers. These equations have been used to describe the scattering of many systems. The only problem is to evaluate the different terms of the matrix eqn (7.131) especially if the x_{ij} and the v_{ij} depend on concentration.

Equation (7.131) has been obtained in the context of what the theoreticians call a 'mean field approximation' which assumes that the probability of any contact is proportional to the average concentration. This is evidently not rigorous, especially near a critical point, where there is a tendency for the molecules of the same type to form clusters and large concentration fluctuations. This effect has to be taken into account in more refined theories. As we have shown, this theory is rigorous at $q = 0$ (eqn (7.69)) and thermodynamically correct. This modification of the random phase approximation which does not require incompressibility but which adjusts the values of the parameters x and v in order to take the concentration effects into account, has been quite successful up to the present.

Part 3 Systems existing in more than one phase

7.10 INTRODUCTION

In the preceding parts of this chapter we have carefully studied the scattering by systems which exist in one phase. Unfortunately, this is often not the case and, as soon as the polymer is partially crystalline or presents immiscibility, the theories which have been developed cannot be applied. One has to find other ways to interpret the data and this is particularly difficult. For instance, at the present time, even though a considerable amount of work has been devoted to the problem of ionomers, the interpretation of the neutron scattering, as well as of the X-ray data from these systems is far from satisfactory. This means that the interpretation is difficult and explains why this part of Chapter 7 will not be as extensive as the preceding ones. In the first section, in order to maintain a sound basis we shall approach the multiphase systems of mixtures by considering the evolution of the scattering when one reaches the critical point, coming from the solution side. In a second section we shall give some rules known as the Porod invariants and show different examples of the treatment of random heterogeneous systems and micellar solutions.

7.11 CRITICAL OPALESCENCE

7.11.1 The Ornstein–Zernike approach

We have seen in Section 7.2 how it is possible to explain the scattering by a one-component system or a solution made of small molecules. Since in such a system the excluded volume is small, there is, in the domain of q in which we are interested, no angular dependence on the scattering. It has been known for a long time that when decreasing the temperature one approaches the critical point the scattering increases drastically and becomes angle dependent. In fact the purpose of the Ornstein–Zernike theory was to explain this experimental fact. Let us go back to the O–Z formula (eqns (7.99) and (7.99′)) which we write again

$$S^{-1}(q) = \frac{1}{N} - c(q) \tag{7.137}$$

using, as in the literature devoted to small molecules, the letter c for the excluded volume ($c = -v$).

We have assumed that the quantity c or $-v$ is independent of q since the interactions between molecules are of very short range. This is not true in the neighbourhood of the critical point. The molecules make clusters of larger and larger size and the scattering depends on the vector $\boldsymbol{q}$. In order to explain this experimental fact we assume that $c(q)$ can be put in the form

$$c(q) = c_0 + c_1 q^2 \tag{7.138}$$

where c_0 and c_1 depend on the system. This means that we expand c as a function of q keeping only the first term and neglecting the others. Introducing eqn (7.137) in eqn (7.138) leads to

$$S^{-1}(q) = \left(\frac{1}{N} - c_0\right)\left[1 + \frac{Nc_1}{1 - Nc_0} q^2\right]. \tag{7.139}$$

We now define $Nc_1/(1 - Nc_0)$ as ξ^2 and nc_1 as K^2 obtaining

$$\frac{1}{S(q)} = \frac{K^2}{\xi^2}[1 + q^2\xi^2] = \frac{K^2}{\xi^2} + K^2 q^2. \tag{7.140}$$

Since at the critical point the zero angle scattering is infinity (see Appendix 2) the quantity $1 - Nc_0$ goes to zero; the quantity K^2 on the other hand has no reason to show anomalous behaviour and we shall assume that it is a constant. In these conditions the quantity ξ^2 also diverges and goes to infinity at the critical point. The quantity ξ has the dimension of a length and is called the correlation length. This definition can be justified by the following arguments. In Chapter 4 we have seen that $S(q)$ can be written as the Fourier transform of a function $G(r)$ which is called the pair distribution (it is equal to $g(r) - 1$ if we forget the factor $1/N$ in eqn (7.137)). This can be written as

$$S(q) = \iiint G(r)\exp(-\mathrm{i}\boldsymbol{q}\cdot\boldsymbol{r})\mathrm{d}\boldsymbol{r} \tag{7.141}$$

and by inverse Fourier transform one can evaluate $G(r)$ from $S(q)$ in eqn (7.140) (see Appendix 1)

$$G(r) = \frac{1}{(2\pi)^3} \iiint S(q)\exp(+\mathrm{i}\boldsymbol{q} \cdot \boldsymbol{r})\mathrm{d}\boldsymbol{q}. \tag{7.142}$$

Using the value of $S(q)$ in eqn (7.140) and forgetting a normalization factor unimportant for our discussion, $G(r)$ becomes

$$G(r) \approx \frac{1}{r} \exp\left(-\frac{r}{\xi}\right) \tag{7.143}$$

this explains clearly the meaning of ξ as a correlation length since it shows that the correlations disappear exponentially with a scale ξ. If the mean field approximation cannot be used to describe the system one has to use other methods like scaling laws and renormalization theories. They slightly modify the Ornstein–Zernike calculation; at $T = T_c$ and at large r for instance it is suggested that

$$\left.\begin{aligned} S(q) &\approx q^{-2+\eta} \\ G(r) &\approx r^{-(d-2+\eta)} \end{aligned}\right\} \tag{7.144}$$

where η is a small exponent of the order of 0.041 (Stanley 1971).

7.11.2 Geometrical representation

With an equation like eqn (7.140) the natural way of plotting the results is to use what is called a Zimm plot in polymer science or an Ornstein–Debye plot in the physics of gases and liquids. Such a schematic plot is given in Fig. 7.14. If K^2 is not independent of T the lines would not be parallel contrary to what the experiments usually show. It is difficult, when a diagram of this kind is experimentally obtained for polymer solutions or mixtures, to recognize if one is in the range where the O–Z theory applies or if one is dealing with an ordinary solution of polymer with low polydispersity. In this last case what we have called ξ^2 would be 1/3 of the square of the radius of gyration. Let us look at what happens when starting from eqn (7.123) one approaches the critical point. For simplicity we assume that the two polymers have the same degree of polymerization and the same form factor obtaining

$$S^{-1}(q) = \frac{1}{\varphi(1-\varphi)zP(q)} - 2\chi. \tag{7.145}$$

Since the system is symmetrical, the critical point corresponds to $\varphi = 0.5$. The critical temperature is obtained when $S^{-1}(0) = 0$. Using these values of $S^{-1}(0)$ and φ_c gives for χ_c the value $z\,\chi_c = 2$.

At $\varphi = 0.5$, $S^{-1}(q)$ can be written as

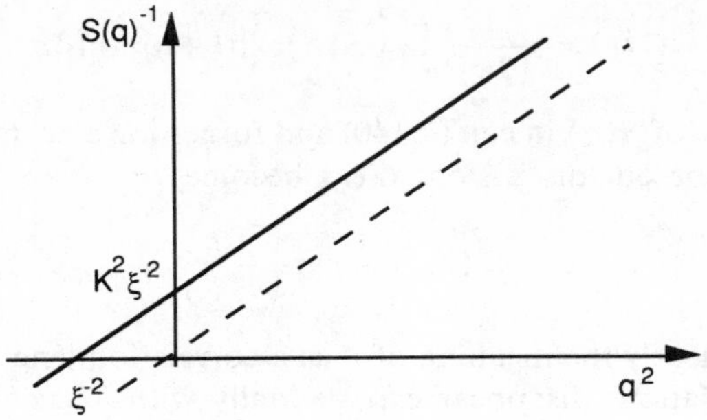

Fig. 7.14 The solid line corresponds to a scattering experiment made slightly above the critical point; the dashed line to an experiment at the critical point. The slope after normalization of the intercept to unity gives the correlation length ξ which can also be obtained from the intercept of the scattering line with the q^2 axis.

$$S^{-1}(q) = \frac{4}{z}\frac{1}{P(q)} - 2\chi \tag{7.146}$$

we can also, using eqn (7.146) write this equation in the form

$$S^{-1}(q) = \frac{4}{z}\left[\frac{1}{P(q)} - 1\right] + \frac{4}{z} - 2\chi = \frac{4}{z}\left[\frac{1}{P(q)} - 1\right] + 2(\chi_c - \chi) \tag{7.147}$$

If we limit ourselves to the first term of the expansion of $S^{-1}(q)$ as function of q^2 we obtain, calling R^2 the radius of gyration of the chains

$$S^{-1}(q) = 2(\chi_c - \chi) + \frac{4q^2R^2}{3z} \tag{7.148}$$

which, written in the form of eqn (7.140) gives for the length ξ

$$\xi^2 = \frac{2R^2}{3z}\frac{1}{(\chi_c - \chi)} \tag{7.149}$$

and for the parameter K

$$K^2 = \frac{4R^2}{3z}. \tag{7.150}$$

The parameter ξ^2 can be measured from the data, either normalizing the intercept to unity and measuring the slope as in the classical Zimm plot, or taking the intercept of the straight line with the x-axis which gives ξ^{-2}.

As suggested previously, the parameter K does not change rapidly with temperature. On the other hand the parameter ξ^2 varies as $(\chi_c - \chi)^{-1}$; if, as it is frequently assumed, χ is a linear function of $1/(T_c - T)$, ξ will be proportional to $1/\sqrt{(T_c - T)}$. This shows that, in the case of large molecules, one does not need to have an excluded volume parameter or a χ coefficient depending on q. These exponents are the classical exponents obtained using a mean field theory but more refined theories may give different values. The problem of the determination of critical exponents has been the subject of many experimental studies. To summarize the preceding discussion the mean field theories (O–Z theory for small molecules and the generalization of Flory theory by the random phase approximation theory for macromolecules) give essentially the same results. The only difference being that, for polymers one does not have to assume q dependence of the excluded volume parameter or of the interaction parameter χ.

7.11.3 Screening length in polymer solutions

In the de Gennes (1973) formula, regardless of the concentration, $S(q)$ decreases as q^{-2} at large q. This suggest the use of the O–Z plot and the interpretation of the quantity ξ in terms of polymer solutions behaviour. Let us first obtain the expression for ξ. We start from eqn (7.94′) and use the expansion of $P(q)$ as function of q^{-1} (eqn (6.108)) and we obtain

$$\frac{\bar{b}^2}{i(q)} = \frac{1}{z\varphi^2 \dfrac{2}{x}\left(1 - \dfrac{1}{x}\right)} + v \qquad (7.151)$$

valid when $x = zq^2l^2/6 = q^2R^2 \gg 1$. Simplifying this expression in order to put it in the O–Z form gives

$$\frac{\bar{b}^2}{i(q)} = \frac{l^2}{12\varphi}\left[q^2 + \frac{1}{R^2} + \frac{12\varphi v}{l^2}\right]. \qquad (7.152)$$

If the molecular weight is large the term $1/\overline{R^2}$ can be neglected and, when we are in semidilute solutions

$$\xi^{-2} = \frac{12\varphi v}{l^2}. \qquad (7.153)$$

In polymer physics, this quantity is called the screening length. When two monomers are separated by a distance larger than ξ there are no more interactions between them: the excluded volume has no influence and the chain behaves like a Gaussian chain made of segments of length ξ (called blobs by de Gennes (1979)). The preceding relationship, obtained by Edwards (1966), leads if one uses the des Cloizeaux's law (1975) $v \approx \varphi^{1/4}$, to $\xi \approx \varphi^{-5/8}$ which is slightly different from the de Gennes (1979) result obtained without using the mean field approximation $\xi \approx \varphi^{-3/4}$.

7.12 THE TWO-DENSITY MODEL

In many cases we could treat the system as a two-density model. We assume that there is a matrix of density ρ_1 in which are imbedded particles of density ρ_2. More generally the density in the system is either ρ_1 or ρ_2 depending on the vector $\boldsymbol{r}$. If the two phases have sharp boundaries there are general rules, frequently called the Porod law and the Porod invariant which give some useful results and which we shall justify in the following section.

7.12.1 The Porod law

In our discussion of the form factors we have seen that, if the boundary between the scattering objects and the matrix is sharp, the scattering varies as q^{-4} and must be proportional to the total interface S_T between the two homogeneous media. We were able to obtain a formula (eqn (6.123)) using as model a solution of identical hard spheres obtaining

$$q^4 I(q) = b_v^2 2\pi NS = b_v^2 2\pi S_T \tag{7.154}$$

where S_T is the total surface of the spheres, i.e. $N4\pi R^2$ (N is the number of spheres) and b_v the contrast factor per unit volume between the two media. Since we know that for large q the result depends only on the behaviour at short distances this result can be generalized to any two-phase system with sharp boundaries giving the general law

$$\text{Lim.}_{(q\to\infty)} q^4 I(q) = b_v^2 2\pi S_T \tag{7.155}$$

where S_T is the total interface. This relationship is rigorous for any structure of a two-phase system as long as its interface is sharp and the value of q^{-1} is smaller than the defects of the surface and much larger than its radius of curvature; it is the Porod law which was discussed earlier (eqn (6.123)).

7.12.2 The Porod invariant

General scattering equation

Let us go back to the general definition of the scattering and write eqn (7.4) in a slightly different manner

$$I(q) = \int_V\int_V \langle \Delta b_v(r)\Delta b_v(r')\rangle \exp[-i\boldsymbol{q}\cdot(\boldsymbol{r}-\boldsymbol{r}')]\mathrm{d}\boldsymbol{r}\mathrm{d}\boldsymbol{r}' \tag{7.156}$$

introducing the fluctuations $\Delta b(r)$ of the local density $b(r)$ of scattering length per unit volume. Using the translational invariance and the isotropy of the system we write that $\langle \Delta b(r)\Delta b(r')\rangle = \langle \Delta b(0)\Delta b(r)\rangle$. This gives, for eqn (7.153)

$$I(q) = V\int_V \langle \Delta b(0)\Delta b(r)\rangle \exp(-\mathrm{i}\boldsymbol{q}\cdot\boldsymbol{r})\mathrm{d}r = V\int_V \langle \overline{\Delta b(0)^2}\rangle \Gamma(r)$$

$$\exp(-\mathrm{i}\boldsymbol{q}\cdot\boldsymbol{r})\mathrm{d}\boldsymbol{r} \tag{7.157}$$

introducing the Debye correlation function defined by the relationship (very similar to eqn (4.46))

$$\langle \Delta b(0)\Delta b(r)\rangle = \overline{\Delta b(0)^2}\Gamma(r). \tag{7.158}$$

The normalization factor fixes $\Gamma(0) = 1$.

From eqn (7.154), $I(q)/V$ is the Fourier transform of $\overline{\Delta b(0)^2}\Gamma(r)$. Reciprocally, following eqn (A1.17) of Appendix 1, one can say that $\overline{\Delta b(0)^2}\Gamma(r)$ is the inverse Fourier transform of $I(q)/V$ and write

$$\frac{1}{(2\pi)^3}\iiint I(q)\exp(\mathrm{i}\boldsymbol{q}\cdot\boldsymbol{r})\mathrm{d}\boldsymbol{q} = V\overline{\Delta b^2}(0)\Gamma(r) \tag{7.159}$$

which is the general equation relating the scattered intensity to the correlation function. We now make $\boldsymbol{r} = 0$ in the preceding equation obtaining

$$\frac{1}{(2\pi)^3}\iiint I(q)\mathrm{d}\boldsymbol{q} = V\overline{\Delta b^2}(0). \tag{7.160}$$

Since by hypothesis we are working in an isotropic system we can immediately integrate over the angular variables, obtaining

$$2\pi^2\overline{\Delta b^2}(0)\,V = \int_0^\infty I(q)q^2\mathrm{d}q. \tag{7.161}$$

This equation relates the integral of the experimental signal $I(q)$ to the average mean square of the scattering length fluctuations . We must be cautious about the definition of $\overline{\Delta b^2}(0) = b^2\overline{\Delta N^2}(0)$ which is a quantity very different from the quantity $\overline{\Delta N^2}$ used in Part 1 corresponding to the scattering at zero angle. The former is the fluctuation for a distance $r = 0$, the latter is the integral of $\Delta N(0)\Delta N(r)$ over the all-space.

Calculation of $\overline{\Delta b^2}(0)$

Let us go back to our two-density medium. We assume that the component 1 of scattering length density b_1 per unit volume occupies the volume fraction φ_1, the second, with scattering length density b_0 the volume fraction $\varphi_0 = 1 - \varphi_1$. The average density in the system is therefore

$$\langle b \rangle = b_1' \varphi_1 + b_0' \varphi_0 .$$

In the region 1 of the system

$$\Delta b_1 = b_1' - \langle b \rangle = (b_1' - b_0')(1 - \varphi_1) \qquad (7.162)$$

and in the region 0

$$\Delta b_0 = b_0' - \langle b \rangle = (b_0' - b_1')\varphi_1 .$$

The average value of the square of the fluctuations of $\overline{\Delta b^2}(0)$ (eqn (7.158)) becomes

$$\overline{\Delta b^2}(0) = \Delta b_1^2 \varphi_1 + \Delta b_0^2 (1 - \varphi_1) = (b_1' - b_0')^2 \varphi (1 - \varphi) \qquad (7.163)$$

calling φ either φ_1 or φ_0.

We introduce the correlation function $\Gamma(r)$ by the relationship

$$\overline{\Delta b^2}(r) = \overline{\Delta b^2}(0)\Gamma(r) = b_v^2 \varphi (1 - \varphi)\Gamma(r) \qquad (7.164)$$

where we have used, as in Appendix 3, the relationship $b_v = b_1/v_1 - b_o/v_0$ and we obtain, putting this value in eqn (7.159), after writing $r = 0$ and $\Gamma(0) = 1$

$$2\pi^2 b_v^2 \varphi (1 - \varphi) = \frac{1}{V} \int_0^\infty I(q) q^2 \mathrm{d}q . \qquad (7.165)$$

This relationship (Porod 1951) is very important since, for instance, it allows verification of whether a system existing under two phases is mixed or not. All quantities appearing in the LHS of eqn (7.165) are known and by integrating the experimental signal multiplied by q^2 one should obtain an identity. If the identity is not satisfied, either the system is partially mixed (the phases are not made of pure 1 and pure 2) or the interfaces are broad. Extensions of this formula to systems with broad interfaces have been proposed but we will not discuss them. We would like to stress that, contrary to what has been done in the literature, this equation cannot be applied to polymeric solutions or gels. In these cases we know that $I(q)$, at large q, varies as q^{-2} or a power less than two in a good solvent (eqn (6.93)). The integral $I(q)q^2$ diverges which means that the interface is too broad and eqn (7.165) has no meaning. This equation has been obtained in a continuous medium approximation which cannot be extrapolated to systems like chains where the lateral dimensions of the monomers have not been taken into account.

7.13 THE DEBYE–BUECHE EQUATION

It is tempting to look at the kind of signal one should observe for $I(q)$ when a simple expression is used for $\Gamma(r)$. This is what has been done by Debye and Bueche (1949) and Debye *et al.* (1957) who proposed for $\Gamma(r)$ the form

$$\Gamma(r) = \exp - r/\xi \tag{7.166}$$

where ξ is a kind of correlation length. Putting this value in eqns (7.161) and (7.154) we recognize an integral which is calculated in Appendix 1 (eqn A1.56) and obtain

$$\frac{I(q)}{V} = b_v^2\varphi(1-\varphi)\iiint_V \exp(-r/\xi)\exp(-i\boldsymbol{q}\cdot\boldsymbol{r})\mathrm{d}\boldsymbol{r}$$

$$I(q) = Vb_v^2\varphi(1-\varphi)8\pi\xi^3\left\{\frac{1}{1+q^2\xi^2}\right\}^2. \tag{7.167}$$

As we should have guessed we obtain for large q the Porod law $I \approx q^{-4}$. Plotting $I^{-1/2}$ as function of q^2 gives the classical and well-known linear variation.

This gives a very simple way of obtaining the correlation length ξ using the methods which were established studying the scattering in the vicinity of the critical point (Section 7.11.1).

This radial distribution function, which has been introduced by Debye and Bueche, seems to be very simple and has been justified quantitatively. In order to justify it qualitatively let us use the following argument. We consider a scattering point M and a straight line going through it. On this line are segments which are made of substance 1 and segments made of 2. The model assumes that the lengths of these segments are randomly distributed. This problem is similar to the problem of the random degradation of a polymer of infinite length. It is known by all physical chemists that for such a polymer the molecular length distribution is the most probable distribution which for the number of molecules or segments gives an exponential function (see Chapter 6). It is clear that this exponential function is related to $g(r)$ and allows us to understand why the exponential function corresponds to a medium where the two species are randomly distributed. From a physical point of view it is difficult to understand why there is such a striking difference between the O–Z model and the Debye and Bueche model. In the first one has a molecular dispersion of one species in the other, while, in the second one has randomly distributed domains with sharp interfaces. When, in a compatible mixture, the temperature decreases below T_c at the critical composition in a classical liquid, the power law at high q should go from q^{-2} above T_c to q^{-4} below T_c. Again this model requires sharp interfaces and the number of cases where it can be applied without modification remains rather limited in the polymer field.

7.14 SCATTERING BY SPHERICAL PARTICLES

In this last paragraph we would like to discuss the case of a suspension of spheres of sufficient concentration (Cabane 1987) to prohibit the use of the dilute solution approximation. It is evident that we could use the Porod invariant and the Porod law to characterize the scattering of these systems, but we can do more than study the large angle behaviour and the integrated intensity, which are difficult to measure. Splitting as before, the scattering into its intramolecular part and its intermolecular part, gives

$$I(q) = b_v^2\left[NV^2P(q) + N(N-1)\sum_{i_1}\sum_{j_2}\langle\exp(-\mathrm{i}\boldsymbol{q}\cdot\boldsymbol{r}_{i_1j_2})\rangle\right] \tag{7.168}$$

where we have to sum over all the points of molecules 1 and 2 and average over all their relative positions. Using the diagram of Fig. 7.15 we write

$$\boldsymbol{i}_1\boldsymbol{j}_2 = \boldsymbol{i}_1\boldsymbol{C}_1 + \boldsymbol{C}_1\boldsymbol{C}_2 + \boldsymbol{C}_2\boldsymbol{j}_2.$$

Now again the three vectors $\boldsymbol{i}_1\boldsymbol{C}_1$, $\boldsymbol{C}_1\boldsymbol{C}_2$, and $\boldsymbol{C}_2\boldsymbol{j}_2$ are independent and the exponential of eqn (7.168) can be split into the product of three exponentials

$$\langle\exp(-\mathrm{i}\boldsymbol{q}\cdot\boldsymbol{r}_{i_1j_2})\rangle = \langle\exp(-\mathrm{i}\boldsymbol{q}\cdot\boldsymbol{r}_{i_1C_1})\rangle\langle\exp(-\mathrm{i}\boldsymbol{q}\cdot\boldsymbol{r}_{C_1C_2})\rangle \langle\exp(-\mathrm{i}\boldsymbol{q}\cdot\boldsymbol{r}_{C_2j_2})\rangle. \tag{7.169}$$

The summation over i and j in the first and the last terms of eqn (7.169) gives the amplitude scattered by a sphere with the origin at its centre and we know that the square of this quantity is proportional to the form factor of the sphere. We can therefore transform eqn (7.165) into

$$I(q) = b_v^2V^2P(q)\,[N + N^2\langle\exp(-\mathrm{i}\boldsymbol{q}\cdot\boldsymbol{r}_{C_1C_2})\rangle] \tag{7.170}$$

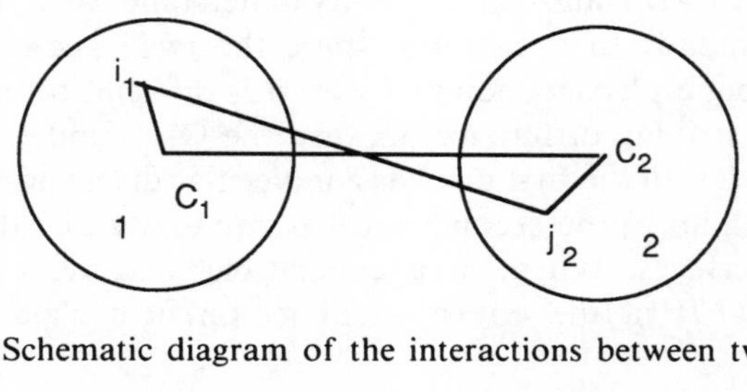

Fig. 7.15 Schematic diagram of the interactions between two spheres.

$$I(q) = b_v^2 N V^2 P(q) [1 + Nh(q)] \tag{7.171}$$

with

$$h(q) = \langle \exp(-\mathrm{i}\boldsymbol{q} \cdot \boldsymbol{r}_{C_1 C_2}) \rangle. \tag{7.172}$$

There is decoupling between the form factor of the spheres and the interaction of their centres. The term in the square brackets is identical to the term we obtained in the case of small molecules (eqn (7.100)) and could be called $S(q)$ because it is the scattering by the centre of the spheres. The dependence on q of $P(q)$ and $S(q)$ has to be taken into account when the radius of the spheres is of the order of q^{-1} and (or) the interactions like electrostatic interactions are long range.

The form of eqn (7.170) allows the theories developed in the statistical mechanics of gases and liquids to be applied. These theories are essentially based on the O–Z equation coupled to another equation called the 'closure relationship'. These latter are necessary in order to be able to evaluate the functions $h(q)$ and $c(q)$ (which is equal to $-v(q)$). Different types of relationships have been proposed and are well-known (Hayter and Penfold 1883; Hansen and McDonald 1986) but will not be discussed here.

As a simple example of the effect of concentration on the structure factor we shall consider the case of hard spheres of radius R without any other interaction than their excluded volume (Cebula *et al.* 1982). The form factor of these spheres is given by the classical expression (eqn (6.50))

$$P(q) = A_0^2(qR) = \left[\frac{3}{u^3}(\sin u - u\cos u)\right]^2 \tag{7.173}$$

with $u = qR$.

The excluded volume parameter is the amplitude scattered by a sphere of radius $2R$

$$v(q) = 8VA_0(2qR) = V\left[\frac{3}{(2u)^3}(\sin 2u - 2u\cos 2u)\right]. \tag{7.174}$$

So we obtain

$$I(q) = b_v^2 N V^2 P(q) [1 - Nv(q)]. \tag{7.175}$$

If the volume fraction occupied by the spheres exceeds 0.125 the scattering intensity at zero q becomes negative which is impossible. This reduces the range of validity of this approximation to very low values of the volume fraction of the spheres. Since the use of the O–Z formula (eqn (7.99′)) has been so successful for polymer solutions even using the dilute solution value for $\boldsymbol{v}$ it is tempting to use the same approximation

$$I(q) = b_v^2 V^2 P(q) \frac{N}{1 + Nv(q)} = b_v^2 N V^2 \frac{P(q)}{1 + 8\varphi A_0(2qR)} \tag{7.176}$$

where, following the notation of Section 7.10, N is the number of particles per unit volume and NV is the volume fraction φ occupied by the particles. Of course this is an approximation which becomes poorer and poorer when the concentration increases but it is interesting to look qualitatively at the influence of the concentration on the shape of the diagrams $I(q)$ vs q^2R^2. This is what has been done in Fig. 7.16(a) where we have plotted $I(q)/\varphi$ as a function of q^2

$$\frac{I(q)}{\varphi} \approx \frac{P(qR)}{1 + 8\varphi A_0(2qR)} \text{ vs } q^2$$

for different concentrations and in Fig. 7.16(b) where a three-dimensional plot is given.

As expected with repulsive forces, the increase of concentration diminishes the fluctuations and decreases the quantity $I(q)/\varphi$ at zero angle. What is more striking is the appearance of a peak in the low q range or of what has been called the correlation hole at low q. This 'hole' can be explained qualitatively by the following argument: interactions between two particles are at larger distances than the intramolecular distances; they decrease faster with q than the form factor $P(q)$. We subtract a quantity which is large at small q and decreases rapidly at larger q. If this effect is large enough it can give a maximum such as the one observed in Fig. 17.16. If we want to include in this model other types of interactions (van der Waals, electrostatic, . . .) we can increase the radius of the excluded volume sphere and replace it by $2R + \Delta R$. More sophisticated methods can be used but we cannot discuss them here.

One problem which could be interesting to solve is to apply a similar technique to particles which are no longer spherical. Even for ellipsoids this introduces serious difficulties. Let us go back to Fig. 7.15; the decomposition of $\boldsymbol{i}_1\boldsymbol{j}_2$ in three vectors is still possible but these vectors are no longer independent; they depend on the orientations of the ellipsoids relatively to the vector $\boldsymbol{C}_1\boldsymbol{C}_2$; the number of variables to introduce increases and, until now, no simple solution has been found.

7.15 THE CORRELATION HOLE

As we just said the appearance of a correlation hole is quite frequent and the subsequent peak has often been attributed to a kind of Bragg reflection. We just want to stress here that this is a quite general feature in the small angle scattering range and can happen frequently without implying the existence of a special correlation distance. If we go back to the general equation of small angle scattering we can always write it as

$$I(q) = b_v^2[Nz^2P(q) - N^2Q(q)] \tag{7.177}$$

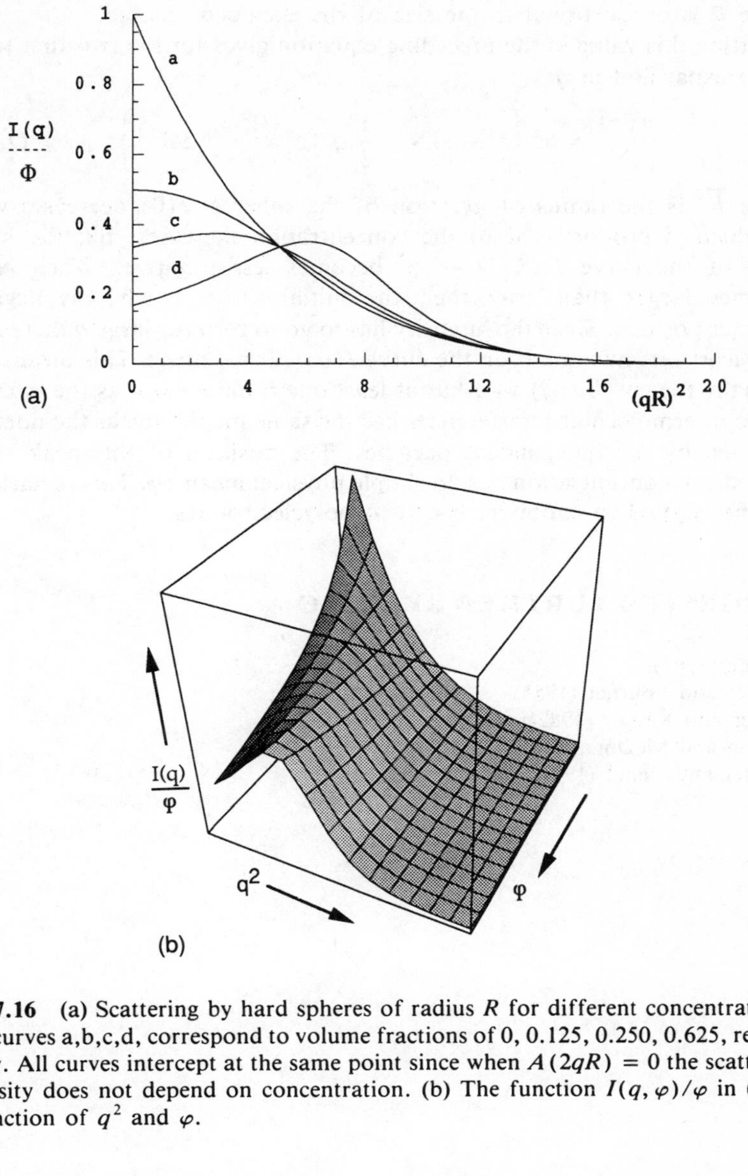

Fig. 7.16 (a) Scattering by hard spheres of radius R for different concentrations. The curves a,b,c,d, correspond to volume fractions of 0, 0.125, 0.250, 0.625, respectively. All curves intercept at the same point since when $A(2qR) = 0$ the scattering intensity does not depend on concentration. (b) The function $I(q, \varphi)/\varphi$ in (a) as a function of q^2 and φ.

where $Q\ (q)$ is a quantity proportional to the excluded volume parameter. From our discussion in Section 7.2 this quantity is always positive for a repulsive particle and can therefore be written as

$$Q(q) = A - Bq^2R^2/3 \tag{7.178}$$

where R is proportional to the size of the excluded volume.

Putting this value in the preceding equation gives for the two first terms in the expansion in q^2

$$\frac{I(q)}{N} = b_v^2 \left\{ z^2 - AN - \frac{1}{3} q^2 [z^2 \overline{R^2} - NBR^2] \right\} \tag{7.179}$$

where $\overline{R^2}$ is the radius of gyration of the spheres. $I(0)$ decreases when N, which is proportional to the concentration increases, but the initial slope of the curve $I(q)/N$ vs q^2 becomes less negative. When NBR^2 becomes larger than $z^2\overline{R^2}$ then the initial slope which was negative becomes positive. Since the intensity has to go to zero for large q there must be at least one maximum on the curve $I(q)/N$ vs q or q^2. This means that normally the curve $I(q)$ vs q has at least one peak as soon as the intensity due to intermolecular interferences has the same magnitude as the intensity scattered by the independent particles. The position of this peak which depends on concentration has no simple physical meaning. This remark can also be applied to copolymers and to polyelectrolytes.

SUGGESTED FURTHER READING

Cabane (1987).
Guinier and Fournet (1955).
Glatter and Kratky (1982).
Hansen and McDonald (1986).
Lindner and Zemb (1991*a*, *b*).

8

Experimental examples of structural studies

8.1 INTRODUCTION

In the earlier chapters of this book we introduced the experimental methods of small angle scattering, and explained how the raw data can be reduced to the point where the quantity remaining is the scattering law, $S(q)$. Subsequent chapters have been devoted to deriving the mathematical forms for $S(q)$ for various models. We now want to demonstrate by using some examples from the literature how these models can be used to extract useful physical parameters from the data. This is in no way to be interpreted as a review of the literature, which is both extensive and growing rapidly.

Small angle neutron scattering was first applied to polymers in order to obtain information about the conformation of individual molecules in bulk samples. It was of course the unique possibility of the deuterium labelling that attracted the experimenters. The determination of molecular size and shape in crowded environments is still the main field of application. We consider situations in which the environments increase in complexity—from one component in one phase, for example melts or glasses, through one-phase multicomponent systems such as solutions or gels to two-phase systems such as crystalline polymers. The information sought may be as simple as the radius of gyration or the molecular weight, or more complex, such as the conformation as determined through fitting a model form for $P(q)$ to the experimental curves. As well as this molecular information, scattering from these more complex systems provides thermodynamic and morphological information, and subsequent sections give some examples. Finally the interpretation of scattering from anisotropic systems, such as stretched polymers or oriented liquid crystal polymers, will be considered.

8.2 SINGLE CHAIN CONFORMATION

8.2.1 H–D mixtures of identical polymers in melts or glasses

The basic equation for describing a mixture of deuterated and hydrogenous polymers identical in all other respects is given in Chapter 5 (eqn (5.31))

$$I(q) = (b_D - b_H)^2 x(1-x) N z^2 P(q).$$

In this formula x is a volume fraction, which is dimensionless. Experimenters, however, more usually work in terms of the concentration, c_D or c_H. In Appendix 3 we show in detail how the physics formulae of Chapters 6 and 7 are altered when real experimental quantities are included. If we take eqn (A3.18) and set $M_H v_H^S = M_D v_D^S$ and $P_H(q) = P_D(q)$ we obtain (with $\chi = 0$)

$$\frac{I(q)}{V} = b_v^2 c_H c_D (v_D^S)^2 (v_H^S) \frac{M_D}{N_A} P(q) \tag{8.1}$$

with $b_v = \dfrac{b_D}{v_D} - \dfrac{b_H}{v_H}$.

The only effect of varying c_D, the concentration of the deuterated species, is to vary the intensity as $c_D c_H$. In principle therefore it is only necessary to carry out a measurement at one value of c_D. As described in the introduction to Chapter 5, the early experiments were carried out before

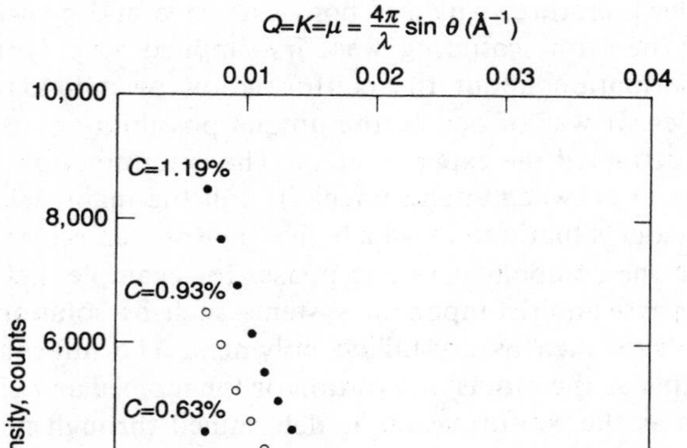

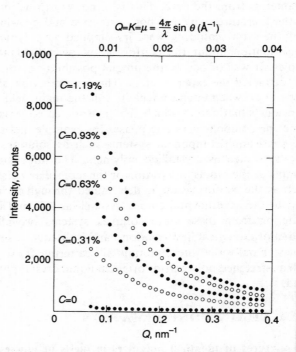

Fig. 8.1 Intensity vs Q ($\equiv q$) for concentrations C of hydrogenous polymethylmethacrylate dissolved in deuterated polymethylmethacrylate (Kirste *et al.* 1975). Reprinted with permission from Wignall (1987).

the theoretical basis of the equation had been developed. These early experiments were seen as an analogy to light scattering from solutions with one of the species replacing the solvent and the other the solute. The fact that as a consequence they were carried out with low levels of the labelling (either c_H or c_D) can now be seen to be an advantage given the potential problems of high deuteration levels discussed in Section 5.6, while the fact that they involved several values of c_D allows us to demonstrate the validity of eqn (8.1). Figure 8.1 shows some beautiful data for low concentration mixtures of normal polymethylmethacrylate (PMMA) in a matrix of perdeutero PMMA. The samples had been very carefully prepared with very low levels of impurity or of residual inhomogeneities left behind by the sample preparation techniques. The quality of the samples can be seen from the extreme flatness and very low level of the signal from the pure deuterated sample. Many scattering experiments show considerably more scattering from the so-called background or matrix sample.

The simplest question to ask of the data in Fig. 8.1 is the value of the radius of gyration R_g. In Chapter 6 we introduced two methods of plotting the data in order to obtain this quantity. Equations (6.105) and (6.106) are the basis of the Guinier (log $I(q)$ vs q^2) and the Zimm ($kc/I(q)$ vs q^2) plots respectively. As explained in Section (6.5.1), because the Zimm plot of a polydisperse system of Gaussian chains obeying the most probable distribution is linear even outside the low q limit, this is the method of plot-

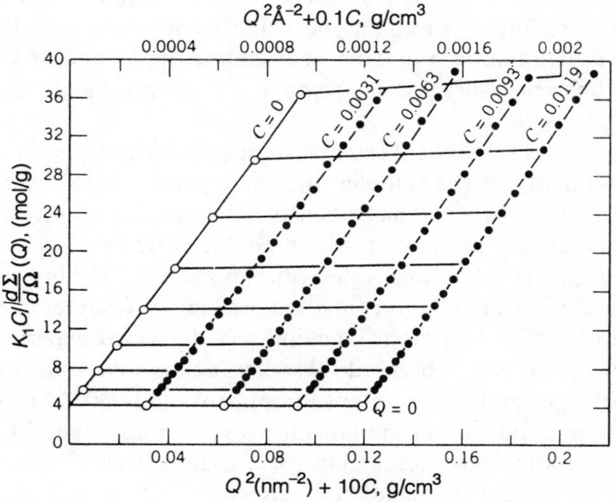

Fig. 8.2 Zimm plot of data in Fig. 8.1. The open circles are the result of extrapolation to zero C and zero Q ($\equiv q$) (Kirste *et al.* 1975). Reprinted with permission from Wignall (1987).

ting the data favoured by polymer scientists. In Fig. 8.2 the same data as in Fig. 8.1 have been plotted in this form for each of the four values of c_D, the scattering from the unlabelled matrix sample having been subtracted, in order to eliminate the incoherent scattering. We will discuss later the form of the abscissa ($q^2 + \lambda c_D$, where λ is a constant). For the moment it is a convenient way of displacing the data for different values of c_D to demonstrate the validity of eqn (8.1), since the curves are rigorously parallel. In view of eqn (8.1) the vertical axis should actually be $kc_D c_H/I(q)$, giving an error in M of order c_D. Equation (6.106) shows that from the ratio of slope to intercept of any of the lines, the radius of gyration can be determined. In this case it is 128 ± 2 Å.

It is important to note that the Zimm approximation for small q (eqn (6.106) was developed for light scattering for which the requirement that $qR_g \ll 1$ is always obeyed even for large molecular weights. Its application to neutron scattering needs care unless one is sure that the polydispersity follows the most probable distribution and that the chains obey Gaussian statistics. In this case, as discussed in Section 6.6.4, the Zimm plot is linear over a wide q range. The data in Fig. 8.2 continue well outside the so-called low q range, but since the polydispersity was reasonably close to the most probable distribution, the linearity continues as high as $qR_g = 5$. The use of this method of plotting the data also depends on being able to extrapolate to zero q values in order to obtain the molecular weight. Scattering from artefacts at low q can make this difficult. Note however that while the value of the molecular weight depends on good absolute calibration of the data the value of $\overline{R^2}$ does not. It should be remembered here that in a polydisperse sample the R_g and M values correspond to different averages, R_g being the z average and M the weight average.

The data in Figs 8.1 and 8.2 remain 'competitive' despite their age because of the high quality of the samples and the careful analysis. Following the development of the theory described in Chapter 5 it was realized that a single measurement at a relatively high labelling level was all that is required, since the scattering is described by eqn (8.1). Thus considerable shortening of the experimental time is possible. In Chapter 5 we also discussed the effect of deuterium labelling and described experiments where phase separation was observed between deuterated and hydrogenous polymers of high molecular weight in approximately 50/50 mixtures. It is therefore now usual to confine labelling levels to less than 20 per cent or so, and good practice to check that there is no effect of varying c_D. The full factor $c_D c_H$ must be included of course.

We should now return to Fig. 8.2 and examine why the abscissa was plotted as $q^2 + \lambda c_D$. This is a result of the historical development of SANS from light scattering. Equation (8.1) can be reorganized to give

$$\frac{c_{\mathrm{D}} V}{I(q)} = \frac{1}{b_{\mathrm{v}}^{2}} v_{\mathrm{H}}^{\mathrm{S}} v_{\mathrm{D}}^{\mathrm{S}} \frac{N_{\mathrm{A}}}{MP(q)} (1 - \varphi) \tag{8.2}$$

$$= \frac{1}{b_{\mathrm{v}}^{2}} v_{\mathrm{H}}^{\mathrm{S}} v_{\mathrm{D}}^{\mathrm{S}} N_{\mathrm{A}} \left(\frac{1}{MP(q)} - \frac{\varphi}{MP(q)} \right)$$

since. $c_{\mathrm{H}} v_{\mathrm{H}}^{\mathrm{S}} = 1 - \varphi$ where φ is the volume fraction of deuterated molecules.

The coefficient of φ for low molecular weight solvents is normally called A_2 and in that case A_2 is independent of q. (Careful examination shows that the $P(q)$ in this term belongs to the solvent and is, of course, unity.) For the present case, since M^{-1} is small we ignore the q variation in A_2. This is the basis of the famous Zimm plot of the light scatterers, developed in order to-interpret data from dilute solutions of polymers. In the low q limit $P^{-1}(q) \rightarrow 1 + q^2 R_{\mathrm{g}}^2/3$. Extrapolation at constant q to zero φ gives a line whose slope to intercept ratio gives R_{g}. Extrapolation at constant φ to zero q gives a line with slope A_2. The zero φ, zero c intercept gives the molecular weight M. Correct determination of the molecular weight is one of the most important checks on the validity of the scattering data which may be affected by the difficulties of correct extrapolation to zero q and the elimination of artefacts. The molecular weight of a well-characterized sample should be obtained with an absolute accuracy of better than 10 per cent if the incoherent scattering has been correctly subtracted. In this case, the value obtained, 220 000, was in good agreement with other measurements.

Figure 8.2 has been plotted according to eqn (8.2). This way of plotting the data presented in Fig. 8.1 was argued to demonstrate the validity of the Flory hypothesis of unperturbed dimensions in the melt. It is often stated that this hypothesis demands that A_2 is zero in a melt. In fact comparison of eqns (8.1) and (8.2) shows that this is not precisely true. As we saw above if we expand eqn (8.1) into the form of eqn (8.2) and take only the zero q limit we obtain $A_2 = M^{-1}$. This is very small but not exactly zero. The crucial criterion for the Flory hypothesis of Gaussian conformation in the melt, which was confirmed by all the SANS experiments is that $(\overline{R^2})$ varies as M when the data are corrected so that the same averages over the polydispersity are used for each parameter. This latter point is a reminder that for polydisperse samples eqns (8.1) and (8.2), apply if R_{g} becomes the z average, $R_{\mathrm{g}z}$, and M the weight average, M_{w}.

The radius of gyration and the molecular weight are important parameters, but by concentrating on these low q parameters alone we are in danger of wasting some of the information contained in the scattering data. As discussed in Chapter 6, Sections 6.3 and 6.5, further information about the molecular conformation is available if the shapes of the scattering curves are considered over a wide q range (not just the low q slope and the

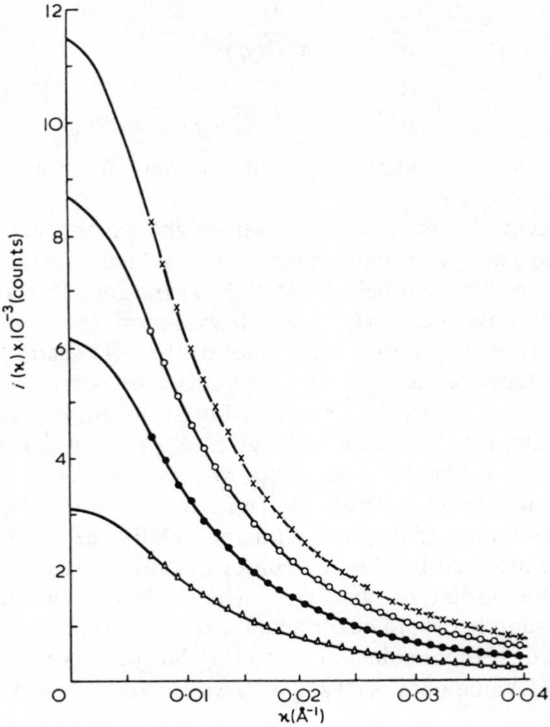

Fig. 8.3 Data as in Figs 8.1 and 8.2. The solid lines are the least squares fit of eqn (6.73) with polydispersity as in eqn (6.131). Note $\kappa \equiv q$. Reprinted from Kirste *et al.* (1975) with permission from the author and the publishers, Butterworth Heinemann ©.

intercept). In Fig. 8.3 the data in Fig. 8.1 have been fitted to the Debye scattering curve (Gaussian chains) using eqn (6.73) modified to include the polydispersity expressed as in eqn (6.131). The indications from Fig. 8.3 are that within the observed q range and hence over distances longer than q_m^{-1} (where q_m is the maximum observed q) which is about 25 Å in this case, the molecules are behaving as Gaussian chains. Figure 8.4 shows similar good agreement when the form factor for stars given in eqn (6.80) is compared over a similar q range with the scattering data from 3- and 18-armed polybutadiene molecules. As far as possible it is useful to confirm the information obtained by fitting a model $P(q)$ to the data by also exploring the relationship between $\overline{R^2}$ and M.

As we have remarked already the Debye curve for Gaussian chains reaches a plateau when plotted as $q^2 I(q)$ against q^2. (This is the Kratky

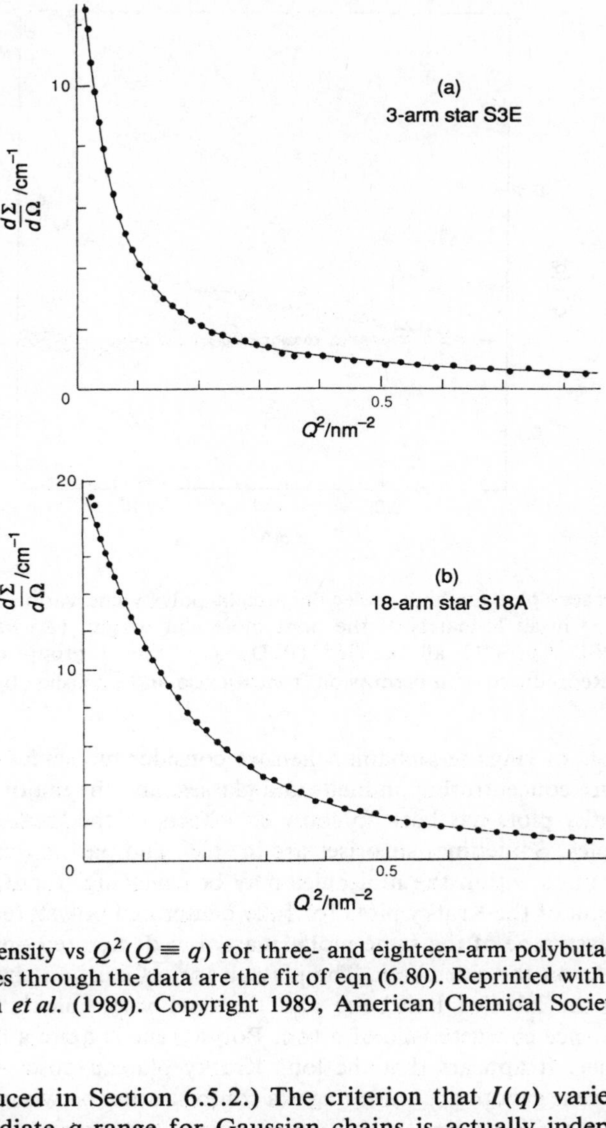

Fig. 8.4 Intensity vs $Q^2(Q \equiv q)$ for three- and eighteen-arm polybutadiene stars. The solid lines through the data are the fit of eqn (6.80). Reprinted with permission from Horton *et al.* (1989). Copyright 1989, American Chemical Society.

plot introduced in Section 6.5.2.) The criterion that $I(q)$ varies as q^2 in the intermediate q range for Gaussian chains is actually independent of polydispersity and can therefore be a very useful check on the chain statistics. However, the detailed effect of the local chain structure comes into play in this q range when distances shorter than about 30 Å are explored. The effect of chain architecture is well demonstrated in a Kratky plot as shown in Fig. 6.20 where the curves for Gaussian chains, rings, and stars were compared. We shall see examples of the differences between the

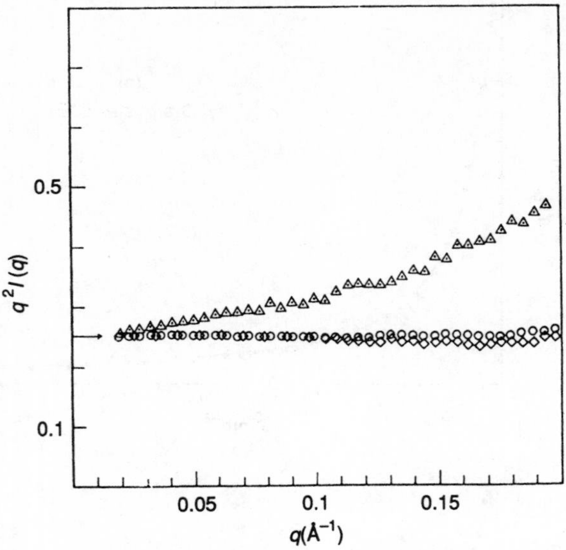

Fig. 8.5 Kratky plot for high molecular weight polystyrene variously deuterated and dissolved in an H-matrix of the same molecular weight. (△) backbone only labelled (PSD_3H_5), (○) all labelled (PSD_8, (◇) phenyl group only labelled (PSD_5H_3). Reproduced with permission from Rawiso and published by Bastide and Benoît 1990.

Kratky plots of rings and chains when we consider molecules in solution. Here, we are concentrating on melts and glasses, and the major application of the Kratky plots has been to show up effects of the local structure of the molecules. Sometimes surprises are in store and partial deuteration of the repeat units within the molecules may be revealing. Figure 8.5 shows a comparison of the Kratky plots for fully deuterated polystyrene with that from a polystyrene of the same molar mass, but deuterated only along the backbone or on the side chain. The plateau for a Gaussian chain is barely visible for the labelled backbone and the data very quickly assume the q^{-1} dependence characteristic of a rod. Polystyrene is quite a rigid unflexible polymer. It appears that the long Kratky plateau observed when all the polystyrene groups are scattering is a compensation between the rod-like signal and the side groups. It should not come as a surprise that the cross section of the chain is as important as its local rigidity on this distance scale and, of course, this compensation may not occur for all polymers. The fact that the monomer structure affects the scattering of polystyrene beyond q values as low as $0.03\,Å^{-1}$ was first observed by neutron scattering on partially deuterated chains in a good solvent (Rawiso *et al.* 1987). Clearly Kratky plots have to be interpreted with care. In Fig. 8.6 scattering data from polycarbonate chains are compared with calculations.

Large deviations from the Debye curve for Gaussian chains are seen at

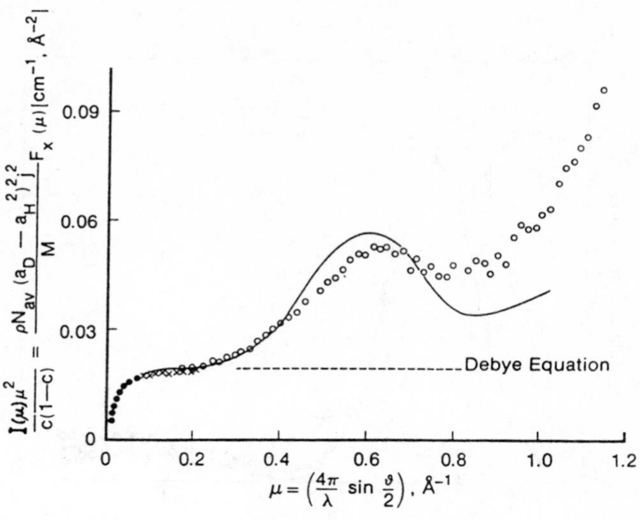

Fig. 8.6 Kratky plot for polycarbonate chains compared to calculations based on the rotational isomeric state model (—). (○ ○ ○) data from Gawrisch *et al.* (1981). Reprinted with permission from Yoon and Flory (1981). Copyright Springer.

distances shorter than about 5 Å. In this range it is possible to compute the form factor (eqn (6.8)) for particular models of the molecule by calculating the factors $\langle \sin(qr)/qr \rangle$ for each pair of scattering centres. The calculated curve in Fig. 8.6 is based on the rotational isomeric state model of the polycarbonate chain, taking each deuterium atom in the benzene rings as a scattering locus, but averaging the three deuterium atoms in the methyl groups to a one point location. The agreement is remarkably good, and such comparisons with neutron data allow the computations to be carefully tested and the models to be refined.

The final type of experiments on H–D mixtures we wish to mention, takes us out of the strictly small angle scattering q range, into intermediate or even wide angles. Here the scattering is often termed 'diffuse'. In this range the very local inter- and intramolecular arrangements dominate. As for the polystyrene case above, it is often fruitful to label parts of the repeat unit on each chain, rather than whole chains. In this way, data which complement X-ray experiments are obtained. Figure 8.7 shows such data for partially deuterated polymethylmethacrylate of two different tacticities. In each case the α-methyl and backbone protons are deuterated. The data of much lower intensity in the figure are from a fully hydrogenous sample. Clear differences are introduced at low q by enhancing the scattering with deuterium, while subtle effects of the tacticity on molecular packing can also be seen. The large incoherent background scattering from hydrogen

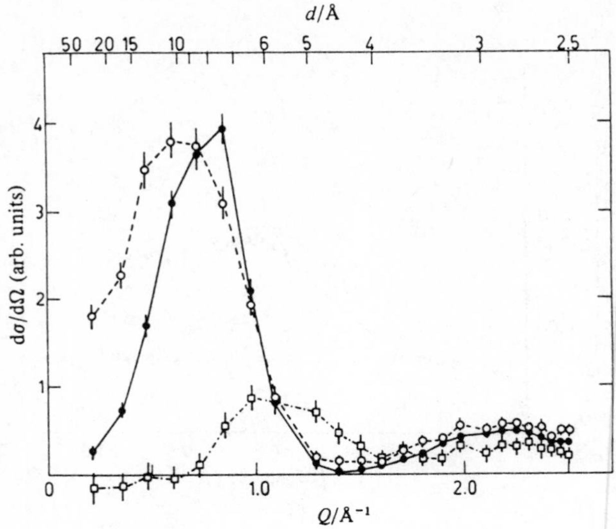

Fig. 8.7 Elastic coherent scattering against Q ($\equiv q$) from two different tacticity samples of polymethylmethacylate: (○) isotactic, (●) syndotactic with the α-CH_3 and methylene hydrogens replaced by deuterium, (□) fully hydrogenous sample. Reproduced with permission from Gabryś *et al.* (1986).

which would occur at these q values has been removed, not by subtraction, but by the use of polarization analysis as described in Section 3.1.3.

As a final remark in this section it is important to note that while the theoretical discussion has centred on melt samples many experiments are carried out on glasses. This is clearly more convenient experimentally, and unless some mesophases (such as crystals, liquid crystals, micelles, or coexisting compositions in phase separated domains) appear during the quenching process, no difference in the results has ever been observed.

8.2.2 H–D mixtures in the presence of a third component—solutions and blends

As discussed in Chapters 5 and 7, as soon as the system under consideration contains more than one type of molecule, the scattering intensity will contain contributions not only from molecular shape but also from both density and concentration fluctuations. In most systems it is the latter which will dominate and make extraction of molecular properties difficult. The great potential of the neutron is to allow this separation of $P(q)$ and $Q(q)$ to be undertaken at finite concentrations by exploiting the use of deuteration. The experiments are essentially all based on the use of eqn (5.41), or variants of it. Since we shall be quoting it many times it is worth reiterating this important equation here.

$$I(q) = x(1-x)(b_D - b_H)^2 N z^2 P(q) + \langle b \rangle^2 [N z^2 P(q) + N^2 z^2 Q(q)] \tag{8.3}$$

where $\langle b \rangle = x b_D + (1-x) b_H - b_0$ and x is the volume fraction of deuterated molecules.

To remind the reader: the first term corresponds to the fluctuations between labelled and unlabelled molecules and the second to the fluctuations of the overall concentration when there is no difference between labelled and unlabelled species. The first term in this equation is the scattering from the labelled molecules while the second is that from a sample with no labelling. One of the important variants is obtained by rearranging the terms in eqn (8.3) to give

$$I(q) = [x(b_D - b_0)^2 + (1-x)(b_H - b_0)^2] N z^2 P(q) + \langle b \rangle^2 N^2 z^2 Q(q). \tag{8.4}$$

Now the first term is still the scattering from the single chain form factor, though with a different coefficient, but the second term is just the interparticle factor, $Q(q)$.

The experimental applications of these equations fall into two groups. In the first group, use is made of the fact that the coefficients of $P(Q)$ and $Q(q)$ depend in different ways on x. The value of x is changed in experiments at fixed N, so that by weighted subtractions the contributions from inter- and intramolecular terms can be separated. In the second group the value of x is fixed, but the value of the solvent scattering length, b_0, is adjusted so that $\langle b \rangle$ is zero and only the first term in eqn (8.3) or (8.4) has a non-zero coefficient. (Note that in these conditions the coefficients of $P(q)$ become identical.) It is of course assumed that $P(q)$ and $Q(q)$ do not vary with x. An alternative method is used when the hydrogeneous polymer is 'matched' to the solvent, so that $b_H = b_0$ giving

$$I(q) = (b_D - b_0)^2 \{x N z^2 P(q) + x^2 N^2 z^2 Q(q)\}. \tag{8.5}$$

Since the coefficients of $P(q)$ and $Q(q)$ vary as x and x^2 respectively, $P(q)$ can be obtained in the classical Zimm plot by extrapolation to zero x.

The study of polymer solutions in the semidilute and concentrated regimes of concentration has been of interest in order to follow the changes in the dimensions and statistics of the molecules as they pass from good solvent dilute solution (with excluded volume) to the melt (Gaussian chains). During the 1970s such studies were undertaken by a number of research groups world-wide and there is a considerable literature. Here we

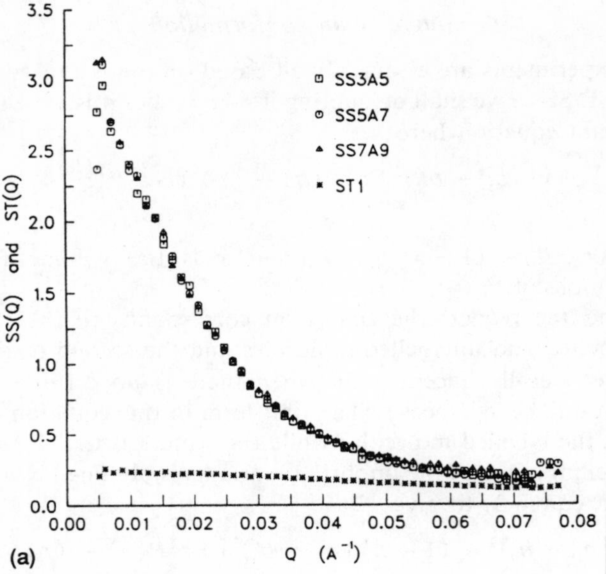

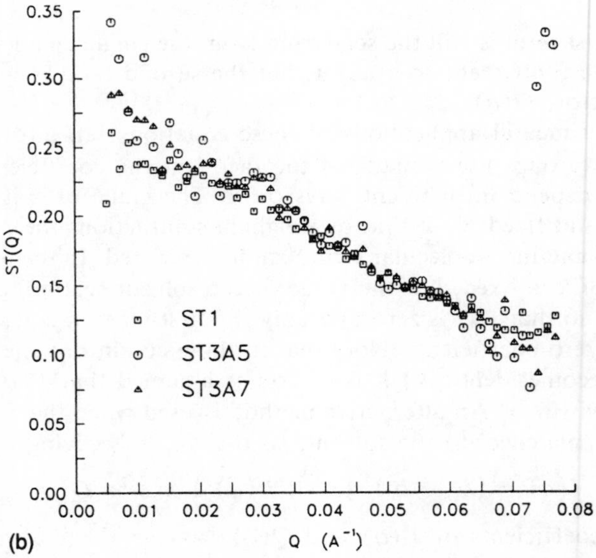

Fig. 8.8 Data for mixtures of HPS and D-PS dissolved in deuterated toluene at 8% overall polymer concentration. (a) Weighted subtractions to give single chain term (first term in eqn (8.3). (⊡) 0.2 and 0.4 H:D ratio; (⊙) 0.4 and 0.6 H:D ratio, (▲) 0.6 and 0.85 H:D ratio, (■) pure hydrogenous sample for comparison. (b) The total scattering term (second term in eqn (8.3)) also obtained by weighted subtractions against Q ($\equiv q$). (⊙) 0.2 and 0.4 H:D ratio, (△) 0.4 and 0.6 H:D ratio, (⊡) pure hydrogenous sample. Reprinted with permission from King *et al.* (1985). Copyright 1985, Americal Chemical Society.

pick just one example. Figure 8.8 shows data from solutions of polystyrene and its deuterated homologue in deuterated toluene. In the study the concentration of the solutions ranged from 0 to 22 per cent. In the figure the concentration is 8 per cent. The molecular weights are around 65 000 and toluene is a good solvent for polystyrene so that the scattering from this solution would contain a large contribution from intermolecular correlations (the virial term would be large). The experimenters made five different measurements all at 8 per cent total polymers concentration, but with the ratio of deuterated to hydrogenous polystyrene at 0, 0.2, 0.4, 0.6, and 0.85. In Fig. 8.8(a) a weighted subtraction of three sample pairs 0.2 and 0.4, 0.4 and 0.6, and 0.6 and 0.85, has produced three measurements of the first term in eqn (8.3) or (8.4), $P(q)$, the single chain term. There is very good agreement between the three subtractions as there should be if they have been done correctly. From the curves for $P(q)$ values of R_{gz} and of M_w were obtained. R_{gz} is 80 Å and M_w agrees within a few per cent with the GPC value. Quite often in such experiments it proves difficult to exactly match the molecular weights of the deuterated and the hydrogenous polymers. This problem was discussed in Section 5.5.1 and eqn (5.31) should be replaced by eqn (5.52).

Figure 8.8(b) shows the data obtained from the weighted subtractions for the second term in eqn (8.3). As we know, this is the scattering from the whole system without any labelling and this is seen clearly since it is identical to the scattering from a solution without deuterated molecules shown also in Fig. 8.8(a). There is more information to be obtained from these total scattering curves, especially the screening length ξ, but this is discussed later when we consider thermodynamic parameters.

The solution to the problem of obtaining $P(q)$ adopted by the light scatterers has already been mentioned. The extrapolation to zero concentration used in the Zimm plot allows the various contributions to be separated, *but only in very dilute solution limits*. Occasionally this approach has been applied to the study of dilute polymer solutions by SANS. An example of a Zimm plot for a solid solution has already been given in the previous section. There is really nothing different in such a plot from a solution in a low molecular weight solvent except that, unless the sample is at the theta temperature, the virial coefficient, A_2, will be much larger. In fact it can be argued that only under exceptional circumstances is it worthwhile studying a solution at low concentrations by neutron scattering, as opposed to the usually more accessible light or X-ray scattering. Such circumstances might centre on the q range or the available contrast. An example would be a low molecular weight hydrocarbon polymer (needs, small q values) in a hydrocarbon solvent (needs deuterium to increase contrast).

It would appear that polymer blends are one further example of solutions and that eqn (8.3) or (8.4) should be applicable in order to observe

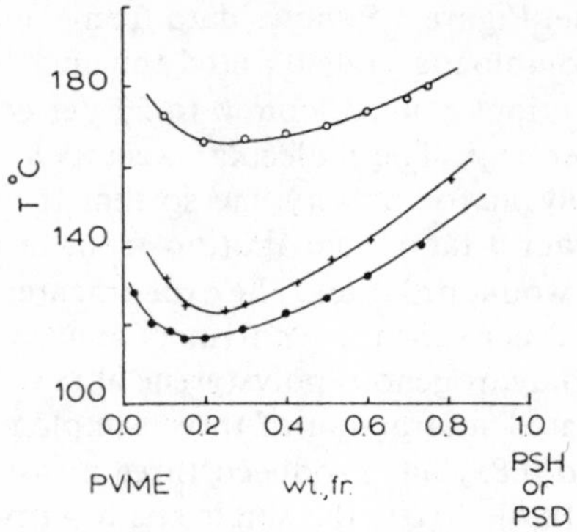

Fig. 8.9 Cloud point curves for mixtures of polyvinyl methyl ether with polystyrene. PVME ($M_w = 99\,000$) with: ○ ○ ○ PSD ($M_w = 119\,000$); ●●● PSH ($M_w = 233\,000$); and + + + PSH ($M_w = 100\,000$). Reprinted with permission from Yang *et al.* (1983). Copyright 1983, John Wiley.

the behaviour of individual components. Here considerable caution is necessary. All the manipulations on solutions were made assuming there is no interaction between H and D species. These assumptions are only valid when small changes in the polymer–solvent interaction, introduced by deuteration do not substantially change the state of the system. For this to be true the sample must be far away from any critical point. Figure 8.9 shows the effect on the phase boundary of a mixture of polystyrene and polyvinylmethylether when the polystyrene is replaced by its deuterated homologue. There is a shift to higher temperature of the cloud point by about 40°. Thus an attempt to extract the single chain form factor $P(q)$ by weighted subtraction of measurements with different values of x means that the data are not obtained for samples in the same thermodynamic state. The term $Q(q)$ may not be the same and the subtractions are suspect. In the literature there are some carefully monitored examples of such subtractions, or more reliable contrast match experiments, but the effect of observations such as Fig. 8.8 has been to make this application rather rare.

At higher q values, as in the bulk samples described in the previous section, the Kratky plot gives detailed information about the chain architecture. A comparison of two Kratky plots such as those shown in Fig. 8.10 shows different architectures—ring vs chain—for the same chemical structure. The 'hump' predicted for the ring architecture in Chapter 6 is clearly visible in Fig. 8.10.

Returning to less complex architecture we mention finally in this section a few other examples. Gels and networks, whether end-linked molecules

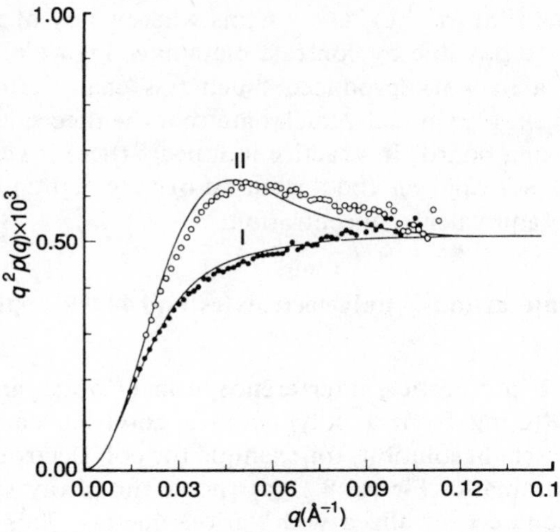

Fig. 8.10 Kratky plot for I, linear and II cyclic, polystyrene in cyclohexane D_{12}. Solid lines are calculated curves as in Chapter 6, Section 6.5. Reprinted with permission from Hadziioannou *et al.* (1987). Copyright 1987, American Chemical Society.

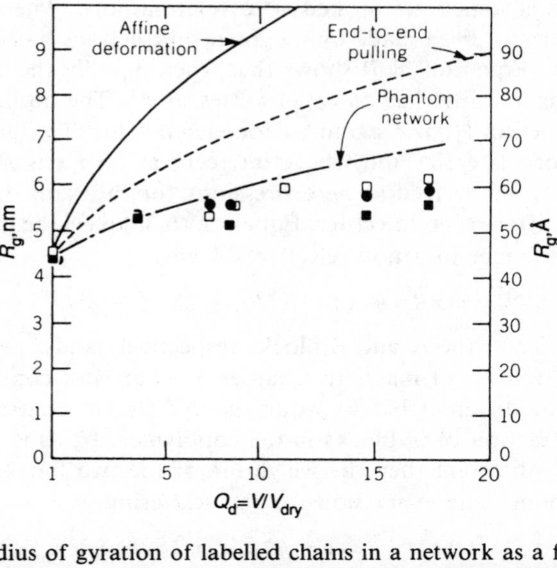

Fig. 8.11 Radius of gyration of labelled chains in a network as a function of the degree of swelling, $Q_d = V/V_{dry}$, compared to various models. The different symbols refer to different sample-to-detector distances (● 2.83 m, □ 5.66 m, ■ 10.66). Reprinted with permission from Bastide *et al.* (1984). Copyright 1984, American Chemical Society.

or randomly cross-linked ones, are systems where study of the chain conformation is made possible by contrast matching. Figure 8.11 shows how the swelling of a network produced much less chain deformation than predicted by the existing models. Such data sent the theoreticians scurrying back to the drawing board. In practice it appears there is very little difference between an isotropic gel and a solution of very high molecular weight polymers of the equivalent concentration.

8.2.3 Strong interactions – polyelectrolytes and block copolymer solutions

The effect of the interparticle interference term $Q(q)$ is not very obvious in the total scattering from a polymer in a good solvent. For strongly interacting polymers in solution, for example for polyelectrolytes, the effect is much more dramatic. Figure 8.12(a) shows the X-ray scattering from polyvinyl sulphonate neutralized with various metals. This is identical to what would be seen in a neutron experiment without any labelled chains. In Fig. 8.12(b) the functions $P(q)$ and $Q(q)$ have been separated in a SANS experiment by deuterating some of the molecules. The origin of the peak in the total scattering can be seen in the strongly negative contribution of $Q(q)$. This arises from the very strong repulsive forces between the molecules and is sometimes termed a 'correlation hole'. The separation in Fig. 8.12(b) has not been made by weighted subtractions but by a series of extrapolations. Equation (8.5) shows that when $b_H = b_0$ the coefficient of $P(q)$ varies as x while that of $Q(q)$ varies as x^2. The authors therefore plotted the function $I(q)/x$, against x for each q value. This gave a straight line whose slope is $Q(q)$ and whose intercept at zero x is $P(q)$.

In Chapter 6 we derived some expressions for the radii of gyration of molecules of different architecture. Equation (6.36) gave the expression for a diblock copolymer in a non-selective solvent.

$$\overline{R_T^2} = x\overline{R_A^2} + (1-x)\overline{R_B^2} + 2x(1-x)\overline{L^2} \tag{8.6}$$

$\overline{R_A^2}$ and $\overline{R_B^2}$ refer to the A and B blocks respectively and L is the distance between their centres of mass. In Chapter 6 we did not consider the possibility that the different blocks would have different contrast factors. x was just the fraction of A blocks in the copolymer. If, as is highly likely, the contrast is different then the weighting of the two blocks has to take this into account. The expression for the scattering is

$$I(q) = \bar{b}_A^2 S_{AA} + \bar{b}_B^2 S_{BB} + 2\bar{b}_A \bar{b}_B S_{AB}$$

where

$$\bar{b}_A = \left[b_A \frac{v_0}{v_A} - b_0\right].$$

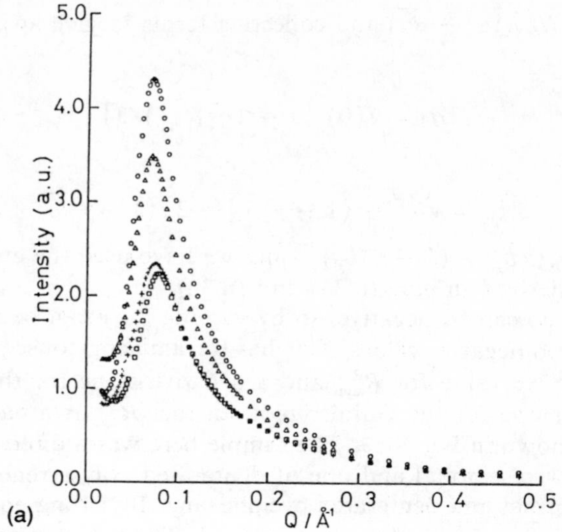

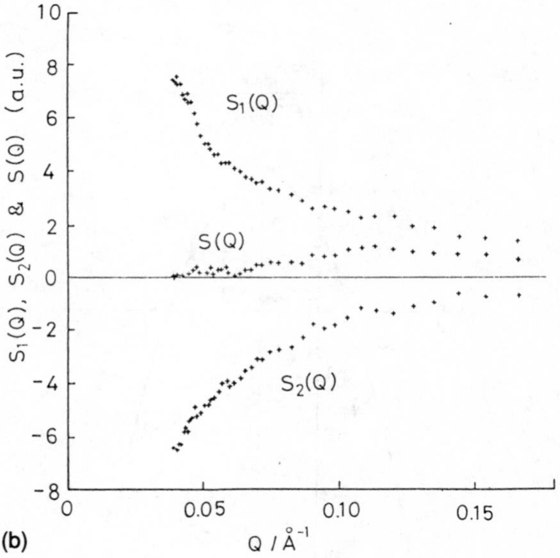

Fig. 8.12 (a) Small angle X-ray scattering curves for aqueous solutions of polyvinylsulphonate unneutralized and neutralized with (+)Li, (△)Na, and (○)K. Reprinted with permission from Kaji *et al.* (1984). Copyright 1984, American Chemical Society. (b) Separation of SANS data from the Na sample in (a) into $S_1(Q)$ ($\equiv P(q)$) and $S_2(Q)$ ($\equiv Q(q)$). Reprinted with permission from Ono *et al.* (1986).

Letting $S_{AA} = Nz_A^2(1 - \frac{q^2}{3}\overline{R_A^2})$ and collecting terms to give an expression of the form

$$I(q) = I(0)\ \{1 - (q^2\overline{R^2_{\text{app}}})/3\}$$

we find

$$\overline{R^2_{\text{app}}} = y\overline{R^2_A} + (1-y)\overline{R^2_B} + 2y(1-y)\overline{L^2} \tag{8.7}$$

where $y = x\bar{b}_A(x\bar{b}_A + (1-x)b_B)^{-1}$ and we have used the expressions for $\overline{R^2_{AB}}$ and $\overline{R^2_T}$ derived in eqns (6.33) and (6.36).

Now $\bar{b}_A$ or $\bar{b}_B$ can be negative, so by varying b_0, y can be made to vary from positive to negative values. This has the amusing consequence of producing a negative value for R^2_{app} and a negative slope for the Zimm plot or a positive slope for the Guinier plot. In fact $\overline{R^2_{\text{app}}}$ is a parabolic function of y as shown in Fig. 8.13. The sample here was a diblock copolymer with one block of normal and one of deuterated polystyrene in a varying mixture of normal and deuterated cyclohexane. By fitting eqn (8.7) to the data values of $\overline{R^2_A}$, $\overline{R^2_B}$, and $\overline{L^2}$ are obtained. The values determined from

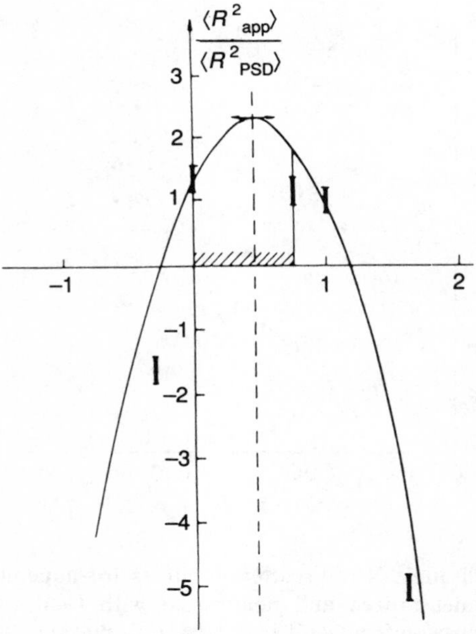

Fig. 8.13 Parabolic dependence of $\bar{R}^2_{\text{app}}$ on contrast y, for a solution of a diblock copolymer PSD–PSH in a mixture of C_6H_{12} and C_6D_{12} at 35°C. Reprinted with permission from Duval *et al.* (1976). Copyright 1976, John Wiley.

these data show that the A and B blocks are probably compatible because $\overline{L^2}$ has the value expected for the distance between the two halves of a single molecule obeying Gaussian statistics.

8.2.4 Systems with mesomorphic phases

In two phase systems the signal generated by any neutron scattering contrast between the two phases will generally dominate the scattering from individual molecules. This is because the effective molecular weight of a domain in such structures is orders of magnitude higher than that of the molecules within it. As an example Fig. 8.14 shows the scattering from a diblock copolymer. The two blocks are polystyrene and polybutadiene. The ratio of hydrogen ($b = -0.33 \times 10^{-12}$ cm) to carbon ($b = 0.65 \times 10^{-2}$ cm) differs in the two polymers. In polystyrene it is approximately $(CH)_n$, while in polybutadiene it is nearer $(CH_2)_n$. This means that even in the unlabelled species there is a very strong signal from the domain structure of the copolymer in the bulk state arising from the natural contrast between the blocks. In this case the two blocks were chosen to have rather different molecular weights, that of the polystyrene being 380 000 while that of the polybutadiene was 46 000. With this ratio the copolymer forms spherical domains of polybutadiene in a polystyrene matrix. The domains are fairly monodisperse in size as can be seen from the oscillating scattering curve in the figure. The data are very similar to the curve for polydisperse spheres given in Fig. 6.25. The intense peak at low q arises from the interparticle interference. From such data information about the morphology can be obtained which is in essence identical to that obtained from small angle X-ray scattering. In fact the radius of the domains is 218 Å from these data, but this type of scattering will be discussed in more detail later on. Here we are more interested in the question of whether, if we were to incorporate some deuterated molecules in the sample, we could obtain information about the conformation of the individual molecules. Leaving aside for the moment the method by which it was obtained, compare the signal for individual blocks in Fig. 8.14 with that from the spherical domains. Clearly if would be difficult to separate the single chain from the spherical domain scattering. We remarked above on the 'natural' contrast between the two polymers which is giving rise to the signal from the hydrogenous diblock system. Included in the figure is the signal obtained if the butadiene segments are deuterated. As expected, the contrast is much enhanced. Deuterium has a positive value of b ($= 0.66 \times 10^{-12}$ cm). The scattering length per unit volume of the deuterated blocks is relatively large and positive, in contrast to that of the hydrogenous blocks which is small but negative. Mixtures can be made, therefore, with any value of the scattering length per unit volume between

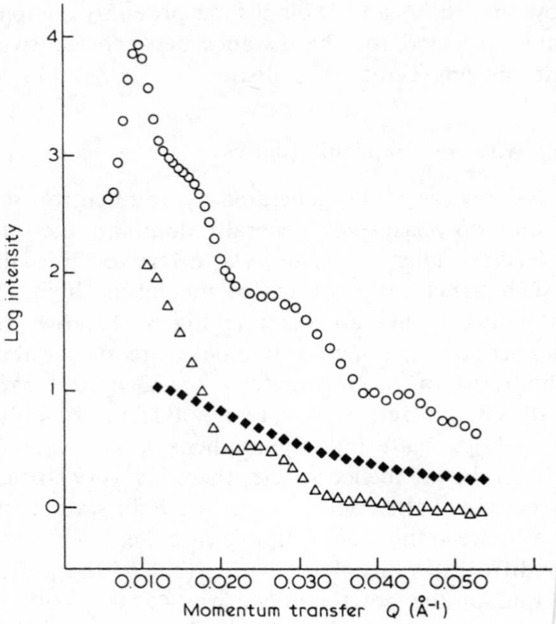

Fig. 8.14 SANS data for polystyrene–polybutadiene diblock copolymers with (○) deuterated butadiene block, (△) hydrogenous butadiene block, and (♦) a mixture chosen to eliminate the second term in eqn (8.3) ($Q \equiv q$). Reprinted from Bates *et al.* (1983) with permission of author and publishers, Butterworth Heinemann.

these two limits by varying the amount, x, of deuterated material. Remembering that the scattering length per unit volume of polystyrene is small but positive it can be seen that a judiciously chosen value of x ought to be able to exactly match the values of the styrene and butadiene segments. This is how the data for the butadiene blocks in Fig. 8.14 have been obtained. The method corresponds to setting the coefficient of the second term in eqn (8.3) to zero considering the styrene segments as the solvent, b_0. From the data the radius of gyration of the butadiene segments is 114 Å in comparison to the value obtained for butadiene of the same molecular weight in a homopolymer which is 86 Å.

The experiment is very cleverly designed and it is almost the only such example in the literature. Why? Well firstly let us apply our check that all has gone well in the contrast matching, and ask about the value of the molecular weight. Unfortunately the data in Fig. 8.13 give a value of M_w for the butadiene segments which is 3.5 times larger than that from the original characterization. This results from a scattering intensity which can already be seen in Fig. 8.14 to be substantial, whereas in contrast-matched samples it is usually quite weak. In several other samples with much lower molecular weights an acceptable agreement was found between the neutron value for M_w and the original characterization value. The authors suggest that the errors arise in high molecular weight samples because the deuterated butadiene segments are not randomly distributed. Unfortunately this throws severe doubts on the reliability of the value obtained for R_g. The non-random distribution could have a thermodynamic origin, in the same way as discussed for homopolymer mixtures of deuterated and hydrogenous polymers in Chapter 5. Alternatively it could be structural since the two diblocks had neither identical block nor identical overall molecular weights.

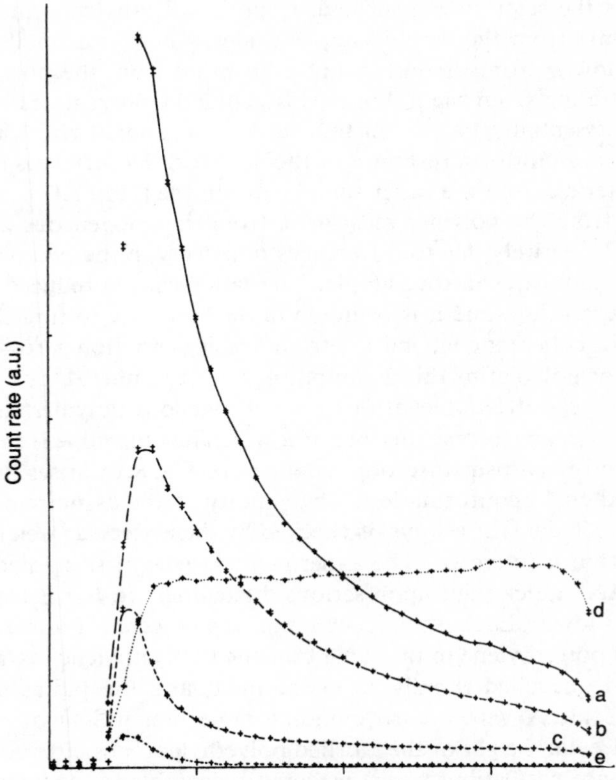

Fig. 8.15 Count rate vs radius on an area detector for isotopic blends of (a) high molecular weight polyethylene, (b) low molecular weight PE, (c) hydrogenous PE, (d) H_2O standard, and (e) empty container. Reprinted with permission from Sadler (1984).

This latter is probably the main reason that the technique of contrast matching to remove domain scattering is not more widely applied. It is very difficult to synthesize molecules with closely matched molecular weights of the deuterated and hydrogenous blocks. Another point worth making is that contrast matching is only possible if the b values of one of the blocks can be changed from smaller to larger than the second block by deuteration. In practice this is usually only the case if the hydrogen content in the two blocks is very different.

All crystalline polymers are effectively two-phase because of the different densities of the crystalline and amorphous phases. However, since any part of any molecule might exist in either phase, contrast matching of the phases such as in the diblock copolymer example is not an option. Figure 8.15 shows the scattering from hydrogenous polyethylene. Underneath is the scattering from the empty sample holder. There is a small amount of scatter at low q from the latter, but even more from the polymer. Some of the scattering from the polymer is isotropic incoherent scattering. This should be essentially flat, or in this case show a small rise with q which is due to a non-uniform response of the detector. The effect is seen clearly in the scattering from the water sample also in the figure. The q dependent scattering from the polymer must arise from inhomogeneous density, but this could be entirely due to differences in density in the two phases, or to voids and impurities in the sample. The latter can be reduced by careful sample preparation, and it is common in the literature to find the assumption that the coherent contribution from density variation across the phases is small. For polyolefins this assumption is not too unrealistic, because the small negative scattering length of the hydrogenous polymer can be made close to zero in an isotopic mixture of low x. Thus the hope is that the scattering from the two such isotopic mixtures in Fig. 8.15 arises largely from the labelled chain conformation. The validity of the assumptions made in analysing such data, as always, is checked by the molecular weight obtained from the zero q intercept. The experimenters who first studied polyethylene by SANS quickly hit upon serious difficulties. In Fig. 8.16 two Zimm plots are shown. Each sample contains 6 per cent of the same linear deuterated polyethylene in the same branched polyethylene matrix. Sample a has been quenched rapidly from the melt, and the radius of gyration obtained is 132 Å with a corresponding molecular weight of 46 000. The actual molecular weight of the labelled polyethylene was 60 000, so the data analysis is not perfect but was considered acceptable in the light of sample b. This identical sample has been slowly cooled from the melt. The Zimm plot looks quite normal, but R_g is 368 Å and M_w is 7×10^5. Such data clearly reveal a non-uniform distribution of the deuterated molecules. There has been considerable discussion about the exact description of this non-uniformity but it is generally agreed to be caused by the different

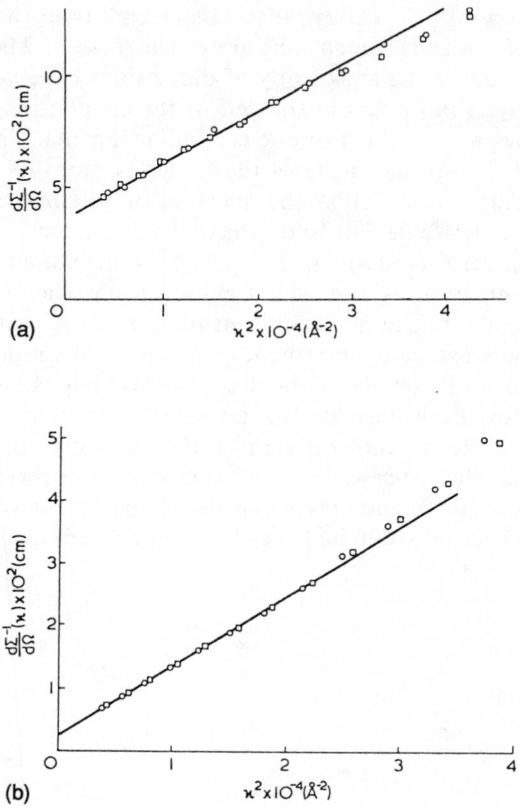

Fig. 8.16 Zimm plots for two mixtures of deuterated and hydrogenous polyethylene ($\kappa \equiv q$). (a) Quenched from the melt and (b) slow cooled. Reprinted from Schelten *et al.* (1976) with permission of the author and the publishers, Butterworth Heinemann ©.

crystallization temperatures of the two species. Reducing the time allowed for segregation by fast quenching or reducing the degree of crystallinity by introducing branches in the polymers all reduce the effect. Clearly such ploys are not always possible or desirable. Attempts to separate the contribution from the single chains have been made by, for example, modelling the signal from the non-uniform distribution at low q, extrapolating to higher q, where the single chain signal is to be analysed, and subtracting it from the data. Such manipulation of the data is fraught with dangers and the molecular weight check is vital.

The scattering at low q from crystalline samples gives the radius of gyration, to be compared with the corresponding value in the melt. For solution grown crystals R_g varies much less steeply than the $M^{1/2}$ dependence of the melt and the dimensions are much smaller. These and other observations, including the anisotropy of dimensions observed in oriented samples, which we will discuss at the end of the chapter, suggest strongly that the molecules are confined to one crystalline lamella. The question of the regularity of the arrangements of the folded sections of the molecules within the lamellae, in particular the question of whether there is a high probability that a molecule will fold directly back on itself is addressed at intermediate q values via the Kratky plots. The problem of adjacent re-entry caused raised temperatures and voices among some scientists for a considerable period, and it is in no small measure due to the neutron scatterers that there is now reasonable agreement over the correct model. This is no place to enter into the details of these arguments, but one example as in Fig. 8.17 will show again how useful the Kratky plot is in differentiating between models of local conformation. The models may be simple mathematical formulae, such as described in Chapter 6, rods or sheets for example, or more complex calculations involving the shape of the stems and their repeat pattern. The most detailed are the Monte Carlo calculations, such as shown in Fig. 8.17.

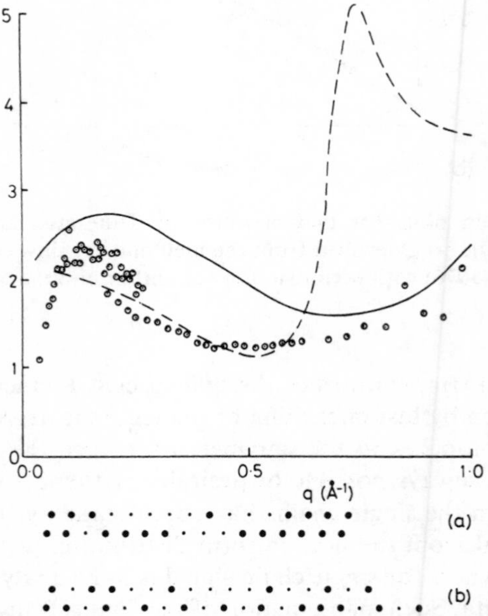

Fig. 8.17 Kratky plot for polyethylene compared to various model calculations. (a) — Adjacent re-entry, (b) — — next but one folding. Reprinted with permission from Sadler (1984). The patterns labelled (a) and (b) indicate the patterns of stems in the crystal corresponding to the models.

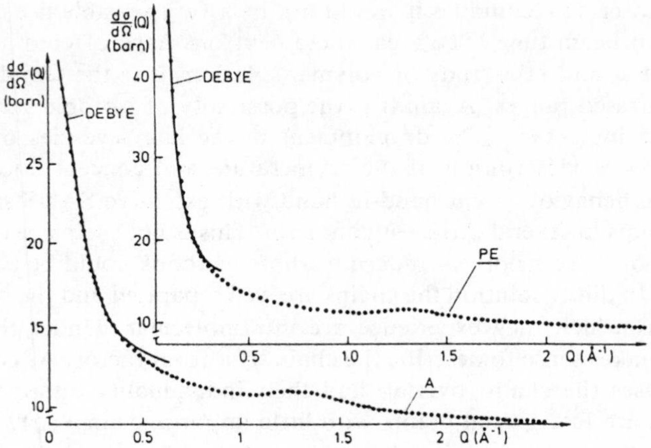

Fig. 8.18 Intensity vs Q ($\equiv q$) for 50 per cent mixtures of D-PE and H-PE (PE) and *n*-alkane melts (A). The solid lines are the Debye function (eqn (6.73)) fitted to the low q data. Reprinted with permission from Schelten and Stamm (1981). Copyright (1981), American Chemical Society.

If the data are extended to still higher q values as described for amorphous systems in Section 8.2.1 further details of the local molecular arrangements may be revealed. In this intermediate range local packing is important. The scattering from both polyethylene and from *n*-alkane melts (with thirty-six carbons) shows a peak at about 1.2 Å^{-1} arising from interchain spacing. The data in Fig. 8.18 are interesting because of the wide q range covered. The relationship between the Debye scattering curve at low q, the Kratky region where deviations from Gaussian behaviour intervene, and the so-called diffuse scattering peaks at very high q are all visible. (The samples were 50/50 H–D mixtures in each case.) By careful analysis of the peak in the high q data which arises from nearest neighbour distances, the authors concluded there is no evidence for back folding of the polyethylene chains in the melt.

8.3 THERMODYNAMIC PARAMETERS

8.3.1 Solutions

The use of scattering to study the thermodynamics of polymer solutions, in particular the virial coefficient A_2, has a long tradition. It is of course possible to use SANS to measure A_2 but since this is excellently well done by light scattering techniques it would not be a very sensible use of expensive neutron beam time. The area where neutrons have offered an important contribution to the study of polymer solutions is in the semidilute and the concentrated ranges. Again it is the possibility of isotopic substitution which is so important. The development in the late seventies of a complete theoretical description of the temperature and concentration effects on solution behaviour went hand-in-hand with extensive SANS studies by several groups in several different countries. This is not the place to go into the details of these theories—indeed a whole textbook could be devoted to this cause. In dilute solution the chains are well separated and the segmental interactions which they experience are intramolecular. Under these conditions it makes sense to describe the chain by a form factor. As concentration increases the chains overlap and their individuality disappears. The chain ends are just special points with little importance in $S(q)$. It now is much more appropriate to describe the sample by a screening length similar to that which was introduced in Chapter 7, Part 3 (in the Ornstein–Zernicke theory). As discussed in Section 7.11.3 this screening length, ξ, is the average distance along the chain between intermolecular contacts with other chains. One of the properties of ξ is that the excluded volume interaction for two points further apart than this distance is negligible. Therefore, the screening length is a measure of the weakening of the excluded volume effect.

In order to observe these effects it is necessary to be in the intermediate q range $qR_g > 1$ and $ql < 1$ where l is the persistence length. Within this q range, the point $q\xi = 1$ divides two regions of behaviour. The scattering function for a single chain behaves as in a dilute solution with excluded volume effects for $q\xi > 1$, and as in the bulk with Gaussian statistics (or theta solvent) for $q\xi < 1$. In Chapter 6 (eqns (6.91) and (6.93)) we showed that in this intermediate q range the dependence of $S(q)$ on q is a power law. When excluded volume is important the power is 5/3, when it is removed in the theta solvent and the chains become Gaussian the power is 2. ξ decreases with increasing concentration and it is also temperature dependent. In principle the value of ξ can be measured in a solution in which most chains are contrast matched with the solvent, i.e. $\bar{b}_H = 0$. ξ is then q^{*-1} where q^* is the crossover from dilute solution behaviour ($S(q) \simeq q^{-5/3}$) to screened behaviour ($S(q) \simeq q^{-2}$). In practice this crossover may be rather diffuse and difficult to define. An alternative method is to observe the many chain scattering in a solution where all the chains have the same contrast. The scattering is then of the form given in eqn (7.140)

$$S(q) = K^2/(q^2 + \xi^{-2}) \tag{8.8}$$

so that ξ is obtained from the slope and the intercept of a plot of $S^{-1}(q)$ against q^2. Alternatively an extrapolation to the negative intercept on the

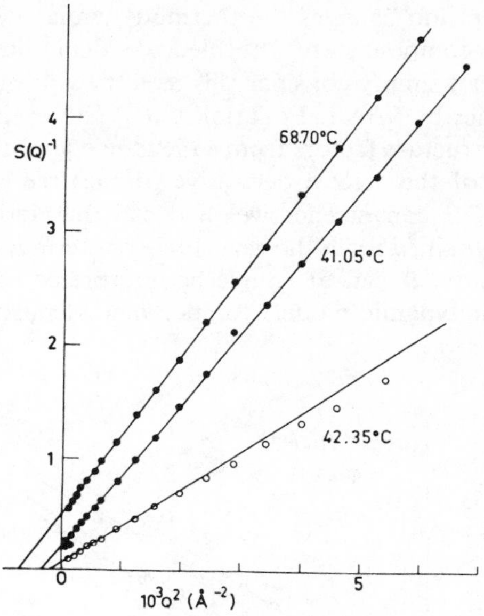

Fig. 8.19 Inverse scattering law vs q^2 ($q \equiv Q$) for two concentrations of PS in C_6D_6. (●) 0.092 g cm^{-3} and (○) 0.058 g cm^{-3}. The values of $S^{-1}(Q)$ for the latter have been divided by 5. Reprinted with permission from Cotton *et al.* (1976).

q^2 axis as shown in Fig. 7.14 gives $-q^2 = \xi^{-2}$. Some data for two concentrations of polystyrene in deuterated cyclohexane are shown in Fig. 8.19. The values of $S^{-1}(q)$ at the lowest concentration have been divided by a factor of five for clarity. At the lower concentration the effect of the crossover to dilute solution behaviour at high q can be seen, since $q\xi = 1$ occurs in the middle of the observed q range for this concentration. To be precise, for this lowest concentration at the low q values the behaviour is dominated by the many chain scattering and is described by eqn (8.8). At higher q the single chain scattering with excluded volume gives rise to a $q^{5/3}$ dependence. For the higher concentration the data follow eqn (8.8) over the whole q range since $q\xi = 1$ occurs outside the range of observation. From the slope and the intercept of these data in the range where eqn (8.8) applies a value of ξ can be obtained. The effect of temperature on ξ can be seen in the same figure as the change in the slope and the intercept of the higher concentration data when the temperature is raised.

8.3.2 Blends

As we have already seen in Fig. 8.9, determining the conformation of individual molecules within a mixture by contrast match is often not possible because the deuteration changes the thermodynamic state of the system. If, however, one component of the blend is deuterated and the other hydrogenous we can simply consider this as a 'new' blend and investigate the thermodynamics by SANS. Equation (7.63) shows that the zero angle scattering partial structure factors from a blend in equilibrium are just (kT) times the inverse of the second derivative of the free energy of mixing, $(\partial^2 g_c/\partial\varphi_i\partial\varphi_j)^{-1}$. For convenience we shall call this derivative g_c'' in this section. If g_c'' is obtained from the zero angle scattering, then it is a model independent quantity. It can of course be interpreted in terms of any of the current thermodynamic models for polymer blends. The form in the

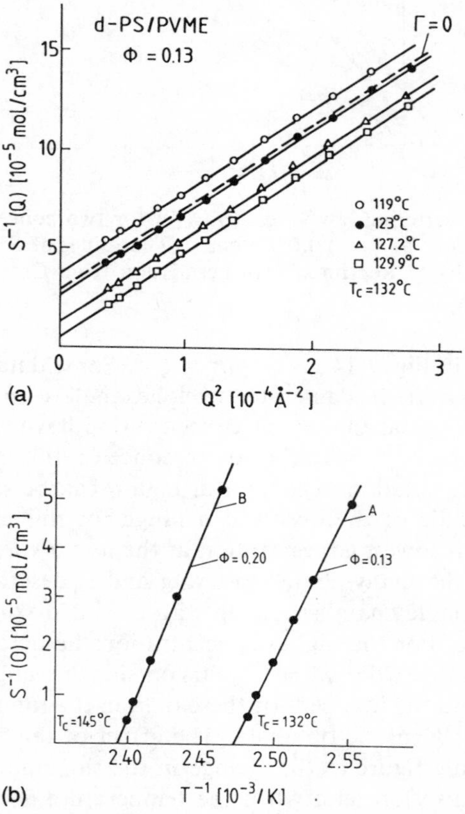

Fig. 8.20 (a) Inverse scattered intensity vs q^2 ($\equiv Q^2$) for a blend of D-PS with polyvinylmethylether at various temperatures as shown. (b) $S^{-1}(\circ)$ obtained from zero q intercepts of data in (a) for this blend and one of lower M_w. Reprinted with permission from Schwahn *et al.* (1992).

Flory–Huggins theory is particularly simple, as shown in Appendix 2, eqn (A2.62).

$$\frac{g_c''}{k_B T} = \frac{1}{z_1\varphi_1} + \frac{1}{z_2\varphi_2} - 2\chi \tag{8.9}$$

the χ parameter in this formalism actually contains all contributions to the free energy of mixing apart from the combinatorial entropy in the first two terms. We shall use this equation as the definition of χ.

Figure 8.20(a) shows some data for a blend of polyvinylmethylether with d-polystyrene. φ_{PS} is 0.13. In Fig. 8.20(b) the zero angle scattering as a function of the temperature for this blend is compared to the behaviour of a second blend with a lower molecular weight d-PS and $\varphi = 0.2$. At the spinodal temperature g_c'' is zero (see Appendix 2) and the values can be read from the extrapolation in Fig. 8.20(b). The spinodal temperature of the lower molecular weight PS occurs at higher temperature as is to be expected from the increase in combinatorial entropy. The linearity of the plots implies that χ varies as T^{-1} as in the simple Flory–Huggins theory.

It is important to re-emphasize that the values of g_c'' or of χ obtained from the neutron scattering data depend on the value chosen for the reference volume when calculating b_v, the contrast factor. To emphasize this point, let us examine the general form of the Flory–Huggins models in the de Gennes formalism (see also eqn (7.123)) which was expressed in Appendix 3 as eqn (A3.18)

$$\frac{V}{I(q)} = \frac{1}{b_v^2}\left\{\frac{1}{\dfrac{c_1 M_1}{N_A} P_1(q)(v_1^S)^2} + \frac{1}{\dfrac{c_2 M_2}{N_A} P_2(q)(v_2^S)^2} - \frac{2\chi}{v_0}\right\}. \tag{8.10}$$

If we substitute the low q expansion of $P_i(q)$ (eqn (6.106)) and reorganize the terms, we obtain

$$\frac{V v_0 b_v^2}{I(q)} = \left\{\frac{1}{\varphi_1 z_1} + \frac{1}{\varphi_2 z_2} - 2\chi\right\} + Aq^2 \tag{8.11}$$

$$A = \frac{1}{3}\left[\frac{\overline{R_1^2}}{\varphi_1 z_1} + \frac{\overline{R_2^2}}{\varphi_2 z_2}\right]. \tag{8.12}$$

Equation (8.12) shows that the slopes of the lines in Fig. 8.20(a) are determined by the values of $\overline{R_i^2}$. The fact that the lines in Fig. 8.20(b) are not precisely parallel reflects a small change with temperature of the values of $\overline{R_i^2}$.

Clearly the determination of thermodynamic parameters in polymer blends by SANS is not generally possible since it requires deuteration in order to provide a suitable contrast. In a few cases there is sufficient natural contrast arising from a different proportion of hydrogen in the two polymers (polyethylene with polystyrene would be an example). Such measurements do allow careful testing of some of the models applied to describe the physics of polymer–polymer miscibility, particularly near the spinodal temperature and, as such, are an important contribution to the understanding of polymer blends. The straight lines in Fig. 8.20(a) and (b) verify the mean field theories in this case but while the Ornstein–Zernicke and Flory–Huggins theories described the data for this blend rather well, this has not always been found to be the case (Brereton *et al.* 1987)

8.3.3 Block copolymers

The situation with copolymers is much more complex. In the bulk we have to take account of three form factors, $P_1(q)$, $P_2(q)$, and $P_{12}(q)$, while in solution we also have three interactions to account for, χ_{01}, χ_{02}, and χ_{12}. The expressions for copolymers in the bulk and solution are given in eqns (7.133) and (7.128), respectively. In solution the requirement of obtaining independent measurements of at least two interaction parameters severely limits the use of eqn (7.128) as a 'thermodynamic' tool for the study of copolymer solutions.

In the bulk on the other hand provided that a reasonable estimate can

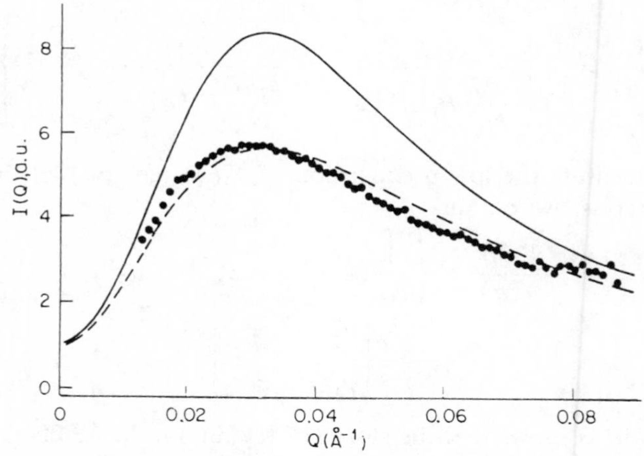

Fig. 8.21 Scattered intensity vs q ($\equiv Q$) for a diblock copolymer of deutero 1,4-polybutadiene with hydrogenous 1,2-polybutadiene. The solid and dashed lines are calculated from eqn (7.133) with $\chi = 9.5 \times 10^{-3}$ and -2×10^{-3}, respectively. Reprinted with permission from Bates (1985). Copyright 1985, American Chemical Society.

be made of the $P(q)$ eqn (7.133) can be applied and provides simply a good estimate of χ. The striking difference with the blend case is that if the copolymer is really monodisperse in composition, it is always true that $I(q = 0) = 0$. Thus as the microphase separation is approached the intensity tends to infinity at a finite q value corresponding to $qR \approx 2$. In fitting the data the usual assumption is that the segments are Gaussian and that the radii of gyration are the same as for homopolymers of the same molecular weight. Data for a diblock copolymer of 1,4 with 1,2-polybutadiene, where the 1,4 block is deuterated, are fitted by the theoretical curve (eqn (7.133)) in Fig. 8.21. All the parameters except χ were obtained from the literature, and polydispersity was also included. In this case a very good fit to the data is obtained.

8.3.4 Phase separation in polymer blends

When a polymer blend with a lower critical solution temperature such as in Fig. 8.8 is heated inside the spinodal to the region where g_c'' becomes negative, the concentration fluctuations responsible for the scattering in Fig. 8.20 become unstable and phase separation follows by a process called spinodal decomposition. Since this is driven by g_c'' observation of the phenomenon can provide an alternative measure of the thermodynamic properties of a blend. The original theoretical development by Cahn and Hilliard (Cahn 1986) was applicable to mixtures of small molecules (for example metallic alloys). It was subsequently extended to polymers and the effects of density as well as concentration fluctuations were included. If the temperature of a sample is suddenly changed to a fixed temperature within the spinodal all the concentration fluctuations grow in amplitude but the

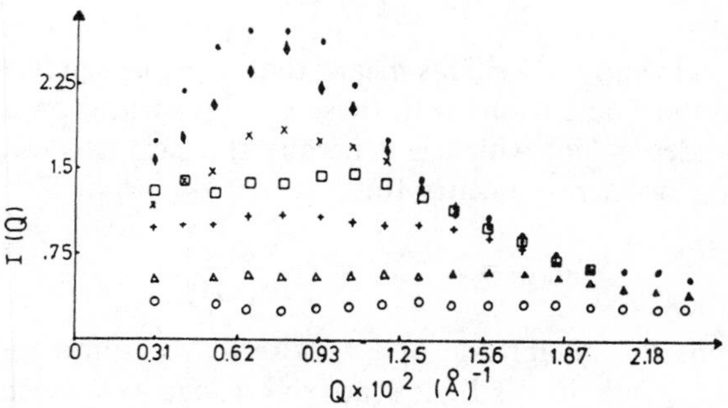

Fig. 8.22 Normalized scattered intensity vs Q ($\equiv q$) for a 26/74 blend of polymethylmethacrylate with poly(α-methylstyrene-co-acrylonitrile) after a temperature jump to within the spinodal at $T = 154.0°C$. The curves are at 10 s intervals, with the lowest intensity curve at 19 s after the jump. (Higgins *et al.* 1989*b*.)

growth rate is wavelength dependent. Rather quickly one wavelength begins to dominate and the scattering function develops a maximum corresponding to a preferred length scale in the phase separated morphology. This maximum is clearly seen in Fig. 8.22 for a blend of poly(α-methylstyrene-co-acrylonitrile) with deuterated polymethylmethacrylate. It can also be seen that the maximum moves to lower q values as time continues. This corresponds to a 'ripening' of the phase separated morphology, where the theoretical models for spinodal decomposition no longer apply.

For the very early stages of spinodal decomposition, the C–H theory can be developed in the same random phase approximation as lead to eqns (8.10) and (8.11). There are three structure factors to consider. That of the blend at T_0, the temperature just before the jump into the spinodal, we designate as $S_{T_0}(q)$. Then there is also the time varying structure factor of the blend after the jump, which we designate as $S(q,t)$, and the structure factor that the blend would have at the temperature inside the spinodal which we designate as $S_T(q)$. In the C–H theory which ignores thermal fluctuations

$$S(q,t) = S_{T_0}(q)\exp(2tR(q)) \tag{8.13}$$

$R(q)$ is the q dependent growth rate which is

$$R(q) = -q^2 M(q)/S_T(q). \tag{8.14}$$

$M(q)$ is a mobility term which we have allowed to be q dependent, although in the simple C–H theory this is not the case. Both $S_T(q)$ and $S_{T_0}(q)$ are given by eqn (8.11). This poses no problems for T_0, but at T remember that the system is inside the spinodal, thus, $(g_c'')^{-1}$ is negative and therefore $S_T(q)$ is actually negative for $q < q_c$ where

$$q_c = \left\{\frac{-g_c''}{k_B TA}\right\}^{1/2} \tag{8.15}$$

and A is defined in eqn (8.12). This means that $R(q)$ in eqn (8.14) is positive for $q < q_c$ and the fluctuations with these q values will grow in amplitude. If $M(q)$ is q-independent which is generally the case it is easy to see that the maximum growth rate occurs for

$$q_{max} = 2^{-1/2} q_c = \left\{\frac{-g_c''}{2k_B TA}\right\}. \tag{8.16}$$

In many systems the magnitude of g_c'' close to the spinodal boundary is so small that q_{max} falls in the light scattering range and much of the published data testing the theories is from light scattering experiments. Some systems have a g_c'' which varies strongly with temperature so that even close to the spinodal g_c'' is large enough for q_{max} to be observed in SANS. The system in Fig. 8.22 is one such. The advantage is that from data inside the spinodal a value of g_c'' is obtained for comparison with that obtained from measurements such as those in Fig. 8.20. The experiments shown in Fig. 8.22 were obtained at 10 s intervals! Even taking for granted that the source was a high flux reactor, such data show how very intense is the signal from a sample when phase separation occurs. We shall see more of this in Section 8.4 when we look at examples of scattering from two phase systems.

8.3.5 Transesterification

Transesterification is the random scission and recombination of polyester molecules induced by heating. The recombination does not necessarily occur between fragments of the same molecule, so that transesterification is a randomization process, which is frequently observed in polyethers and polyimides as well as in polyesters. If the initial sample is a mixture of a deuterated and a hydrogenous polyester molecules then the transesterification process results in copolymers containing segments of deuterated and hydrogenous monomers. The apparent molecular weight determined from the zero angle scattering will decrease, and if this can be followed in a SANS experiment then information on the kinetics of the process becomes available. Such kinetic data is rather rare in the literature so that, in this case SANS is, somewhat unusually, valuable as a 'chemical' rather than a structural tool.

In Chapter 6 it was shown that the SANS data in the intermediate q range lead to values of the number average molecular weight (eqn (6.130)). When the random phase approximation is applied to this q range, we obtain

$$S^{-1}(q,t) = \frac{q^2 l^2}{12x(1-x)} + \frac{1}{2x(1-x)}\left\{\frac{1}{z_D(t)} + \frac{1}{z_H(t)} - \frac{1}{z_T^0}\right\} - 2\chi \tag{8.17}$$

where $z_i(t)$ is the number average degree of polymerization of segments of species i at time t, and z_T^0 is the number average degree of polymerization of the whole sample. This formula is valid for any polydisperse system containing a mixture of homopolymers and copolymers. It has the usual q^2 dependence for this q range with a slope depending on x, but not on the molecular weights

$$\frac{1}{z_T^0} = \frac{x}{z_0(0)} + \frac{1-x}{z_H(0)}. \tag{8.18}$$

The segment length, l, has been assumed to be the same for the two species, and since the molecular weights are small we let χ be zero. In fact χ will later be removed by subtracting the scattering at zero time. The intercept at zero q of the asymptotic line described by eqn (8.17) is

$$S^{-1}(0,t) = \frac{1}{2x(1-x)}\left\{\frac{1}{z_D(t)} + \frac{1}{z_H(t)} - \frac{1}{z_T0}\right\}. \quad (8.19)$$

Each time a scission occurs the number of blocks of H and D increase by unity so that z_H and z_D decrease.

It is easy to establish a relationship between z_D, z_H, and S_c where $S_c(t)$ is the number of effective scissions.

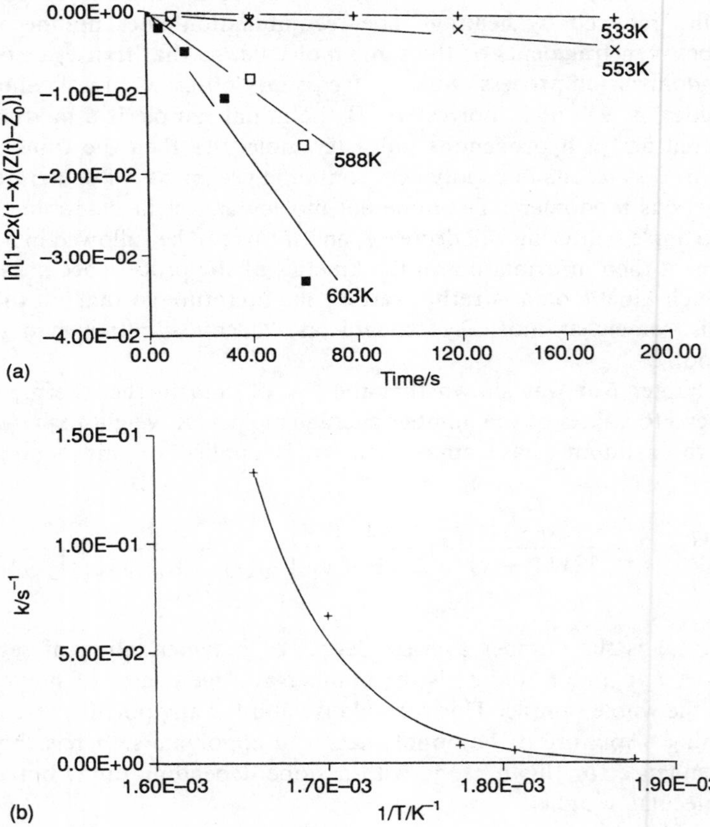

Fig. 8.23 (a) Variation of molecular weight with time during transesterification of a main chain liquid crystal terpolyester. Data have been plotted according to eqn (8.23). (b) Rate constants from the slopes of the lines in (a) as a function of inverse temperature. Reprinted with permission from MacDonald *et al.* (1991). Copyright 1991, American Chemical Society.

$$S^{-1}(0,t) = S^{-1}(0,0) + \frac{1}{2x^2(1-x)^2}\frac{S_c(t)}{N_T} \tag{8.20}$$

$S_c(t)$ can be determined by simple kinetic arguments.

$$S_c(t) = N_T x(1-x)[1-\exp(t/\tau)] \tag{8.21}$$

where N_T is the total number of monomer segments per unit volume and τ is related to the rate constant, k per monomer segment by

$$k = 2\tau^{-1}N_T \tag{8.22}$$

whence

$$S^{-1}(0,t) - S^{-1}(0,0) = \frac{\{1-\exp(-t/\tau)\}}{2x(1-x)}\,. \tag{8.23}$$

Plotting the function $\ln[1 - 2x(1-x)(S^{-1}(0,t) - S^{-1}(0,0))]$ against t yields τ^{-1} as the slope from which k is obtained.

Figure 8.23 shows some data for the transesterification of an aromatic main chain random terpolyester. In Fig. 8.23(a) the increasing rate at higher temperature is clear from the increasing slope (slope $= \tau^{-1}$). In Fig. 8.23(b) the rate constants per segment mole are shown as a function of temperature. The line is the best fit of the relation $k = A\exp(-E_a/k_b T)$ and the activation energy obtained is $E_a = 150 \pm 8\,\mathrm{kJ\,mol^{-1}}$.

8.4 STRUCTURE AND MORPHOLOGY IN MULTIPHASE SYSTEMS

8.4.1 Form factors of particles of known shape

In Chapter 6 we developed formulae for the scattering from discrete particles of different shapes including the sphere (eqn (6.50)), the shell (eqns (6.55) and (6.57)), the disc (eqn (6.58)), and the rod (eqn (6.63)). Unless the nonspherical particles are oriented in some way, their scattering patterns are orientationally averaged and hence very similar and it would be difficult to distinguish between models using simply the $S(q)$ functions. This is seen in the similarity of Figs 6.8 and 6.9. Structures with spherical symmetry, however, show very clear characteristic oscillations in their scattering patterns (Fig. 6.6). The appearance of such oscillations is a clear indication of the existence of such spherical structures in the sample, and (as seen from Fig. 6.25) of a reasonably narrow size distribution. Generally speaking, small angle neutron scattering experiments from particulate dispersions have focused interest on the internal structure of the particles. In the polymer field the investigation by SANS of polymers adsorbed onto the surface of such particles in order to stabilize their dispersions has been well developed by a number of research groups. In order to reduce effects of interparticle interference (Sections 7.14 and 8.4.2) such experiments are carried out with relatively low volume fractions of dispersed particles. In fact at high q, where the effects of the surface structure is best observed the interparticle interference effects are small. In Chapter 6 we considered (Section 6.4.3) the case of a spherical particle with a surface layer of uniform density. The scattering contains three contributions as shown in eqn (6.98). These are the scattering due to the shape of the particles themselves, that due to the surface layer, and a cross term between the two

$$I(q) = \bar{b}_p^2 S_p(q) + \bar{b}_l^2 S_l(q) + \bar{b}_p \bar{b}_l S_{lp}(q) \tag{8.24}$$

$\bar{b}_p$, $\bar{b}_l$ are the contrast between the solvent and the particle or the layer respectively, and $S_p(q)$, $S_l(q)$, are the scattering laws for particle and layer respectively. In Chapter 6 we remarked that experiments designed to investigate a surface layer must focus on a q range of the order of $ql \sim 1$. It is also important that the particles are large ($qR \gg 1$ where R is the radius of the particle) so that the surfaces are effectively flat. In these circumstances $S_p(q)$ can be approximated by the Porod formula, eqn (6.119). Even in this q range the scattering from $S_p(q)$ can be very intense and it has become common in experiments on adsorbed polymers to adjust the scattering length of the solvent until $\bar{b}_p \equiv 0$. This is another contrast match condition and examination of eqn (8.24) shows that the scattering depends only on the scattering from the surface layer. This can be related to the density profile, $\rho_b(r)$, as in Section 6.4. The term we require is the second term on the right-hand side of eqn (6.98).

$$I_l(q) = N \frac{16\pi^2}{q^2} |\widehat{r\rho_b(r)}|^2 \tag{8.25}$$

where $(\widehat{r\rho_b(r)})$ is the one-dimensional FT of $r\rho_b(r)$

$$(\widehat{r\rho_b(r)}) = \int_R^{R+z} r\rho_b(r) \sin qr \, dr. \tag{8.26}$$

Assuming that $\rho_b(r)$ is zero outside the range R to $R+z$ and letting $r = R+z$ (z is the thickness of the surface layer)

$$(\widehat{r\rho_b(r)}) = \int_0^z \rho_b(z) \{\sin q(R+z)\} (R+z) dr. \tag{8.27}$$

There are two alternatives to analysis of the data. Either a model can be chosen for $\rho(z)$ so that $I(q)$, can be calculated from eqns (8.27) and (8.25) and compared to the data, or the data can be transformed through the same two equations to give $\rho(z)$ directly. This latter process is fraught with difficulties. In particular there is a mathematical difficulty because of the presence of the parameter R in eqn (8.27) and also the data may not extend over a wide enough q range to obtain a reliable FT.

Nevertheless with care some authors estimate it possible to obtain $\rho(z)$ from SANS data of adsorbed polymer layers. Figure 8.24(a) shows logarithmic plots of data at contrast match for high molecular weight polyethylene oxide adsorbed onto fully deuterated polystyrene latex spheres dispersed in D_2O. The problems of Fourier transforming these 'real' data

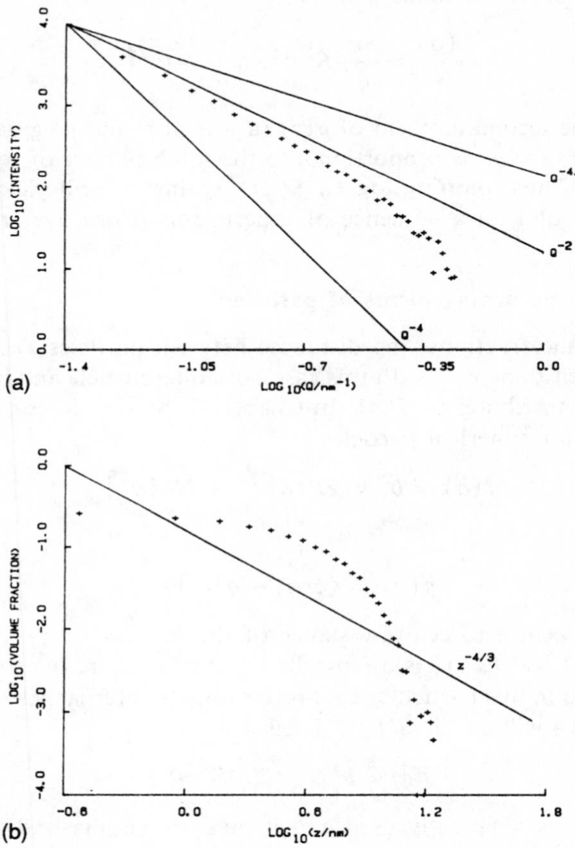

Fig. 8.24 Log–log plot of scattered intensity vs Q ($\equiv q$) for polyethylene oxide (M_w = 280 000) adsorbed on deuterated polystyrene latex in D_2O. Lines showing q^2 and q^4 dependencies are indicated. (b) Log–log plot of the adsorbed polymer fraction $\varphi(z)$ as a function of the distance z from the surface, obtained from the data in (a). The solid line is the prediction of scaling theory. Reprinted with permission from Cosgrove *et al.* (1987). Copyright 1983, American Chemical Society.

are obvious. There are insufficient data at low q, and the residual noise from removal of the background scattering at high q is large. By smoothing data and by using a dispersion integral technique to cope with the mathematical difficulties described above the data in Fig. 8.24(a) were transformed to give the profile for $\rho(z)$ shown in Fig. 8.24(b), where the profile is also compared to a prediction of scaling law theories.

An alternative to fitting models or transforming the data is to use approximate forms for $S_1(q)$ and extract some simple parameters.

In the range $qR \gg 1$ but $qz \ll 1$ eqn (8.25) becomes (in analogy with the development in Section 6.2.2)

$$\frac{I_1(q)}{N} = \frac{8\pi^2}{q^2} R^2 m^2 \exp(-\sigma^2 q^2) \tag{8.28}$$

where σ^2 is the second moment of $\rho(z)$ (a sort of radius of gyration of the polymer layer) and m is proportional to the total number of segments per unit area, Γ. Thus, plotting log $(q^2 S(q))$ against q^2 will yield σ and the relative value of Γ in a sequence of experiments (Cosgrove *et al.* 1990).

8.4.2 Structural arrangements of particles

At high volume fractions, the distances between particles becomes comparable to their dimensions. In this case both interparticle and intraparticle interference contributes to $S(q)$. In Chapter 7, Section 3 (eqn (7.171)) we showed that for spherical particles.

$$I(q) = \bar{b}_v^2 N V^2 P(q)\,[1 + Nh(q)] \tag{8.28$'$}$$

where

$$h(q) = \langle \exp(-\mathrm{i}\boldsymbol{q}\cdot\boldsymbol{r}') \rangle \tag{8.29}$$

and $\boldsymbol{r}'$ is the centre to centre distance of the spheres.

The term $[1 + Nh(q)]$ is often called in the literature the structure factor, $S(q)$, and in this particular case refers only to interparticle effects. For clarity here we will call it $S'(q)$. Then

$$I(q) = \bar{b}_v^2 N V^2 P(q) S'(q).$$

This equation is frequently used and it must be emphasized that it only applies to spherical particles. In eqn (7.78) $S'(q)$ is related to $g(r)$, the radial distribution function, and then in eqn (7.79) to the interparticle potential, $U(r)$. Observation of $S'(q)$ can thus be used, in principle, to test models of $U(r)$. In practice, however, the mathematical manipulations involved are not trivial, and calculations only exist for a number of simple forms of $U(r)$.

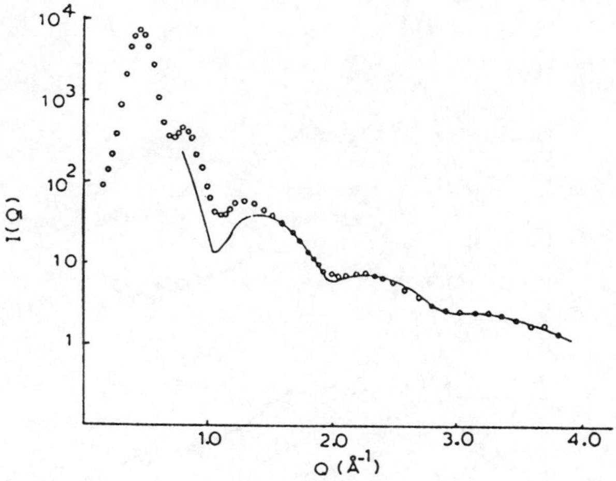

Fig. 8.25 Scattered intensity plotted logarithmically against Q ($\equiv q$) for a diblock copolymer of polystyrene with polyisoprene which has formed cylindrical domains. The solid line is the theoretical calculation for $P(q)$. Reprinted with permission from Richards and Thomason (1983). Copyright 1983, American Chemical Society.

An example of this will be given shortly, but first, observe the effect of $P(q)$ and $S'(q)$ in Fig. 8.25. The sample is a diblock copolymer of styrene with isoprene and the length ratios of the two blocks are such that cylindrical domains of styrene are embedded in an isoprene matrix (confirmed by electron microscopy). In this case some of the cylinders oriented with the long axis perpendicular to the plane of the sample so that a characteristic series of oscillations is seen in the data at higher q values arising from the circular cross section (very similar to those from spheres). The solid line is the theoretical curve for $P(q)$. At low q, there are serious deviations from the theoretical curves because of the contribution of $S'(q)$. Remembering the logarithmic scale on which the data have been plotted the dominant effect of $S'(q)$ at low q is evident.

In reasonably dilute systems $P(q)$ and $S'(q)$ can be dealt with separately, since $P(q) \to 1$ for $qR \ll 1$ but $S'(q) \to 1$ for $qd \gg 1$ where d is the interparticle spacing. Figure 8.26(a) shows data from a diblock copolymer polystyrene—poly(ethylene-co-propylene) in dodecane solution at four weight per cent concentration. In this solvent spherical micelles with polystyrene cores are formed by the copolymer and, since there is no contrast between the ethylene-co-propylene (PEP) chains in the corona, and the dodecane) the SANS experiment only observes the cores as well separated polystyrene spheres (of diameter about 200 Å and centre to centre distance around 800 Å). Nevertheless, the spheres strongly repel each other

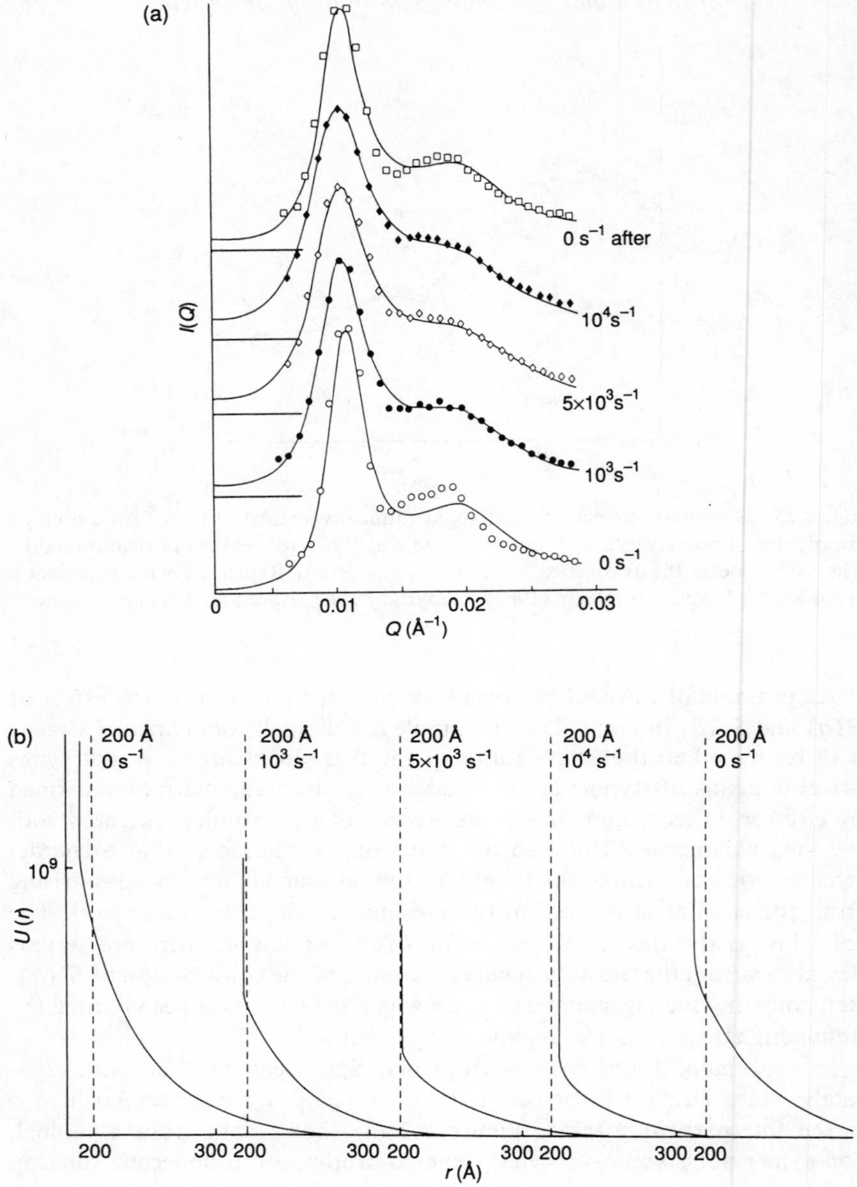

Fig. 8.26 (a) Scattered intensity vs Q ($\equiv q$) for a 4 per cent sample of diblock copolymer polystyrene-b-(ethylene-co-propylene) in dodecane. The experiments were carried out in a Couette cell at the shear rates indicated. The solid lines are fits to the data. (b) The interparticle potential $U(r)$ used to fit the data in (a). (Higgins *et al.* 1988.)

via the excluded volumes of the soluble PEP chains. At low q, as shown in the figure, $S'(q)$ dominates the scattering. In this case the solid lines have been calculated from a potential, $U(r)$, of the form shown in Fig. 8.26(b). The sample was being subjected to shear flow, in a Couette cell, and the subtle effects observed on $S'(q)$ as the shear rate, $\dot{\gamma}$, changes have been related to small changes in the shape of $U(r)$. The theoretical curves in Fig. 8.26(a) include a contribution from $P(q)$ as well as $S'(q)$.

8.4.3 Complex structures

As a general rule, the use of SANS to investigate morphology in random two-phase structures is not to be recommended. There are several other, more readily available techniques, such as electron microscopy, light scattering, and small angle X-ray scattering which would normally be first choice. Nevertheless, there are a few experimental situations where the neutron might be called for, and a few others where scattering from random structures interferes with an experiment designed to investigate something quite different. These latter effects often involve residual inhomegeneities in samples which have been quenched from the melt to the glassy state. They may be due to density fluctuations in amorphous polymers, concentration fluctuations in blends, or domain structures in liquid crystal polymers. Having done his or her very best to reduce the spurious scattering by careful sample preparation, the experimenter's problem is usually to give a satisfactory explanation of intense scattering at low q, and then to find some way of removing it from the desired signal at higher q. This procedure may involve fitting the signal to one of the model scattering laws for random structures which we will discuss shortly. Although a great deal of effort has been and will continue to be, devoted, to such problems, experiments have to be dealt with on a case by case basis. It is not appropriate to give a general recipe for dealing with intense contamination to the scattering at low q except to reiterate that extreme care in sample preparation will help to reduce it.

Returning now to the experiments designed to investigate two-phase morphology, what are the situations requiring neutrons? As always, the answer to this question arises from contrast and from penetration. The neutron can be used for relatively thick, opaque samples, and isotopic labelling can be used to change the contrast and high-light different features in the structure. There were two main models discussed in Section 7.3 for random two-phase structures. These are the Porod law (eqn (7.155)) and the Debye–Bueche eqn (7.167). In both cases, the boundaries between the two phases must be sharp. The Debye–Bueche formula applies best when the average correlation length, ξ between two phases is of the order of q^{-1}

$$I(q) = Vb_v^2\varphi(1-\varphi)8\pi\xi^3\left(\frac{1}{1+q^2\xi^2}\right)^2. \tag{8.30}$$

A plot of $I^{-1/2}(q)$ against q^2 should be linear and yield a value of ξ from the slope/intercept. The high q limit of eqn (8.30) yields the familiar Porod q^4 law (eqn (7.155)).

$$q^4\frac{I(q)}{V} = (b_v^2)\ 2\pi S/V \tag{8.31}$$

with $2\pi S/V = \varphi(1-\varphi)\ 8\pi/\xi$.

Figure 8.27 shows an example of the use of these two equations in the same experiment on a polymer blend. Ethylene-propylene-diene elastomer (EPDM) is immiscible with a copolymer of styrene and vibylpyridine (PSVP). It forms a very 'crumbly' fragile blend. Addition of a few zinc neutralized sulphonate groups to the EPDM (Zn-S–EPDM) causes a coordination interaction between the zinc sulphonate and vinyl pyridene groups, thus forcing the incompatible polymers to mix at a much more intimate level than hitherto. These blends are tough and transparent. In the experiment the PSVP was deuterated (D-PSVP) to give enhanced contrast. Figure 8.27(a) shows a Porod plot for EPDM–D-PSVP blends of several compositions. The surface to volume ratios would correspond to dimensions of the order of 6 to 20 μm if the shape were spherical. Figure 8.27(b) shows a Porod plot for the same mixtures when the interacting Zn–S groups are added. Even a very small level (0.65 mol per cent) is sufficient to dramatically change the scattering, which no longer obeys the Porod law (i.e. there are no longer large particles with sharp interfaces). It does, however, obey the Debye–Bueche equation quite well, as shown in Fig. 8.27(c). The correlation lengths are around 50 to 100 Å. To complete the comparison, in Fig. 8.27(d) the Debye–Bueche plot clearly will not fit the unfunctionalized blend, indicating that a simple two-phase model cannot describe these data satisfactorily.

8.4.4 Fractal systems

As we remarked in Section 6.4.4 the fractal idea is beginning to be used to interpret scattering data, though, as yet, its applications to polymer systems are relatively few. In Section 6.4.4 we showed that the fractal dimension, D, of self-similar objects can be determined in two ways. The mass, $M(R)$ is related to the radius R by $M(R) \propto R^D$, while the scattering is described by eqn (6.102), $I(q) \propto q^{-D}$. A typical example of the use of these fractal ideas for understanding scattering data is in the curing of epoxy structures. An example of such data is shown in Fig. 8.28. The epoxy monomer is a diglycylether of bisphenol A (DGEBA) cured with a triamine.

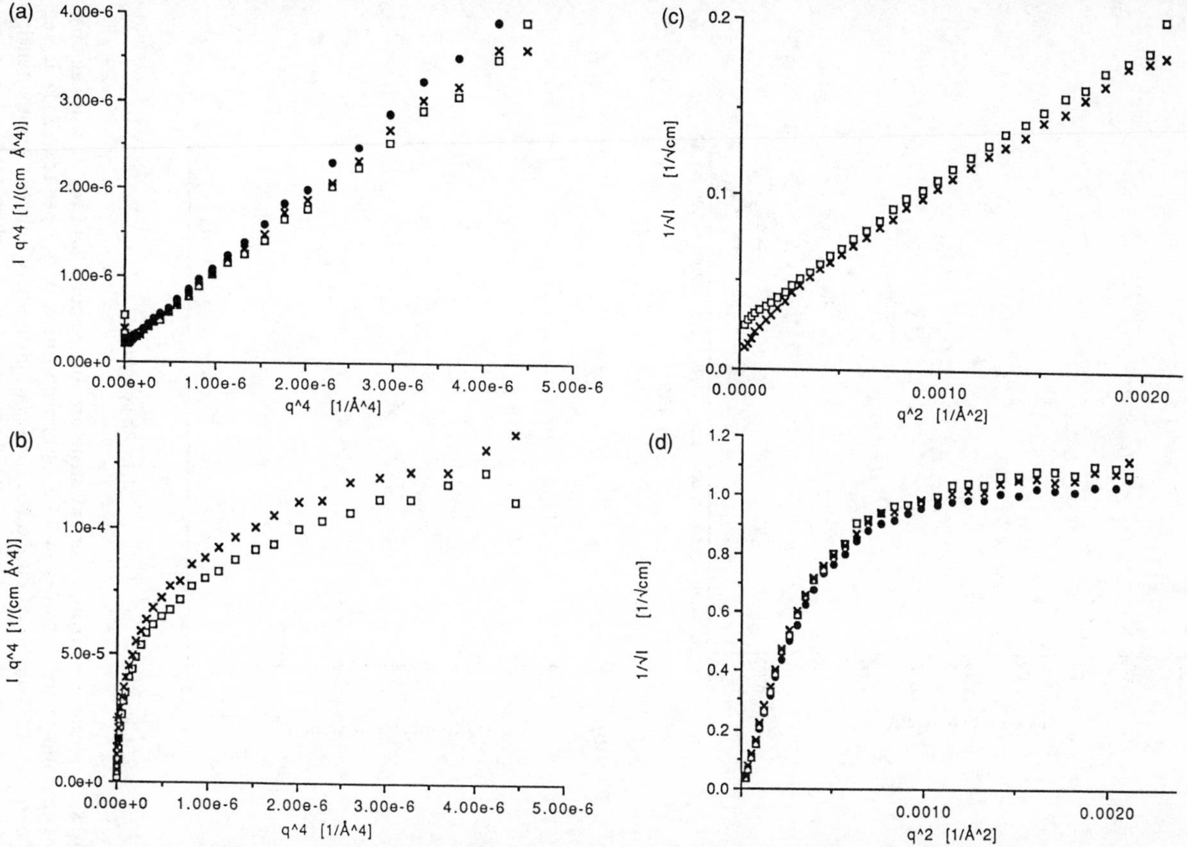

Fig. 8.27 Mixtures of ethylene-propylene-diene elastomer (EPDM) with a deuterated copolymer of styrene and vinyl pyridine (D-PSVP). (a) Porod plot for three concentrations: (●) 90/10, (□ 82/18, and (×) 70/30. (b) Porod plot for two of the same concentrations when interacting Zn–S groups are added to 0.65 mol per cent (same symbols as (a)). (c) Debye–Bueche plot for data as in (b). (d) Debye–Bueche plot for data as in (a). (Clark *et al.* 1991.)

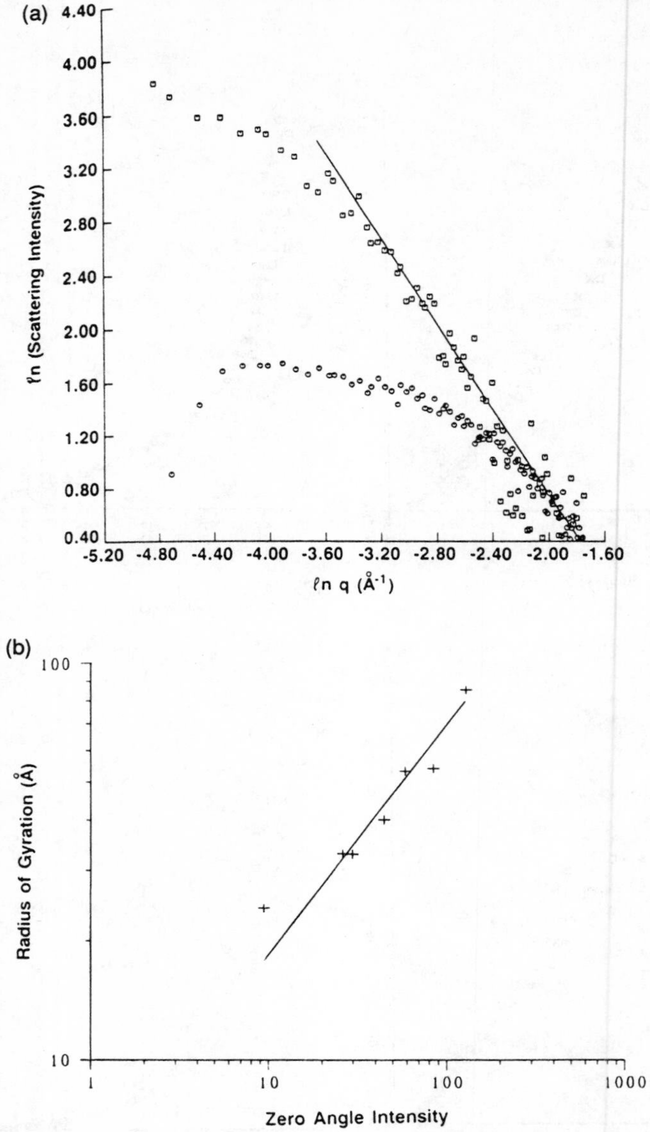

Fig. 8.28 Scattering from an epoxy structure during cure. (a) Log intensity against q. The samples are dissolved in THF at concentration of 1.5 per cent. (□) degree of cure 3.5, (○) degree of cure 0.09. (b) Relationship between zero angle intensity and z-average R_g obtained from a Zimm analysis of data such as that in (a) at low q. Reprinted from Wen-Li Wu *et al.* (1989) with permission from the authors and publishers, Butterworth Heinemann ©.

Samples were extracted at different stages in the curing process and neutron scattering experiments were carried out. Figure 8.28(a) shows data for two different degrees of cure, plotted logarithmically. Only for the higher degree of cure is an extended linear range observable, from which $D = 1.6$ is obtained. Figure 8.28(b) shows a plot of M vs R for a sequence of samples, where the M and R values have been obtained by making a Zimm plot at low q values. The value of D from Fig. 8.28(b) is 1.67, so the two methods are in good agreement. The samples were not fractionated after the curing was arrested, so they are highly polydisperse. When polydispersity is included, a critical percolation model of gelation predicts $D = 1.59$ in good agreement with the results.

8.5 ANISOTROPIC STRUCTURES

When the first small angle neutron scattering apparatuses were being built in the early seventies it was felt essential to increase count rates by using, as far as possible, large area detectors. This was in contrast to light and X-ray scattering where, because of the much higher photon intensity, single angle or at most, linear detectors had been acceptable. The almost automatic inclusion of an area detector in a SANS spectrometer had other benefits. It made information on scattering from directionally oriented or stretched samples straightforward to obtain in a single experiment, if a loss in count rate could be tolerated. Partly as a result of the exciting (and sometimes unexpected) results of SANS, more X-ray and light-scattering spectrometers now also include area detectors. In the following sections we will highlight a few examples of what, in many ways is the most interesting use of the technique.

8.5.1 Effect on the scattering formulae of orienting the samples

Let us consider the simple case of the scattering by two points separated by a fixed distance r. If they rotate freely the intensity scattered by the pair will be proportional to $\langle \exp(-\boldsymbol{q}\cdot\boldsymbol{r}) \rangle = \sin qr/qr$ and the scattering by the pair, as a function of q will be $1 - q^2r^2/6$. If we use Cartesian coordinates u, v, and w and call the components of r, r_u, r_v, and r_w (Fig. 8.29(a)) this can be written as $1 - q^2r_u^2/2$ since $r_u^2 = r_v^2 = 1/3r^2$. Now, if r is oriented the scattering depends strongly on the orientation: for instance if r is oriented along the u direction $r_u^2 = r^2$ and r_v^2 and r_w^2 are equal to zero. This can be generalized to any case if we know the distribution of the orientations. In order to show how it works for a simple example we shall evaluate the first term of the scattering function (the radius of gyration) for one molecule placed in an orienting field. We have to evaluate

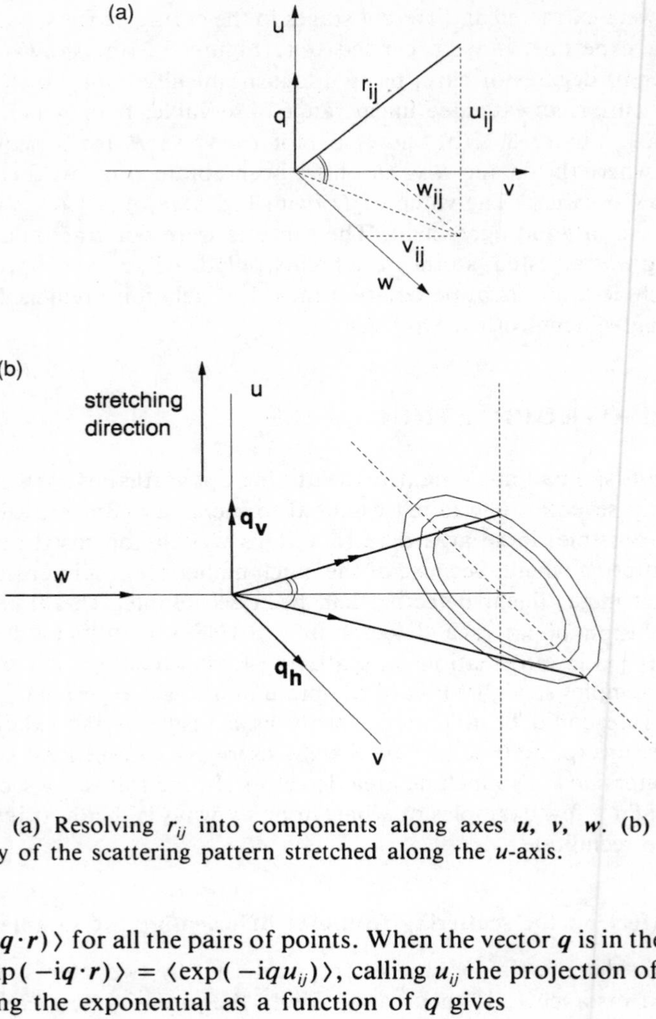

Fig. 8.29 (a) Resolving r_{ij} into components along axes *u, v, w*. (b) Elliptical anisotropy of the scattering pattern stretched along the *u*-axis.

$\langle \exp(-\mathrm{i}\boldsymbol{q}\cdot\boldsymbol{r})\rangle$ for all the pairs of points. When the vector $\boldsymbol{q}$ is in the u direction: $\langle \exp(-\mathrm{i}\boldsymbol{q}\cdot\boldsymbol{r})\rangle = \langle \exp(-\mathrm{i}qu_{ij})\rangle$, calling u_{ij} the projection of $\boldsymbol{r}_{ij}$ on $\boldsymbol{u}$. Expanding the exponential as a function of q gives

$$I_z(q) = \sum\sum \langle \exp(-\mathrm{i}qu_{ij})\rangle = z^2 - \frac{q^2}{2}\sum\sum\langle u_{ij}^2\rangle \qquad (8.32)$$

$$= z^2\left(1 - \frac{q^2}{2z^2}\sum\sum\langle u_{ij}^2\rangle\right).$$

This defines a mean square radius of gyration $\overline{R_u^2}$ in the direction $\boldsymbol{u}$, remembering the definition of the radius of gyration of a molecule is

$$\bar{R}^2 = \frac{1}{2z^2}\sum\sum\langle r_{ij}^2\rangle = \frac{1}{2z^2}\sum\sum\{\langle u_{ij}^2\rangle + \langle v_{ij}^2\rangle + \langle w_{ij}^2\rangle\}. \qquad (8.33)$$

If the orientation is isotropic the three quantities $\overline{R_u^2} = \frac{1}{2z^2}\sum\sum\langle u_{ij}^2\rangle$ are equal and equal to one-third of the radius of gyration. If the system is oriented $\overline{R_u^2}$, $\overline{R_v^2}$, and $\overline{R_w^2}$ are different; they can be called radii of gyration with respect to a plane since they measure the average value of the square of the distance to the plane going through the centre of mass. This shows that the scattered intensity will be different depending on the orientation of the particles relative to $\boldsymbol{q}$.

Two things are important:

1. The orientation of independent particles does not affect the extrapolated zero angle intensity. This intensity can only be modified if the interactions along the different directions are not isotropic.

2. It is the orientation with respect to the vector $\boldsymbol{q}$ which is important. At small angles $\boldsymbol{q}$ is practically in the plane of the sample. If the sample is stretched in the direction u the intensity will decrease faster in this direction since $\overline{R_u^2}$ increases and the isointensity curves will be similar to ellipses having their larger axis horizontal (Fig. 8.29(b)). If the particles are rigid $R_u^2 = R_w^2$. Since for rigid particles we always have $R_u^2 + R_v^2 + R_w^2 = R^2$ we see that, compared to the isotopic case, the intensity decrease ΔR_{u0}^2 is twice the intensity increases ΔR_{v0}^2 or ΔR_{w0}^2.

If the particles can be deformed, which is always the case for chain molecules, change of the radius of gyration in the u direction with no change in directions v and w becomes possible (this implies a change in the volume of the sample). In the case of a rubber we assume that the volume of the rubber does not in fact change during the deformation (rubbers are practically incompressible). Therefore, if λ is the elongation ratio and assuming affine deformation in the u direction we obtain

$$\overline{R_u^2} = \lambda\overline{R_{u0}^2};\quad R_v^2 = \lambda^{-1/2}\overline{R_{v0}^2} \text{ and } \overline{R_w^2} = \lambda^{-1/2}\overline{R_{w0}^2}. \tag{8.34}$$

8.5.2 Aligned or stretched single chains

If the observation of single polymer molecules in crowded environments was the first and most basic experimental gap that SANS filled (by the use of H–D substitution) then the observation of the deformation and relaxation of these molecules was a close second. The question of affine deformation in stretched melts or rubbers has been of long-standing and vital importance to the understanding of viscoelasticity. In a melt, deformation is followed by stress relaxation, and generally the requirement for adequate statistics from an anisotropic sample, where the signal cannot be radially averaged over the detector, prohibits real time experiments. Samples have to be heated to the melt, deformed, and quenched, with a known time period allowed for stress relaxation between the two processes. Figure 8.30 shows the isointensity contour plots obtained on an area detector from a polystyrene sample M_w 7.8 × 10^5, 15 per cent deuterated in a hydrogenous matrix) which has been subjected to a fast stretching deformation with overall extension ratio, $\alpha = 3$ and then quenched. Note the similarity to Fig. 8.29(b). The elliptical pattern has a ratio of long to short axes of 5.3. If the polymer molecules have deformed affinely with the sample deformation then their radii of gyration parallel to the stretching direction should be $R_g^{\prime\prime} = 3R_{g0}$, while the value perpendicular to the stretch, $R_g^{\perp} = 3^{-1/2}R_{g0}$ (see eqn (8.34)). If we assume we are in the low q range so that the scattering can be described by eqn (6.106), then at isointensity points on the major and minor axes, $q_1^2 R_g^{\prime\prime 2} = q_2^2 R_g^{\perp 2}$. Thus, the expected ratio of axes will be $R_g^{\prime\prime}/R_g^{\perp}$ which is $3^{3/2}$ (=5.2) for affine deformation, very close to the observed values of 5.3. In this and many other experiments, the interest centred not only on whether deformation was initially affine, but on the details of the recovery of the Gaussian chain conformation during the subsequent relaxation. Such problems are attacked by waiting for measured recovery periods in the melt state before quenching the samples. The whole procedure of course depends on both quenching and initial stretching being fast compared to the relaxation. Moreover, the assumption that eqn (6.106) holds is unlikely to be true for the high molecular weight polymers needed to test relaxation theories. In general $qR_g^{\prime\prime} > 1$ and eqn (6.106) can only be applied to the $R_g^{\perp}$ data. This is clear in Fig. 8.30—much of the intensity in the parallel direction is lost behind the beamstop.

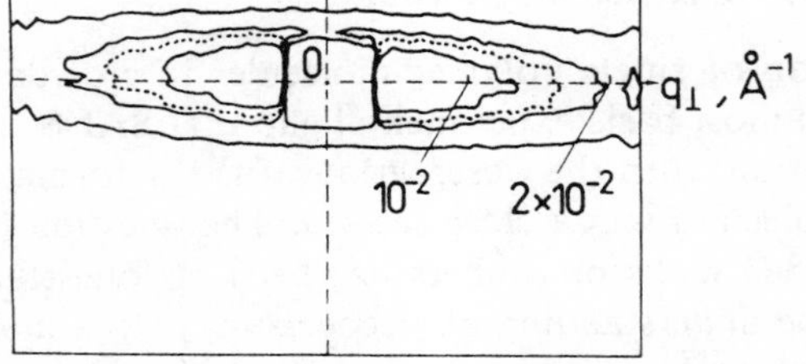

Fig. 8.30 Curves of equal intensity in the $q_\perp - q_\parallel$ plane for a sample of 15 per cent D-PS in H-PS with $M_w = 7.8 \times 10^5$ immediately after stretching with an extension ratio, $\alpha = 3$ and quenching. Reprinted with permission from Boué *et al.* (1982).

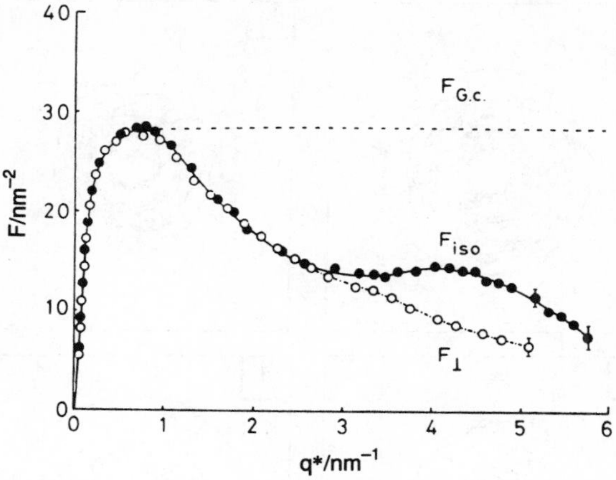

Fig. 8.31 Kratky plots for polymethylmethacrylate containing 50 per cent deuterated molecules with $M_w = 2.3 \times 10^5$. (•) Unstretched, (∘) cold drawn with extension ratio of $\alpha = 2$ and measured along $q_\perp$. $F_{G.c.}$ is the Debye function for comparison. Reprinted with permission from Dettenmaier *et al.* (1986). Copyright 1986, American Chemical Society.

Not only the low q, but also the high q data may carry information about molecular deformation. Figure 8.31 shows Kratky plots for polymethylmethacrylate which has been cold drawn – i.e. stretched in the glassy state. The molecular weight is 2.5×10^5, the extension ratio 2 and 50 per cent of the molecules are deuterated. In the Kratky plot the data perpendicular to the stretch direction are compared to those from an unstretched sample, and to the curve for a Gaussian chain. As already mentioned in Section 8.1 for other polymers, deviations from Gaussian behaviour are clearly seen. What is particularly of interest is the fact that the stretched sample shows clear effects of a very local deformation, on a scale of less than 25 Å.

8.5.3 Anisotropic structures

In the experiments described in the previous section anisotropy of the scattering data was expected, and the experiments were designed to observe it. In this section we just mention briefly two 'surprises' – where anisotropy was either unexpected, or of an unexpected form. In Fig. 8.26(a) the scattering from a solution of polystyrene – poly(ethylene-co-propylene) micelles was compared to calculations based on a model interparticle potential. It can be seen in the figure that the data, especially before shear was applied show more structure than predicted by the $S'(q)$ calculated for a liquid-like arrangement of spheres. The data in Fig. 8.26(a) have been obtained

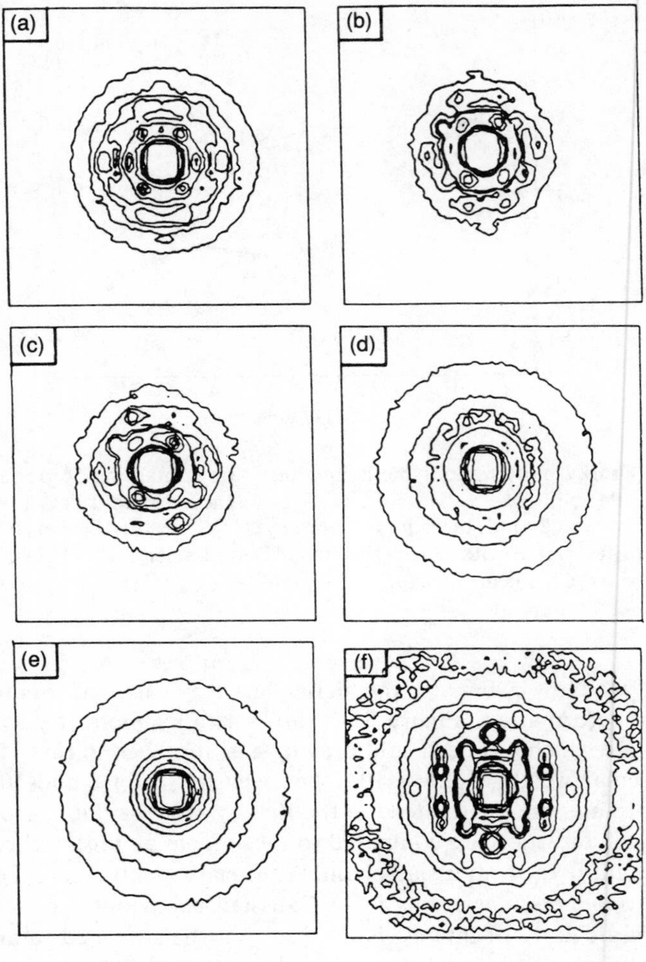

Fig. 8.32 Contour plots of intensity of an area detector from diblock copolymers as in Fig. 8.26 but at 6 per cent in deuterated dodecane. The sample conditions or the shear rate in the Couette cell were: (a) static; (b) after lowering stator; (c) after rotating stator once; (d) at 25 s^{-1}; (e) at 70 s^{-1}; (f) static again. (Phoon 1992).

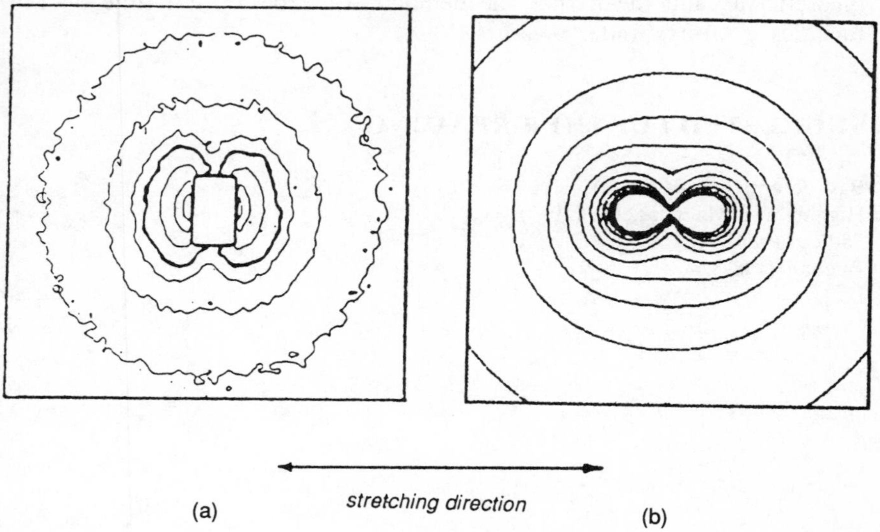

Fig. 8.33 (a) Scattering from a polymethylsiloxane network containing labelled free chains which has been stretched with an extension ratio of 2.43. Reprinted with permission from Oeser *et al.* 1987. (b) Calculated isointensity curves to compare to (a). Reprinted with permission from Bastide *et al.* (1990). Copyright 1990, American Chemical Society.

by radially averaging data on an area detector. In Fig. 8.32 similar data for a 6 per cent sample are shown for the full area detector. The scattering clearly shows long range organization of the micelles. They are not really liquid-like, but actually forming a three-dimensional crystal-like array. During shear this 3-D order disappears, which is why the radial averages in Fig. 8.26(a) fit the liquid-like model better at higher shear.

For the final surprise we take one example of what has been called 'butterfly' patterns. The origin of the name is clear in Fig. 8.33. These are data for labelled chains trapped in a deformed network and at first one might suppose the data belong in the previous section. However, the pattern in Fig. 8.33(a) cannot be explained by the scattering from the labelled chains. There is no change perpendicular to the stretch and extra intensity in the direction parallel fo the stretch at low q. It has been argued that in this case the intensity arises from deformation of concentration fluctuations in an inhomogeneous network. Figure 8.33(b) shows a theoretical calculation, which at least predicts the observed form of the scattering.

Results such as these continue to arise from the SANS experiments being carried on around the world. Such surprises are the delight of the theoreticians and mean that the demand for experimental time on the facilities is always under pressure.

SUGGESTED FURTHER READING

Bastide and Benoît (1990).
Higgins and Maconnachie (1987).
Sadler (1984).
Wignall (1987).

9

Dynamics

9.1 INTRODUCTION

9.1.1 Types of motion

Motion of polymer molecules occurs over many decades of frequency from the slow reorientation of whole molecules in melts just above their glass transition temperatures (T_g) which may take hours, to the torsions and vibrations of side groups at around 10^{14} Hz. As will be seen, the neutron technique samples only the upper end of this frequency range with a limit which has currently been pushed down by the ultrahigh resolution spin echo spectrometer to around 10^7 Hz. Within this range fall the molecular vibrations of side chains, the rotations of such side groups, the travelling waves in crystalline samples, and the higher frequency backbone motions of polymers in solution and in melts at temperatures well above T_g.

There is considerable overlap in the information obtained from neutron scattering with that from techniques such as NMR relaxation measurements, infrared and Raman spectroscopy, high frequency viscoelastic and ultrasonic measurements. However, the unique feature of the neutron scattering technique is that the relatively heavy neutron has large momentum at modest energies. The wavevector changes associated with energy transfer are important and neutron experiments explore regions of momentum transfer space away from the $q \approx 0$ region of infrared and Raman spectroscopy. Because of this neutron scattering spectra inherently convey direct information about spatial correlations as well as energy fluctuations in a way not available from the other techniques mentioned. In addition, because the neutron interacts directly with the scattering nuclei, there are no optical selection rules and the difference between hydrogen and deuterium allows labelling techniques to be used.

Despite these advantages, neutron scattering has not been extensively applied to the study of polymer dynamics. This is partly because the highest resolution spectrometers are scarce and not widely available and partly because the necessity of taking the sample to the neutron source has meant these spectrometers have appeared very inaccessible to polymer scientists. It remains true that because of inevitable experimental limitations (resolution and flux) the information from neutron spectroscopy needs to compliment that from other techniques and to be part of wider studies. The applications in polymer science will always be limited. Nevertheless, in a volume such as this it is important to make clear both what is possible and what is not possible. Taking the basic theory developed in Chapter 4 and showing how this has been used to interpret a number of experiments we hope to demonstrate both the advantages and the limitations of neutron spectroscopy for the study of macromolecular motions.

In Chapters 5–8 we have only taken note of the incoherent scattering as a background to the structurally-dependent coherent scattering. In inelastic and quasielastic measurements, both coherent and incoherent scattering are affected by the neutron energy changes, so both have to be considered in theory and experiments. At the end of Chapter 4 we showed that the inelastic scattering arising from vibrational, rotational, and translational motion appears in different parts of the energy transfer spectrum. Translational motion broadens the elastic scattering peaks (Fig. 4.4) while rotational motion gives this peak structure with a sharp central peak and a broadened 'foot' (Fig. 4.5). Vibrational motion, on the other hand, gives rise to sharp peaks each shifted away from the elastic scattering by an amount corresponding to a vibrational transition (Fig. 4.6).

The techniques available for studying inelastic and quasielastic scattering are described in Chapter 3. The simplest is the time-of-flight spectrometer where the neutron *velocity* is measured before and after scattering by a sample into a range of detectors. Given the velocity and the angle of scatter, the energy transfer $\hbar\omega$, and momentum transfer q can be calculated. Data are, however, usually displayed as intensity against time-of-flight (inverse velocity) as seen in Fig. 9.1. Since these experiments are carried out at constant θ, but $k_i \neq k_f$, q varies with $\hbar\omega$ across each spectrum. Also, since $\hbar\omega$ varies as τ^{-2}, the resolution tends to deteriorate at higher value of $\hbar\omega$. Figure 9.1 shows the time-of-flight spectra obtained for an amorphous polymer, polymethylmethacrylate (PMMA), a semicrystalline polymer, polytetrafluorethylene (PTFE or Teflon) and a rubber, polydimethylsiloxane (PDMS) all at room temperature. The incident neutron wavelength is about 4 Å, the energy, 5 meV. With this incident energy, the energy transfer range for reasonable resolution reaches to about 50 meV (400 cm^{-1}). The intense peak at the right-hand side of each spectrum corresponds to the incident energy and to neutrons which have been approximately elastically scattered. To the left of this peak are neutrons with shorter flight times, which have gained energy from the sample. The data correspond to only one-half of Fig. 4.6. In order to observe the other half, the neutron energy loss spectrum, much higher incident energies are required. A further difference to Fig. 4.6 is that there are clearly many inelastic transitions, corresponding to the many vibrational modes in the sample, rather than the single transition shown there.

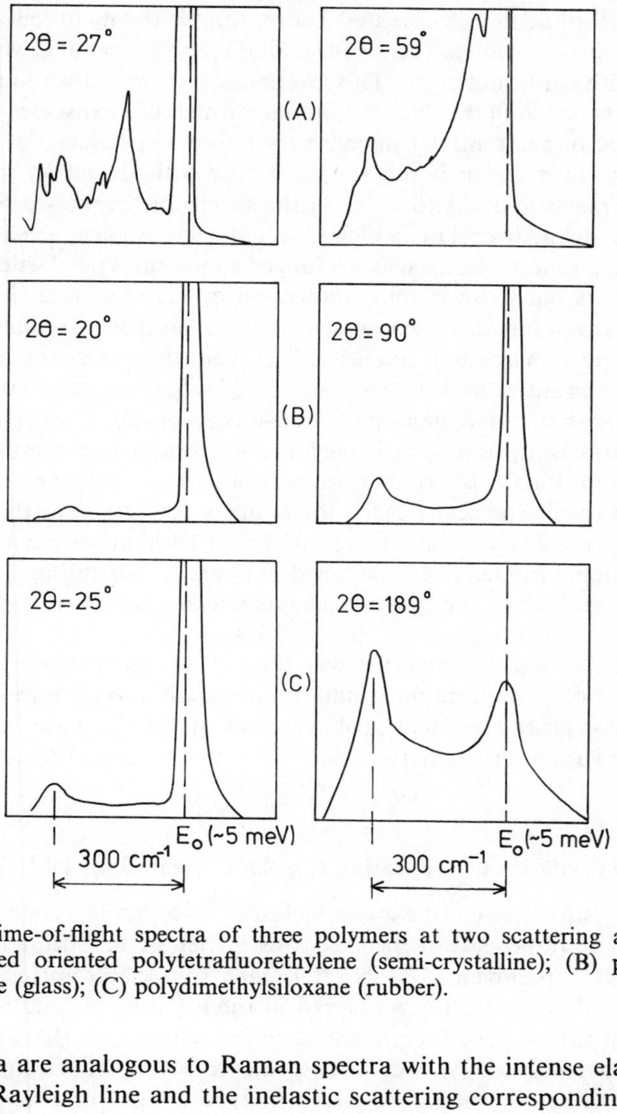

Fig. 9.1 Time-of-flight spectra of three polymers at two scattering angles (2θ). (A) Stretched oriented polytetrafluorethylene (semi-crystalline); (B) polymethylmethacrylate (glass); (C) polydimethylsiloxane (rubber).

The spectra are analogous to Raman spectra with the intense elastic peak being the Rayleigh line and the inelastic scattering corresponding to anti-Stokes lines.

Although different techniques may be necessary to analyse the detail, the effects of nearly all the different types of polymer motion to be discussed in this chapter can be seen in the time-of-flight spectrum in Fig. 9.1. The fluorine in the Teflon means that from this sample coherent scattering is dominant, while for the other two polymers the hydrogen ensures that the incoherent contribution is observed. In the spectrum of the polycrystalline stretch-oriented Teflon sample the main features are an intense, unbroadened elastic peak and inelastic structure which varies strongly with scattering angle. This inelastic structure is due to propagating modes in the crystalline regions—phonons which obey dispersive relationships. Some of these modes have a very strong dependence of energy on the wavevector $\boldsymbol{q}$ and it is the variation of $\boldsymbol{q}$ with scattering angle which causes the inelastic peaks to move in the spectrum. In the scattering from the polymethylmethacrylate, which is an amorphous glass, the sharp structure of the phonon spectrum is no longer apparent. The elastic scattering is still intense but there is some indication of broadening at its foot and there is a broad inelastic peak which does not shift with scattering angle, but which does increase in intensity. This behaviour is characteristic of a molecular vibration, and in the case of PMMA this particular feature is associated with torsional motion of the α-methyl group. The slight broadening at the foot of the quasielastic peak comes from the rotations of the ester methyl group. Finally the rubber not only shows very intense inelastic scattering, but the elastic scattering is increasingly broadened as the scattering angle increases. The inelastic scattering from PDMS arises from the methyl group torsional mode. The broadened quasielastic scattering is due to the 'wriggling' motion of the polymer chains which characterizes the rubbery (liquid) state of polymers.

In the following sections, we will discuss the three different types of molecular motion—vibration, rotation, and translation—both in terms of experimental procedures and problems and of the theoretical framework required to interpret the data.

9.1.2 Practicalities of separating translation, rotation, and vibration

The formalism required to express inelastic scattering has been introduced in Chapter 4. In general, if the scattering nuclei are in motion on the time scale of the experiment, $k_\mathrm{i} \neq k_\mathrm{f}$. It is then the double differential cross section $\partial^2\sigma/\partial\Omega\,\partial\omega$ which is measured in the experiment, and as has been mentioned already, the incoherent term is no longer a flat background, since it carries spatially uncorrelated information about motion of the scattering nuclei. It should be noted that, although for convenience the coherent and incoherent terms are discussed separately, experimentally they may be very difficult to separate. The scattering cross sections are related to two space–time correlation functions $G(\boldsymbol{r}, t)$ and $G_\mathrm{s}(\boldsymbol{r}, t)$. $G(\boldsymbol{r}, t)$ expresses the probability that if there is a nucleus at position $\boldsymbol{r}_i$ at time $t = 0$, there will be another nucleus at position r_j at time t. $G_\mathrm{s}(\boldsymbol{r}, t)$ is the self-correlation function, expressing correlations between the same nucleus at time 0 and t. (See eqns (4.73) and (4.77).) The cross sections are:-

$$\left(\frac{\partial^2\sigma}{\partial\Omega\,\partial\omega}\right)_{\text{coh}} = \frac{k_{\text{f}}}{k_{\text{i}}}\frac{b^2V}{2\pi}\iint \mathrm{d}\boldsymbol{r}\,\mathrm{d}t\exp\{-\mathrm{i}(\boldsymbol{q}\cdot\boldsymbol{r}-\omega t)\}G(\boldsymbol{r},t) \tag{9.1}$$

$$\left(\frac{\partial^2\sigma}{\partial\Omega\partial\omega}\right)_{\text{inc}} = N\frac{k_{\text{f}}}{k_{\text{i}}}\frac{\overline{\Delta b^2}}{2\pi}\iint \mathrm{d}\boldsymbol{r}\,\mathrm{d}t\exp\{-\mathrm{i}(\boldsymbol{q}\cdot\boldsymbol{r}-\omega t)\}G_{\text{s}}(\boldsymbol{r},t) \tag{9.2}$$

where

$$\hbar\omega = \Delta E = E_{\text{f}} - E_{\text{i}} = \frac{\hbar^2}{2m}(k_{\text{f}}^2 - k_{\text{i}}^2). \tag{9.3}$$

The integrals in eqns (9.1) and (9.2) contain all the information about the scattering system and are usually called the scattering laws with (eqn (4.84))

$$S_{\text{coh}}(\boldsymbol{q},\omega) = \frac{V}{2\pi N}\iint \exp\{-\mathrm{i}(\boldsymbol{q}\cdot\boldsymbol{r}-\omega t)\}G(\boldsymbol{r},t)\mathrm{d}\boldsymbol{r}\,\mathrm{d}t \tag{9.4}$$

and an analogous definition for $S_{\text{inc}}(\boldsymbol{q},\omega)$ in terms of $G_{\text{s}}(\boldsymbol{r},t)$. If only the spatial Fourier transforms (in eqn (9.4) and its coherent analogue) are performed, the quantities obtained are the so-called intermediate scattering functions $S_{\text{coh}}(\boldsymbol{q},t)$ and $S_{\text{inc}}(\boldsymbol{q},t)$ (eqns (4.89) and (4.90)).

Although, in principle, the experimentally observed cross sections can be Fourier transformed to obtain the correlation functions, in practice the $\boldsymbol{q}$ and ω ranges are usually limited so it is more usual to calculate scattering laws based on models of the molecular motion and compare these with the data.

In some cases it is the cross sections (9.1) or (9.2) which are calculated, in others it is the scattering laws $S(\boldsymbol{q},\omega)$ or $S(\boldsymbol{q},t)$. In every case, of course, the experimentally determined data must be corrected for background scattering and detector efficiency in analogy to the small angle experiments already discussed. Resolution is almost always an important factor and usually varies from spectrum to spectrum as well as across the energy range in a given spectrum. The only really satisfactory way to deal with this is to convolute a realistic resolution function with the model before comparison with the data.

In the analysis of inelastic scattering from polymers it is assumed that the effects of translation, rotation, and vibration can be separated in the spectra so that $S(q,\omega) = S_{\text{tr}} \otimes S_{\text{rot}} \otimes S_{\text{vib}}$. Two factors determine whether this assumption is valid. The first is the purely practical one of resolving different effects within data obtained at finite resolution. The second is more fundamental. Are the three types of motion intrinsically separable? In Chapter 4, we saw that to separate the motion we describe the time-dependent position vector of a scattering nucleus in terms of three vectors, $\boldsymbol{c}_i(t)$, $\boldsymbol{b}_i(t)$, and $\boldsymbol{u}_i(t)$ describing centre of mass, its rotation about the centre of mass, and its vibrations with respect to the centre of mass, respectively. The fundamental assumption is that each of these three vectors can change independently of the other two. Although this assumption is nearly always made it is not necessarily valid. Consider the case of a small molecule in a liquid. It may well have interactions with its neighbouring molecules which are orientation-dependent. In this case translation and rotation of the molecule are likely to be coupled. For a polymer, to a good approximation fast rotations of side groups such as methyl groups are unlikely to be coupled to main chain motion. However, rotation of bulky side groups or segments of the main chain clearly are coupled to the overall diffusive motion we call translation for a polymer. If we do assume the three vectors $\boldsymbol{c}_i$, $\boldsymbol{b}_i$, and $\boldsymbol{u}_i$ are uncoupled, then in Chapter 4 we showed that the scattering is still the *convolution* of three scattering laws. Thus, for example, each of the delta functions representing vibrational energy transitions in Fig. 4.6 is broadened by the effect of translation and rotation. Equally, the central delta-function in Fig. 4.5, representing the rotation of a side group, will be broadened by translational motion (to look like Fig. 4.4) if the polymer is molten, and make the second, broadened component, at the foot of the elastic peak very difficult to resolve.

Generally speaking, therefore, it is safest to separate a particular motion by working at a temperature, or in a frequency range where the other motions cannot be observed. For example, rotational motion of side chains is best observed below the glass transition temperature where translational motion does not occur, and vibrational spectra are sharpest at low temperatures where the other motion is very slow or stopped.

9.2 VIBRATIONS

9.2.1 The scattering laws for vibrational motion

In Chapter 4 we derived expressions for the coherent and incoherent scattering from a molecule with a single vibrational frequency, $\omega(\kappa)$ (eqns (4.124) and (4.125)). While the real situation is complicated by the existence of many vibrational modes, these simple 'one phonon' formulae contain all the essential features which allow interpretation of the neutron inelastic spectra. Reiterating these formulae we have

$$\left(\frac{\partial^2\sigma}{\partial\Omega\partial\omega}\right)_{\text{inc}} = \frac{k_\text{f}}{k_\text{i}}\overline{\Delta b^2}N\exp(-2W)\left\{\delta(\omega_0) + \frac{(\boldsymbol{q}\cdot\boldsymbol{u})^2}{2}\,\delta(\omega+\omega_0) + \frac{(\boldsymbol{q}\cdot\boldsymbol{u})^2}{2}\,\delta(\omega-\omega_0)\right\} \tag{9.5}$$

$$\left(\frac{\partial^2\sigma}{\partial\Omega\partial\omega}\right)_{\text{coh}} = \frac{k_\text{f}}{k_\text{i}} b^2 \sum \exp(-2W)\left\{\delta(\omega_0)\delta(\boldsymbol{q}-\boldsymbol{\sigma}) + \frac{(\boldsymbol{q}\cdot\boldsymbol{u})^2}{2}\delta(\omega_0+\omega)\delta(\boldsymbol{q}+\boldsymbol{\kappa}-\boldsymbol{\sigma}) + \frac{(\boldsymbol{q}\cdot\boldsymbol{u})^2}{2}\delta(\omega_0-\omega)\delta(\boldsymbol{q}-\boldsymbol{\kappa}-\boldsymbol{\sigma})\right\}. \tag{9.6}$$

In each case, we have an elastic peak which, in the absence of translation and rotation is as sharp as the resolution function, and peaks shifted in energy by the creation or annihilation of a phonon. We consider first the coherent scattering. As described in Chapter 4 the delta functions in $\boldsymbol{q}$ and ω_0 in eqn (9.6) place very strong restrictions on where neutron scattered intensity will be observed. These conditions

$$\left.\begin{aligned} \hbar\omega_0 &= E_\text{i} - E_\text{f} = \pm\hbar\omega \\ \boldsymbol{q} &= \boldsymbol{k}_\text{i} - \boldsymbol{k}_\text{f} = \boldsymbol{\sigma} \pm \boldsymbol{\kappa} \end{aligned}\right\} \tag{9.7}$$

mean that a proper choice of $\hbar\omega_0$ and $\boldsymbol{q}$, usually made using a triple axis spectrometer (see Chapter 3), allows the relationship between $\omega(\kappa)$ and κ to be traced out for phonons travelling along different lattice directions in a crystal. Polymers are usually polycrystalline, and not very highly oriented (compared to the single crystal samples usually used for such experiments). There are, moreover, many propagating modes. It is usually possible, however, to distinguish between those modes where the displacement is perpendicular to the propagation direction (transverse modes) and those where it is parallel to this direction (longitudinal modes) and we will see an example of such observations for stretch oriented polyethylene later on in this chapter.

Looking at the incoherent scattering we see again an elastic and two inelastic peaks but the delta function in q has disappeared. This means that for a given $\hbar\omega_0$ *all* possible transitions $\hbar\omega_0 = \pm\hbar\omega$ are observed, irrespective of the value of q, in a function called the density of states, $z(\omega)$. This is not precisely true in fact, because we must consider the factor $(\boldsymbol{q}\,.\,\boldsymbol{u})^2$ which multiplies all the inelastic terms in both eqns (9.5) and (9.6). This is a sort of selection rule, in that if $\boldsymbol{q}$ is perpendicular to the vibrational displacement $\boldsymbol{u}$ there will be no signal. In stretch oriented samples, even in the incoherent scattering, the orientation of $\boldsymbol{q}$ relative to the sample can be important, as we see subsequently, but most amorphous or polycrystalline samples are not oriented so that what this factor does is to modify the intensity of the elastic peaks by their corresponding vibrational amplitude u. The resulting spectrum of energy transitions, extrapolated to zero q, is called the amplitude-weighted density of states, $g(\omega)$. It is this quantity which is frequently presented as the result of inelastic incoherent neutron scattering experiments.

Finally, in considering eqns (9.5) and (9.6) we note that both inelastic and inelastic terms are multiplied by the Debye–Waller factor $\exp(-2W)$ (where $W = \langle(\boldsymbol{q} \cdot \boldsymbol{u})^2\rangle$ which was discussed in Chapter 4. This factor causes the intensity to drop off as q increases, and to do so more rapidly for large amplitude motion.

As discussed in Section 9.1, the sharp vibrational spectra are broadened by the effect of rotational and translational motion. For this reason, it is desirable to observe vibrational spectra from samples at low temperatures, where these other motions have less importance. There are other advantages as well. The Boltzmann factor will ensure that most transitions are between the ground state and the first excited state, so that broadening due to anharmonicity of the vibrational levels ($\hbar\omega_{0-1} \neq \hbar\omega_{1-2}$) is avoided. In polymers, however, the vibrational spectra are frequently broadened even at low temperatures, especially in the glassy state. This is most probably due to local inhomogeneities in the sample so that the interactive potential energy of the same chemical group with its surroundings varies from place to place within the sample. Low temperature samples can only be observed with neutrons which have enough energy to excite the vibrational modes. The incoming neutrons have relatively high energies and the spectra are observed in 'down scattering' or 'neutron energy loss'. Since such spectrometers often have poorer resolution than those with low incoming energies, vibrational spectra from polymers have often been observed from room temperature samples in neutron energy gain or 'up scattering' despite the broadening effects discussed above.

9.2.2 Vibrational motion in polymeric samples

The vibrational modes of motion can be grouped as acoustic and as optic phonons. The former correspond to travelling waves, with a strong $\omega - \boldsymbol{q}$ correlation. The latter correspond to localized modes, including the motion of side groups, and usually show rather small effects of dispersion (i.e. $\omega(\kappa) =$ constant). For these reasons, acoustic phonons are more usually observed from oriented samples, using coherent scattering techniques able to vary $\boldsymbol{q}$ and ω independently (specifically the triple axis spectrometers), while optic phonons are often observed in a density of states experiment using incoherent scattering. Given the very highly advanced state of Fourier transform infrared spectroscopy and Raman spectroscopy and the very poor resolution (and high cost) of inelastic neutron scattering experiments it is not unreasonable to look very closely at the advisability of attempting inelastic neutron spectroscopy on any samples (leave alone polymers which have, as we shall see, their own special problems).

The unique information available from the neutron experiment is the relatively wide range of $\boldsymbol{q}$, ω pairs available compared to FTIR or Raman where $\boldsymbol{q}$ is always close to zero. This allows the dispersion curves of acoustic modes to be tracked over a wide range and fitted to model calculations of the inter- and intramolecular force fields with the corresponding elastic moduli. This not only tests the structural models and force field information but provides data for calculation of materials parameters. Unfortunately polymeric samples are at best polycrystalline and often only partially crystalline, so that the techniques applied very successfully to analysing the crystal dynamics of perfect crystal small molecule samples are of only very limited applicability to polymers.

The advantage that incoherent scattering has over other forms of spectroscopy is that hydrogen modes may be much more visible. Neutron spectroscopy has thus been useful in filling in some gaps left by these other spectroscopies, particularly in identifying motion of torsional side groups and testing densities of states calculated from model force fields.

As will be demonstrated, applications of neutron inelastic scattering to the study of vibrational modes of polymers are strictly limited in usefulness and should only be undertaken after a fairly exhaustive use of more conventional spectroscopy. Nevertheless there are useful contributions to be made by the neutron and this book would be incomplete without a brief discussion of them.

9.2.3 Inelastic scattering from torsional vibrations

Examination of eqn (9.5) shows that large amplitude vibrational modes involving nuclei with large cross sections will give intense peaks in the inelastic neutron spectra. The incoherent scattering from hydrogen is an order of magnitude larger than that from other nuclei. Torsional vibrations of methyl or phenyl side chains of polymers with their large amplitudes are, therefore, good candidates for neutron investigation, especially since these bands are often weak and difficult to identify from infrared and Raman spectra. Deuteration of the side group, where this is possible, reduces the cross section by a factor of around 20 and the peak drops out of the spectra, thus confirming the assignment. Particularly in the early days of neutron spectroscopy, such identification of methyl or phenyl torsions in polymeric samples was quite common and useful to the spectroscopists. In many of these examples, partial deuteration of the molecule has allowed an unambiguous assignment of the torsional mode. In the cases of polydimethylsiloxane and polypropylene a more complicated situation emerged, in the former case, because deuteration of the methyl group reduces all low frequency modes and is therefore unhelpful and in the latter, because of complications arising from the crystalline structure. These two examples illustrate both the advantages and the problems of neutron scattering spectroscopy and will be described in some detail.

Henry and Safford (1969) used a time-of-flight spectrometer to observe the inelastic scattering from PDMS under varying degrees of cross-linking and filler. A broad peak at $160\,\text{cm}^{-1}$ for the room temperature sample was assigned to the methyl group motion which was considered to be almost free rotation because of the breadth of the peak. Allen *et al.* (1975) observed a similar peak at $165\,\text{cm}^{-1}$, made the same assignment, and calculated the barrier to rotation assuming a threefold potential well. This barrier is about $8\,\text{kJ}\,\text{mol}^{-1}$ which is about $3\,kT$ at room temperature and argues against the rotation being completely free though it will certainly be fast. A free rotation being non-quantized gives no inelastic delta functions in eqn (9.5) and the elastic peak becomes broadened. Thus $\partial^2\sigma/\partial\Omega\partial\omega$ (and $S(\boldsymbol{q},\omega)$) are broad distributions centred around $\hbar\omega = 0$. The time-of-flight cross section, $\partial^2\sigma/\partial\Omega\partial\tau$, however, is related to $\partial^2\sigma/\partial\Omega\partial\omega$ by the factor τ^3 (eqn (3.17)). Multiplication by this factor distorts the spectrum and increases intensity towards $\hbar\omega \neq 0$. It can even produce the appearance of an inelastic peak. An example of this effect is shown in Fig. 9.2(a) where the spectrum of molten polyethylene oxide for which $S(\boldsymbol{q},\omega)$ is a smooth function centred at $\hbar\omega = 0$ shows a peak in the time-of-flight spectrum around $200\,\text{cm}^{-1}$. Figure 9.2(b) shows the same pair of functions for PDMS. (Both the time-of-flight spectra were obtained at 45° scattering angle with $\lambda_i = 4.2\,\text{Å}$.) In this case, the peak at $160\,\text{cm}^{-1}$ is clearly still visible when the τ^3 effect is removed and corresponds to a genuine inelastic feature. Henry and Safford did not attempt to present their data converted from $\partial^2\sigma/\partial\Omega\partial\tau$ to $\partial^2\sigma/\partial\Omega\partial\omega$ and fitted a calculation for free rotation to the time-of-flight data. The confusion about freedom of rotation of the methyl group in PDMS was extended by Amaral *et al.* (1976) using temperature dependence of total scattering cross section measurements who intepreted the Debye–Waller factor to arrive at a barrier to rotation of less than $1.6\,\text{kcal mol}^{-1}$. This measurement, by its nature, is much less accurate than the direct inelastic measurement of the torsional transitions. Although these authors suggest that the interpretation of the time-of-flight inelastic spectra in terms of free or hindered rotation is a matter of choice, the above discussion shows that the removal of the τ^3 effect leaves no ambiguity. The breadth of the inelastic peak is not due solely to the nature of the methyl rotation itself, but also to the very fast backbone motions in PDMS which broaden both elastic and inelastic delta functions in eqn (9.5). These motions will be discussed further in Section 9.4, but it is interesting to note that the inclusion of fast rotations of backbone segments might well account for the low apparent barrier to methyl rotation given by the total scattering experiment. Moreover, many of the effects of filler, of cross-linking and of temperature in narrowing the methyl peak, which were

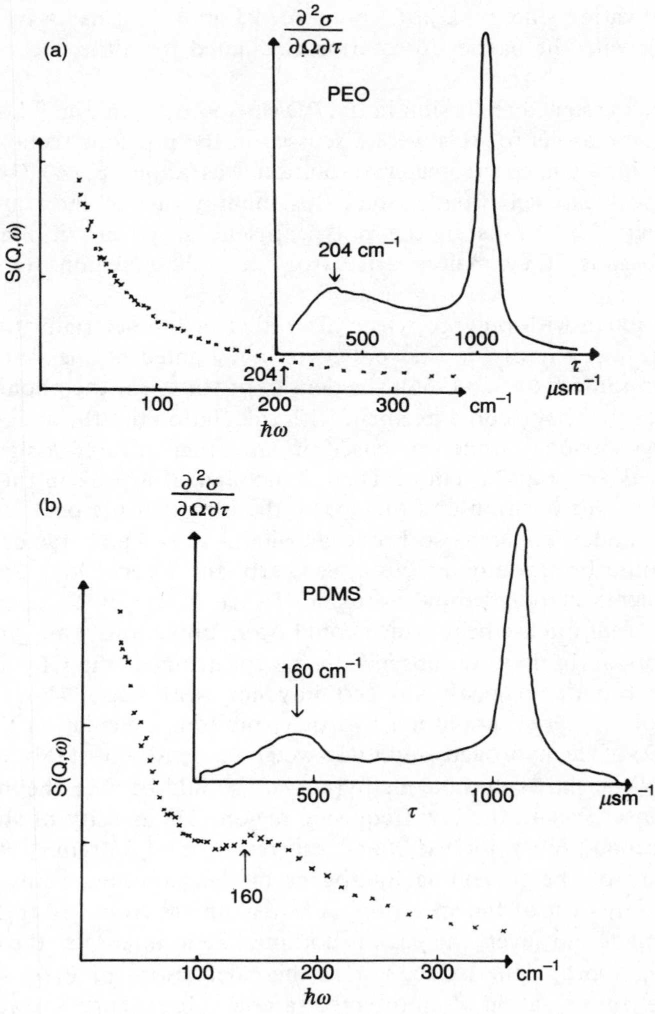

Fig. 9.2 Time-of-flight spectra, $\partial^2/\partial\Omega\partial\tau$ and corresponding scattering laws $S(Q,\omega)$ $(Q \equiv q)$ for (a) polyethylene oxide (PEO) and (b) polydimethyl siloxane (PDMS) melts.

interpreted by Safford and Henry in terms of their effects on the methyl rotation itself may also be attributed to the effects of slowing down the backbone motion. Finally, unpublished measurements of NMR spin lattice relaxation frequencies as a function of temperature for the relaxation mode associated with the methyl motion arrive (Pellow, private communication) at an activation energy E_a of about 6.7 kJ mol^{-1}. This is in excellent agreement with the barrier to rotation calculated from the neutron elastic peak.

The peak assigned to torsion in the PDMS spectrum in Fig. 9.2(b) is very broad. The reasons for this were discussed in the previous section. In this case, the inherent spectrometer resolution was about $E_f/10$ (~25 cm^{-1}) but the peak is even broader than this, mainly due to the translational broadening. The peaks in the polypropylene spectrum in Fig. 9.3 are sharper because they mainly arise from crystalline regions in a highly oriented sample.

The problem with polypropylene has rather different origins to that for PDMS. It was one of the first polymers investigated by inelastic neutron scattering. (Safford *et al.* 1964) The density of states for the phonon lattice modes was in fairly good agreement with calculation but the assignment of the methyl torsional mode was based on an earlier infrared assignment of a very weak band at 200 cm^{-1}. There is no sign of a peak in the neutron spectrum at this position and this led to the idea that the peak due to the torsional modes would be so broad and its Debye–Waller factor so large that it would be virtually undetectable (Zerbi and Piseri 1968). Subsequent measurements in poly(propylene oxide) (Allen *et al.* 1972), however, did show one clear intense peak which could be unambiguously assigned to the methyl torsion by its total absence in the spectrum of the CD_3 analogue. Although broad, the peak was certainly not weak—as it should not be because of the large amplitude hydrogen motion. Calculation (Takeuchi *et al.* 1982) of the hydrogen amplitude-weighted density of states, $g(\omega)$, for polypropylene shows that the methyl torsion should be overwhelmingly the most intense peak in the low frequency region. The density of states does show a broad, fairly intense, band centred around 230 cm^{-1} which, on deuteration of the remaining hydrogens in the molecule, gains intensity relative to the rest of the spectrum. A reassignment from 210 to 230 cm^{-1} must be made; however, the peak is nothing like as intense as the torsional mode in polypropylene oxide, nor as the calculations of $g(\omega)$ suggest it should be. Investigation of an oriented sample subsequently showed strong dispersion effects for this mode, unexpected for a methyl torsion. The experimental results are shown in Fig. 9.3. A highly stretch-oriented sample was prepared, and effects of the factor $(\boldsymbol{q}\cdot\boldsymbol{u})^2$ in eqn (9.5) were clearly seen. When the $\boldsymbol{q}$ vector pointed along the chain axis, the torsional mode was relatively intense and sharp around 240 cm^{-1}. When $\boldsymbol{q}$ was perpendicular to the axis the band was much broader and apparently shifted to 220 cm^{-1}. In this molecule the crystalline form is a helix with the methyl groups sticking out sideways. The explanation of the data in Fig. 9.3 is that coupling between chains may be enhanced by the methyl groups on neighbouring chains leading to a large frequency dispersion for the methyl

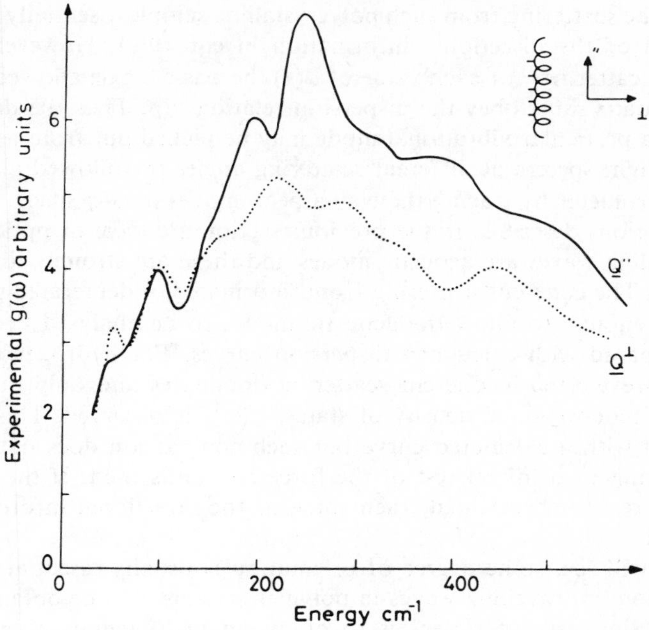

Fig. 9.3 Experimental hydrogen amplitude–weighted density of states, $g(\omega)$ for stretch-oriented polypropylene with $Q(\equiv q)$ parallel and perpendicular to the helical axis. Reprinted with permission from Takeuchi *et al.* (1982).

torsion centred around 220 cm^{-1} in the direction perpendicular to the helical axis and a small dispersion around 240 cm^{-1} for the parallel direction. In unoriented samples, a broad feature around 230 cm^{-1} results.

These two examples illustrate the importance of the large incoherent cross section of hydrogen in assigning side chain modes for polymers, together with the necessity of careful interpretation of the time-of-flight spectra which ought to be transformed to $g(\omega)$ if possible. The directional properties of the neutron energy transfer arising from the sizeable momentum transfer are also important. Another example of the use of sample orientation will be discussed in the next section.

9.2.4 Acoustic phonons in crystalline samples

The observation of travelling waves in polymer crystals offers the attractive possibility of obtaining both the inter- and intrachain force constant. The drawback is the near impossibility of obtaining single crystals of polymers. Most polymers contain only small crystallites embedded in an amorphous matrix. The scattering from such polycrystalline sample essentially averages out much of the directional information in eqn (9.6). However, in the coherent scattering, for each value of $\omega(\boldsymbol{\kappa})$ there is a single allowed $\boldsymbol{q}$ value and ω–$\boldsymbol{q}$ pairs must obey the dispersion relationship. Thus the dispersion curve of a particular vibrational mode may be picked out from a series of time-of-flight spectra at different scattering angles or followed on a triple axis spectrometer by tracing the way a peak moves in ω–$\boldsymbol{q}$ space. The torsional motions described in the previous section are local or optic modes, the travelling waves are acoustic modes and these are strongly directional in nature. The coherent scattering from fluorinated or deuterated polymers is strong enough to allow the acoustic modes to be analysed in this way and compared with calculated dispersion curves. For hydrogenous polymers, however, the incoherent scattering dominates and only the hydrogen amplitude-weighted density of states, $g(\omega)$, is observed. This may be compared with a calculated curve but such comparison does not provide a very stringent or direct test of the force constants used. If the polymer samples are stretch oriented, then some of the directional information is recovered.

As we will see, some degree of orientation is usually essential if useful information on travelling waves in polymer crystals is to be obtained. The first and simplest experiment is to orient an hydrogenous polymer and observe the incoherent inelastic scattering.

Figure 9.4 shows the amplitude-weighted density of states, $g(\omega)$, for stretch-oriented polyethylene determined using a triple axis spectrometer so that in one case, the $\boldsymbol{q}$ vector was oriented always along the chain axis (longitudinal) and in the second case it was perpendicular (transverse). The data were obtained at finite $\mathbf{q}$ and the q^2 terms in eqn (9.5) divided out. The two curves have been normalized at 190 cm^{-1}. The two main peaks at 525 and 190 cm^{-1} have been assigned to the longitudinal stretch bend (accordion) mode of the C–C–C skeleton and to the out-of-plane torsion of the methylene groups about the C–C bond respectively. The disappearance of the 525 cm^{-1} band from the $q_{\perp}$ spectrum confirmed its assignment to the longitudinal mode —in this configuration there is no way for the neutron to excite the motion. The fact that the transverse mode shows up in both spectra (though without the normalization mentioned, it is, in fact, reduced in intensity by a factor two in the $\boldsymbol{q}_{\parallel}$ spectrum) is due to crystalline field mixing of the modes.

Although for this hydrogenous polymer the incoherent cross section of the hydrogen means that the incoherent scattering dominates it is notable that directional information can be obtained. Such information is however considerably enhanced if the coherent cross section can be made to dominate as in a fluorinated or deuterated polymer.

This is the case for the Teflon sample whose spectrum is shown in

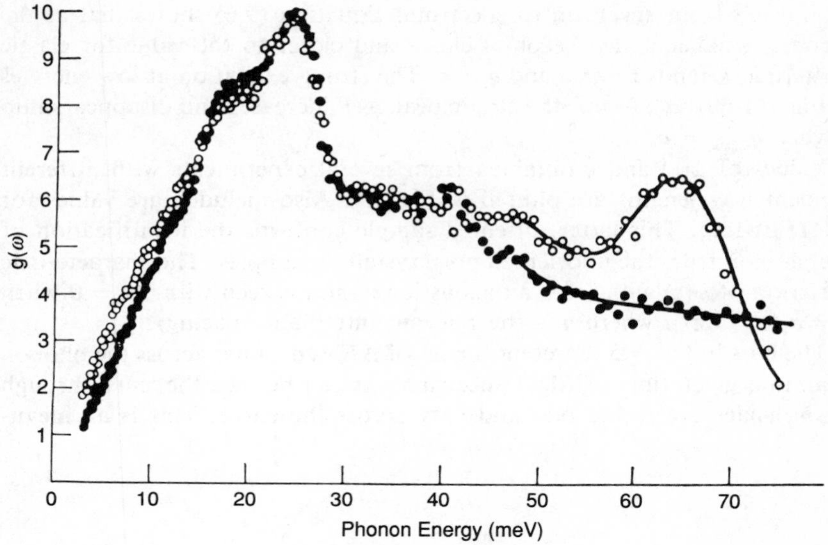

Fig. 9.4 Amplitude-weighted density of states, $g(\omega)$ for a stretch-oriented sample of polyethylene at 100 K. (○) longitudinal, (●) transverse. Reprinted with permission from Myers *et al.* (1966).

Fig. 9.1. The sample was polycrystalline hexagonal PTFE for which the unit cell is very anisotropic and highly symmetric with the c direction at 19.5 Å very much larger than the equivalent a direction at 5.66 Å. This crystal structure has only one strong Bragg reflection at small q values. It arises from the $[1\,0\,\bar{1}\,0]$ planes with an interchain spacing of 4.96 Å. As shown by eqn (9.6) the excitation of acoustic phonons along directions close to a strong Bragg reflection gives rise to intense peaks in the neutron spectrum. In this case the phonons are giving their energy *to* the neutrons (the spectra in Fig. 9.1 are neutron energy gain spectra) and being annihilated in the scattering process. For acoustic phonons, $\hbar\omega$ varies strongly with $\boldsymbol{\kappa}$. The conservation conditions (eqn (9.7)) have to be obeyed simultaneously. This is why the phonon peak corresponding to the acoustic phonon occurs at different energy transfer values (different neutron time-of-flight) for different scattering angles (different q values). In Fig. 9.1 there is one such peak at low $\hbar\omega$ which is intense and moves with scattering angle. This is the acoustic phonon. The peaks at higher $\hbar\omega$ are optic phonons (little q variation). In time of flight spectroscopy there are usually detectors at many scattering angles. When, as in this case, there is only one strong Bragg reflection and one strong acoustic phonon excitation, the peak can be tracked from spectrum to spectrum. Equation (9.6) shows that as $\hbar\omega$ becomes smaller and $\boldsymbol{q}$ becomes closer and closer to the value for elastic scattering, $\boldsymbol{\kappa}$ tends to zero and $\boldsymbol{q} = \boldsymbol{\sigma}$. The strong excitation at low energies in Fig. 9.1 moves *towards* the elastic peak as θ increases and disappears into it when $\boldsymbol{q}_{\text{elastic}} = \boldsymbol{\sigma}$.

Values of $\omega(\kappa)$ and κ obtained from several experiments with different incident wavelengths are plotted in Fig. 9.5. Also included are values for a PTFE fibre. This latter, oriented sample confirms the identification of the phonon from the unoriented polycrystalline samples. The characteristic variation of $\omega(\kappa)$ with κ for an acoustic phonon is seen with $\hbar\omega = 0$ when $\boldsymbol{q} = \sigma$ ($\sigma = 2\pi/a$ where a is the relevant interplane spacing).

The bars in Fig. 9.5 represent the set of ΔE vs $\boldsymbol{q}$ values across the phonon peak in a given time of flight spectrum. As can be seen the 'cuts' through $[q, \hbar\omega]$space are rather odd and vary across the curve. This is an inevit-

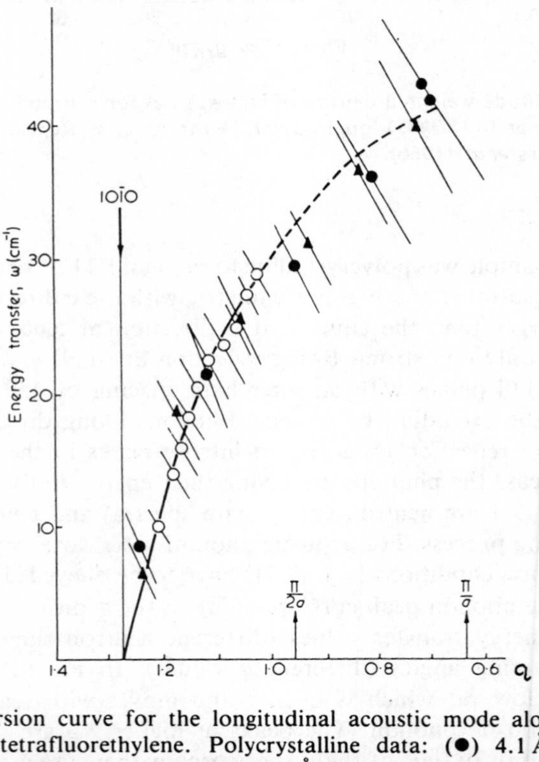

Fig. 9.5 Dispersion curve for the longitudinal acoustic mode along [1 0 $\bar{1}$ 0] in hexagonal polytetrafluorethylene. Polycrystalline data: (●) 4.1 Å, (▲) 5.0 Å incident neutrons. Fibre specimen: (○) 5.8 Å incident neutrons. Reprinted from Twisleton and White (1972) with permission from the authors and the publishers, Butterworth Heinemann ©.

able result of scanning energy at fixed angle. For optic phonons this is not so great a problem since $\hbar\omega$ varies little with q. However, for acoustic phonons, the fixed angles do cause difficulties and the triple-axis spectrometer described briefly in Chapter 3 provides considerable advantages because it allows $\hbar\omega$ and q to be varied independently. Thus $\hbar\omega$ can be varied at constant q and the resolution or the phonon peaks greatly improved. Normally, however single crystal sample textures are essential because it is only rarely the case that a single Bragg reflection dominates (as for PTFE). Thus from a polycrystalline sample many such reflections, and their associated acoustic phonons occur in any spectrum making it impossible to identify the modes. Such triple-axis experiments are extremely time consuming, and, except for very rare cases where highly oriented single crystal-like polymer samples of a reasonable size can be prepared, they are frankly not worthwhile.

In order not to finish this section on a negative note and to indicate the prize to be won, if good, large, oriented, deuterated or fluorinated crystals are available we include the set of phonon dispersion curves displayed in Fig. 9.6. These have been calculated for deuteropolyethylene (D-PE) from a particular model of the inter- and intramolecular forces. The neutron data in this figure were obtained from a large D-PE sample which was highly crystalline and essentially single crystal-like. As can be seen two acoustic phonons have been tracked and some optic phonons observed both by neutrons, Raman, and infrared spectroscopy. High quality neutron data could therefore provide a stringent test of model forces fields in polymeric

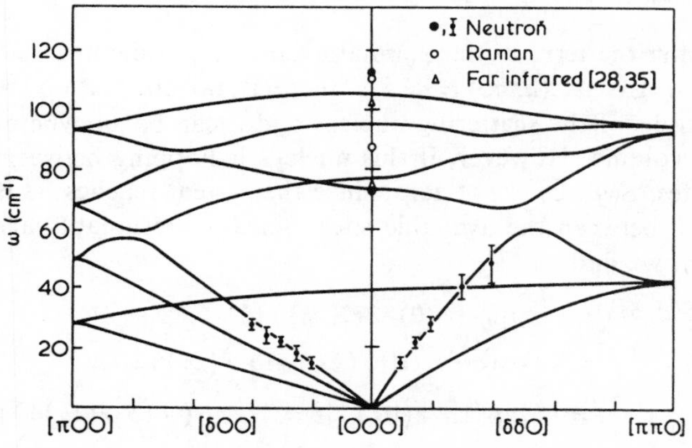

Fig. 9.6 Phonon-dispersion curves for deuterated polyethylene along [ξ 0 0] and [ξ ξ 0] calculated from model potential parameters together with experimental data at 77 K. Reprinted from Twisleton *et al.* 1982 with permission from the authors and the publishers, Butterworth Heinemann ©.

samples. From the slopes at low $\boldsymbol{\kappa}$ of the $\omega(\boldsymbol{\kappa})$ vs $\boldsymbol{\kappa}$ plots for acoustic phonons the velocity of sound is obtained and hence the elastic stiffness constants in the crystals. From the data in Fig. 9.6 the value of C_{11} (which is the stiffness perpendicular to the chains) is $18.2 \pm 0.2 \times 10^9\ \mathrm{Nm}^{-2}$.

Clearly such information is potentially very valuable in understanding polymeric material properties—if only the necessary samples could be prepared.

9.3 ROTATIONAL MOTION OF SIDE GROUPS

9.3.1 The scattering laws

In Chapter 4 we considered the very simple case of a rotation between two fixed sites, 1 and 2. The elastic scattering becomes separated into two components, a central delta function and a broadened 'foot'. The expresssion derived in eqn (4.111) for the simple two-site jump can be extended for the general case to

$$S^{\mathrm{rot}}_{\mathrm{inc}}(\boldsymbol{q}, \omega) = A_0(q)\delta(\omega) + \frac{1}{\pi}(1 - A_0(\boldsymbol{q}))F(\omega). \tag{9.8}$$

It is instructive to see how this form arises. To do this, we return to the general formula for the incoherent scattering in eqn (4.76)

$$\left(\frac{\partial^2\sigma}{\partial\Omega\partial\omega}\right)_{\mathrm{inc}} = \frac{k_{\mathrm{f}}}{k_{\mathrm{i}}} N \frac{\overline{\Delta b^2}}{2\pi} \int_{-\infty}^{\infty} \mathrm{d}t \exp(\mathrm{i}\omega t)\, \langle \exp \mathrm{i}\boldsymbol{q}\cdot(\boldsymbol{r}_1(0) - \boldsymbol{r}(t))\rangle$$

remembering the term in the thermal average $\langle\ \rangle$ is defined as $S_{\mathrm{inc}}(\boldsymbol{q}, t)$. In a liquid or gas $S_{\mathrm{inc}}(\boldsymbol{q}, \infty) = 0$, i.e. there is no correlation between the final positions of the scattering nucleus and it can be anywhere within the scattering volume. However, if this nucleus is hopping between a number of fixed sites, $S(\boldsymbol{q}, \infty)$ is not zero, since the system reaches its equilibrium distribution between the available sites. Since it presumably also starts in equilibrium we find

$$\begin{aligned} S(\boldsymbol{q}, \infty) &= \langle \exp[\mathrm{i}\boldsymbol{q}\cdot(\boldsymbol{r}_1(0) - \boldsymbol{r}_1'(\infty))]\rangle \\ &= \langle \exp \mathrm{i}\boldsymbol{q}\cdot\boldsymbol{r}_1(0)\rangle\langle \exp -\mathrm{i}\boldsymbol{q}\cdot\boldsymbol{r}(\infty)\rangle \\ &= |\langle \exp(\mathrm{i}\boldsymbol{q}\cdot\boldsymbol{r}(0))\rangle|^2 = |\langle \exp(-\mathrm{i}\boldsymbol{q}\cdot\boldsymbol{r}(\infty))\rangle|^2. \end{aligned}$$

Now when we make the Fourier transform to obtain $\left(\frac{\partial^2\sigma}{\partial\Omega\partial\omega}\right)_{\mathrm{inc}}$, this non-zero (but constant) value for $S(\boldsymbol{q}, \infty)$ leads as usual to a delta function (as in eqn (9.8)) whose coefficient, $A_0(\boldsymbol{q})$ is just the Fourier transform of the shape of equilibrium distribution between the available sets for the scattering nuclei. The existence of an elastic component in the scattered intensity is the signature of scattering centres which are essentially located in space. Correspondingly, the absence of such a component is the signature of a system with dynamical translational disorder. In our discussion we will concentrate on the relatively simple but common case of a methyl group side chain on a polymer. For a hydrogen nucleus in a methyl group rotating about a fixed axis between three equivalent sites

$$A_0(q) = \frac{1}{3}\left(1 + 2j_0(qr\sqrt{3})\right) \tag{9.9}$$

where r is the radius of the motion, and $j_0(x)$ is again a zero order spherical Bessel function. $A_0(q)$ is called the elastic incoherent structure factor, EISF.

The broadened component, $F(\omega)$, is Lorentzian in form

$$F(\omega) = \frac{2\tau_r/3}{1 + \omega^2(2\tau_r/3)^2} \tag{9.10}$$

with a full width at half maximum $4\tau_r/3$, where τ_r is the time between jumps, or mean residence time.

The model for a threefold symmetrical potential of a methyl side group is shown in Fig. 9.7. V_3 is the height of the potential barrier between equivalent positions of the methyl group. Equilibrium values occur when the group is oriented in one of these positions and it is the inter- and intramolecular forces of neighbouring atoms which will determine the height of V_3 (i.e. how difficult is it for the group to rotate? How well is the equilibrium minimum located?) Except at zero kelvin there will be small

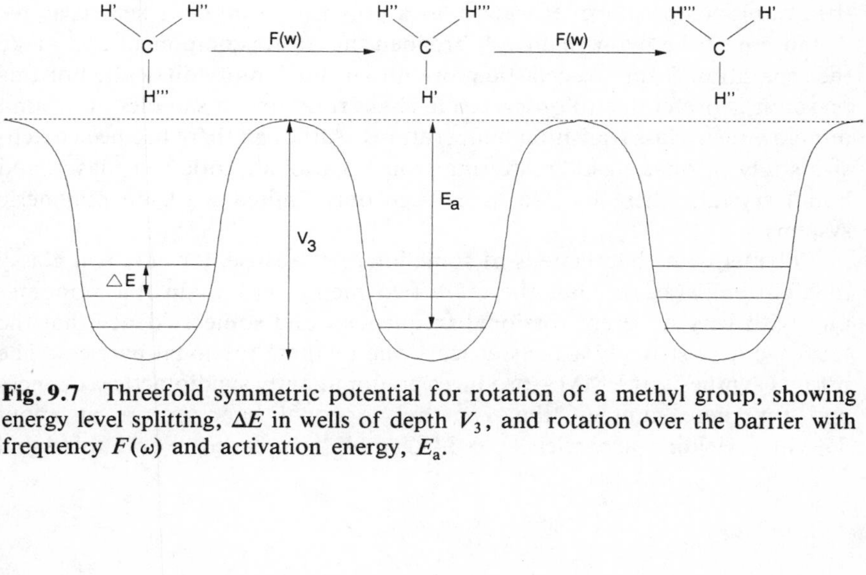

Fig. 9.7 Threefold symmetric potential for rotation of a methyl group, showing energy level splitting, ΔE in wells of depth V_3, and rotation over the barrier with frequency $F(\omega)$ and activation energy, E_a.

vibrations about this minimum—the torsional oscillations already discussed in Section 9.2.2. These are quantized motions and a few of the quantized energy levels are indicated. $\hbar\omega$ is the energy of the torsional peak in the inelastic neutron spectrum. By using a threefold symmetric potential such as that in Fig. 9.7, V_3 can be calculated from the measured $\hbar\omega$ values. The Boltzmann factor ensures that at temperatures around T_g and below, most of the groups occupy the ground state. Except at zero kelvin, there is a chance of a rotational jump to the next equivalent position. The Boltzmann factor determines this probability as $\exp(-E_a/kT)$. The activation energy E_a is less than V_3 by the value of the ground state energy (usually very small). The model assumes instantaneous jumps between sites and a residence time of τ_r. $(3\tau_r)^{-1}$ is thus the rotational frequency, and $F(\omega)$ the function which describes the scattering from this motion. In practice, the width of the central delta function will be governed by the resolution of the spectrometer and, if there is translational motion, also by the effect of $S_{tr}(\boldsymbol{q}, \omega)$. The overall intensity will also be modified by the Debye–Waller factor.

9.3.2 Neutron quasielastic scattering from rotating side groups

As we have seen the rotation of a side group between the positions of the potential minima gives rise to a broadened component underneath the elastic scattering (see eqn (9.8)) and the width of this quasielastic component is governed by the rotational frequency. Separation of the elastic from the quasielastic components depends crucially on the incident energy spread (the resolution of the apparatus), and on the shape of the resolution function. To some extent, since the rotational motion is an activated process, its frequency can be adjusted to match the available resolution by changing the sample temperature. However, as a polymeric sample is heated above T_g motion of the main chain will broaden the elastic component and make the separation from quasielastic motion that much more difficult. For this reason it is preferable to observe side chain rotations in samples at around or below their glass transition temperatures. Although there has been extensive study of quasielastic scattering from rotational motion in plastic and liquid crystals, there has, to date, been only limited work on polymeric systems.

Polymethylmethacrylate is of some interest because not only is it glassy to high temperatures, but there are two methyl groups in the monomer unit with very different torsional frequencies and some evidence that the local tacticity strongly affects at least one of these torsional barriers. The normal synthesis of PMMA produces predominantly syndiotactic sequences and for these samples, the α-methyl torsional band occurs at about $350\,\text{cm}^{-1}$ yielding a barrier $V_3 \sim 34\,\text{kJ mol}^{-1}$. In earlier work, this band was assigned by comparing the α-chloro analogue with PMMA itself, but deuteration of the α-methyl group has recently reconfirmed this assignment. In a predominantly isotactic sample this barrier is reduced to 23 kJ mol^{-1}. The ester methyl group has a torsional band at rather lower energy (around 100 cm^{-1}) leading to a value for V_3 of between 3 and 10 kJ/ mol^{-1}. The band is difficult to assign with more precision even by specific deuteration because it occurs in a region of other backbone motions which are also reduced in intensity in the neutron spectra after deuteration of the methyl group. From the inelastic spectra, it is not possible to determine the effect of the tacticity on this barrier. The barriers for the two methyl groups are different enough so that even without the aid of specific deuteration, it would be possible to observe separately the effects of their rotational motion in the quasielastic spectra for samples below T_g. At room temperature, the α-methyl group is rotating at less that 10^9 Hz while the ester methyl moves much faster at 10^{11} Hz. Specific deuteration is helpful, however, in that it reduces the background and the elastic scattering from the other hydrogen nuclei (immobile in the glass on this time scale apart from vibrations which are at such high frequency they average out). Using a time-of-flight spectrometer, as described in Chapter 3, with its resolution adjusted to match the appropriate rotational motion frequency and a back

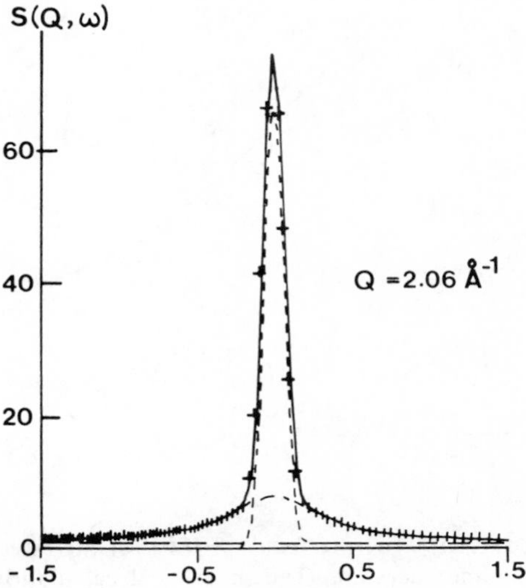

Fig. 9.8 Quasielastic scattering from polymethylmethacrylate at 25°C showing the rotational motion of the ester methyl group as a broadened component under a resolution broadened delta function. $q = 2.06$ Å^{-1}. (Gabryś *et al.* 1984) ($q \equiv Q$).

scattering spectrometer (also described in Chapter 3) for the slowest rotational motion, it was possible to separate the elastic and quasielastic contributions for both methyl groups and to extract the rotational frequency ν_{rot} and the elastic incoherent structure factor, the EISF (see eqn (9.9)). Figure 9.8 shows $S(\boldsymbol{q}, \omega)$ in the quasielastic scattering region from a sample of predominantly syndiotactic PMMA in which all the hydrogens apart from the ester methyl have been deuterated. The time-of-flight spectrometer has an approximately triangular resolution function which makes the two-component nature of the scattering evident. The fits to the data of a delta function and a broadened Lorentzian, both resolution broadened, are also shown. Figure 9.9 shows the value of $A_0(q)$ at 315 K as a function of q extracted using two different methods of separating the data and from two different experiments. The curve $A_0'(q)$ is the calculated EISF for the rotating methyl group adjusted for the extra incoherent scattering from the remaining nuclei in the molecule. There is a mismatch between the calculated and experimental values. However, the coherent scattering is not inconsiderable from this sample in which five out of the eight hydrogen nuclei have been deuterated. The X-ray scattering from this polymer shows

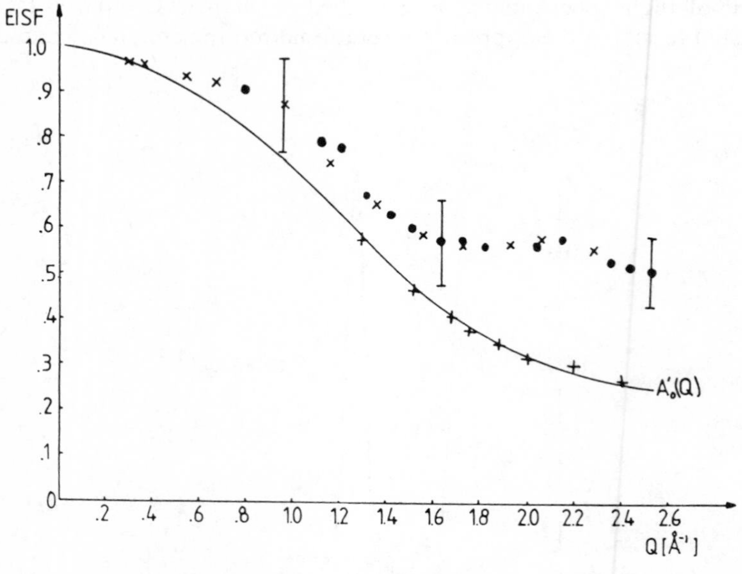

Fig. 9.9 Elastic incoherent structure factor (EISF) for the ester methyl rotation in syndiotacic polymethylmethacrylate. The theoretical calculation including corrections for elastic incoherent scattering is shown as the solid line $A_0'(Q)$ $(Q \equiv q)$. (×) (●) data from two time-of-flight spectometers, (+) data corrected as described in the text. (Gabryś *et al.* 1984.)

peaks at around $q = 0.8$ and 2.2 Å^{-1}. The neutron diffraction from the same sample as that used for the rotational measurements also showed peaks around these values at the points of maximum deviation of the data from the calculated $A_0'(q)$. Correction of the data for this coherent elastic contribution brings the data into excellent agreement with the calculations as shown by the crosses in Fig. 9.9. Since $A_0(q)$ is the Fourier transform of the rotational volume, it is clear that the alternative strategy of modifying the proposed rotational motion to achieve the shape of the experimental $A_0(q)$ would demand a reduction in the rotational radius—which is not possible for a methyl group rotating around its axis unless we imagine shortening the C–H bonds or drastically changing the bond angles!

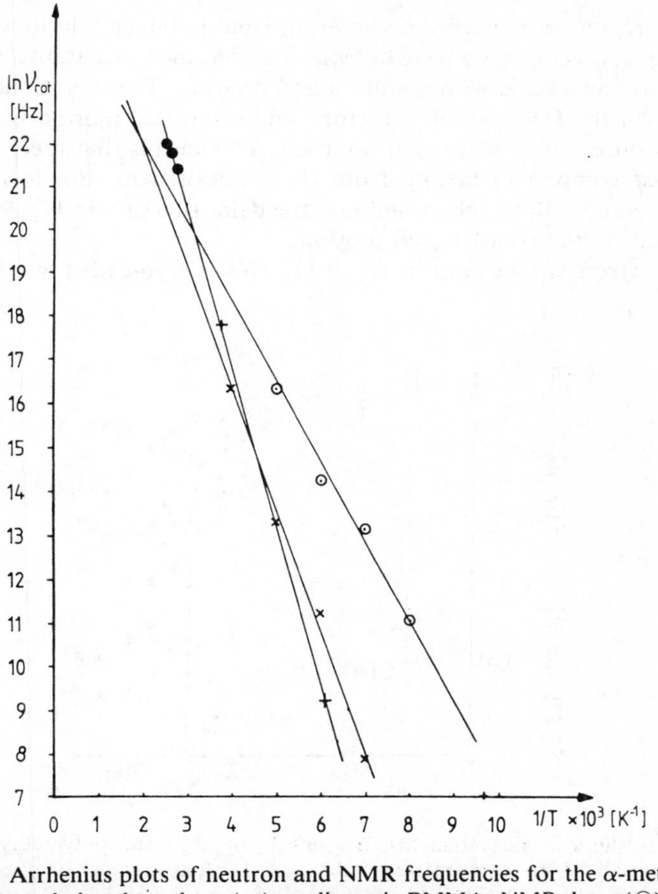

Fig. 9.10 Arrhenius plots of neutron and NMR frequencies for the α-methyl rotation in polymethylmethacrylate. (×) syndiotactic PMMA, NMR data; (⊙) isotactic PMMA, NMR data; (+) atactic PMMA, NMR data; (●) predominantly syndiotactic PMMA, neutron data. (Ma 1981.)

The question now arises as to how to correlate measurements such as these with relaxation times obtained from other techniques. This is usually done by comparing the temperature dependence. Figure 9.10 shows a plot of the rotational frequencies obtained from neutron quasielastic measurements on a α-methyl rotation in a predominantly syndiotactic sample, together with NMR relaxation values (Ma 1981). The slope of this Arrhenius plot gives values of E_a; see Fig. 9.7. Excellent agreement is obtained between the values of E_a from NMR and from neutron experiments as evidenced by the continuity of the neutron and the NMR data.

9.3.3 Window scans and quasielastic scattering

When interpreting quasielastic scattering from polymer solutions or melts in subsequent sections, we have to remember that the translational scattering function is convoluted with vibration and rotation. The vibrational motion appears via the Debye–Waller factor. Any rotational motion is generally very fast on the time scale of translation. This means that the Lorentzian broadened component arising from this rotational motion forms a low intensity, almost flat background and the delta function in Fig. 9.8 is now broadened by the translational motion.

These effects can be seen in Fig. 9.11. This is a so-called 'window scan'

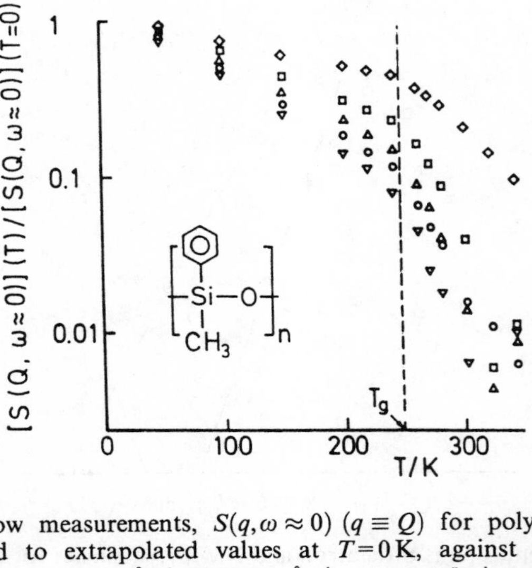

Fig. 9.11 Window measurements, $S(q,\omega \approx 0)$ $(q \equiv Q)$ for polymethylphenylsiloxane normalized to extrapolated values at $T = 0\,\mathrm{K}$, against temperature for selected values of q. (◇) 1.6 Å^{-1}, (□) 2.6 Å^{-1}, (△) 3.4 Å^{-1}, (○) 4.2 Å^{-1}, (▽) 4.8 Å^{-1}. Reprinted with permission from Meier *et al.* (1989). Copyright 1989, American Chemical Society.

performed on a high resolution (back-scattering) spectrometer. The sample is hydrogenous methylphenylsiloxane whose glass transition temperature is at 247 K (from differential scanning calorimetry, DSC). The window scan consists in setting a narrow energy window – in this case equal to 8 μeV, which is the spectrometric resolution at zero energy transfer – and counting the integrated intensity within this narrow window as a function of the sample temperature. If the sample were a crystalline or glassy solid, then the change in intensity would be given just by the Debye–Waller factor in eqns (9.5) and (9.6). Rotations of molecular side groups would give rise to an extra loss of intensity from this near elastic region as the Lorentzian in eqn (9.8) becomes broader at higher temperatures and intensity in its wings falls outside the narrow defined window. Thus the whole temperature dependence below T_g may not be associated with a Debye–Waller factor for vibrations alone. From the q-dependence at fixed temperature a value of the root mean square amplitude $\langle u^2 \rangle^{1/2}$ can be obtained since the intensity should decrease as $\exp(q^2\langle u^2 \rangle/3)$ (Section 4.9.3). For the data in Fig. 9.11 $\langle u^2 \rangle^{1/2}$ is 0.37 Å at 150 K and 0.5 Å at 240 K. This is extremely large for harmonic vibrations and implies a large scale motion, which might, in this case be, rotation of the methyl group.

At about 240 K, close to T_g, quite clearly a second process comes into play, which in this case is the rotation of the phenyl group. Intensity is being lost from the elastic window as the quasielastic peak broadens. In this range the faster motion of the methyl rotation is a very broad background. Window scans such as these are often used to identify the temperature ranges where noticeable quasielastic broadening can be detected with a particular spectrometer resolution. Although the α-process of the main polymer chain might be expected to begin at T_g, the associated motion is too slow for the resolution of the neutron spectrometers until much higher temperatures are reached. Typically such temperatures are at least $T_g + 100$.

9.3.4 Effects of multiple scattering

In all the discussion so far it has been assumed that all the detected neutrons have been scattered by one nucleus and then travelled unhindered to the detector. Except for a sample one atom only in thickness, this assumption is not true. Neutrons may be absorbed (though the absorption cross section of most nuclei in polymeric materials are low), or, more importantly, they may be scattered again, and even again and again. This process is called multiple scattering, and since at each scattering event the q, ω profile will be governed by $S(q, \omega)$, multiple scattering has the potential to severely modify the resulting observed scattering. Moreover, if this is interpreted in terms of single scattering, the parameters extracted may be unreliable.

We do not have the time here to enter into detailed discussion of multiple scattering. The reader is referred to the extensive literature, and in particular to a very detailed discussion of the subject in Beé (1988). The only way to deal with multiple scattering is to calculate it for the sample configuration and dynamic model in question, and then to correct for it. Generally speaking, multiple scattering is least important for near elastic scattering (which is why we were not concerned by it in Chapter 8). For a rotating group, however, the second scattering varies with $\boldsymbol{q}$ and ω differently to the first scattering, thus modifying the apparent EISF. Beé shows, however, that the quasielastic width may not be much affected in the example of a rotating methyl group.

The first approach to this problem is to minimize multiple scattering by working with thin samples with greather than 90 per cent transmission. The second is to check that any deviations from expected simple models are not caused by multiple scattering (for example if the EISF does not go to unity as $\boldsymbol{q}$ tend to zero). The third is to do a full multiple scattering calculation with all the difficulty this entails. For polymeric samples, where rather simple models are fitted, such heroic efforts have not so far been deemed worthwhile.

9.4 MAIN CHAIN MOTION

9.4.1 The correlation time and its q dependence

In liquids the elastic components in eqns (9.5) and (9.8) are replaced by Doppler broadened quasielastic peaks. A set of data for polydimethyl siloxane, obtained from a time-of-flight spectrometer were shown in Fig. 1.16. Although a polymer solution or melt is a true liquid in the sense that there is no fixed centre of mass, the experimental $\boldsymbol{q}$ value defines whether the motion of all or only part of the molecule is observed. The exponents $(\boldsymbol{q} \cdot \boldsymbol{r})$ in eqns (9.1) and (9.2) ensure that the scattering from self- or pair-correlations over a distance r will be maximum when $q = 1/r$ (or multiples of $1/r$ for a periodic array such as a crystal). The q value will thus 'pick out' different scales of molecular motion. For $q < 1/R_g$ where R_g is the radius of gyration of the polymer molecule, overall centre of mass diffusion is dominant. If this is (as is usual) Fickian diffusion, both the scattering law, $S(\boldsymbol{q}, \omega)$, and the intermediate scattering laws $S(q, t)$ have particularly simple dependence on q and the diffusion coefficient, D, for both coherent and incoherent scattering (see Chapter 4, Section 4.9.1).

$$\frac{S(\boldsymbol{q}, \omega)}{S(\boldsymbol{q})} = \frac{\Gamma}{\Gamma^2 + \omega^2} \tag{9.11}$$

$$S(q, t) = S(q, 0) \exp(-\Gamma t) \tag{9.12}$$

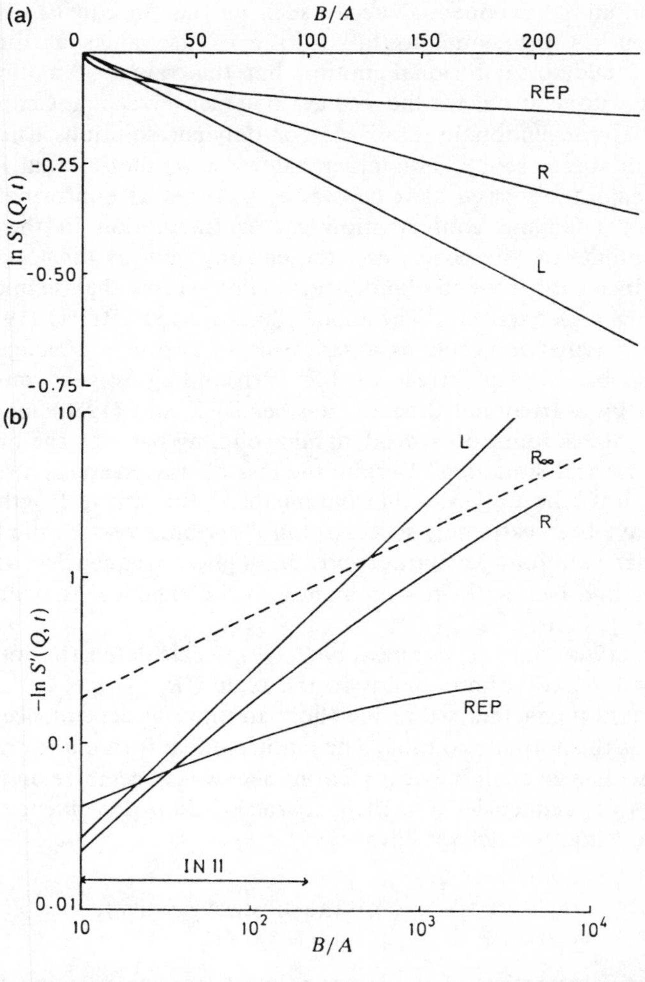

Fig. 9.12 Scattering functions L (eqn (9.12)), R (eqn (9.18)) and REP (eqn 9.24)) and the asymptotic limit for the Rouse model (eqn (9.23)) plotted against the experimental variable in spin-echo, the field in the main coils (to convert to a time scale, $1\text{A} = 5.33 \times 10^{-11}\text{s}$). $\Gamma_L = \Gamma_R = 5 \times 10^7 \text{s}^{-1}$, $qD_R = 3$.

where Γt is the normalized time. The $S(\boldsymbol{q}, \omega)$ are broadened elastic peaks as shown in Fig. 1.16. The $S(\boldsymbol{q}, t)$ are decay functions—examples can be seen in Figs 9.12 and 9.15.

The simplest parameters which can be extracted from the experimental data are the half width at half maximum of $S(\boldsymbol{q}, \omega)$, called $\Delta\omega$ (or $\omega_{1/2}$), and the slope of a plot of ln $S(\boldsymbol{q}, t)$ against t, called the inverse correlation time

Γ. (For a simple exponential decay such as that in eqn (9.12) Γ is the time taken for the signal to fall to $1/e$ of its values at time $t = 0$.) $\Delta\omega \equiv \Gamma \equiv Dq^2$ for diffusional motion but this range of motion is only accessible with neutrons for the smallest polymer molecules. Centre of mass diffusion is conventionally observed for polymer solutions using photon correlation spectroscopy. For larger values of q, the internal motion of the molecule is observed as it constantly changes its conformation. This continually changing conformation causes fluctuation in the distances between points on the chain, and the entropy gain as the length of any segment increases from its equilibrium value causes that segment to act as a spring under tension. The model developed by Rouse (1953) represents the polymer molecule as a series of such springs of length a connected by beads. The effect of the surrounding solvent molecules is described by a frictional drag on the beads. Zimm (1956) modified the model to take account of hydrodynamic coupling between the beads existing in many real situations. Despite the loss of any chemical structure for Rouse chains (the only variable parameter is the spring length a) these models have been extremely successful in describing viscoelastic behaviour of polymer solutions excepting only the highest frequencies where local vibrations and rotations are dominant and the chemical structure can no longer be ignored.

Pecora (1968) and de Gennes (1967, 1971) calculated the intermediate scattering laws for Zimm chains in the region $R_g^{-1} < q < a^{-1}$. Both the coherent and incoherent scattering show an unusual dependence on $(\Gamma t)^{2/3}$ where Γt is the normalized time. The mathematical formula for the scattering laws will be given in the next section; here we concentrate on the relaxation times or frequencies and their characteristic q dependence.

For the Zimm model we have

$$\Gamma_Z \propto \frac{k_B T}{\eta_s} q^3 \,(\eta_s \text{ is the solvent viscosity}). \tag{9.13}$$

Fourier transformation of the intermediate functions does not lead to an analytical form for $S(\boldsymbol{q}, \omega)$ and $S_s(\boldsymbol{q}, \omega)$ but the width function $\Delta\omega_{inc}$ and $\Delta\omega_{coh}$ show the same q variation as Γ.

In circumstances where the hydrodynamic interactions may be screened out (i.e. a polymer melt) the original Rouse description may be more valid. This leads to an even slower (square root) variation with normalized time and

$$\Gamma_R \propto \frac{k_B T a^2}{\xi_0} q^4 \tag{9.14}$$

ξ_0 is the friction factor per segment. Table 9.1 summarizes the numerical

Table 9.1 Numerical prefactors for the width function calculated for several models of molecular motion

Model	*Form of q dependence of width function*		*Numerical coefficients for* Γ	Δω	
Rouse $R_g^{-1} < q < \alpha^{-1}$	$\frac{k_B T \alpha^2}{\xi_0} q^4$	coherent	0.106	0.02	θ solvent conditions
		incoherent	0.106	0.03	
Zimm $R_g^{-1} < q < \alpha^{-1}$	$\frac{k_B T}{\eta_s} q^3$	coherent	0.0375*	0.0303	
		incoherent	0.0953	0.0413	
Fickian diffusion	Dq^2	coherent	1	1	
		incoherent	1	1	

*Assuming pre-averaged Oseen tensor. (Benmouna and Akcasu 1980.)

coefficients in eqns (9.13) and (9.14) for Γ and Δω for both coherent and incoherent scattering.

For distances shorter than α, local vibrational fluctuations are explored. There are no analytical models for this region but both qualitative discussion in terms of a Brownian motion and numerical calculation (Akcasu *et al.* 1980*a*) suggest a return to a q^2 dependence of Δω and Γ. The Rouse model depends on the Gaussian nature of the conformation of segments within one spring length α. Realistically then, even for the most flexible polymers, α is expected to be of the order 10 Å or more. The lowest q values obtained with the high resolution quasielastic spectrometers described in Chapter 3 are around 0.03 Å^{-1} or more. The neutron experiments inherently explore therefore the transition region $q\alpha \approx 1$ and the short range local motions. For this reason, the model calculations by Akcasu *et al.* (1980*a*) has been particularly valuable. These authors calculated the correlation functions $S(q, t)$ continuously over the range from $q < R_g^{-1}$ to $q > \alpha^{-1}$ for the Rouse and Zimm models. Clearly at the upper inequality at which motion of a single 'bead' dominates, the model becomes unrealistic but as analysis in Section 9.4.3 shows, this theory offers the best objective method of assessing the neutron scattering results. Their work suggests a model independent parameter should be extracted from the correlation functions (or intermediate scattering functions), which should be the first cumulant (or initial slope) of the normalized function $S(q, t)/S(q, 0) = S'(q, t)$

$$\Omega = \lim_{t \to 0} \frac{d}{dt} S'(q, t). \tag{9.15}$$

For Fickian diffusion $\Omega = \Gamma$, for the Rouse model in the range $R_g^{-1} < q < a^{-1}$, $\Omega \equiv 2^{1/2}\Gamma$.

Figure 9.13 shows the variation of Ω with q over the whole range of motion described for the Zimm model of a polymer in a good solvent. Since T and η_s are fixed, there is only one variable parameter, the spring length a. Akcasu *et al.* (1980*a*) were able to develop the discussion of Ω in some detail, looking at the effect of solvent interactions, solution concentrations and the strength of hydrodynamic interactions. As will be seen, this approach now forms the basis for interpretation of many of the neutron results from polymer solutions.

In a polymer melt, apart from a general slowing down of the motion of a single chain, two qualitative changes from the behaviour in dilute polymer solutions are expected. First, the hydrodynamic effects are screened out, so that the correlation function in the 'universal' regime is in the Rouse limit and the inverse correlation time varies as q^4. Secondly, a new length scale may come into play: the distance between entanglements of the polymer chains. In the reptation model, introduced by de Gennes (1971) and extended by Doi and Edwards (1979) into a full theory of polymer viscoelasticity, the distance between entanglements is expressed as a tunnel diameter, D_R. Within this 'tunnel' the polymer molecule executes free Rouse motion, but when distances larger than D_R are explored a very slow 'creep' intervenes. The relevant correlation functions, calculated by

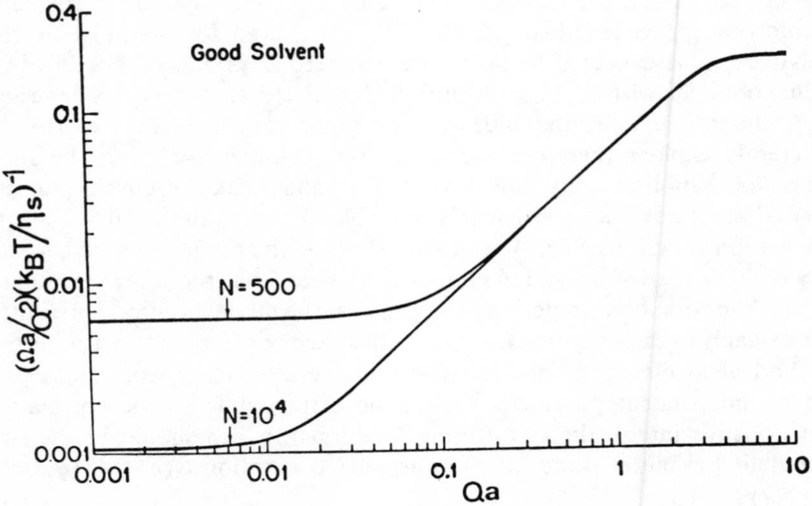

Fig. 9.13 Calculated values of $(\Omega a/Q)$ $(k_B T/\eta_s)$ vs Qa $(Q \equiv q)$ for a good solvent and for two molecular weights, N. Reprinted from Akcasu *et al.* (1980*a*) with permission from authors and the publishers, Butterworth Heinemann ©.

de Gennes (1980) are Rouse-like at short times (at least the inverse correlation time has the same q dependence) and have a very slowly decaying tail at long times as entanglements come into play. An example can be seen in Fig. 9.12(a).

Typical values for the various lengths introduced above are: $R_g \approx 200$ Å, $a \approx 10$ Å and $D_R \approx 50$ Å. In order to explore the long range internal modes characteristic of polymer molecules and to detect effects of entanglements, experiments must thus be carried out at very low q values (<0.1 Å^{-1}). The corresponding values of Ω range from 10^7 s^{-1} in a melt up to 10^9 s^{-1} in a solution. These values are equivalent to half widths of quasielastic scattering ranging from about 10^{-8} to 10^{-6} eV and thus require very high resolution neutron scattering experiments.

9.4.2 The scattering laws

The scattering laws defined by eqn (9.4) are calculated from space–time correlation functions, $G(\mathbf{r}, t)$ via spatial and time Fourier transformation. It turns out that while analytical forms result from the spatial Fourier transformation to the intermediate scattering functions, $S(\mathbf{q}, t)$ the second Fourier transform can often only be obtained numerically. Conveniently, however, the highest resolution quasielastic neutron scattering (the spin-echo technique) results in the normalized intermediate scattering function.

$$S'(\mathbf{q}, t) = \frac{S(\mathbf{q}, t)}{S(\mathbf{q}, 0)}. \tag{9.16}$$

We will therefore only discuss in detail some of the formulae developed for $S'(\mathbf{q}, t)$ while leaving their derivations to more advanced texts. The quantity often extracted is the first cumulant, Ω which was defined in eqn (9.15).

In the early part of this section we took the low q limit of $\Omega = \Gamma = Dq^2$. At finite q, we have to remember the term $S(q)$ in the denominator of eqn (9.16) so that for coherent scattering

$$\Omega = \frac{q^2 D}{S(q)}. \tag{9.17}$$

For polymers, for $q \ll R_g$, $S(q) \to 1$ and we recover $\Omega = \Gamma = Dq^2$. For the intermediate q range $S^{-1}(q) \to \frac{q^2 l^2}{12}$ (eqn (6.109)) where l is a length of an elementary step. If we equate l to $\mathbf{a}$, then we recover the q^4 dependence of eqn (9.14), noting also that $k_B T/\xi_0$ is the diffusion coefficient of a monomer.

A more general expression than eqn (9.17) can be obtained by replacing D by a generalized mobility term $k_B T\mu(\mathbf{q}, c)$ which allows for interactions, both thermodynamic and hydrodynamic within the system. This simply shows a more general result, that for coherent scattering the inverse correlation times, Γ, are dependent on the structure factor, $S(q)$.

Equations (9.12) and (9.17) are frequently used as a first approximation to interpretation of the dynamic scattering. The more rigorous approach is to model the particular system in terms of $G(r, t)$. This is not the place to undertake the detailed mathematics. We simply quote the important results for $S'(q, t)$ for polymeric systems.

For the Rouse model the coherent scattering is given by

$$S'_{\rm R}(q, t) = \int_0^\infty \mathrm{d}u \exp\left\{-u - (\Gamma_{\rm R} t)^{1/2} g\left[\frac{u}{(\Gamma_{\rm R} t)^{1/2}}\right]\right\} \tag{9.18}$$

$$g(y) = \frac{2}{\pi}\int_0^\infty \mathrm{d}x \left(\frac{1-\exp(-x^2)}{x^2}\right)\cos yx$$

where the inverse correlation time $\Gamma_{\rm R}$ is given by

$$\Gamma_{\rm R} = \frac{k_{\rm B} T a^2}{3\pi\xi_0} q^4. \tag{9.19}$$

In solution, hydrodynamic interactions transmitted via the free solvent molecules are important, thus modifying the Rouse normal modes, as shown by Zimm. In this case, du Bois Violette and de Gennes (1976) calculated the scattering law as

$$S'_{\rm Z}(q, t) = \int_0^\infty \mathrm{d}u \exp\left\{-u\left(h\left[\frac{u}{(\Gamma_{\rm Z} t)^{2/3}}\right]\right)\right\} \tag{9.20}$$

$$h(y) = \frac{4}{\pi}\int_0^\infty \mathrm{d}x \frac{\cos x^2}{x^3}\left[1 - \exp(-y^{-3/2}x^3)\right]. \tag{9.21}$$

There is a particularly simple form for $\Gamma_{\rm Z}$ which now depends only on solvent viscosity and on temperature

$$\Gamma_{\rm Z} = 0.0375 \frac{k_{\rm B} T}{\eta} q^3. \tag{9.22}$$

Both eqns (9.18) and (9.20) reach simple asymptotic forms at long times, where the scattering functions become similar to their incoherent counterparts and

$$S'_{\rm Z}(q, t) \to \mathrm{e}^{-(\Gamma_{\rm Z} t)^{2/3}}, \; S'_{\rm R}(q, t) \to \mathrm{e}^{-(\Gamma_{\rm R} t)^{1/2}}. \tag{9.23}$$

These limits are only reached at times of order 10 Γ^{-1} which is well beyond the range explored even in the neutron spin-echo experiments. The term Γ_R (or Γ_Z) may be obtained by fitting eqn (9.18) or (9.20) to the scattering function directly, or by simply taking the corresponding initial slopes Ω_R or Ω_Z of these functions since it has been shown that $\Omega_Z = 2^{1/2}\Gamma_Z$ and $\Omega_R = (\pi/4)\Gamma_R$.

As mentioned in Section 9.4.1 the time Fourier transform of eqns (9.18), (9.20) or even the simpler incoherent laws, are not analytical functions. Nevertheless the full width at half maximum of the scattering laws $\Delta\omega_{inc}$ and $\Delta\omega_{coh}$ can be calculated and the q variations and numerical coefficients were listed in Table 9.1.

For the entangled melt ($R_g^{-1} < q \leq D_R^{-1}$) in the framework of the reptation model, de Gennes (1980) calculated an overall correlation function of one chain which can be written as:

$$S'(q,t) = \frac{q^2 D_R}{36} S'_{local}(q,t) + \left(1 - \frac{q^2 D_R}{36}\right) S'_{creep}(q,t) \tag{9.24}$$

$$S'_{local}(q,t) = \exp(\Omega_R t)\left[1 - \text{erf}(\Gamma_R t)^{1/2}\right] \tag{9.25}$$

$$S'_{creep}(q,t) = \frac{8}{\pi^2}\exp\left(-\frac{t}{T_{rep}}\right)\left\{1 + \sum_{n=1}^{odd}\frac{1}{n^2}\left(\frac{\pi^2}{8}\right)^{n^2-1} \times \left[\frac{8}{\pi^2}\exp\left(-\frac{t}{T_{rep}}\right)\right]^{n^2-1}\right\} \tag{9.26}$$

where T_{rep} is the reptation time.

An alternative description of entangled chains has been given by Ronca (1983). The elastic interactions between the polymer chains are incorporated into the equation of motion through a memory term (the kernel is related to the relaxation modulus). The calculated scattering function for single-chain motion shows a continuous change from Rouse-like behaviour to a flat response with decreasing q. In the range $q^2\langle r_E^2\rangle \leq 1$ where $\langle r_E^2\rangle$ is the mean-square end-to-end distance of the entanglement molecular weight, the predicted long-time limit is a plateau value

$$S'(q,\infty) \approx 1 - \frac{q^4\langle r_E^2\rangle^2}{496}. \tag{9.27}$$

This suggests a weaker decay of $S'(q,t)$ than predicted by the reptation model. In the short-time limit, Rouse-like behaviour is expected to hold.

Figure 9.12(a) compares on a semilogarithmic scale the scattering functions in eqns (9.12), (9.18), and (9.24) and Fig. 9.12(b) and includes the asymptotic long-time behaviour of the Rouse-type motion, eqn (9.23). Typical values of Γ_R and T_{rep} have been used in the calculations. Note that on the double logarithmic plot in Fig. 9.12(b), eqn (9.24) exhibits similarities to the long-time asymptotic behaviour of the Rouse chain. Furthermore, the short-time behaviour of eqn (9.24) (see Fig. 9.12(a)) is reminiscent of that shown by incoherent scattering functions in that the function approaches $t = 0$ as an asymptote to the ordinate. This arises because a model based on a continuous rather than a discrete description of chain motion necessarily fails at short times.

9.4.3 Motion in polymer solutions

In practice, the intermediate scattering functions $S'(\boldsymbol{q}, t)$ have been more generally explored for polymer solutions than the $S(\boldsymbol{q}, \omega)$. This is for three reasons. First, as has been mentioned, the slow, main chain motion requires the high resolution of the neutron spin-echo technique, which automatically provides time–domain data. Second, the theoretical curves are calculated from the space–time correlation functions $G(\boldsymbol{r}, t)$ and, while $S'(\boldsymbol{q}, t)$ may have simple analytical forms, the second Fourier transformation to frequency space can often only be performed numerically. Third, removal of resolution effects is a simple division process in the time–domain results from spin-echo, while in the frequency domain it entails an unreliable deconvolution process. Generally, in this latter case, as described above, model scattering laws are convoluted with the observed resolution function and are fitted to the experimental data.

The problem of obtaining information about diffusive polymer motion from $S(\boldsymbol{q}, \omega)$ data is clearly seen in Fig. 9.14 (Allen *et al.* 1976). A 3 per cent by weight solution of polytetrahydrofuran (PTHF) in CS_2 was observed by high resolution time-of-flight spectroscopy. At $q = 0.25$ Å^{-1} a small amount of broadening beyond the instrument profile is seen. By convoluting this resolution profile with the Fourier transforms of eqns (9.18) and (9.20) the data are compared with the Rouse and Zimm models (labelled q^4 and q^3 in the figure respectively). The full width at half maximum of the resolution function in Fig. 9.14 is 25 μeV. Given that the q variation of the width function, $\Delta\omega$, is at least q^2 for simple diffusion, and as fast as q^4 for the Rouse model, it is clear that the time-of-flight spectrometer could not be used for much lower q values than those in the figure. The best resolution achieved by a back-scattering spectrometer is about 1 μeV, and the effective resolution of the spin-echo spectrometers is about 0.01 μeV. It thus becomes clear why spin-echo spectrometry, limited though its availability is, has become the desirable technique for observing long-range polymer motion.

Another complication with data such as that in Fig. 9.14 is the difficulty of separating incoherent and coherent scattering. Both coherent and incoherent scattering are affected by the quasielastic broadening. The intensity of the incoherent scattering, $S_s(\boldsymbol{q}, \omega)$ is almost q independent. That of the coherent scattering varies with q as $S(\boldsymbol{q})$ and also depends on the contrast factors. In particular for a polymer solution intensity increases with

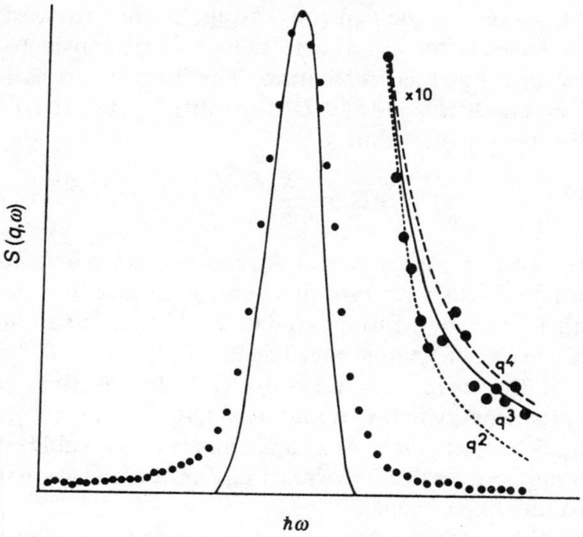

Fig. 9.14 Quasielastic scattering from a 3 per cent solution of polytetrahydrofuran in CS_2 at $q = 0.25$ Å^{-1} together with the instrumental resolution shown as a solid line. Enlarged wing profile shows the best fit calculated for eqn (9.12) labelled q^2, eqn (9.20) labelled q^3, and eqn (9.18) labelled q^4. (Allen *et al.* 1976.)

decreasing q and tends to be small at high q. For an hydrogenous polymer in a deuterated solvent, therefore, the observed quasielastic scattering will be a mixture of $S(\boldsymbol{q}, \omega)$ and $S_s(\boldsymbol{q}, \omega)$. At high q, $S_s(\boldsymbol{q}, \omega)$ will tend to dominate, while at low q $S(\boldsymbol{q}, \omega)$ may become more important. This problem cannot be overcome by using an hydrogenous solvent with deuterated polymer because the background incoherent scattering from the solvent swamps the signal. The solvent CS_2 in Fig. 9.14 was chosen advisedly. It has no incoherent scattering and it almost exactly contrast matched to the hydrogenous PTHF so that $S(q) \approx 0$. In a second experiment on deuterated PTHF dissolved in CS_2 the coherent scattering dominates because of the scattering length difference between polymer and solvent but the incoherent scattering becomes negligible because there is no hydrogen present. For most polymer solutions this contrast matching is not possible and the observed scattering is an inextricable mixture of coherent and incoherent terms.

In neutron spin-echo, as explained in Chapter 3, the tracking of the neutron spin means the $S_{coh}(\boldsymbol{q}, t)$ is measured unambiguously. It is then usual in spin-echo experiments to choose as samples hydrogenous polymers dissolved in deuterated solvents in order to enhance $S(q)$ as much as possible and to reduce the neutron wastage to (not counted) incoherent scattering. To illustrate the use of spin-echo data we consider the centre of mass diffusion of a polymer in solution. The inverse correlation time Γ is just Dq^2 for simple diffusion, and D is the diffusion coefficient, given by the Stokes–Einstein relationship as

$$D = \frac{k_B T}{6\pi\eta_s R_H}, \tag{9.28}$$

where η_s is the solvent viscosity and R_H the hydrodynamic radius.

In an organic solvent with typical viscosity around 5×10^{-4} Nsm^{-2}, a molecule with radius of gyration around 20 Å (and assuming R_H of the same order) has a diffusion coefficient of $\sim 2.5 \times 10^{-6}$ cm^2 s^{-1}. At $q = 0.025$ Å^{-1}, therefore, $\Gamma = 1.5 \times 10^7$ s^{-1}. In the time range of the spin-echo experiment (which is about 1.3×10^{-8} s), ln $S'(q, t)$ will have dropped from 0 at time 0 to -0.2, and thus be reasonably well resolved. This would require a resolution of 0.01 μeV in a conventional frequency-domain scattering experiment.

Figure 9.15 shows (a) the normalized intermediate scattering functions $S'(q, t)$ for a solution of polydimethylsiloxane of molecular weight 6400 ($R_g \approx 21$ Å) in deuterated benzene at 3.5 per cent concentration and 25°C and (b) the inverse correlation time Γ/q^2 plotted against q^2 for a series of molecular weights of the same polymer under the same experimental conditions (Higgins *et al.* 1983). The $S'(q, t)$ have had the spectrometer *resolution* simply removed by division (compare the difficulty of convoluting models with the experimental resolution function to analyse the data in Fig. 9.14).

The plateaux at low q in Fig. 9.15(b) correspond to the regime $q < R_g^{-1}$ and to simple diffusive motion with $\Gamma = Dq^2$. The variation with molecular weight of the transition to the region $q > R_g^{-1}$ is clearly seen. (Compare Fig. 9.13.)

All the curves in Fig. 9.15(b) converge for $q > R_g^{-1}$. In this range eqn (9.12) no longer applies, and the data should be better described by eqn (9.20). This can be seen in Fig. 9.15(a). For the high q values ln $S'(q, t)$ no longer varies linearly with time. The values of Γ shown in the figure for these higher q values were obtained by fitting eqn (9.20) to the data.

Figure 9.13 shows that for small stiff molecules, where R_g is small and α is large, the region where eqn (9.20) applies may be quite small or even non-existent. Since α in particular, may not be known independently, it is better not to presuppose the mathematical form of $S'(q, t)$, in order to obtain Γ, but to take the short time asymptotic slope, $\Omega = (d/dt)(\ln S'(q, t))$. As discussed in Section 9.4.1 Ω is simply related to Γ for each of the models of interest, but its extraction from the data is model independent. For the Zimm model

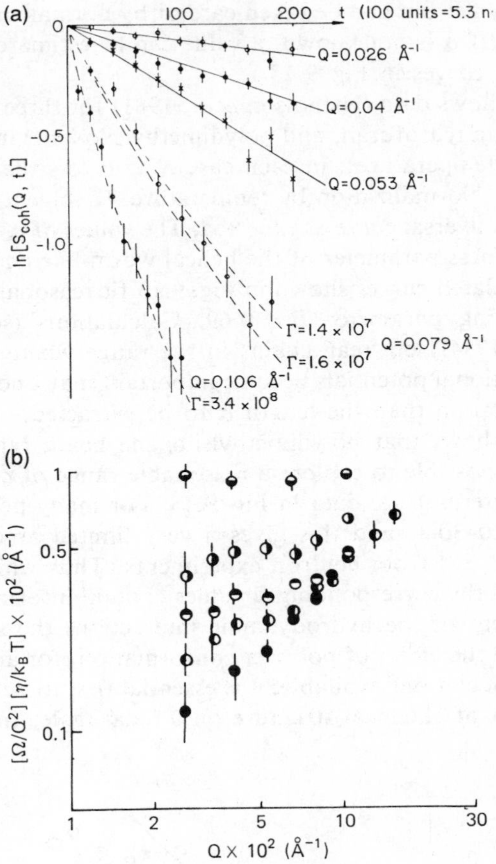

Fig. 9.15 (a) $S(q, t)$ against t for 3.5 per cent polydimethylsiloxane (PDMS) chains ($M_n = 6400$) in C_6D_6 for various values of $Q(\equiv q)$. (b) Plot of Ω/Q^2 against Q for PDMS of various molecular weights in C_6D_6 at 30°C. (◒) 1100, $R_g = 7$ Å; (◑) 2700, $R_g = 12.4$ Å; (◒) 6400, $R_g = 21.2$ Å; (◐) 15 100, $R_g = 40$ Å; (●) 174 000, $R_g > 100$ Å. (Higgins *et al.* 1983).

$$\Omega \equiv (2)^{1/2}\Gamma = (2)^{1/2} \times \left(0.0375 \frac{k_B T}{\eta_s} q^3\right). \tag{9.29}$$

Data from different solutions and different temperatures can be normalized to each other by multiplying Ω by η_s/T. Moreover the results should be independent of the chemical structure of the polymer. However, when $q \sim \alpha^{-1}$, Ω will deviate from q^3 behaviour. This value of q will vary from polymer to polymer but can be taken care of by normalizing the Ω and the q values by a. (If a is not known, a value can be estimated by comparing the data to the curves in Fig. 9.13.)

Figure 9.16 shows data (Nicholson *et al.* 1981) for three polymers (polystyrene, polytetrahydrofuran, and polydimethylsiloxane) in two solvents at three different temperatures; in each case R_g^{-1} is less than the smallest q value observed. Normalization by temperature T, solvent viscosity η_s and a, produced a universal curve as expected. The values of a used correspond to λ^{-1}, the stiffness parameter of the helical worm-like chain. (Yamakawa 1977) The calculated curves shown in Fig. 9.13 fit reasonably to these data with the draining parameter $B = 0.69$. Calculations (see for example Allegra *et al.* (1984)) on 'real' chains in the range where details of bond angles and rotational potentials become important may one day allow more detailed information than the length a to be extracted.

Figure 9.16 shows that polydimethylsiloxane has a rather small value of a. It is thus possible to explore a reasonable range of $q \leq a^{-1}$, making it possible to interpret the data in Fig. 9.15. For many polymers however a is typically 20–30 Å and this gives a very limited amount of data in the q range $q < a^{-1}$ from neutron experiments. Thus while, in principle, the $S'(q, t)$ and the corresponding Ω values contain information about the nature and extent of the hydrodynamic interactions the solvent–polymer interactions and the effect of polymer conformation, for all of which there are extensive calculations available, it is essential first to be absolutely clear about the effect of chemical structure on a local scale. In other words it

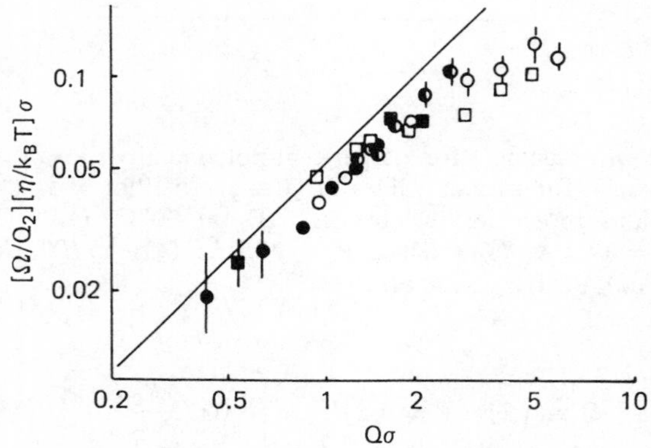

Fig. 9.16 $(\Omega\sigma/Q^2)$ (η/kT) $(\sigma \equiv a$ and $Q \equiv q)$ against $Q\sigma$ for polymer solvent systems as follows (values of σ in brackets). (○) PS/C_6D_6 (36 Å) at 30°C; (□) PS/C_6D_6 (36 Å) at 70°C; (■) PTHF/C_6D_6 (16 Å) at 30°C; (◐) PDMS/C_6D_6 (16 Å) at 30°C; (●) PDMS/C_6D_6 (16 Å) at 70°C. The solid line is eqn (9.22). (Nicholson *et al.* 1981.)

is no use interpreting data in terms of Gaussian chains when on the scale of observation the chains are no longer behaving as Gaussians!

9.4.4 Contrast matching in dynamic studies of polymer solutions

The study of dilute polymer solution, as described so far, only uses deuteration where necessary to enhance the contrast between polymer and solvent. The particular features neutron spectroscopy add to, for example dynamic light scattering, is the time vs distance scale explored. If we, instead, make a mixture of deuterated and hydrogenous molecules in a solvent which can either be D or H or a mixture, then we have the dynamic analogy of some of the SANS experiments described in Chapter 8. There are two types of fluctuations to consider, $\Delta(C_H + C_D)$ and $\Delta C_H (= -\Delta C_D)$. Deuteration can be used in an exactly analogous way to small angle neutron scattering in order to contrast match solutions and observe single chain motion in complex surrounding by setting $\Delta(C_H + C_D) = 0$. To illustrate this point, we choose two examples.

The first concerns the dynamics of a semidilute polymer solution (Csiba *et al.* 1991). In such solutions a new characteristic length comes into play, intermediate between R_g and α. It is the so-called dynamic screening length ξ_h. For distances less than ξ_h the chains are observed as simple Gaussian chains, with the dynamic behaviour described in the previous section. For distances longer than ξ_h cooperative motion will be observed as well.

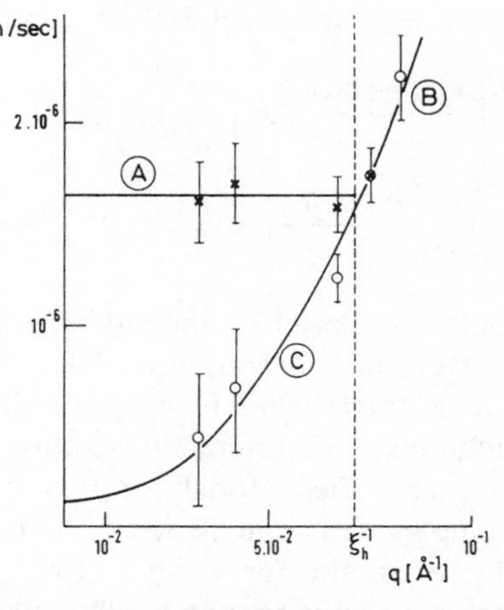

Fig. 9.17 Γ_q/Q^2 against q for (×) PDMS-H in deuterated toluene and (○) a mixture of PDMS-H with PDMS-D in a mixed solvent of deuterated and normal toluene. A, B, guides to the eye, C, calculated values from eqn (9.19). Reprinted with permission from Csiba *et al.* (1991).

Figure 9.17 shows data for Ω from two experiments on polydimethylsiloxane in toluene at 20 wt per cent concentrations. In one experiment the PDMS was hydrogenous and the toluene deuterated. These data fall on the curves A–B in Fig. 9.17. In the second experiment the PDMS was a 50/50 mixture of hydrogenous and deuterated polymer, while the toluene was also an H–D mixture, calculated to exactly match the contrast of the polymer. This is the condition described in eqn (8.3) or (8.4) where the second term disappears and $S(q, 0) \equiv P(q)$.

These data correspond to the curve C–B in the figure. The point at which the two curves part company is the point at which the interchain cooperative effects become important. In the sample with only hydrogenous chains these effects dominate and Ω varies as q^2 for $q\xi_h < 1$

$$\Omega_{\text{coop}} = \frac{k_B T}{6\pi\eta_s\xi_h} q^2. \tag{9.30}$$

From the data in Fig. 9.17, the value of ξ_h is 22 Å. The data from the contrast-matched sample corresponds to the motion of single chains and the results are similar to those in Fig. 9.16. However, the hydrodynamic effects are screened out in semidilute solutions, and now Ω varies as q^4 rather than q^3 (eqn (9.18) rather than eqn (9.20)).

The second example of the use of contrast matching concerns diblock copolymers in solution (Duval *et al.* 1991) i.e. it is the dynamic analogue of the small angle scattering experiments described in Chapter 8, Section 8.2.3.

The scattering law for a symmetrical diblock copolymer can be written as

$$I(q,t) = \left\{\frac{(b_D - b_H)}{2}\right\}^2 N\left(\frac{z}{2}\right)^2 (P_{DD} - P_T)\mathrm{e}^{-\Gamma_1 t}$$
$$+ \left(\left(\frac{b_D + b_H}{2}\right) - b_0\right)^2 N^2 z^2 \frac{P_T}{1 + \nu\phi N P_T} \mathrm{e}^{-\Gamma_c t}. \tag{9.31}$$

Equation (9.31) should be viewed in analogy with the static equation for a two component mixture in solution, eqn (5.41). The first term is essentially the scattering from the labelled blocks, and the second that from the whole molecules (including their interactions, normally expressed as $\Omega(q)$ but here written in a simplified model form in the denominator of the second term). In analogy with eqn (9.12) each term is multiplied by a dynamic structure factor of the form $\exp(-\Gamma t)$.

P_{DD} is the particle form factor for the labelled block while P_T is that for the whole molecule. The interesting form of the first term in eqn (9.31) was derived in Chapter 5, eqn (5.57).

In contrast match conditions the second term vanishes. This is the term which describes the motion of the copolymer molecules as a whole, i.e. it arises from fluctuations in $\Delta(C_D + C_H)$. We then have $S'(q,t) = \exp(-\Gamma_1 t)$ where Γ_1 is the intramolecular relaxation of the two halves of the molecule with respect to each other, i.e. fluctuations in $\Delta C_D(= -\Delta C_H)$.

As described in Section 9.4.2, Γ_i can be expressed in terms of $S(q)$ and a general mobility function, $\mu(q, c)$.

$$\Gamma_1(q) = q^2 k_B T \frac{\mu(q,c)}{S(q)}. \tag{9.32}$$

If there are no interactions (in this case hydrodynamic *or* thermodynamic) $\mu(q, c) = \xi_0^{-1}$ and eqn (9.32) reduces to eqn (9.14) since in this q range $S^{-1}(q)$ is proportional to $q^2 a^2$.

Figure 9.18 shows data for a symmetrical diblock copolymer of deuterated and hydrogenous polystyrene in contrast matched mixture of H and D benzene. For such a diblock copolymer it is reasonable to expect the thermodynamic interactions to be small between the H and D blocks, i.e. $\chi = 0$. The total molecular weight was 10 800 and $S(q)$ was obtained from the variation of intensity with q using the spin-echo spectrometer. In Fig. 9.18 $\mu(q, c)$ varies both with q and with c, indicating that hydrodynamic (or other) interactions may not be ignored.

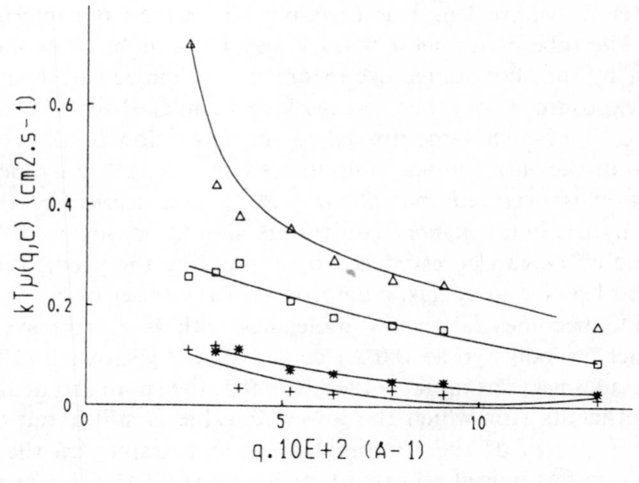

Fig. 9.18 Mobility $\mu(q, c)$ as a function of q for polystyrene diblock copolymers. (+), (*), (□), and (△) correspond to concentrations 48.4×10^{-2}, 41.1×10^{-2}, 24.6×10^{-2} and $15.4 \times 10^{-2}\,\mathrm{g\,cm^{-3}}$. Reprinted with permission from Duval *et al.* (1991). Copyright 1991, American Chemical Society.

9.4.5 Dynamics in the melt

As already mentioned in the previous section, when concentration increases, hydrodynamic interactions are screened out, and the single chain dynamics revert to the Rouse behaviour in eqn (9.18). This model, however, ignores the entanglements between chains which are responsible for viscoelastic behaviour and which also affect the single chain correlation functions. In the introduction to this chapter we mentioned the reptation theory of Edwards and de Gennes. One of the points of discussion about this theory has always been to ask just how accurately does it describe the motion of the polymer molecules in crowded surroundings. Since spin-echo neutron scattering offers the possibility of observing individual molecules in such a situation because of the deuteration labelling techniques, it has been the goal of one or two groups of physicists to use neutron spin-echo to check the validity of the reptation model at a molecular level. (Most experimental checks are at a macroscopic level, i.e. the prediction of the temperature or molecular weight dependencies of viscosities or moduli.)

The reptation model proposes that a set of entangled molecules would move most easily along their own contours, like snakes. As well as their frictional drag, the surrounding molecules now introduce a third length scale into the description of motion of a single chain. In Section 9.4.1, we described the physical restraints imposed by the surroundings in terms of the diameter D_r whose length is certainly larger than the intermolecular distances. The tube or tunnel it defines may be thought of as the blurred outline left by the short-range excursions of the molecular segments in a long-time exposure. (Long here means long compared to local relaxation times but of course short compared to the relaxation of the constraints by removal of the surrounding molecules.) For $q > D_r^{-1}$, only local Rouse —like motion is expected, but for $q < D_r^{-1}$, modification of the Rouse behaviour by the entanglement constraints should be observed.

The value of D_r can be estimated by comparing the predictions of the model with observed viscoelastic data. It is of the order of 30 Å for polyethylene and becomes larger for molecules with larger transverse cross sections such as polystyrene (80 Å) or polydimethylsiloxane (50 Å). For many molecules it seems unlikely therefore that the spin-echo neutron scattering experiments (for which the lowest q value is still about 0.02 Å^{-1}) would show effects of these entanglements constraints on the polymer motion, and it was indeed reported that for polydimethylsiloxane no such effects were observed.

Polytetrahydrofuran $[(CH_2)_4O]_n$ has a molecular structure fairly close to polyethylene. There was until recently, however, little viscoelastic data available to confirm that it also has a small value of D_r. Nevertheless, neutron scattering data indicate this by showing effects that are difficult to explain except in terms of chain entanglements. In the experiment (Higgins and Roots 1985) samples of high- and low-molecular weight materials were mixed so that in one sample the labelled PTHF molecules were surrounded by low molecular weight (unentangled molecules) (H/L) while in a second experiment the surroundings were high molecular weight (entangled) molecules (H/H). In each case the surroundings were deuterated molecules so that $S(q)$ was significant and $S(q, t)$ could be observed. Figure 9.19 compares the observed intermediate scattering functions in the two experiments at one value of q. The lower curve from the unentangled molecules (H/L) is well fitted by the Rouse model calculations, but the upper curve from the entangled molecules (H/H) cannot be fitted by this model because of the very slow long time decay. If these data are fitted to calculations for the reptation model (eqn (9.24)) then the best value of D_r obtained is 30 Å. Viscoelastic measurements give a value of $D_R = 31.5$ Å from the plateau modulus of this polymer.

If, instead of eqn (9.24), eqn (9.27) is used to interpret the long time behaviour of the correlation functions, then a value of the root mean square entanglement distance $\langle r_E^2 \rangle^{1/2}$ can be extracted. Not surprisingly, this is essentially the same as that for D_r. The Ronca theory (Ronca 1983), however, also allows a clear prediction of whether the effects of entanglements can be observed within the spin-echo time scale. Ronca predicted serious derivations from Rouse behaviour when

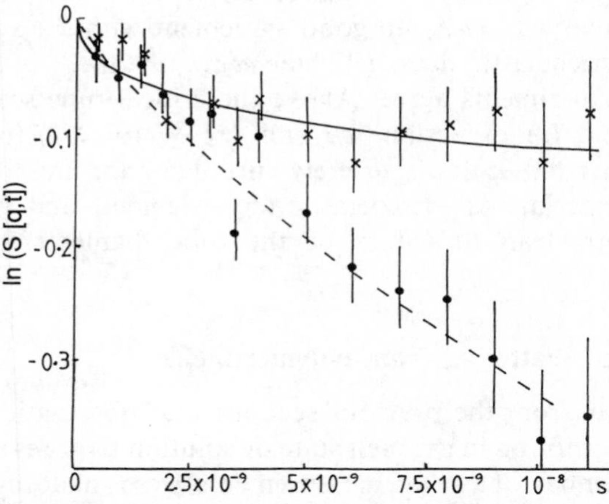

Fig. 9.19 $Ln(S(q, t))$ as a function of time for a mixture of high molecular weight polytetrahydrofuran (PTHF) in: (×) a high molecular weight deuterated PTHF matrix and (●) a low molecular weight deuterated PTHF matrix (----) eqn (9.18) with $\Gamma_R = 5.55 \times 10^{-7} s^{-1}$ and (—) eqn (9.24) with the same value of Γ_R and $D_R = 30$ Å. (Higgins and Roots 1985.)

$$t > [\langle r_E^2 \rangle q^2/24]^2 \Gamma_R^{-1}. \tag{9.33}$$

For the PTHF used in these experiments $\langle r_E^2 \rangle^{1/2} = 30$ Å and $\Gamma_R = 5.55 \times 10^7\,s^{-1}$ for the data in Fig. 9.19. In eqn (9.33) this gives $t \approx 2 \times 10^{-9}$ s, which compares well with the point at which the H/H data deviate from the H/L data in Fig. 9.19; on the other hand, PDMS which has often been studied because of its high flexibility, has values of Γ_R perhaps a factor 2.5 times larger than those of PTHF, but the value of D_R is 50 Å (from the plateau modulus). In eqn (9.33) these values lead to times longer by a factor of three than those for PTHF and correspondingly more difficult to observe (Richter *et al.* 1981). In fact eqn (9.33) shows that the possibility of observing the slowing down due to entanglements is a delicate balance of two parameters, Γ_R and $\langle r_E^2 \rangle^{1/2}$.

In recent years the resolution of the spin-echo spectrometer has been considerably improved to give a current maximum observation time of 40×10^{-9} s. The group that worked originally on PDMS has extended their studies to a polymer with a somewhat smaller value of $\langle r_E^2 \rangle^{1/2}$. An alternating copolymer of ethylene–propylene was prepared with $\langle r_E^2 \rangle^{1/2}$ of order 45 Å. At 292 K deviations from Rouse motion were observed at about 10^{-8} s, as predicted by the Ronca calculations. Moreover a much longer section of the plateau regime was detected and it was possible to test the details of the model. Excellent agreement with Ronca's model was obtained, as well as satisfactory agreement with the long-time behaviour of the de Gennes function eqn (9.24). In these experiments it was again possible to fit data at several q values simultaneously to the model, and the value of $\langle r_E^2 \rangle^{1/2}$ obtained was 47 Å, in good agreement with the value of 43 Å obtained from viscoelastic data. (Richter *et al.* 1990).

Thus all the experiments agree. Above the critical time scale t_c defined by eqn (9.33) and for excursions beyond $\langle r_E^2 \rangle^{1/2}$ (or D_R) the motion of entangled polymer molecules is severely curtailed and, moreover, the two very different experiments—viscoelastic measurements and neutron spin-echo spectroscopy lead to values of the tube diameter in very close agreement.

9.4.6 Incoherent scattering from polymer melts

It should be clear from the previous sections that observation of characteristic polymeric motion in the melt state or solution requires the spin-echo technique. A number of earlier measurements were made using incoherent scattering and the highest resolution time-of-flight or back-scattering spectrometers, but these have been superseded for the long range Rouse or Zimm type motion by the spin-echo data. It remains true, however, that the incoherent scattering functions do contain useful information. This is particularly true at higher q values where effects of the molecular structure become dominant. There remains the difficulty that fitting procedures are required, involving convoluting spectrometer resolution functions with model scattering laws, so it is necessary to have available suitable models before embarking on the scattering experiment. An area where there is a considerable interest is the motion responsible for the α-relaxation (associated with the glass transition temperature). This may be rather local motion and analysis of the incoherent quasielastic scattering may provide insight alongside other relaxation techniques into the mechanisms and processes involved. Most of the experimental investigations using neutron scattering have concentrated on small molecule glass forming liquids such as ortho ter phenyl (Bartsch *et al.* 1989) but there has been recent interest in relatively simple polymers, such as polybutadiene (Frick *et al.* 1988, Kanaya *et al.* 1991). Well above the glass transition, the details of the quasielastic broadening can be analysed and compared to numerical simulations. The analysis is similar to that used in Section 9.3 for side groups, now applied to rotations of the main chain. For polybutadiene, isolated conformational transitions were shown to occur alongside cooperative motion.

A much discussed recent theory of the glass transition is the mode coupling theory. (Frick *et al.* 1988; Bartsch *et al.* 1989). This theory predicts a critical temperature (usually about 30°C above T_g) at which there is a singularity in the density fluctuations (much as there is a singularity in the concentration fluctuations of a mixture near its critical point as discussed in Chapter 8). The theory has attracted the attention of neutron scatterers because the density–density correlation function is measured directly in incoherent quasielastic neutron scattering experiments and thus these measurements can be used to check directly the predictions of the theory. The observations involve window scans (Section 9.3.3) and divergences in the vibrational amplitude obtained from the Debye–Waller factors. So far much of the predicted behaviour has been observed but it is not clear yet that this model is the best one for all glasses and it still leaves some questions unanswered. Nevertheless, it is an example of a case where a model is precise enough in its predictions to be tested by the neutron experiments and further models already exist or are being developed. The original 'simple' incoherent scattering experiments on hydrogenous polymers, are as a result of this, enjoying a comeback as sensitive tests of the models.

SUGGESTED FURTHER READING

Beé (1988).
Higgins (1984).

10

Neutron reflection for studying surfaces and interfaces

10.1 INTRODUCTION

A recent addition to the family of neutron scattering techniques, and one of particular importance for polymeric samples is neutron reflection. This is a technique which explores the variation in composition normal to a surface reflecting the neutrons. In essence it belongs with the small angle scattering experiments, and not a few of the results already published have been obtained on spectrometers constructed for small angle scattering. However, both the formalism adopted to interpret the data and the conceptual design of the experiments are very different from those for small angle scattering, so reflection enjoys the privilege of a short chapter of its own. Although we will be relatively brief, the explosion of interest in the applications of neutron reflection to polymer surfaces and interfaces and to polymers adsorbed at surfaces and interfaces almost rivals the similar explosive growth which occurred when the small angle scattering technique began to be available in the seventies. Again, it is likely to be the usual inhibiting factors of restricted availability and access which limit the applications of the technique. In this case, though, we will see that the actual form of the samples is very important. They must be flat and, usually, thin.

Throughout this book the dual nature of the neutron as both a wave and as a particle has been a recurring theme. We have frequently shown how the interference of the neutron waves scattered by a sample gives rise to scattering patterns related to the sample structure. We have had cause often enough to remark on the analogies to light (or to X-ray) scattering phenomena. Up to this point, however, the scattering properties of the sample have been discussed in terms of scattering lengths, b, or cross sections, σ.

The neutron refractive index was introduced briefly in Chapter 2 during the discussion on neutron guides. When considering neutron reflection measurements it is this refractive index which is the important property of the reflecting sample and which allows us to pursue the analogy to the reflection of light. The neutron refractive index, is:

$$\nu = 1 - \lambda^2\left(\frac{\rho_b}{2\pi}\right) + \mathrm{i}\lambda\left(\frac{\rho_a}{4\pi}\right) \tag{10.1}$$

ρ_b and ρ_a are the scattering length density (i.e. the average scattering length per unit volume) and the absorption cross section density (with a similar definition) respectively. As can be seen, eqn (10.1) defines a complex quantity. However, for most polymeric materials of interest, ρ_a is negligibly small so that the last term in eqn (10.1) can be dropped. It is this refractive index and in particular its variation normal to a reflecting surface, which governs the reflection of neutrons by surfaces and interfaces. It is worth noting that since ρ_b appears in eqn (10.1), rather than ρ_b^2, the sign of ρ_b will be important. Because it is so fundamental we will give a brief derivation of the real part of ν based on the analogy with the refractive index for light.

Consider a plane wave travelling from left to right in the x-direction as shown in Fig. 10.1(a). The phase of this wave at any point along x is $\exp(-\mathrm{i}kx)$ where $k = 2\pi/\lambda$. In Fig. 10.1(a) a film of thickness e is

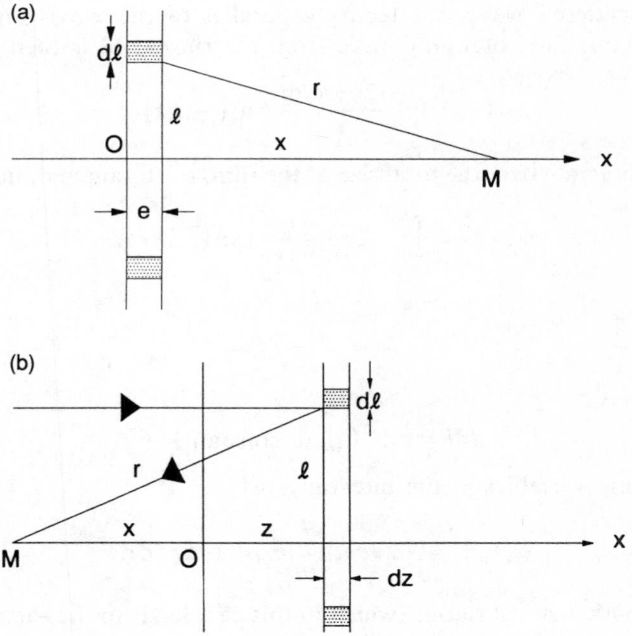

Fig. 10.1 (a) A plane wave impinging on a film of thickness e and scattering length density ρ_b and being transmitted. (b) The same situation, but the wave is reflected by the film.

introduced into the beam. Since the optical refractive index, ν, is the ratio of the velocities of light inside and outside the film respectively, the difference in phase introduced by the film (which is the difference in effective distance travelled) is $(\nu e - e)$ so that the phase of the wave after passing through the film is

$$\psi(x) = \exp(-\mathrm{i}k(x + (\nu - 1)e)). \tag{10.2}$$

We now have to make a calculation of the phase of the neutron wave passing through the film, in order to compare it with eqn (10.2).

Suppose the film is composed of neutrons of scattering length b, giving a scattering length density ρ_b and consider the phase of the neutron beam a long way from the film at the point M on the x-axis. The amplitude at M is the sum of the direct beam and the contributions scattered by the nuclei in the film. In Fig 10.1(a) a ring of width $\mathrm{d}l$ and radius l is drawn in the film. The volume of this ring is $2\pi le\mathrm{d}l$. Each nucleus within this element scatters an amplitude at M of $-(b/r)\exp(-\mathrm{i}kr)$ where r is the distance from the ring to M. The negative sign comes from the Fermi pseudopotential and the conventions used for the definition of scattering length in Chapter 4. If M is far enough from the film we can consider that this scattered wave is effectively parallel to the x-axis and to the incident beam. The total amplitude from the ring at M is then

$$\psi(x_{\mathrm{M}}) = \frac{-2\pi\rho_b e l\mathrm{d}l}{r}\exp(-\mathrm{i}kr) \tag{10.3}$$

and if we integrate over the total size of the film, the total amplitude at M is

$$\psi(x_{\mathrm{M}}) = \int_0^\infty -2\pi\rho_b e\frac{l\mathrm{d}l}{r}\exp(-\mathrm{i}kr). \tag{10.4}$$

Now

$$l^2 = r^2 - x_{\mathrm{M}}^2$$

so

$$l\mathrm{d}l = r\mathrm{d}x\ (x_{\mathrm{M}}\text{ is constant}) \tag{10.5}$$

and changing variables in the integral gives

$$\psi(x_{\mathrm{M}}) = -2\pi e\rho_b\int_{x_{\mathrm{M}}}^\infty \exp(-\mathrm{i}kr)\,\mathrm{d}r. \tag{10.6}$$

Ignoring mathematical rigour (which in this case leads to the same answer) we can simply integrate eqn (10.6) and obtain

$$\psi(x_{\mathrm{M}}) = -2\pi e\rho_b\left(\frac{-\mathrm{i}}{k}\right)\exp(-\mathrm{i}kx_{\mathrm{M}}). \tag{10.7}$$

We now have to add the intensity in the unscattered beam at M giving

$$\psi_{\text{total}}(x_{\text{M}}) = \exp(-\mathrm{i}kx_{\text{M}})\left[1 + \frac{2\pi \mathrm{i}e\rho_b}{k}\right]. \tag{10.8}$$

We aproximate the term in square brackets to an exponential (since $2\pi \mathrm{i}e\rho_b/k \ll 1$) to give

$$\psi_{\text{total}}(x_{\text{M}}) = \exp\left(-\mathrm{i}kx_{\text{M}} + \frac{2\pi \mathrm{i}e\rho_b}{k}\right) \tag{10.9}$$

$$= \exp\left(\mathrm{i}k\left\{x_{\text{M}} - \frac{2\pi\rho_b e}{k^2}\right\}\right)$$

and comparison with eqn (10.2) shows that

$$\nu = 1 - \frac{2\pi\rho_b}{k^2}$$

or, remembering $k = 2\pi/\lambda$

$$\nu = 1 - \frac{\lambda^2\rho_b}{2\pi}$$

as required.

When they reflect light, films of oil-on-water or coatings on lenses give rise to interference effects with which we are all familiar. The neutron reflection technique exploits the same phenomenon, but with the much smaller distance scale arising from the smaller neutron wavelength, and with the potential of manipulating ν by varying ρ_b through deuteration. Total reflection of slow neutrons by surfaces was first reported in 1946 by Fermi and Zinn (1946) and the phenomenon has been exploited for many years in the design and construction of neutron guides (see Chapter 2). The application to study of surfaces only really began in the early eighties, and the first dedicated spectrometers were built even more recently than that. We shall not present the theory in great detail—in any case many ideas for data interpretation are still being developed. We will describe the basic ideas, and show what information is potentially available from the experiments and illustrate this with a few examples.

10.2 NEUTRON SPECULAR REFLECTION

In this section we recall some simple laws of optics defining the critical angle and explore the differences between light and neutron reflection. Figure 10.2(a) shows an example of a wave passing from one medium to another. Since $\theta_1 < \theta_0$ it is clear from the Snell–Descartes law* (for refraction of light) that in this case $\nu_1 < \nu_0$. Now, examination of eqn (10.1) shows that, for materials not containing a large amount of hydrogen, ρ_b will be positive and ν will normally be less than unity. For a vacuum (and essentially for air) ν will be unity and therefore greater than ν of the sample, so that Fig. 10.2(a) applies to the general case of a neutron beam entering a sample from air. Again, from the Snell–Descartes law, there will be a critical angle of incidence, $\theta_0 = \theta_c$ when θ_1 becomes zero. This is defined by

$$\cos\theta_c = \nu. \tag{10.10}$$

Neutrons are totally externally reflected. Not only are most neutron refractive indices less than that for air (or vacuum) they are also very small. Typically, $1 - \nu$ is of the order of 10^{-6} so that θ_c is very small indeed ($\leq 1°$) and the total external reflection occurs at glancing angles of incidence. Since θ_c is small, a series expansion can be used for $\cos\theta_c$ in eqn (10.10), giving for small θ_c:

$$\theta_c/\lambda = (\rho_b/\pi)^{1/2}. \tag{10.11}$$

When $\theta < \theta_c$ there is still some penetration of the medium by an imaginary component with an exponentially decaying amplitude. This is called the evanescent wave and is well-known in optical experiments. This imaginary wave penetrates to a small depth determined by θ_c into the medium and in optical experiments it has been exploited by scattering it off, for example, fluorescent labelled molecules near the surface (Ausserre *et al.* 1985). There are pilot experiments afoot to exploit the evanescent neutron wave in surface diffraction studies, but the phenomenon is extremely low in intensity and looks unlikely to be very widely applicable until even more powerful neutron sources become available.

Finally, in this section, we ask why, in the heading, we have allowed an apparent tautology to appear in the use of the adjective 'specular'. This is normally used just to distinguish the specular beam from the scattering which occurs from the surface if it contains fluctuations in scattering length density. The specular beam occurs at a well defined angle following the laws of optics just discussed. Off-specular reflection is diffuse scattering away from this angle and arises from the surface fluctuations in flatness, in density or in composition. Its exploitation in terms of studying in-plane morphology of a surface is just beginning. It is normally very much weaker in intensity than the specular reflection and needs careful experimentation. Nevertheless, it is possible that surface scattering or non-specular reflection may become an important tool in the future.

* This book has been written by an Anglo-French team. Though each nationality is used to calling this law by the name of only one of these scientists, we prefer to remember both of them.

10.3 REFLECTIVITY PROFILES

10.3.1 General

The reflectivity R is defined as the reflected neutron intensity divided by the incident neutron intensity. For $\theta < \theta_c$, R is unity. For the interface between two bulk media R was calculated by Fresnel and the results can be demonstrated as follows.

At a sharp boundary between two media, as in Fig. 10.2(a), if the only variation in potential or scattering length is in the z-direction, we need consider only the perpendicular component of the wave, $\psi(z) = \exp(ikz\sin\theta)$. The wave must be *continuous* and smoothly varying across the interface, which requires that both $\psi(z)$ and $d\psi(z)/dz$ are equal for components on either side of the interface. We define the intensity, T, of the transmittal wave in the same way as the reflectivity. Clearly $R + T = 1$. The total amplitude of the perpendicular component in medium 0 is then the sum of the incident and reflected waves

$$\psi(z) = \exp(ik_0 \sin\theta_0 z) + R^{1/2} \exp(-ik_0 \sin\theta_0 z)$$

and of the perpendicular component of the wave in medium 1 is,

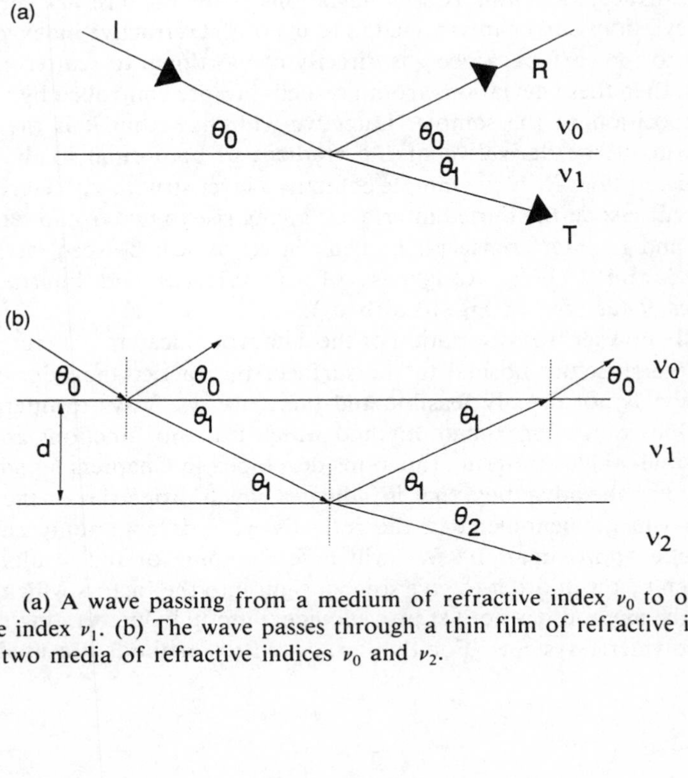

Fig. 10.2 (a) A wave passing from a medium of refractive index ν_0 to one with refractive index ν_1. (b) The wave passes through a thin film of refractive index ν_1 between two media of refractive indices ν_0 and ν_2.

$$\psi(z) = T^{1/2} \exp(\mathrm{i}k_1 \sin\theta_1 z)$$

equating the values for $\psi(z)$ and those for $\mathrm{d}\psi(z)/\mathrm{d}z$ for the transmitted and reflected components at the boundary and solving for R gives

$$R = \left| \frac{k_0 \sin\theta_0 - k_1 \sin\theta_1}{k_0 \sin\theta_0 + k_1 \sin\theta_1} \right|^2 .$$

Now since $k = 2\pi/\lambda = mv/\hbar$, we have $k_0/k_1 = v_0/v_1 = \nu_0/\nu_1$ so that an alternative form for R is:

$$R = \left| \frac{\nu_0 \sin\theta_0 - \nu_1 \sin\theta_1}{\nu_0 \sin\theta_0 + \nu_1 \sin\theta_1} \right|^2 \tag{10.12}$$

and its shape as a function of q can be seen in Fig. 10.3.

Remembering that ν is a function of λ so that R is a function of both θ_0 and λ, the reflectivity profile can be scanned by varying θ_0, as in a conventional spectrometer, or by varying λ. As described in Chapter 3 (Section 3.3), the neutron pulsed sourced spectrometers, which vary λ at fixed θ have a number of advantages over the spectrometers which use fixed λ and vary θ. The basic theory is, of course, the same in either case.

If there is any variation in ρ_b and hence in the refractive index, normal to the surface, then this causes deviations from the Fresnel curve, and these deviations can be interpreted in terms of the refractive index variation normal to the surface. Since ν is directly proportional to scattering length density, then these deviations from Fresnel's law are controlled by variation in composition of the sample. Later we will show that it is the Fourier transform of the derivative of the gradient of ρ_b normal to the surface which determines R. If the sample contains a layer structure, then reflection may occur also at the buried interfaces, giving rise to two or more reflected beams and to interference phenomena in R, as will be seen later on, in Figs 10.5 and 10.6(d). Roughness of the surfaces and interfaces also modifies R (as seen in Fig. 10.6 (b–d)).

Ideally Fourier transformation of the data would lead to a direct measure of sample structure normal to the surface. In practice the Fourier transformation is not usually feasible and there are two ways to interpret the data. One is an approximate method which leads to functions analogous to the small angle scattering functions developed in Chapters 6 and 7. This method has the advantage that it indicates which properties of the surface lead to which phenomena in the reflectivity. It is commonly called the 'kinematic approximation'. We will describe some of the results of this approach below in order to give some insight into the factors which control R, but the method has not yet proved widely useful in interpreting real data from polymeric systems. For this, exact matrix methods are used, which

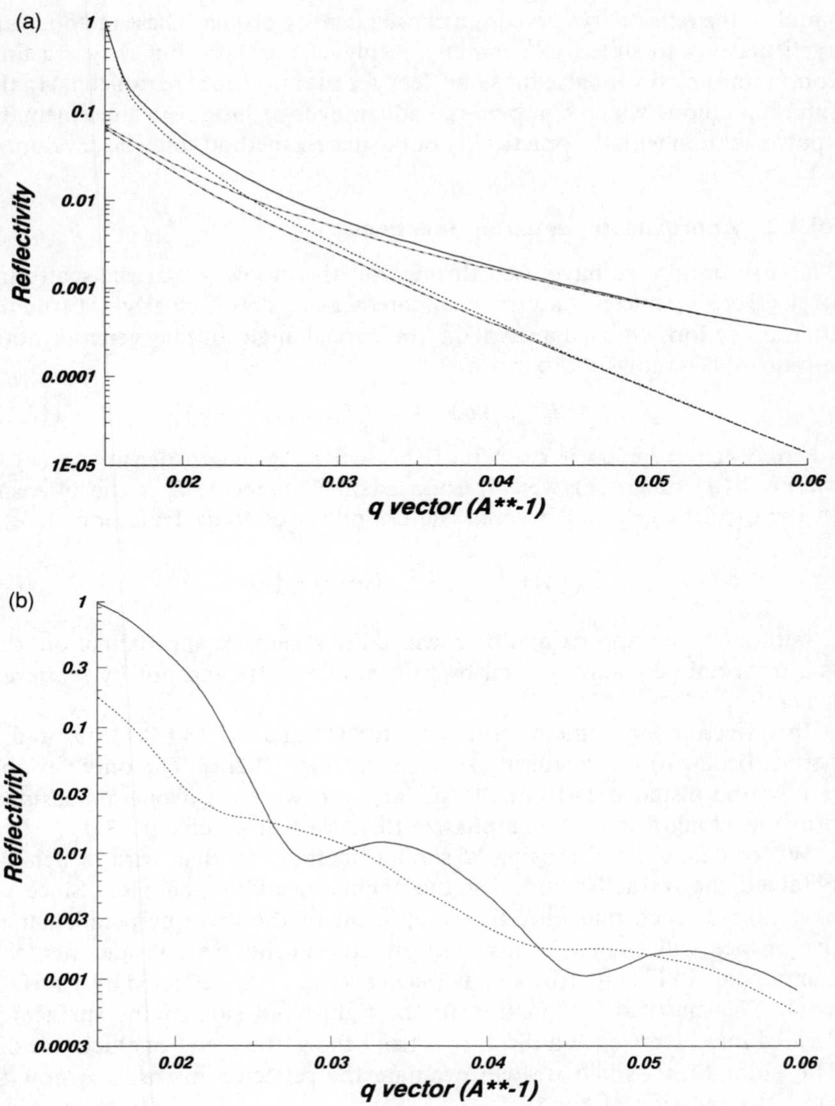

Fig. 10.3 Comparison of exact calculations (optical matrix method) and kinematic approximation for reflectivity of several different surfaces. (a) The Fresnel curve for a sharp interface: — exact; —·— kinematic approximation. The Fresnel curve modified by a diffuse interface with $\sigma = 30$ Å: — — — exact; —·— kinematic approximation. (b) A thin film of thickness 300 Å and scattering length density 0.7×10^{-5} Å^{-2} on a substrate with $\rho_b = 0.05\ 10^{-5}$ Å^{-2}. — exact; — — — kinematic approximation.

simulate the reflectivity curve for a chosen density profile. These methods are particularly well suited to computer fitting of the data, but they are time consuming and do not give the same 'feel' for the important parameters. In the future, methods which combine the advantages of insight in the kinematic approximation with the practicality of the matrix methods may be developed.

10.3.2 Approximate scattering functions

The assumption we have used throughout this book is that the scattering of neutrons by nuclei is a very weak interaction. This is clearly not true for total reflection, but well away from the critical angle an easy generalization of eqn (4.49) allows us to write.

$$I(\boldsymbol{q}) = I_0\langle|\hat{\rho}_b(\boldsymbol{q})|^2\rangle = I_0\langle\hat{\rho}_b(\boldsymbol{q})\hat{\rho}_b(-\boldsymbol{q})\rangle \tag{10.13}$$

where $\hat{\rho}_b(\boldsymbol{q}) = b\hat{n}(\boldsymbol{q})$ is the FT of the scattering length density, $\rho_b(\boldsymbol{r}) = bn(\boldsymbol{r})$. $\hat{n}(\boldsymbol{q})$ and $n(\boldsymbol{r})$ were introduced in Chapter 4. I_0 is the intensity in the incident beam. We recall the definition of $\hat{n}(q)$ from eqn (4.48).

$$\hat{\rho}_b(\boldsymbol{q}) = \int_{-\infty}^{\infty} \exp - (\mathrm{i}\boldsymbol{q}\cdot\boldsymbol{r})\,\rho_b(\boldsymbol{r})\,\mathrm{d}\boldsymbol{r}.$$

Although, this approximation is called the kinematic approximation, this is a term only commonly used by reflection experts and not by scatterers generally.

In reflection experiments, the $\boldsymbol{q}$ vector is equal to $(4\pi/\lambda)\sin\theta_0$ and is perpendicular to the scattering surface; see Fig. 10.2(a). The only variable is now the distance z, from the surface and we have a one-dimensional problem along this q. To emphasize this, we will specify q as $q_\perp$.

We calculate $R(q_\perp)$ using a similar method to that with which we obtained the refractive index at the beginning of this chapter. Since we have already seen that only the component of the wave perpendicular to the surface will change in a reflection experiment, we consider just this component. In Fig. 10.1(b) we have such a wave being reflected by a surface at O. The material behind (i.e. to the right-hand side of the surface) is divided into layers each a distance z from the surface and of thickness $\mathrm{d}z$. The point M at which we will evaluate the reflected intensity is now to the left-hand side of the surface.

If the origin of phases is at O, then the path of the wave to M is $x_\mathrm{M} + 2z$ along the axis and (provided M is far from the surface) this is also true for any point off the axis. We approximate $r = x_\mathrm{M} + z$.

The amplitude at M of this perpendicular component is

$$\psi(x_\mathrm{M}) = -\frac{b}{r}\exp(-\mathrm{i}k(2z + x_\mathrm{M})). \tag{10.14}$$

We again divide each layer into rings of radius l and width $\mathrm{d}l$, so that the number of scatterers within the ring is $2\pi l \mathrm{d} l \mathrm{d}z$. Then for each layer the reflected amplitude at M is

$$\psi(x_{\mathrm{M}}) = \int_0^\infty 2\pi\rho_b \mathrm{d}z \frac{l\mathrm{d}l}{r} \exp - [\mathrm{i}k(r+z)].$$

Changing variables from l to r as in eqn (10.5) and making the lower integration limit $x_{\mathrm{M}} + z$ leads to

$$\psi(x_{\mathrm{M}}) = \frac{-\mathrm{i}}{k} 2\pi\rho_b \mathrm{d}z \exp(-\mathrm{i}k(x_{\mathrm{M}} + 2z)). \tag{10.15}$$

If we assume the material in the surface has uniform density ρ_b, then we can integrate over all z to give

$$\psi_{\text{total}}(x_{\mathrm{M}}) = \frac{-2\pi\rho_b \mathrm{i}}{k} \int_0^\infty \exp(-\mathrm{i}k(x + 2z))\,\mathrm{d}z. \tag{10.16}$$

As before, we substitute $u = x_{\mathrm{M}} + 2z$, to obtain

$$\psi_{\text{total}}(x_{\mathrm{M}}) = \frac{-2\pi\rho_b \mathrm{i}}{R} \int_{x_{\mathrm{M}}}^\infty \exp(\mathrm{i}ku) \frac{\mathrm{d}u}{2} \tag{10.17}$$

and

$$\psi_{\text{total}}(x_{\mathrm{M}}) = \frac{\pi\rho_b}{k^2} \exp(-\mathrm{i}kx_{\mathrm{M}}). \tag{10.18}$$

The intensity is just the complex square of this amplitude $= \pi^2\rho_b^2/k^4$. Now since this was just the perpendicular component of the wave, k should be the perpendicular component and $q_\perp = 2k$. The intensity of the incident wave was assumed to be unity and the reflectivity is the ratio of the reflected to incident intensities giving

$$R(q_\perp) = \frac{16\pi^2\rho_b^2}{q_\perp^4} \tag{10.19}$$

which is just the Porod law for a sharp interface and the Fresnel law away from the critical edge. We should now compare this approximate curve to the exact Fresnel calculation R_{F} (eqn (10.12)). Let us consider two extremes. Firstly, far away from the critical edge θ_0 and θ_1 become large (we approach normal incidence) and, since $\nu \simeq 1$, $\sin\theta_0 \simeq \sin\theta_1$ and $R_{\mathrm{F}}^{1/2} \approx (1-\nu)/(1+\nu) \approx (1-\nu)/2$. Now, we have $\nu = 1 - 2\pi\rho_b/k^2$ and $q_\perp = 2k$ (for normal incidence), leading immediately to $R_{\mathrm{F}}(q_\perp) \approx 16\pi^2\rho_b^2/q_\perp$ which is eqn (10.19). Suppose, however, we approach the critical angle, $\theta_1 = 0$, where eqn (10.12) gives us $R_{\mathrm{F}} = 1$. The kinematic approximation always fails and gives values of R which are too small.

This is shown in Fig. 10.3(a) where the approximation is compared to the exact calculation of the Fresnel curve. Away from the critical edge, the kinematic approximation has been used successfully to interpret reflection data and may have advantages in terms of speed and efficiency.

We now consider the case when ρ_b is not constant and the scattering length density ρ_b is a function, $\rho_b(z)$ of z. We then obtain

$$\psi_{\text{total}}(x_M) = \frac{-2\pi \mathrm{i}}{k} \exp(-\mathrm{i}kx) \int_0^\infty \rho_b(z) \exp(-\mathrm{i}2kz)\, \mathrm{d}z.$$

Since $\rho_b(z)$ is always zero for $z < 0$ we can extend the integral to $-\infty$ and remembering $2k = q_\perp$, the integral becomes the Fourier transform of $\rho_b(z)$

$$\hat{\rho}(q_\perp) = \int_{-\infty}^{\infty} \exp(-\mathrm{i}q_\perp z)\, \rho_b(z)\, \mathrm{d}z. \tag{10.20}$$

The intensity is then the complex square of $\psi_{\text{total}}(x_M)$

$$I(q_\perp) = \frac{16\pi^2}{q_\perp^2} \left| \hat{\rho}_b(q_\perp) \right|^2 \tag{10.21}$$

and the reflectivity

$$R(q_\perp) = I(q_\perp)/I_0 = \frac{16\pi^2}{q_\perp^2} \left| \hat{\rho}_b(q_\perp) \right|^2. \tag{10.22}$$

In Appendix 1 we show that the Fourier transform of a derivative of a function $f(r)$ is just iq times the FT of $f(r)$. This leads to an equivalent form for $R(q_\perp)$

$$R(q_\perp) = \frac{16\pi^2}{q_\perp^4} \left| \hat{\rho}'_b(q_\perp) \right|^2 \tag{10.23}$$

where

$$\hat{\rho}'_b(q_\perp) = \int_{-\infty}^{\infty} \exp(\mathrm{i}q_\perp z) \frac{\mathrm{d}\rho_b(z)}{\mathrm{d}z}\, \mathrm{d}z.$$

We see that $\hat{\rho}'_b(q_\perp)$ (and hence $R(q_\perp)$) is governed by the Fourier transform of the concentration gradient normal to the surface.

To illustrate the sensitivity of $R(q_\perp)$ to features in $\rho(z)$ equations, eqn (10.22) or (10.23) can be applied to three types of surface or interface.

If the interface is a sharp step between materials with different scattering length densities ρ_b^1 and ρ_b^0 then the derivative $\mathrm{d}\rho/\mathrm{d}z$ becomes $(\rho_b^1 - \rho_b^0)$ $\delta(z)$ where $\delta(z)$ is again a Dirac delta function. Substitution into eqn (10.23) gives (defining $\Delta\rho_b = \rho_b^1 - \rho_b^0$)

$$R(q_\perp) = \frac{16\pi^2}{q_\perp^4}\,\Delta\rho_b^2\left\{\int_{-\infty}^{\infty}\delta(z)\exp(-\mathrm{i}q_\perp z)\,\mathrm{d}z\right\}^2 \tag{10.24}$$

and remembering that the FT of a delta function is 1

$$R(q_\perp) = \frac{16\pi^2}{q^4}\Delta\rho_b^2 \tag{10.24'}$$

i.e. we recover eqn (10.19).

Clearly the larger the change in ρ_b at the interface, the stronger is the reflectivity. If the interface is not sharp, then the $\delta(z)$ has to be replaced by a mathematical function describing its shape. Suppose the interface is diffuse and described by the error function

$$\rho_b(z) = \frac{\Delta\rho_b}{\sigma\sqrt{2\pi}}\int_{-\infty}^{z}\exp\left(\frac{z^2}{2\sigma^2}\right)\mathrm{d}z$$

where σ is the standard derivation of the Gaussian function describing the concentration *gradient*, i.e.

$$\frac{\mathrm{d}\rho_b}{\mathrm{d}z} = \frac{\Delta\rho_b}{\sigma\sqrt{2\pi}}\exp-\left(\frac{z^2}{2\sigma^2}\right) \tag{10.25}$$

and $z = \pm\sqrt{2}\sigma$ describes the points where $\mathrm{d}\rho_b(z)/\mathrm{d}z$ drops to $1/e$ of its value at $z = 0$.

Since we know that the Fourier transform of a Gaussian function, such as that in eqn (10.25), is another Gaussian (see Appendix 1) when we substitute eqn (10.25) in eqn (10.23) we obtain

$$\hat{\rho}_b'(q_\perp) = \Delta\rho_b\exp\left(\frac{-q_\perp^2\sigma^2}{2}\right)$$

and hence

$$R = \frac{16\pi^2}{q_\perp^4}\Delta\rho_b^2\exp(-q_\perp^2\sigma^2) = R_{\text{step}}\exp(-q_\perp^2\sigma^2) \tag{10.26}$$

where R_{step} is defined by eqn (10.24′).

Equation (10.26) shows that the reflectivity of a diffuse interface falls away *below* that for a sharp interface and the fall off is steeper for broader interfaces, as usual when we make a Fourier transform. The approximate calculation in eqn (10.26) is compared to the exact matrix calculation in Fig. 10.3(a). It is clear from the Fig. 10.3(a) that detailed information about the interfacial profile is best obtained at the low reflectivity end of the Fresnel curve.

Finally in this section, we consider a thin layer of material of scattering length ρ_b^1 deposited on a substrate of scattering length ρ_b^2. If a layer is of thickness e and the scattering length density of the outer medium in ρ_b^0 then we have two steps $\Delta\rho_b(=\rho_b^1-\rho_b^0)$ and $\Delta\rho_b^1(=\rho_b^2-\rho_b^1)$

$$\frac{\mathrm{d}\rho}{\mathrm{d}z}(z) = \Delta\rho_b\delta(z) + \Delta\rho_b^1\delta(z-e). \tag{10.27}$$

Substitution of this into eqn (10.24) and integration leads to

$$\hat{\rho}_b'(q_\perp) = \Delta\rho_b + \Delta\rho_b^1 \exp(-\mathrm{i}q_\perp e) \tag{10.28}$$

and taking the complex square

$$R = \frac{16\pi^2}{q_\perp^4}\left[\Delta\rho_b^2 + \Delta\rho_b'^2 + \Delta\rho_b\Delta\rho_b'\{\exp(\mathrm{i}q_\perp e) + \exp(-\mathrm{i}q_\perp e)\}\right]$$

$$= \frac{16\pi^2}{q_\perp^4}\{\Delta\rho_b^2 + \Delta\rho_b'^2 + 2\Delta\rho_b\Delta\rho_b'\cos(q_\perp e)\}. \tag{10.29}$$

We see, as expected, interference fringes arising from interference between the waves reflected at the two interfaces. Equation (10.29) is compared to an exact calculation in Fig. 10.3(b). It is clear that the kinematic approximation is very poor in this case and in fact only in rare cases would it be used.

These three simple examples have shown how analytical expressions for the reflectivity are related to the structure of the reflecting samples via Fourier transformation. Unfortunately, as we have seen in Figs 10.3(a) and (b) the kinematic approximation breaks down for total reflection, so it cannot describe the scattering near to the critical angle. For this reason it has not been widely applied to analysis of reflection profiles, and by far the most popular method is the matrix method to be described in the next section. Recently Sun *et al.* (1988) have been working on the problem of describing the shape of the reflectivity curve in the neighbourhood of the critical edge in an attempt to verify scaling laws. Their work is still in progress but has not yet been used to characterize polymeric systems. Tidswell *et al.* (1990) have developed a generalization of eqn (10.29) for several layers which takes account of refraction and absorption at the interfaces. In a recent report (Eastoe and Penfold 1992) it has been shown that this can provide a reasonable approximation for thin layers (<100 Å) and is much more rapid and simple than the optical matrix method to be described in the next section. Recently, Henderson *et al.* (1993) have shown that a combination of the kinematic approximation with contrast variation, when analysing data from polymers adsorbed at a liquid surface, allowed them to dipense with an initial model, thus removing an objection frequently raised to the use of the optical matrix method. Great care is required however if the polymer is highly extended into the subphase as the reflectivity may not be sensitive enough to the tails.

10.3.3 Multilayer optical method for calculating reflectivity

This method is an exact way of calculating the reflection of a sequence of sharp interfaces (or discrete layers). If the boundaries are not sharp, then a good approximation is achieved by dividing the interface into a sequence of steps (i.e. turning the scattering length density profile into a histogram).

The expression for scattering from a film of thickness d, with sharp boundaries and constant scattering length, resting on a substrate (as in Fig. 10.2(b)) can be obtained, as before, by matching both ψ and $d\psi/dz$ across the interfaces. There are now three scattering length densities and two interfaces and by a generalization of eqn (10.12) the reflectivity is

$$R = \left| \frac{r_{01} + r_{12} \exp(2i\beta)}{1 + r_{01} r_{12} \exp(2i\beta)} \right|^2 \tag{10.30}$$

where r_{ij} is defined as

$$\frac{\nu_i \sin\theta_i - \nu_j \sin\theta_j}{\nu_i \sin\theta_i + \nu_j \sin\theta_j}$$

$$\beta = (2\pi/\lambda)\, \nu_1 \mathrm{d}_1 \sin\theta.$$

Examination of eqn (10.21) and Fig. 10.2 shows that β is the optical path length within the film and the exponential term takes care of the diminishing intensity of the multiply reflected waves. The exact calculation can be extended to structures containing several layers if necessary. If the surfaces are rough or the interfaces not sharp, but diffuse, the effect on $\rho_b(z)$ is identical and it is impossible to distinguish between the two cases. In either case, the reflectivity is modulated by an exponential decay (as shown in eqn (10.26)) in the kinematic approximation). When using the optical multilayer method it is usual to multiply r_{ij} for each interface by the term $\exp(q_i q_j \langle\sigma\rangle^2/2)$ where q_i is $4\pi\nu_i/\sin\theta_i$ and $\langle\sigma\rangle$ is the root mean square of a Gaussian curve approximating to the roughness or diffuseness profile of the interface.

A complex $\rho_b(z)$ profile can be approximated by a histogram or sequence of layers. Each layer adds another reflected wave, all of which have to be added together, taking account of their relative phases. R is then the complex square of this total amplitude. A general solution, very suitable for computation of the reflectivity may then be adopted. In this method, which relies on the condition that the wave function of the neutron, and its gradient, is continuous across each boundary, each layer has a characteristic matrix M_i.

$$M_i = \begin{bmatrix} \cos\beta_i & -(i/\kappa_i)\sin\beta_i \\ -i\,\kappa_i \sin\beta_i & \cos\beta_i \end{bmatrix} \tag{10.31}$$

where $\kappa_i = \nu_i \sin\theta_i$.

The reflectivity is then obtained from the product of these matrices

$$M = [M_1]\,[M_2]\,[M_3]\,\ldots\,[M_n]. \tag{10.32}$$

The result is a 2 × 2 matrix

$$M = \begin{bmatrix} M_{11} & M_{12} \\ M_{21} & M_{22} \end{bmatrix}$$

and the reflectivity is given by

$$R = \left| \frac{(M_{11} + M_{12}\,\kappa_s)\,\kappa_a - (M_{21} + M_{22})\,\kappa_s}{(M_{11} + M_{12}\,\kappa_s)\,\kappa_a + (M_{21} + M_{22})\,\kappa_s} \right|^2 \tag{10.33}$$

where the subscript a refers to the outer (air) medium and s to the final (substrate) medium.

As we have said, this method is very well suited to modelling the reflection from samples with complex layer structures and interfaces. It does not of course allow inversion of the data to obtain $\rho_b(z)$. This means that it is very important to have as much supporting information about the structure as possible. Although experimenters outside the field continually query the uniqueness of profiles obtained by fitting data in this way, in practice there is always enough external evidence to rule out most alternatives.

10.4 A MODEL REFLECTION EXPERIMENT

The matrix formalism which is translated into computer codes for fitting reflectivity data also makes easy the demonstration of the effects of various surface properties. In Fig. 10.4 (a–f) we demonstrate *separately* the effects of surface roughness and of experimental resolution on the model scattering from a substrate and from a film with sharp interfaces.

The instrumental resolution is compounded of the spread in scattering angle with that in wavelength. The former is usually the most important because it also includes the effect of non-flatness of the reflecting surface. The perfect Fresnel curve from a flat glass surface in Fig. 10.4(a) is modified by a 'wavy' glass surface in Fig. 10.4(b). The main effect is to 'blur' the critical edge. In Fig. 10.4(c), the surface is again flat but with a high frequency roughness imposed (as in eqn (10.26)). For a Gaussian roughness with mean square deviation of 30 Å the Fresnel curve drops more steeply at higher q. In Fig. 10.4(d) two examples are shown where a thin layer has been superimposed on the flat, smooth substrate. The interference pattern predicted in eqn (10.29) arising from reflection at the two interfaces is clearly seen. The frequency is determined by the layer thickness. In Fig. 10.4(e), the effect of resolution or of a wavy surface on this interference pattern is shown and, finally, in Fig. 10.4(f) the effect of rough interfaces. It is interesting to note that roughness of the interfaces damps the amplitude of the interference fringes while roughness of the surface lowers the intensity at high q.

A number of pointers to good experimental design can be picked out from these simulated results. Firstly, it is of paramount importance that the sample is as flat as possible (always assuming good instrumental resolution). For liquid samples this means suppressing vibrations. For solid samples, the substrates are usually optical flats or polished silicon discs and the films are spin-cast from solution onto the substrate with or without subsequent annealing. Secondly, the effect of roughness was most noticeable in Fig. 10.4(c) at higher q. Generally, interfacial effects are better observed at higher q values. Noting the q^{-4} dependence in eqn (10.22) and the large variation in the vertical scale in Fig. 10.4 it is clear that high neutron intensity and adequate counting times to obtain good statistics at high q are vital if reliable information about interfaces is to be extracted from the data.

Finally, we noted in Fig. 10.4(f) that the effect on the reflectivity profile of equivalent roughness at different interfaces was not the same. This result arises because the strength of the reflection at any interface depends on $\Delta\rho_b$, the change in scattering length. The layer structure used for simulation had a deuterated material on top of a hydrogenous one. Thus, the strongest reflection is at the air surface and a much weaker one comes from the polymer interface. By choosing where to put the deuterium in such a sample, different aspects of the structure may be highlighted.

10.5 EXAMPLES OF REFLECTION DATA FROM POLYMERIC SAMPLES

10.5.1 Polymers adsorbed at a liquid interface

There are two types of experiment in this category. In one the polymer adsorbs to a solid surface under a liquid phase, in the other the polymer segregates at the air–liquid interface of a solution. In each case, while the design of the experiment may differ, the desired quantity is the segment density profile normal to the surface. For a number of organic solvents the hydrogenous species has a net negative scattering length density, while that of the deuterated species is positive. A mixture can be made therefore with zero scattering length density and $\nu = 1$. Such a liquid is a null reflecting liquid, and any reflection comes from the adsorbed polymer

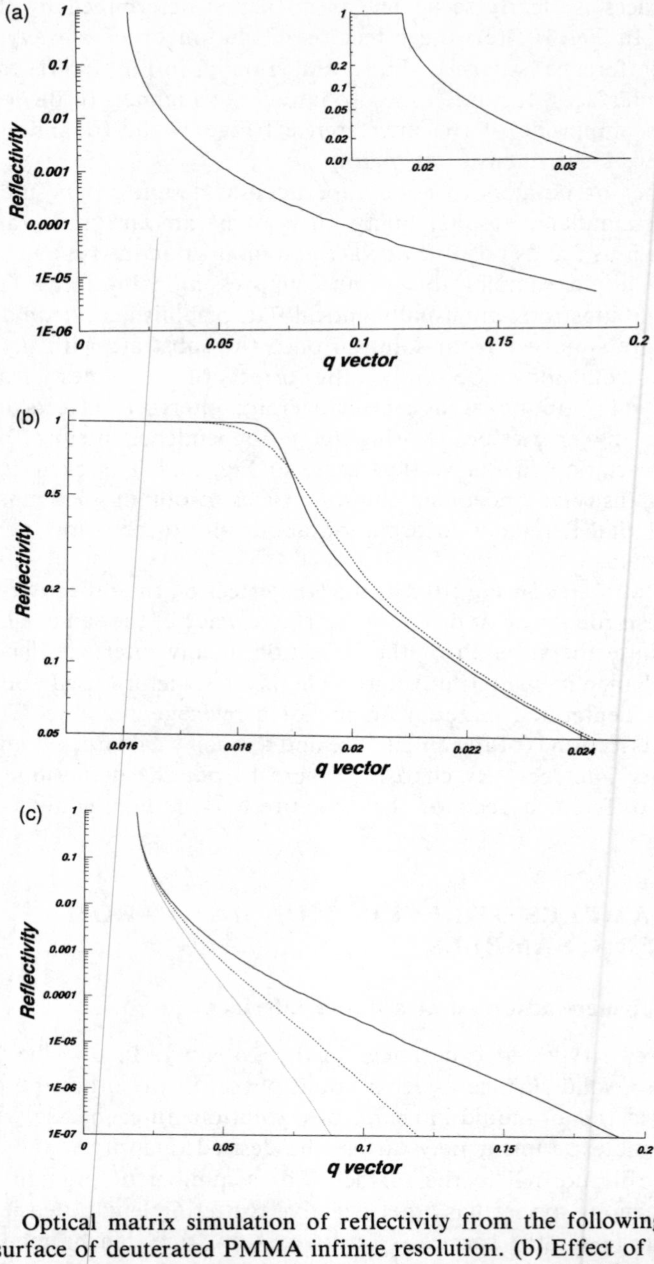

Fig. 10.4 Optical matrix simulation of reflectivity from the following surfaces. (a) Flat surface of deuterated PMMA infinite resolution. (b) Effect of finite resolution on (a): $\Delta\theta/\theta = 3$ per cent (—), 8 per cent (– – –). (c) Effect of surface roughness on (a): $\sigma =$ (—) 10 Å, (– –) 20 Å, (. . . .) 30 Å. (d) A thin film of thickness 400 Å (– – –), 600 Å (—) of D-PMMA superimposed on a substrate

(d)

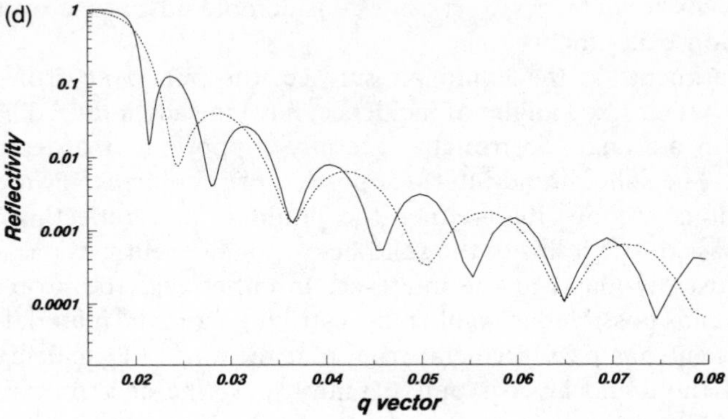

(e)

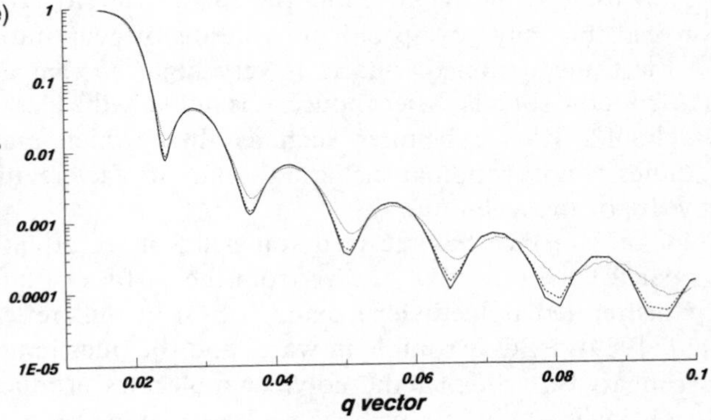

(f)

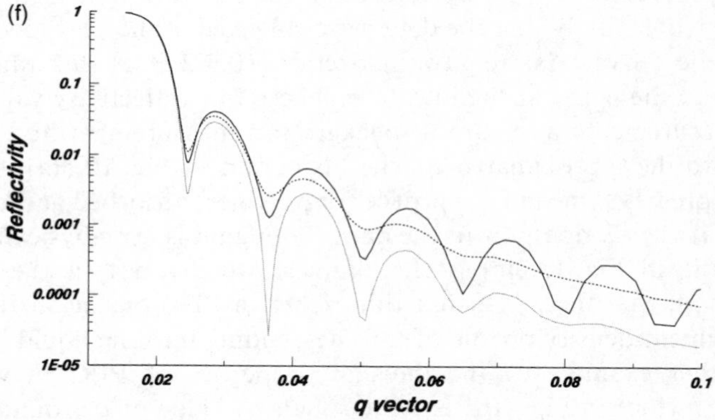

(quartz). (e) The effect of resolution on reflection from a thin film $\Delta\theta/\theta = 0$ (—), $= 3$ per cent (– – –) and 8 per cent (....). (f) The effect on rough surfaces on the reflection from a thin film. – ($\sigma_{01} = \sigma_{12} = 0$). – – – ($\sigma_{01} = 0$, $\sigma_{12} = 30$ Å) .... ($\sigma_{01} = 30$ Å, $\sigma_{12} = 0$).

layer. This gives neutron reflection a considerable advantage over X-rays for reflection experiments.

For experiments at the liquid–air surface, the pulsed neutron spectrometers, with their fixed angles of incidence, have the advantage. The sample is usually in a Langmuir trough, carefully supported in order to avoid vibrations. The solid–liquid interface poses more problems. Two different configurations are possible—either the liquid is null reflecting and the neutrons pass through this to the solid below, or the neutrons pass *through* the solid (usually glass) to the interface. In either case the problem is to reduce as far as possible the incoherent scattering from the material through which the neutrons pass. Remembering that the angle of incidence is only 1°, even if the liquid layer is only 0.1 mm thick, the neutron path length would be 5.7 mm, and the transmission of this thickness of hydrogenous material is less than 10 per cent. Thus the surface liquid layer needs to be very thin and this may pose problems in terms of evaporation. On the other hand the transmission of quartz is very high, so that approach to the interface from the solid is easier though it is unlikely the glass could be as thin as the liquid. It is difficulties such as these which made neutron reflection studies of adsorption at the liquid–solid interface relatively rare in the earlier years of the technique.

The liquid–air interface has received somewhat more attention. Figure 10.5 shows part of the reflectivity curve from the surface of a 0.1 per cent solution of deuterated polyethylene oxide (PEO) in null reflecting water (Rennie *et al.* 1989). PEO is soluble in water and the question investigated in the experiments was whether the polymer molecules produce an excess or a depletion layer at the air–water interface. Using the optical matrix method, values of $R(q)$ were calculated for two different segment density profiles. Note, firstly that the data were collected at the very low reflectivty end of the curve. As we saw in Section 10.3.2 it is this range that is sensitive to the $\rho_b(z)$ at the interface. These low reflectivity values require long measurements and careful background measurement and removal.

The profile approximated by the histogram in Fig. 10.5(a) is the theoretically predicted mean field profile for polymers adsorbed at the air–liquid surface. It clearly does not fit the data. Though it is not physically realistic, the profile in Fig. 10.5(c) is the simplest which does fit the data. The implication, that there is a thin film richer in PEO on the surface, with a weak segment density profile of tails descending into the liquid is probably yet another example of the unusual properties of PEO in water. The polymer surface in Fig. 10.5 is quite rough. A value of σ around 20 Å was needed to fit the data, as compared to 3 Å for the water alone. Twenty angstroms is about the value of the statistical segment length for the polymer, so it is possible that the surface roughness is determined by the local stiffness of the polymer molecules.

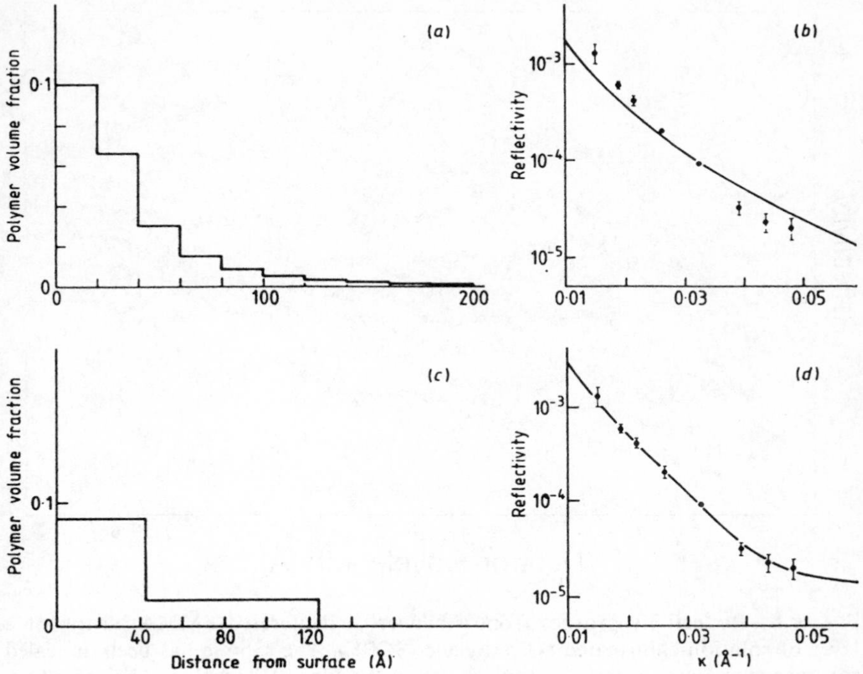

Fig. 10.5 Observed and calculated reflectivity profiles from the air–solution interface of a 0.1 per cent by weight solution of deuterated polyethylene oxide (M_n = 19 600) in null reflecting water (background substracted). (b) Profile calculated for the mean field distribution shown in (a). (d) Profile calculated for the distribution shown in (c). Reprinted with permission from Rennie *et al.* Copyright 1989, American Chemical Society.

10.5.2 Polymer–polymer interdiffusion

This is an area which has received considerable attention in the period since the reflectivity profile technique became routinely available. On the one hand, if the two polymers are of the same species, differing only in that one of them is deuterated, observation of the developing interfacial region as interdiffusion occurs can be used to test current theories of polymer motion. On the other hand, if the two species differ chemically the thermodynamics of polymer–polymer miscibility determines the equilibrium interface and moreover, there are potential applications to the study of polymer–polymer bonding. In order to demonstrate some of the possibilities, we pick an example from our own work (Fernández *et al.*

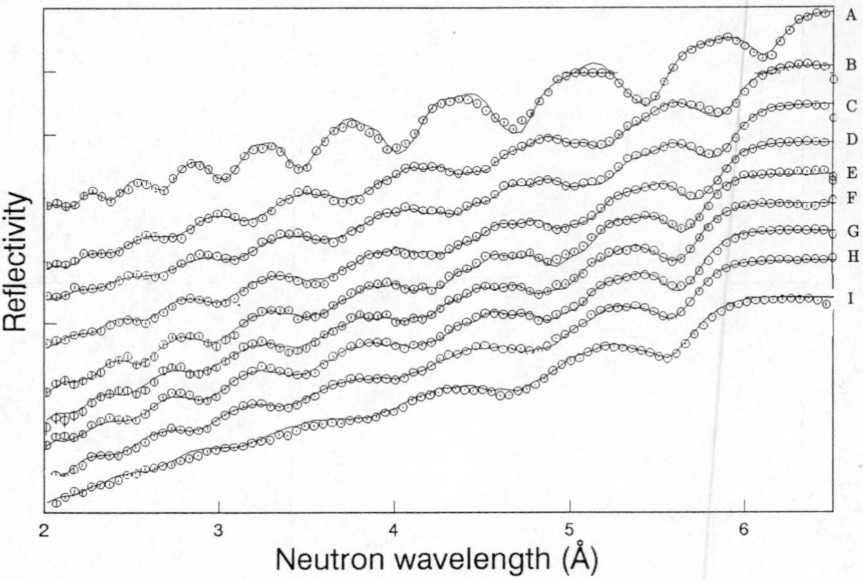

Fig. 10.6 Reflectivity profiles from a bilayer of deuterated PMMA on top of a layer of solution-chlorinated polyethylene (SCPE). The sample has been annealed for the accumulated lengths of time shown in the Fig. 10.7 caption. The error bars enclosed by the circles represent the data points while the solid lines represent the best fits achieved using the interfacial profiles in Fig. 10.7. The data are displaced vertically relative to each other for display purposes. (Fernández *et al.* (1991).)

1991). The data in Fig. 10.6 are the reflectivity profiles from a bilayer of deuterated polymethylmethacrylate (D-PMMA) on solution-chlorinated polyethylene (SCPE). These polymers are partially miscible. The lower SCPE film is $\sim 3.5 \times 10^3$ Å thick and was spin cast from solution directly onto the glass substrate. This SCPE layer gives rise to the high frequency oscillation just visible in the data. The D-PMMA layer was spin-cast separately and then floated on to water and finally picked up on top of the SCPE layer. The D-PMMA layer is 1180 Å thick and because it has the largest scattering length density, gives rise to the dominant periodicity in the data. This sample has then been annealed at 120°C for increasing periods of time and the reflectivity profile measured after each annealing period. The data in Fig. 10.6 correspond to accumulated annealing times of 0(A) to 5217(I) min. The data were fitted using the optical matrix method. (The fits are the solid lines through the data.) As well as the thicknesses and scattering length densities of the two films, the resolution $\Delta\theta$, and the surface and interface roughness, it was necessary to include

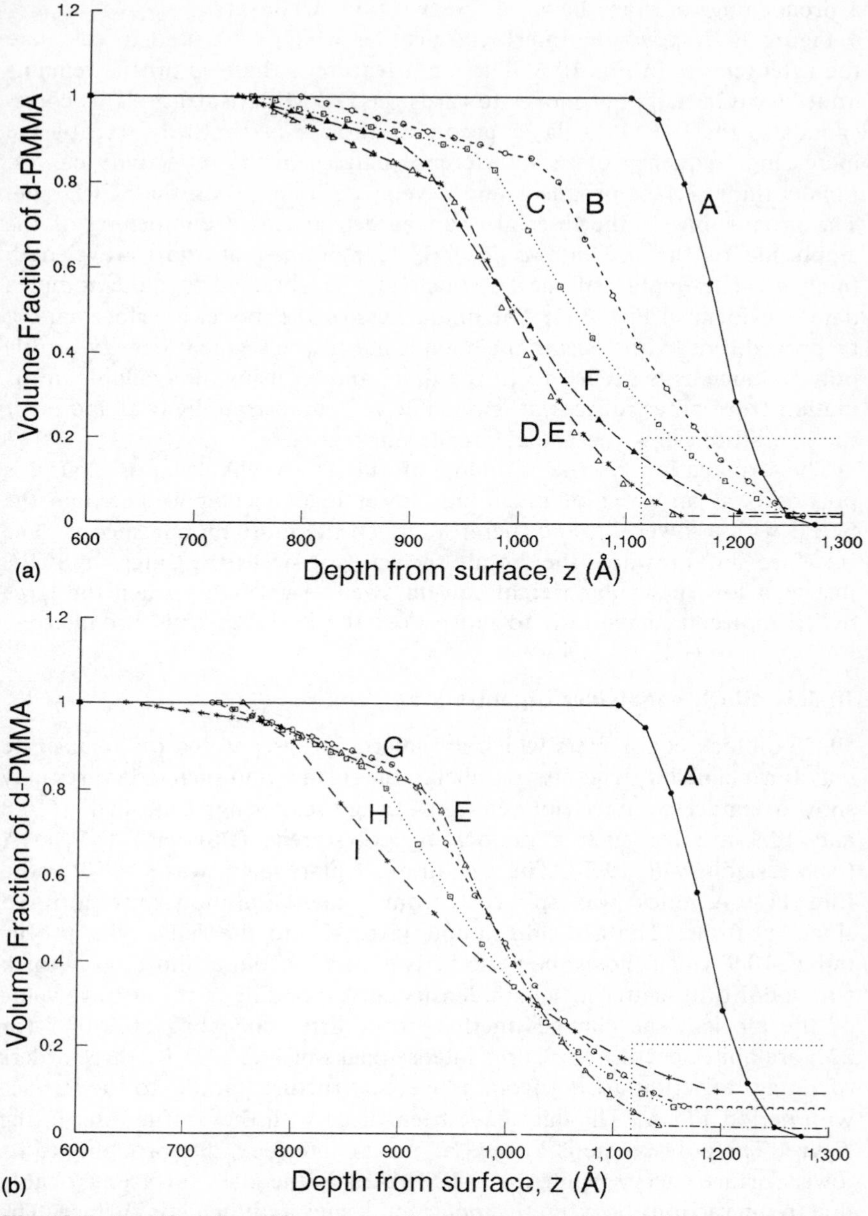

Fig. 10.7 Interfacial profiles used to fit experimental data in Fig. 10.7. (a) Early stages of annealing at 120°C. (b) Late stages of annealing at 120°C. A as made, B = 37 min, C = 141 min, D = 203 min, E = D + 6 months at room temperature, F = 243 min, G = 297 min, H = 807 min, I = 5217 min. (Fernández *et al.* (1991).)

a broadening interface between the two polymer layers.

Figure 10.7 shows the interfacial profiles which were used to calculate the fitted curves in Fig. 10.6. The main feature is that the profile remains quite sharp initially but moves towards the D-PMMA surface. This corresponds to the D-PMMA layer becoming thinner as is clearly seen by the increasing frequency of its interference pattern in the reflectivity curves. Finally the interface broadens and develops a long tail on the SCPE side. The broadening of the interface can be detected as a dampening of the amplitude of the oscillations in $R(q_{\perp})$, especially at short wavelength (high q). The quality of the fit which can be obtained to quite complex data is evident in Fig. 10.6. The uniqueness of the model interface cannot be proved, but by increasing the complexity of the segment density profile only as much as is needed to fit the data, and by using all available information from other sources, it is possible to demonstrate by trial and error the overall features the actual profile must possess.

The explanation of the thinning of the D-PMMA layer is that this polymer was an order of magnitude lower in molecular weight than the SCPE with a lower T_g, and therefore much the more mobile species. The interface shifts towards the mobile species as it swells the 'sluggish' SCPE just as a low molecular weight solvent swells a gel. Only when the large SCPE molecules have time to move does the interface truly broaden.

10.5.3 Block copolymer organization

50–50 diblock copolymers which are microphase separated on a substrate may form lamellar structures parallel to the surface and such structures may show a feature we have not yet seen — Bragg scattering. Data in Fig. 10.8 and 10.9 are for such a diblock of polystyrene (PS) with D-PMMA (Anastasiadis *et al.* 1990). The overall molecular weight was 3×10^4 and a film 1140 Å thick was spin-cast from toluene solution onto a fused silica substrate. The as-made sample gives rise to the reflectivity profile in Fig. 10.9 which shows the periodicity expected from a film 1140 Å thick with a uniform scattering length density corresponding to the average value of the blocks. The changes in the profile after annealing at 170°C for 22 h are quite spectacular. Three intense peaks in Fig. 10.9 are three orders of Bragg reflection arising from a layered structure parallel to the surface with period 175 Å. The data have been fitted with the profile inset in the figure. The polystyrene is located at the air interface, due probably to its lower surface energy, while the D-PMMA is at the glass surface, probably due to interactions between this polar molecular and the glass surface. The data have been fitted with a cosine^2 gradient across each PS–D-PMMA interface. This gives a width of the PS–PMMA interface of 50 Å. The scattering length density in the layers used to fit the data does not corre-

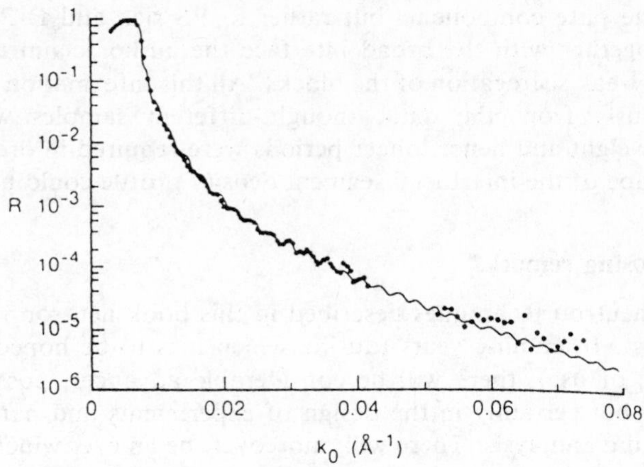

Fig. 10.8 Neutron reflectivity normal to the surface of a thin film of P(S-b-d-MMA) copolymer of $M_W \times 10^4$ on a fused silica substrate prior to annealing. The solid lines are calculated reflectivity profiles ($k_0 = q_\perp/2$). Reprinted with permission from Anastasiadis *et al.* (1990).

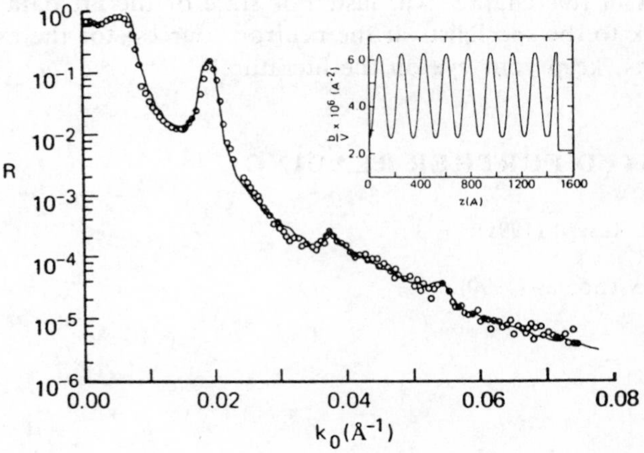

Fig. 10.9 The sample as in Fig. 10.8 after annealing at 170°C for 22 h. The inset shows the scattering length density profile normal to the film surface, used to calculate the reflectivity profile shown as a solid line. Reprinted with permission from Anastasiadis *et al.* (1990).

spond to the pure components but rather to PS rich and D-PMMA rich regions. Together with the broad interface the author claim this implies only rather weak segregation of the blocks. All this information is obtained unambiguously from the data, though different samples with higher molecular weight and hence longer periods were required in order that the detailed shape of the interfacial segment density profile could be explored.

10.5.4 Closing remarks

Of all the neutron techniques described in this book neutron reflection is the youngest. In coming years (during which it is to be hoped this book will still be of use), there will be considerable advances, possibly in the techniques, but certainly in the design of experiments and particularly in the area of data analysis. There will, moreover, be an ever widening spread both of the groups using the technique to study polymeric samples and in the scientific problems being tackled. As well as the questions of surface adsorption, interdiffusion and surface organization mentioned here, problems of surface segregation in blends, plasticizer migration, and location of interfacial agents in polymer mixtures are already being investigated. Others will surely follow. The authors, trying not to be out of date before their book reaches the printers, can only tell the reader that the basic ideas in this chapter will last. For state of the art data treatment, go and talk to the specialists at the neutron sources; for the exciting new applications, keep your eye on the literature!

SUGGESTED FURTHER READING

Felcher and Russell (1991).
Lekner (1987).
Penfold and Thomas (1990).

Appendix 1

The Fourier transform and some of its applications

A1.1 DEFINITIONS IN ONE AND THREE DIMENSIONS

In this appendix we do not intend to give a full mathematical theory of the Fourier transform. We only want to outline some of the properties of these transforms which are used throughout this book.

Let us assume that we have a function $f(x)$ well-defined for all values of $x(-\infty < x < +\infty)$. We assume also that we can integrate it or take its derivatives to any order; this is not absolutely necessary but we do treat the $f(x)$ as 'physical functions', assuming that all the functions we are defining do exist and can be derived or integrated at any order. We define the Fourier transform of $f(x)$ by the relationship

$$g(q) = \int_{-\infty}^{+\infty} f(x)\exp(-\mathrm{i}qx)\mathrm{d}x \tag{A1.1}$$

this transforms a function f of the variable x into a new function g of the variable q. The function g is the Fourier transform of f. In this expression i is the square root of (-1) and $\exp(x)$ the exponential function.

The function f may be imaginary but in this book we have considered mainly real functions. If the function is even $(f(x) = f(-x))$ and it can be seen immediately that its Fourier transform is also real and even: to demonstrate this we change x into $x' = -x$ and expand the exponentials as follows

$$g(q) = \int_{-\infty}^{+\infty} f(x)\exp(-\mathrm{i}qx)\mathrm{d}x = \int_{-\infty}^{+\infty} f(x)\,(\cos qx - \mathrm{i}\sin qx)\mathrm{d}x \tag{A1.2}$$

$$g(q) = \int_{-\infty}^{+\infty} f(-x')\exp(\mathrm{i}\,qx')\mathrm{d}x' = \int_{-\infty}^{+\infty} f(x)\,(\cos qx + \mathrm{i}\sin qx)\mathrm{d}x.$$

Since these expressions must be identical we have:

$$\int_{-\infty}^{+\infty} f(x)\sin qx\,\mathrm{d}x = 0$$

which means that $g(q)$ is real and even.

In the majority of the cases we will consider functions defined in 3-dimensional space, i.e. functions of three variables. This means that, instead of using a scalar x, we shall use a vector $\boldsymbol{r}$ and integrate over all $\boldsymbol{r}$ space writing:

$$g(\boldsymbol{q}) = \iiint_{V\to\infty} f(\boldsymbol{r})\exp(-\mathrm{i}\boldsymbol{q}\cdot\boldsymbol{r})\mathrm{d}\boldsymbol{r} \tag{A1.3}$$

$\mathrm{d}\boldsymbol{r} = \mathrm{d}x\mathrm{d}y\mathrm{d}z$ in Cartesian coordinates where one can write eqn (A1.3) as

$$g(q_x, q_y, q_z) = \iiint_{-\infty}^{\infty} f(x, y, z)\exp - \mathrm{i}(q_x x + q_y y + q_z z)\mathrm{d}x\mathrm{d}y\mathrm{d}z. \tag{A1.3$'$}$$

In polar coordinates using as variables the three classical quantities r, θ, φ

$$g(\boldsymbol{q}) = \int_{r=0}^{+\infty}\int_{\theta=0}^{\pi}\int_{\varphi=0}^{2\pi} f(\boldsymbol{r})\exp(-\mathrm{i}\boldsymbol{q}\cdot\boldsymbol{r})r^2\mathrm{d}r\sin\theta\,\mathrm{d}\theta\,\mathrm{d}\varphi. \tag{A1.4}$$

It is evident by generalization of eqn (A1.2), that if $f(\boldsymbol{r})$ is real and depends only on the length of the vector $\boldsymbol{r}$ the function $g(\boldsymbol{q})$ is also a real function of the length of $\boldsymbol{q}$.

A1.2 FT OF A GAUSSIAN—THE DIRAC DELTA FUNCTION

Before going into a discussion of the properties of the Fourier transformation (FT) it is interesting to evaluate the value of $g(\boldsymbol{q})$ for a practical example. We shall take the case of a Gaussian probability which has been used for the characterization of the probability for the ends of a Gaussian chain to be at a distance r from each other. This probability depends only on the distance r, and in polar coordinates is written as

$$w(\boldsymbol{r}) = \left(\frac{3}{2\pi\bar{l}^2}\right)^{3/2} \exp - \frac{3r^2}{2\bar{l}^2}. \tag{A1.5}$$

This function is normalized in order to have a total probability equal to unity. ($\iiint w(\boldsymbol{r})\,\mathrm{d}\boldsymbol{r} = 1$) and $\bar{l}^2$ is the root mean square distance between the ends of the chain.

Substituting this function for $f(\boldsymbol{r})$ in the definition of the three-dimensional FT in polar coordinates gives

$$g(\boldsymbol{q}) = = \left(\frac{3}{2\pi\bar{l}^2}\right)^{3/2} \iiint_0^{\infty} \exp\left[-\frac{3r^2}{2\bar{l}^2} - \mathrm{i}\boldsymbol{q}\cdot\boldsymbol{r}\right]\mathrm{d}\boldsymbol{r}. \tag{A1.6}$$

Completing the square in the exponential

$$-\frac{3\boldsymbol{r}^2}{2\bar{l}^2} - \mathrm{i}\boldsymbol{q}\cdot\boldsymbol{r} = -\frac{3}{2\bar{l}^2}\left[\boldsymbol{r} + \mathrm{i}\frac{\bar{l}^2}{3}\boldsymbol{q}\right]^2 - \frac{\bar{l}^2}{6}\boldsymbol{q}^2.$$

Introducing this expression in eqn (A1.6) gives

$$g(q) = = \left(\frac{3}{2\pi\bar{l}^2}\right)^{3/2} \iiint_0^{\infty} \exp\left[-\frac{3}{2\bar{l}^2}(\boldsymbol{r} + \mathrm{i})\frac{\bar{l}^2\boldsymbol{q}}{3}\right]^2 \exp - \frac{\bar{l}^2q^2}{6}\mathrm{d}\boldsymbol{r}. \tag{A1.6$'$}$$

We see that the final exponential does not depend on $\boldsymbol{r}$ and can be taken outside the integral sign. We substitute now the new variable $\rho = \boldsymbol{r} + \mathrm{i}\,\bar{l}^2/3\boldsymbol{q}$, $(\mathrm{d}\rho = \mathrm{d}\boldsymbol{r})$ obtaining

$$g(q) = \exp - \frac{\bar{l}^2q^2}{6}\left[\left(\frac{3}{2\pi\bar{l}^2}\right)^{3/2} \iiint_0^{\infty} \exp\left[-\frac{3\rho^2}{2\bar{l}^2}\right]\mathrm{d}\rho\right]. \tag{A1.6$''$}$$

The integral remaining in the square bracket is identical to the integral of eqn (A1.5); its value is 1 and we obtain finally for $g(q)$

$$g(\boldsymbol{q}) = \exp - \frac{q^2\bar{l}^2}{6} \tag{A1.7}$$

with

$$w(\boldsymbol{r}) = \left(\frac{3}{2\pi\bar{l}^2}\right)^{3/2} \exp - \frac{3r^2}{2\bar{l}^2}. \tag{A1.7$'$}$$

This result is well known: the FT of a Gaussian function peaked at the origin is another Gaussian function. There is, however, an important difference: the parameter l which appears in the denominator of eqn (A1.6) is in the numerator of eqn (A1.7). When l decreases the Gaussian curve representing eqn (A1.5) becomes narrower but its FT becomes broader. At the limit when l goes to zero the Gaussian function becomes a function infinite at $r = 0$, having a value equal to zero everywhere except at the origin. Moreover, since the integral of this function is equal to unity, the area between the function and the r axis remains unity. In the classical sense this is not a real mathematical function and is called by mathematicians a 'distribution'. In physics it is very useful, called the improper Dirac delta function and denoted by $\delta(r)$. (See Fig. A1.1.)

We can also write $w(\boldsymbol{r})$ as a function of $g(\boldsymbol{q})$. In order to give a precise definition of this function we can go back to eqns (A1.7) and (A1.7$'$) and after the same elementary calculations establish the relationship

$$w(\boldsymbol{r}) = \left(\frac{1}{2\pi}\right)^3 \iiint g(\boldsymbol{q})\exp(+\mathrm{i}\boldsymbol{r}\cdot\boldsymbol{q})\mathrm{d}\boldsymbol{q}. \tag{A1.8}$$

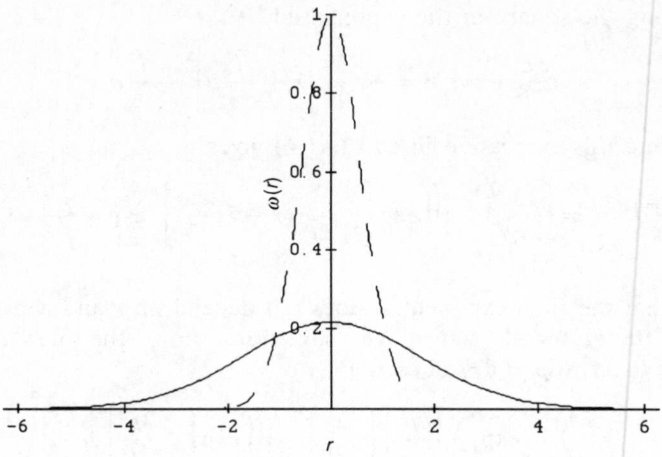

Fig. A1.1 The full line is a Gaussian function and the dotted line its FT. The calculation has been made for $l = 3$ and the scale has been multiplied by 50 in order to make it visible. When l increases the function becomes broader and its FT. narrower.

The plus sign is introduced for reasons which are important when one deals with imaginary functions but, since eqns (A1.5) and (A1.7) are real it has no effect on the result.

As already mentioned, when we approach the limit $l \to \infty$, $g(\boldsymbol{q})$ becomes unity and $w(\boldsymbol{r})$ becomes what we have called the improper Dirac delta function

$$\delta(\boldsymbol{r}) = \left(\frac{1}{2\pi}\right)^3 \iiint \exp(+\mathrm{i}\boldsymbol{r}\cdot\boldsymbol{q})\mathrm{d}\boldsymbol{q}. \tag{A1.9}$$

This equation can be used as a convenient definition of the delta function. In this appendix we are working in a three-dimensional space but it is easy to go back to one dimension and we leave this to the reader, giving simply the definition of a delta function for one variable

$$\delta(t) = \frac{1}{2\pi} \int_{-\infty}^{+\infty} \exp(\mathrm{i}\omega t)\mathrm{d}\omega. \tag{A1.9'}$$

A1.3 PROPERTIES OF THE DIRAC DELTA FUNCTION

(a) By definition, the delta function is zero everywhere except for $\boldsymbol{r} = 0$. It follows that the integration of the product of the delta function with any function $f(\boldsymbol{r})$ over all space gives

$$\iiint_V f(\mathbf{r})\delta(\mathbf{r})\mathrm{d}\mathbf{r} = f(0) \tag{A1.10}$$

if the origin is in the volume V; the value is zero if it is not.

(b) This can be generalized to any point $\mathbf{r}_0$ using, the new variable $\mathbf{u} = \mathbf{r} - \mathbf{r}_0$ in eqn (A1.9)

$$\iiint_V f(\mathbf{r})\delta(\mathbf{r} - \mathbf{r}_0)\mathrm{d}\mathbf{r} = \iiint_V f(\mathbf{r}_0 + \mathbf{u})\,\delta(\mathbf{u})\mathrm{d}\mathbf{u} = f(\mathbf{r}_0) \tag{A1.11}$$

if the volume V includes the point $\mathbf{r}_0$ and is zero otherwise.

(c) If, in a volume V, one has N points at positions $\mathbf{r}_i$, the quantity

$$\sum_i^N \delta(\mathbf{r} - \mathbf{r}_i) \tag{A1.12}$$

is what can be called the local number density $n(\mathbf{r})$ of the points i in the volume V. This can be shown by integrating over V

$$\iiint \sum_i^N \delta(\mathbf{r} - \mathbf{r}_i)\mathrm{d}\mathbf{r} = N = \iiint_V n(\mathbf{r})\mathrm{d}\mathbf{r}. \tag{A1.13}$$

This relationship is rigorous for any value of the volume V and is very useful when one wants to transform a sum into an integral. Let us give an example: the amplitude scattered by a system made of N particles at position $\mathbf{r}_j$ is, in discrete notation

$$A = \sum_{j=1}^N \exp(-\mathrm{i}\mathbf{q}\cdot\mathbf{r}_j). \tag{A1.14}$$

If we introduce $n(\mathbf{r})$, in continuous notation we write

$$A = \iiint_V n(\mathbf{r}) \exp(-\mathrm{i}\mathbf{q}\cdot\mathbf{r})\mathrm{d}\mathbf{r} \tag{A1.15}$$

replacing $n(\mathbf{r})$ by its definition from eqn (A1.13) we obtain

$$A = \iiint_V \sum_j^N \delta(\mathbf{r} - \mathbf{r}_j)\exp(-\mathrm{i}\mathbf{q}\cdot\mathbf{r})\mathrm{d}\mathbf{r} = \sum_j^N \iiint_V \delta(\mathbf{r} - \mathbf{r}_j)\exp(-\mathrm{i}\mathbf{q}\cdot\mathbf{r})\mathrm{d}\mathbf{r}$$

$$= \sum_{j=1}^N \exp(-\mathrm{i}\mathbf{q}\cdot\mathbf{r}_j). \tag{A1.16}$$

This shows that the two descriptions of an 'ensemble' of points are identical.

A1.4 RECIPROCITY IN FTs

It has just been shown that, in a very special case (Gaussian function), the Fourier transform of $g(\boldsymbol{q})$ is proportional to the original function $f(\boldsymbol{r})$. More precisely, in the case of the Gaussian function we have written the following equation (eqn (A1.6))

$$f(\boldsymbol{r}) = \frac{1}{(2\pi)^3} \iiint g(\boldsymbol{q}) \exp(+\mathrm{i}\boldsymbol{r}\cdot\boldsymbol{q})\mathrm{d}\boldsymbol{q}. \tag{A1.17}$$

We shall now show that this equation is quite general, in other words the inverse FT recovers the original function. It should be noted that the number i has a + sign and that a constant term $1/(2\pi)^3$ has been introduced. In order to justify this result let us assume that eqn (A1.17) is correct and replace $g(\boldsymbol{q})$ by its definition from eqn (A1.1) giving

$$f(\boldsymbol{r}_0) = \frac{1}{(2\pi)^3} \iiint_{\boldsymbol{q}} \exp(+\mathrm{i}\boldsymbol{r}_0\boldsymbol{q}) \left[\iiint_{\boldsymbol{r}} \exp(-\mathrm{i}\boldsymbol{r}\cdot\boldsymbol{q}) f(\boldsymbol{r})\mathrm{d}\boldsymbol{r} \right] \mathrm{d}\boldsymbol{q}. \tag{A1.18}$$

Care is needed about the distinction between $\boldsymbol{r}_0$ and $\boldsymbol{r}$; $\boldsymbol{r}_0$ is the value of $\boldsymbol{r}$ we are interested in; $\boldsymbol{r}$ is an integration variable in the second integral; it is a 'dummy' variable which disappears in the final result. By interchanging the order of integration we obtain

$$f(\boldsymbol{r}_0) = \iiint_{\boldsymbol{r}} f(\boldsymbol{r}) \left[\frac{1}{(2\pi)^3} \iiint_{\boldsymbol{q}} \exp[-\mathrm{i}(\boldsymbol{r}-\boldsymbol{r}_0)\cdot\boldsymbol{q}]\mathrm{d}\boldsymbol{q} \right] \mathrm{d}\boldsymbol{r}. \tag{A1.19}$$

The integral over $\boldsymbol{q}$ inside the square brackets, is just the definition of the delta function of $(\boldsymbol{r} - \boldsymbol{r}_0)$, $\delta(\boldsymbol{r} - \boldsymbol{r}_0)$. We write therefore

$$f(\boldsymbol{r}_0) = \iiint_{\boldsymbol{r}} f(\boldsymbol{r})\delta(\boldsymbol{r} - \boldsymbol{r}_0)\mathrm{d}\boldsymbol{r} = f(\boldsymbol{r}_0) \tag{A1.20}$$

and the relationship (A1.16) is demonstrated. This reciprocity means that we quite frequently speak about $\boldsymbol{r}$ and $\boldsymbol{q}$ space as conjugate spaces since we move from one to the other by similar transformations. A curve in $\boldsymbol{r}$ space has a corresponding conjugate curve in $\boldsymbol{q}$ space. As a general rule small distances in $\boldsymbol{r}$ space correspond to large distances in $\boldsymbol{q}$ space and vice versa.

A1.5 EXPANSION OF THE FT AROUND $q = 0$. THE FT AS A SUM OF MOMENTS

If we expand in series form the exponential in the integral of eqn (A1.3) we obtain, assuming that $w(\boldsymbol{r})$ depends only on $r = |:\boldsymbol{r}|$ and is normalized to unity

$$\iiint w(r)\mathrm{d}r = 1$$

$$g(q) = \iiint_V w(r)\left[1 - \mathrm{i}\boldsymbol{q}\cdot\boldsymbol{r} - \frac{(\boldsymbol{q}\cdot\boldsymbol{r})^2}{2} + \mathrm{i}\frac{(\boldsymbol{q}\cdot\boldsymbol{r})^3}{6} + \frac{(\boldsymbol{q}\cdot\boldsymbol{r})^4}{4!} - \mathrm{i}\frac{(\boldsymbol{q}\cdot\boldsymbol{r})^5}{5!} + \ldots\right]\mathrm{d}\boldsymbol{r}. \quad \text{(A1.21)}$$

Due to the fact that $w(r)$ depends only on the distance r, a term like $(\boldsymbol{q}\cdot\boldsymbol{r})^n$ can be written, after averaging over the orientations, as

$$(qr\cos\theta)^n = q^n r^n \overline{\cos^n\theta} = q^n r^n \frac{1}{n+1}$$

if n is even and zero if it is odd, since the average value of $\overline{\cos^n\theta}$, taken over all directions of space is given by $\overline{\cos^n\theta} = 1/(n+1)$ for n even and 0 for n odd. This leads to:

$$g(q) = 1 - \frac{q^2}{6}\int_0^\infty w(r)4\pi r^4 \mathrm{d}r + \frac{q^4}{5!}\int_0^\infty w(r)4\pi r^6 \mathrm{d}r + \ldots. \quad \text{(A1.22)}$$

Thus, this expansion gives the even moments of the function $w(r)$. We define the quantity $\overline{r^n}$ as the nth moment of a distribution $f(r)$ by

$$\overline{r^n} = \frac{\int f(r) r^n 4\pi r^2 \mathrm{d}r}{\int f(r) 4\pi r^2 \mathrm{d}r}. \quad \text{(A1.23)}$$

The second moment $(n = 2)$ is the mean square value of the distance r. This suggests a method of building up a probability law; if all the even moments of the distribution are known, we can build up the function $g(q)$ as

$$g(q) = 1 - \frac{q^2}{3!}\overline{r^2} + \frac{q^4}{5!}\overline{r^4} - \frac{q^6}{7!}\overline{r^6} + \ldots. \quad \text{(A1.24)}$$

and, by inverse FT obtain $w(r)$. The major problem in this type of calculation is the lack of precision in $w(r)$ when the number of moments is limited.

A1.6 SOME GENERAL RULES FOR FTs

From the definition of the FT it is evident that:

1. If one multiplies a function by a constant its FT will be multiplied by the same constant.
2. The FT of the sum of two functions f_1 and f_2 is the sum of the FT of f_1 and f_2.

Another property which is quite useful, arises when we consider the following: if one knows $g(q)$, the FT of $f(r)$ what is the FT of $\partial f(r)/\partial r$ as a function of $g(q)$? In order to solve this problem we express the function $f(r)$ as function of its FT $g(q)$ (eqn (A1.12))

$$f(\boldsymbol{r}) = \frac{1}{(2\pi)^3} \iiint\limits_{V}^{\infty} g(\boldsymbol{q})\exp(+\mathrm{i}\boldsymbol{r}\cdot\boldsymbol{q})\mathrm{d}\boldsymbol{q}. \tag{A1.25}$$

Taking the derivative with respect to $\boldsymbol{r}\cdot$ gives

$$\frac{\partial f(\boldsymbol{r})}{\partial \boldsymbol{r}} = \frac{1}{(2\pi)^3} \iiint\limits_{V}^{\infty} \mathrm{i}\boldsymbol{q}g(\boldsymbol{q})\exp(+\mathrm{i}\boldsymbol{r}\cdot\boldsymbol{q})\mathrm{d}\boldsymbol{q} \tag{A1.26}$$

showing that the FT of $\partial f(\boldsymbol{r})/\partial \boldsymbol{r}$ is $\mathrm{i}\boldsymbol{q}\, g(\boldsymbol{q})$.

This process can be extended to the higher order derivatives of $f(\boldsymbol{r})$ (for instance the FT of $\partial^2 f(\boldsymbol{r})/\partial \boldsymbol{r}^2$ is $-\,q^2 g(\boldsymbol{q})$) as well as to the integrals of $f(\boldsymbol{r})$. This can be very useful for solving linear differential equations. Writing the equation in reciprocal space transforms a linear differential equation into a simple linear equation which can be solved much more easily.

There is a final property of the FT which is useful in scattering and which we have already demonstrated; it is described by the two following equations

$$g_{(q=0)} = \iiint f(\boldsymbol{r})\mathrm{d}\boldsymbol{r}$$

$$f_{(r=0)} = \left[\frac{1}{2\pi}\right]^{3/2} \iiint g(\boldsymbol{q})\mathrm{d}\boldsymbol{q}. \tag{A1.27}$$

The first is obtained by putting $\boldsymbol{q} = 0$ in eqn (A1.1), the second by putting $\boldsymbol{r} = 0$ in eqn (A1.17).

A1.7 THE CONVOLUTION THEOREM

Let us consider the following problem. Suppose we know that $w_1(\boldsymbol{r}_1)$ is the probability to go from the origin to a point M_1. (See Fig. A1.2.) We assume that we have a second probability law $w_2(\boldsymbol{r}_2)$ independent of the first one which takes us from M_1 to M_2. The probability of going from the origin to M_2 through M_1 is evidently $w_1(\boldsymbol{r}_1)w_2(\boldsymbol{r}_2)$ or $w_1(\boldsymbol{r}_1)w_2(\boldsymbol{R}-\boldsymbol{r}_1)$ where $\boldsymbol{R}$ is the quantity $\boldsymbol{R}=\boldsymbol{r}_1+\boldsymbol{r}_2$. If one wants to know the probability $W(\boldsymbol{R})$ to go to M_2, independently of the position of M_1 we have to integrate over all positions of M_1 and we obtain

$$W(\boldsymbol{R}) = \iiint w_1(\boldsymbol{r}_1)w_2(\boldsymbol{R}-\boldsymbol{r}_1)\mathrm{d}\boldsymbol{r}_1 \tag{A1.28}$$

This equation, which is called the convolution equation, is often written as

$$W = w_1 * w_2.$$

Here we have taken as an example a particular problem but relationships of this kind are found in many domains of physics and physical chemistry. Consider for instance the process of making column fractionation: for a monodisperse system the column gives a signal which can be characterized by w_1. The polydispersity curve of the sample being studied is given by w_2; it can be shown that the experimental result is given by W. How it is possible to calculate w_2 knowing W and w_1. The FT provides a very elegant method of solving this problem. If we call $\Omega(q)$, $\omega_1(q)$ and $\omega_2(q)$ the FT of W, w_1, and w_2 respectively, it can be shown that we have the simple relationship:

$$\Omega(q) = \omega_1(q)\omega_2(q). \tag{A1.29}$$

In order to legitimize the result let us start from the definition of $W(\boldsymbol{R})$ and replace w_1 and w_2 by their expressions as functions of their FT. We obtain

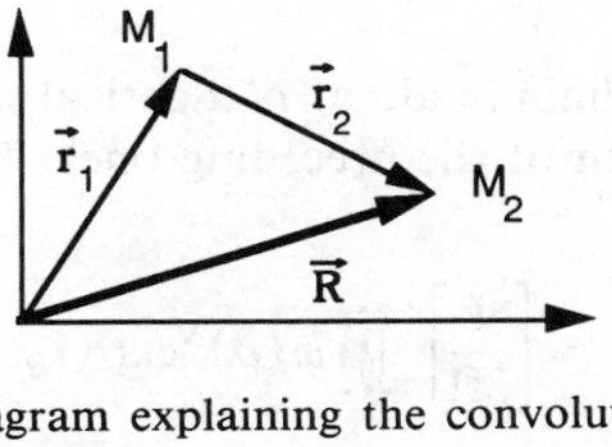

Fig. A1.2 Diagram explaining the convolution theorem.

$$W(\boldsymbol{R}) = \left[\frac{1}{2\pi}\right]^6 \int_{r_1} \mathrm{d}\boldsymbol{r}_1 \int_{q_1} \mathrm{d}\boldsymbol{q}_1 \omega(\boldsymbol{q}_1) \exp(\mathrm{i}\boldsymbol{q}_1 \cdot \boldsymbol{r}_1) \int_{q_2} \mathrm{d}\boldsymbol{q}_2 \omega(\boldsymbol{q}_2) \exp(\mathrm{i}\boldsymbol{q}_2 (\boldsymbol{R} - \boldsymbol{r}_1)). \tag{A1.30}$$

After transforming these three integrals into a multiple integral we see that the integration over $\mathrm{d}\boldsymbol{r}_1$ can be done immediately since (see eqn (A1.9))

$$\left[\frac{1}{2\pi}\right]^3 \int_{r_1} \mathrm{d}\boldsymbol{r}_1 \exp(\mathrm{i}(\boldsymbol{q}_1 - \boldsymbol{q}_2) \cdot \boldsymbol{r}_1) = \delta(\boldsymbol{q}_1 - \boldsymbol{q}_2) \tag{A1.31}$$

and the former expression for $W(\boldsymbol{R})$ becomes

$$W(\boldsymbol{R}) = \left[\frac{1}{2\pi}\right]^3 \int_{q_1} \int_{q_2} \omega_1(\boldsymbol{q}_1) \omega_2(\boldsymbol{q}_2) \exp(\mathrm{i}\boldsymbol{q}_2 \cdot \boldsymbol{R}) \delta(\boldsymbol{q}_1 - \boldsymbol{q}_2) \mathrm{d}\boldsymbol{q}_1 \mathrm{d}\boldsymbol{q}_2 \tag{A1.32}$$

From the definition of the delta function, integrating over $\boldsymbol{q}_2$ gives: $\boldsymbol{q}_1 = \boldsymbol{q}_2$; we then replace $\boldsymbol{q}_1$ by $\boldsymbol{q}$ and obtain finally

$$W(\boldsymbol{R}) = \left[\frac{1}{2\pi}\right]^3 \iiint \omega_1(\boldsymbol{q}) \omega_2(\boldsymbol{q}) \exp(\mathrm{i}\boldsymbol{q} \cdot \boldsymbol{R}) \mathrm{d}\boldsymbol{q}. \tag{A1.33}$$

The right-hand side of this equation is just the FT of $W(\boldsymbol{R})$ which demonstrates the fact that the FT of $W(\boldsymbol{R})$ is the product of $\omega_1(\boldsymbol{q})$ and $\omega_2(\boldsymbol{q})$.

A1.8 APPLICATION OF THE CONVOLUTION THEOREM TO THE STATISTICS OF GAUSSIAN CHAINS

If, instead of having two steps characterized by w_1 and w_2 in $\boldsymbol{r}$ space (or ω_1 and ω_2 in $\boldsymbol{q}$ space) we have N steps $w_1 w_2 \ldots\ldots w_N$ the generalization of the preceding theorem is obvious and the FT of the probability function $W(\boldsymbol{R})$ of going from 0 to $\boldsymbol{R}$ by these N steps will be

$$\Omega(\boldsymbol{q}) = \prod_{i=0}^{N} \omega_i(\boldsymbol{q}). \tag{A1.34}$$

If one knows the $\omega_i(\boldsymbol{q})$, one can evaluate their product, take the inverse FT and obtain $W(\boldsymbol{R})$.

We now consider a chain made of N identical steps without correlations (each step is independent of the preceding one). The $\omega_i(\boldsymbol{q})$ are identical to $\omega(q)$ and

$$W(R) = \left[\frac{1}{2\Pi}\right]^3 \iiint_s \omega(q)^N \exp(\mathrm{i}\boldsymbol{q} \cdot \boldsymbol{R}) \mathrm{d}q. \tag{A1.35}$$

This expression allows the evaluation of the probability distribution for many steps, knowing the probability law for one step provided that their are no correlations between the steps.

It will be shown now that, regardless of the nature of $\omega(q)$, if the number of steps N is large we obtain a Gaussian law for $W(R)$. For this purpose we first expand $\omega(q)$ as a function of q using the general expansion defined in eqn (A1.16)

$$\omega(q) = 1 - \frac{\overline{r^2}}{6}q^2 + \frac{\overline{r^4}}{5!}q^4 - \frac{\overline{r^6}}{7!}q^6 + \dots \tag{A1.36}$$

where $\overline{r^{2n}}$ is given by

$$\overline{r^{2n}} = \int w(r) 4\pi r^{2n+2} \mathrm{d}r.$$

We now have to find an expression for $\omega^N(q)$. For this purpose we write

$$\omega^N(q) = \exp\{N\log\omega(q)\} = \exp\left\{N\log\left[1 - \frac{1}{6}\overline{r^2}q^2 + \frac{1}{5!}\overline{r^4}q^4 + \dots\right]\right\} \tag{A1.37}$$

The first term of this expansion is obtained writing $\log(1 + x) \approx x + O(x)$:

$$\omega(q)^N = \exp\{N\log(q)\} = \exp\left\{-N\left[\frac{1}{6}\overline{r^2}q^2\right]\right\} + O(N). \tag{A1.38}$$

We can easily be convinced that, if N is large, all the other terms can be neglected and we obtain:

$$\Omega(q) = \exp -\frac{1}{6}N\overline{l^2}q^2 \tag{A1.39}$$

where we have defined $\overline{l^2}$ as the root mean square distance between two consecutive steps.

This result is important since we know (see eqn (A1.7)) that it means that the chain is Gaussian, with an end-to-end distance equal to $N\overline{l^2}$ (see eqn (A1.5)). This method leads easily to the properties of a Gaussian chain. A generalization shows that, if the N steps are not identical the same formula can be used writing:

$$N\overline{l^2} = \sum_1^N \overline{l_i^2}. \tag{A1.40}$$

This result can be understood by saying that the superposition of N independent statistical processes which are arbitrary leads to a Gaussian distribution. This is what is called the 'central limit theorem'.

An alternative method of obtaining the same result is the following: the probability law for a given conformation where the position of all the segments is specified is given quite generally by:

$$W(\boldsymbol{r}_1, \boldsymbol{r}_2, \ldots \boldsymbol{r}_i \ldots \boldsymbol{r}_N) = w_1(\boldsymbol{r}_1) w_2(\boldsymbol{r}_2) \ldots w_i(\boldsymbol{r}_i) \ldots w_N(\boldsymbol{r}_N) \tag{A1.41}$$

where w_1, w_2, w_N are the probabilities of finding the vectors $\boldsymbol{r}_1$, $\boldsymbol{r}_2$ ...... $\boldsymbol{r}_N$ in a given position. This function counts all the possible states of the chain. In order to obtain the probability for the two ends to be at a given distance $\boldsymbol{R}$ we must suppress all conformations for which $\Sigma \boldsymbol{r}_i \neq \boldsymbol{R}$. This is easily done by multiplying W by $\delta(\boldsymbol{R} - \Sigma \boldsymbol{r}_i)$ and integrating over all the positions of all subunits. This will leave only the terms corresponding to $(\boldsymbol{R} - \Sigma \boldsymbol{r}_i = 0)$. Using the definition for δ given in eqn (A1.9) we obtain

$$W(\boldsymbol{R}) = \left[\frac{1}{2\pi}\right]^3 \int_q \mathrm{d}\boldsymbol{q} \int_{r_1} \mathrm{d}\boldsymbol{r}_1 w(\boldsymbol{r}_1) \int_{r_i} \mathrm{d}\boldsymbol{r}_i w(\boldsymbol{r}_i) \int_{r_N} \mathrm{d}\boldsymbol{r}_N w(\boldsymbol{r}_N) \exp\{\mathrm{i}\boldsymbol{q} \cdot (\boldsymbol{R} - \Sigma \boldsymbol{r}_i)\}. \tag{A1.42}$$

It is easy to collect all the terms corresponding to the integration over $\boldsymbol{r}_i$; each becomes its own FT $\omega_i(\boldsymbol{q})$ and we are left with

$$W(\boldsymbol{R}) = \left[\frac{1}{2\pi}\right]^3 \iiint_q \exp(-\mathrm{i}\boldsymbol{q} \cdot \boldsymbol{R}) \prod_i \omega_i(\boldsymbol{q}) \mathrm{d}\boldsymbol{q} \tag{A1.43}$$

thus recovering eqn (A1.25). This method is very efficient and a simple extension allows us to obtain simultaneous probabilities; for instance, of having i and j at a distance r_{ij} and the extremities at a distance R. It is also used to study chains subject to an anisotropic constraint as in stretched gels or rubbers.

A1.9 CALCULATION OF THE FT FOR SOME EXAMPLES

In books specializing in Fourier transforms one finds tables of functions with their FT. In this paragraph, a few examples of pairs of functions will be presented together with the method used to obtain them. We shall limit our examples to the case where the function f depends only on the modulus r of the vector $\boldsymbol{r}$.

From the definition of an FT (eqn (A1.1)) we write

$$g(q) = \int_{r=0}^{\infty} \int_{\theta=0}^{\pi} \int_{\varphi=0}^{2\pi} f(r) \exp(-\mathrm{i}\boldsymbol{q} \cdot \boldsymbol{r}) r^2 \mathrm{d}r \sin\theta \, \mathrm{d}\theta \, \mathrm{d}\varphi. \tag{A1.44}$$

We can integrate immediately over φ since the integral is independent of φ. If we use as a variable $\cos\theta = u$ with $\mathrm{d}u = -\sin\theta \, \mathrm{d}\theta$ we obtain

$$g(q) = 2\pi \int_{r=0}^{\infty} \int_{-1}^{1} f(r)\exp(-\mathrm{i}qru)r^2\mathrm{d}r\,\mathrm{d}u \tag{A1.45}$$

then

$$g(q) = 2\pi \int_{r=0}^{\infty} f(r)\left[\frac{-1}{\mathrm{i}qr}(\exp(-\mathrm{i}qr) - \exp(\mathrm{i}qr))\right]r^2\mathrm{d}r. \tag{A1.46}$$

Replacing the exponential by its expression as a function of cosine and sine gives

$$g(q) = \int_{r=0}^{\infty} f(r)\frac{\sin qr}{qr}4\pi r^2\mathrm{d}r. \tag{A1.47}$$

If the distance r is fixed at a length a, i.e. we have a sphere of radius a, we see that the FT reduces $\sin qa/(qa)$ (the front factor disappears since if, we consider that $f(r)$ is $\iiint_V^{\infty} f(r)\mathrm{d}r = 1$, it is known that its FT is equal to unity for $q = 0$).

If the function $f(r)$ is of the form $f(r) \approx r^{-n}$ we use the variable $qr = u$, obtaining

$$g(q) \approx 4\pi \int_{u=0}^{\infty} \frac{q^n}{u^n}\frac{\sin u}{u}\frac{u^2\mathrm{d}u}{q^3} \approx 4\pi q^{n-3}\int_{u=0}^{\infty}\frac{\sin u}{u^{n-1}}\mathrm{d}u. \tag{A1.48}$$

The value of the last integral is a number which can be found in tables of integrals; we shall call it C with

$$C = \Gamma(2-n)\sin\frac{(2-\mathrm{n})\pi}{2} \tag{A1.49}$$

where Γ is the gamma function. We obtain for $g(q)$

$$g(q) = 4\pi C q^{n-3}. \tag{A1.50}$$

As an application of this, we know that a Gaussian chain has, for its structure factor, an asymptotic behaviour in q^{-2}; this means that the pair distribution function is in $1/r$. It should be also in $1/r$ for all two-dimensional objects, in $1/r^2$ for linear molecules and r for a three-dimensional object.

A last example is the FT of a function decreasing like $1/r^n\exp - r/\xi$. To solve this problem we replace $f(r)$ in eqn (A1.47) by this function, obtaining

$$g(q) = \int_0^{\infty} \exp\left(\frac{-r}{\xi}\right)\frac{1}{r^n}\frac{\sin qr}{qr}4\pi r^2\mathrm{d}r. \tag{A1.51}$$

We can say that this integral is the imaginary part of the integral

$$I_1 = \frac{4\pi}{q} \int_0^\infty \frac{1}{r^{n-1}} \exp\left\{-r\left[\frac{1}{\xi} - iq\right]\right\} \mathrm{d}r \qquad \text{(A1.52)}$$

and introduce the new parameter v defined as

$$v = \frac{1}{\xi} - iq$$

using as a new variable $u = rv$ we obtain

$$I_1 = 4\pi \frac{v^{n-2}}{q} \int_0^\infty \frac{\exp - u}{u^{n-1}} \mathrm{d}u. \qquad \text{(A1.53)}$$

Care must be taken over this integral as it is an integral in the complex plane. We shall now consider the two following particular cases:

(a) The case where $n = 0$ or $f(r) = \exp - r/\xi$. In this case eqn (A1.53) can be written as

$$I_1 = \frac{4\pi}{qv^2} \int_0^\infty \exp(-u) u \,\mathrm{d}u = \frac{4\pi}{qv^2} \qquad \text{(A1.54)}$$

after integration by parts.

We take the imaginary part of this quantity by multiplying the numerator and the denominator by v^{*2}; this gives

$$I_1 = \frac{4\pi}{q} \left\{\frac{1}{\xi^1 + iq} \frac{\xi^{-1} + iq}{\xi^{-1} + iq}\right\}^2 = \frac{4\pi}{q} \left\{\frac{\xi^{-1} + iq}{\xi^{-2} + q^2}\right\}^2 \qquad \text{(A1.55)}$$

and from this

$$g(q) = \frac{8\pi}{\xi} \left\{\frac{1}{q^2 + \frac{1}{\xi^2}}\right\}^2. \qquad \text{(A1.56)}$$

(b) The case where $n = 1$ or $f(r) = \frac{1}{r} \exp - r/\xi$. Equation (A1.53) becomes

$$I_1 = \frac{4\pi}{qv} \int_0^\infty \exp(-u) \,\mathrm{d}u = \frac{4\pi}{qv}. \qquad \text{(A1.57)}$$

After extracting the imaginary part we obtain

$$g(q) = \frac{4\pi}{\xi} \frac{1}{q^2 + \frac{1}{\xi^2}} \qquad \text{(A1.58)}$$

$\left(\frac{3}{2\pi\bar{\ell}^2}\right)^{3/2} \exp-\frac{3r^2}{2\bar{\ell}^2}$	$\exp-\frac{\bar{\ell}^2 q^2}{6}$
$\frac{1}{r^n}$	$4\pi\Gamma(2\text{-}n)\sin\frac{(2-n)\pi}{2}\, q^{n-3}$
$\frac{1}{r}\exp-\frac{r}{\xi}$	$4\pi\frac{1}{q^2+\frac{1}{\xi^2}}$.
$\exp-\frac{r}{\xi}$	$\frac{8\pi}{\xi}\left\{\frac{1}{q^2+\frac{1}{\xi^2}}\right\}^2$

Fig. A1.3 Some functions and their FTS.

All the results derived for FT in this Appendix are summarized in Fig. A1.3.

SUGGESTED FURTHER READING

Abramowitz and Stegun (1970).
Bracewell (1965).
Champeney (1973).
Gradshtein and Ryzhik (1965).

Appendix 2

Thermodynamics

In this appendix we derive some of the fundamental thermodynamic relationships which have been used in discussing the scattering problems. We also briefly review Flory's theory of the free energy of mixing of polymers.

A2.1 THE FUNDAMENTAL EQUATIONS OF THERMODYNAMICS

Assuming that the definition and the properties of the functions U and S (respectively the energy and the entropy of the system) are known, the Helmholtz free energy A and the Gibbs free energy (or free enthalpy) G of the system are defined by the relationships

$$A = U - TS \tag{A2.1}$$

$$G = U + pV - TS. \tag{A2.2}$$

The quantity A is a chemical potential at constant volume V and temperature T. This means that for a system in equilibrium at T and V, constant A is a minimum. ΔA, the variation of the free energy at T and V constant, is the work which can be done by the system. The natural variables are V and T and at constant composition one has

$$\mathrm{d}A = -p\mathrm{d}V - S\mathrm{d}T \tag{A2.3}$$

since

$$\mathrm{d}U = -p\mathrm{d}V - \mathrm{d}Q = -p\mathrm{d}V - S\mathrm{d}T.$$

If the system contains p constituents one defines the chemical potential of the species i by: $\mu_i = (\partial A/\partial N_i)_{N_j, V, T}$, where N_i is the number of molecules of species i in the system. For the differential form of A we have

$$\mathrm{d}A = \sum_1^p \mu_i \mathrm{d}N_i - p\mathrm{d}V - S\mathrm{d}T. \tag{A2.4}$$

The partition function Z is given by

$$Z = \frac{1}{\Pi N_i!} \sum \exp - \frac{U}{k_B T} \tag{A2.5}$$

where the summation is extended to all possible states of the system at V and T constant and the factorial term avoids the Gibbs paradox. Statistical thermodynamics tells us that Z is related to the free energy A by the relationship

$$A = - k_B T \log Z. \tag{A2.6}$$

The Gibbs free energy is a potential at constant p and T. It is minimum in these conditions and its derivative can be written as

$$dG = \sum_1^p \mu_i dN_i + Vdp - SdT \tag{A2.7}$$

where μ_i is now defined as

$$\mu_i = \left(\frac{\partial G}{\partial N_i}\right)_{N_j, p, T}. \tag{A2.8}$$

Since G is a first order homogeneous function in the variables N_i and zero order in the variables p and T the Euler theorem gives

$$G = \sum_1^p \mu_i N_i. \tag{A2.9}$$

Differentiating eqn (A2.9) gives

$$dG = \sum_1^p \mu_i dN_i + \sum_1^p d\mu_i N_i$$

and hence, using eqn (A2.7)

$$\sum_1^p N_i d\mu_i = Vdp - SdT \tag{A2.10}$$

which is the well-known Gibbs–Duhem relationship. Applied at constant temperature and pressure it shows that the μ_i are not independent. From eqn (A.7) writing

$$dG = \sum_1^p \frac{\partial G}{\partial N_i} dN_i + \frac{\partial G}{\partial p} dp + \frac{\partial G}{\partial T} dT \tag{A2.11}$$

we obtain

$$\mu_i = \left(\frac{\partial G}{\partial N_i}\right)_{p, T, N_k} \quad V = \frac{\partial G}{\partial p} \text{ and } S = - \frac{\partial G}{\partial T} \tag{A2.12}$$

and via further differentiation

$$\left(\frac{\partial^2 G}{\partial N_i \partial N_j}\right)_{p,T,N_k} = \left(\frac{\partial \mu_i}{\partial N_j}\right)_{p,T,N_k} = \left(\frac{\partial \mu_j}{\partial N_i}\right)_{p,T,N_k}. \tag{A2.13}$$

A2.2 RELATIONSHIP BETWEEN $(\partial\mu_j/\partial N_i)_{p,T,N_k}$ AND $(\partial\mu_i/\partial N_i)_{V,T,N_k}$

From the theory of functions with more than one variable

$$\left(\frac{\partial \mu_i}{\partial N_j}\right)_{p,N} = \left(\frac{\partial \mu_i}{\partial N_j}\right)_{V,N} + \left(\frac{\partial \mu_i}{\partial p}\right)_{N}\left(\frac{\partial p}{\partial N_j}\right)_{V,N} \tag{A2.14}$$

where the index N means that all the N_i, except any which are in the differential form, are constant. From the definition in eqn (A2.12) and exchanging the order of differentiation, we obtain

$$\left(\frac{\partial^2 G}{\partial N_i \partial p}\right)_{p,T,N_k} = \left(\frac{\partial \mu_i}{\partial p}\right)_{N} = \left(\frac{\partial V}{\partial N_i}\right)_{p} = v_i \tag{A2.15}$$

v_i being the partial molar volume of the species i.

In order to evaluate the quantity $(\partial p/\partial N_j)_{V,N}$ we write

$$\mathrm{d}V = \left(\frac{\partial V}{\partial N_j}\right)_{p} \mathrm{d}N_j + \left(\frac{\partial V}{\partial p}\right)_{N_j} \mathrm{d}p \tag{A2.16}$$

since all the other variables do not change; this gives at constant volume ($\mathrm{d}V = 0$):

$$\left(\frac{\partial p}{\partial N_j}\right)_{V} = -\frac{\left(\frac{\partial V}{\partial N_j}\right)_{p}}{\left(\frac{\partial V}{\partial p}\right)_{N_j}} = \frac{v_j}{\beta V} \tag{A2.17}$$

where v_j is the partial molar volume of component j and β the isothermal compressibility $(-1/V)(\partial V/\partial p)$. Substituting the results of eqns (A2.15) and (A2.17) in eqn (A2.14) gives the desired result, i.e.

$$\left(\frac{\partial \mu_i}{\partial N_j}\right)_{V,N} = \left(\frac{\partial \mu_i}{\partial N_j}\right)_{p,N} + \frac{v_i v_j}{\beta V}. \tag{A2.18}$$

A2.3 FLUCTUATIONS IN A GRAND CANONICAL ENSEMBLE

Up to this point we have discussed the ensemble at constant volume, temperature and number of molecules (characterized by the function A) and the ensemble at constant pressure, temperature, and number of molecules (the function G). We shall now introduce the grand ensemble in which the volume, temperature, and the chemical potential of the constituents are fixed. This ensemble is characterized by the letter Ω and is defined by the relationship

$$\Omega = G - A = pV. \tag{A2.19}$$

If we write its differential form we obtain, using eqns (A2.4) and (A2.7)

$$\mathrm{d}\Omega = \sum_{1}^{p} N_i \,\mathrm{d}\mu_i + p\mathrm{d}V + S\mathrm{d}T = \mathrm{d}(\mathrm{pV}). \tag{A2.20}$$

This gives the equilibrium condition for a system with V, T, and μ_i constant. The corresponding partition function is what is called the Grand Partition Function, Ξ, defined by the relationship

$$\Xi = \sum C \exp \frac{1}{k_{\mathrm{B}}T}\left[\sum \mu_i N_i - U\right] \tag{A2.21}$$

where C is a factor which, for this problem, can be considered constant. The summation has to be made over all possible states of the system with constant V, T, and chemical potential μ_i. It is easy to show that

$$\Omega = k_{\mathrm{B}}T \log \Xi. \tag{A2.22}$$

The probability w for one state of the system is given by

$$w(\text{one state}) = \frac{\exp \dfrac{1}{k_{\mathrm{B}}T}\left[\sum \mu_i N_i - U\right]}{\Xi} \tag{A2.23}$$

this gives for the average value of N_i, $\bar{N}_i$

$$\bar{N}_i = \frac{\sum C N_i \exp \dfrac{1}{k_{\mathrm{B}}T}\left[\sum \mu_i N_i - U\right]}{\sum C \exp \dfrac{1}{k_{\mathrm{B}}T}\left[\sum \mu_i N_i - U\right]}. \tag{A2.24}$$

The practical method for obtaining this average value is to take the derivative of $\log \Xi$ with respect to μ_i

$$\frac{\partial \ln \Xi}{\partial \mu_i} = \frac{1}{k_B T} \frac{\sum C N_i \exp \frac{1}{k_B T}\left[\sum \mu_i N_i - U\right]}{\sum C \exp \frac{1}{k_B T}\left[\sum \mu_i N_i - U\right]} = \frac{\bar{N}_i}{k_B T}. \tag{A2.25}$$

Multiplying the two terms on the right-hand side of this equality by $k_B T \Xi$ gives

$$\sum N_i C \exp \frac{1}{k_B T}\left[\sum \mu_i N_i - U\right] = \Xi \bar{N}_i. \tag{A2.26}$$

Differentiating this expression again with respect to μ_k and multiplying by $k_B T/\Xi$ leads to

$$\overline{N_i N_k} - \bar{N}_i \bar{N}_k = k_B T \frac{\partial \bar{N}_k}{\partial \mu_i} = k_B T \frac{\partial \bar{N}_i}{\partial \mu_k} = (\overline{\Delta N_i \Delta N_k})_{V,N} \tag{A2.27}$$

which gives the desired equation

$$(\overline{\Delta N_i \Delta N_k})_{V,N} = k_B T \left(\frac{\partial \bar{N}_i}{\partial \mu_k}\right)_{V,T} = k_B T \left(\frac{\partial \bar{N}_k}{\partial \mu_i}\right)_{V,T} \tag{A2.28}$$

which has been used for the evaluation of the scattered intensity at zero angle in Chapter 7.

A2.4 THE EXCHANGE CHEMICAL POTENTIAL

We assume that we have a system made of a solvent (characterized by the index 0) and p constituents, each being characterized by the number of molecules N_i of partial volume v_i and chemical potential μ_i. We define the number z_i as the ratio of the volume of one molecule of polymer to the volume of one molecule of solvent and write (for $i \neq 0$)

$$\bar{\mu}_i = \frac{\mu_i}{z_i} - \mu_0 .$$

The volume fraction φ_i occupied by the species i is

$$\varphi_i = \frac{z_i N_i}{N_0 + \sum_1^p z_i N_i} .$$

If we imagine the molecules distributed on a lattice we have to divide the large molecules in units having the size of a lattice cell which is identical to the size of a solvent molecule. $N_0 + \sum_1^p z_i N_i$ represents the total number of cells in the system and will be called N_T. The expression for G, following eqn (A2.9) becomes

$$G = N_T\left[\mu_0 + \sum \bar{\mu}_i \varphi_i\right]. \tag{A2.29}$$

Introducing the free energy per cell, g_c, defined as the free energy per volume of one molecule of solvent

$$g_c = \frac{G}{N_T} = \mu_0 + \sum \bar{\mu}_i \varphi_i \tag{A2.30}$$

we obtain by differentiation at T and N_T constant

$$dg_c = \mu_0 \frac{dV}{V} + \sum \bar{\mu}_i d\varphi_i \tag{A2.31}$$

and

$$\left(\frac{\partial g_c}{\partial \varphi_i}\right)_{p,T,N_T} = \bar{\mu}_i. \tag{A2.32}$$

Substituting $\bar{\mu}_i$ for μ_i in the Gibbs–Duhem eqn (A2.10) and assuming z_i constant we obtain

$$d\mu_0 + \sum \varphi_i d\bar{\mu}_i = v dp - s dT \tag{A2.33}$$

where v and s are the volume and the entropy respectively, per cell. The relationship

$$\frac{\partial \bar{\mu}_i}{\partial p} = \frac{1}{z_i}\frac{\partial \mu_i}{\partial p} - \frac{\partial \mu_0}{\partial p} = \frac{v_i}{z_i} - v_0 = 0 \tag{A2.34}$$

explains why

$$\left(\frac{\partial \bar{\mu}_i}{\partial N_i}\right)_{V,T} = \left(\frac{\partial \bar{\mu}_i}{\partial N_i}\right)_{p,T}. \tag{A2.35}$$

All the properties established for μ_i can be easily adapted to the $\bar{\mu}_i$. For instance: if a system is in equilibrium in the presence of different phases the chemical potential of the phases must be identical. This is also true for the exchange potentials: the system is in equilibrium if the $\bar{\mu}$s and μ_0 are the same in the different phases. This is obvious if one writes the $p + 1$ corresponding equations.

When a two-component and two-phase system reaches the critical point the following relationships have to be obeyed. Either

$$\frac{\partial \mu_1}{\partial \varphi_1} = \frac{\partial^2 \mu_1}{\partial \varphi_1^2} = 0 \tag{A2.36}$$

or

$$\frac{\partial \mu_0}{\partial \varphi_0} = \frac{\partial^2 \mu_0}{\partial \varphi_0^2} = 0. \tag{A2.36$'$}$$

They are replaced, using the exchange potentials, by the simple relationship

$$\frac{\partial^2 g_c}{\partial \varphi^2} = \frac{\partial^3 g_c}{\partial \varphi^3} = 0. \tag{A2.37}$$

The curve $\partial^2 g_c / \partial \varphi^2 = 0$ is the spinodal which, as we have seen (Chapter 7) occurs when the scattering becomes infinite and can be generalized for multicomponent systems by

$$\text{Det} \left| \frac{\partial^2 g_c}{\partial \varphi_i \partial \varphi_j} \right| = 0 \tag{A2.38}$$

A2.5 THE THEORY OF REGULAR SOLUTIONS

A.2.5.1 The entropy of mixing of ideal solutions

Let us consider a cell model. In the initial stage we assume that N_0 contiguous cells are filled with the liquid 0, and N_1 with the liquid 1, both liquids having the same molecular volume. If we suppress the partition between 1 and 2 the molecules will occupy the $N = N_0 + N_1$ cells randomly and the mixing entropy will be the logarithm of the total number of possible states multiplied by the Boltzmann constant. The calculation of this number is classical, assuming that the only term which is modified during the mixing is the configurational term

$$\Omega = \frac{N!}{N_1! N_2!} \tag{A2.39}$$

Taking the logarithm and using Stirling's formula we obtain for the entropy of mixing, ΔS

$$\Delta S = -\, k_B \left[N_1 \log \frac{N_1}{N_1 + N_2} + N_2 \log \frac{N_2}{N_1 + N_2} \right]. \tag{A2.40}$$

If there is no energy change during the mixing, the free energy of mixing is just equal to $-T\Delta S$ and the free energy of mixing per cell g_c is

$$g_c = k_B T [\,(1 - \varphi) \log (1 - \varphi) + \varphi \log \varphi] \tag{A2.41}$$

where φ is the volume fraction of the constituent 1. This gives for the exchange chemical potential of 1

$$\bar{\mu}_1 = \frac{\partial g_c}{\partial \varphi} = \log\varphi - \log(1-\varphi). \tag{A2.42}$$

A2.5.2 The energy term in regular solutions

This theory can be improved by taking into account the fact that the energy (or enthalpy) of the molecules is not the same in the pure liquid and in the mixture. For this purpose we assume that the interactions are limited to the nearest neighbours and we assume that each molecule has, in the lattice, s nearest neighbours. We call w_{00} the energy of a pair of molecules of type 0, w_{11} the energy of a pair 1–1 and w_{01} the energy of a pair 0–1. This energy is the difference between the energy of an isolated molecule and the energy of the molecule in the liquid and is always negative. It can be measured approximately, for pure liquids by the latent heat of evaporation. In the pure liquids 0 and 1 one has

$$U_{00} = \tfrac{1}{2} N_0 s w_{00} \quad \text{and} \quad U_{11} = \tfrac{1}{2} N_1 s w_{11} \tag{A2.43}$$

since, if we forget about the effects of the wall, there are $N_0 s/2$ 0–0 pairs and $N_1 s/2$ 1–1 pairs. In the mixture there are three different types of pairs 0–0, 1–1, and 0–1. If we call sX_{00}, sX_{11}, and sX_{00} the number of the corresponding pairs we have

$$U_{\text{mixture}} = sX_{00} w_{00} + sX_{11} w_{11} + sX_{01} w_{01}. \tag{A2.44}$$

The problem is to evaluate the variation of energy on mixing, i.e. the quantity

$$\Delta U = U_{\text{mixture}} - (U_{00} + U_{11}). \tag{A2.45}$$

Counting the number of pairs starting from the molecules 0 we find sN_0 pairs and among these pairs there are sX_{10} joining a molecule 1 and $2sX_{00}$ joining two molecules 0 (each pair is counted twice in the second case). This gives

$$sN_0 = sX_{01} + 2sX_{00}. \tag{A2.46}$$

Interchanging 0 and 1 gives

$$sN_1 = sX_{01} + 2sX_{11}. \tag{A2.47}$$

We eliminate sX_{00} and sX_{11} between eqns (A2.45), (A2.46), and (A2.47) obtaining

$$\Delta U = sX_{01}\left[w_{01} - \tfrac{1}{2}(w_{00} + w_{11}) \right]. \tag{A2.48}$$

Thus the only parameter which has to be taken into account is the quantity $w_{01} - (w_{00} + w_{11})/2$ the difference in energy between a 0–1 pair and the average value of the energy of the 0–0 and 1–1 pairs. Alternatively we could say that, in order to make two 0–1 pairs one has to break one 0–0 pair and one 1–1 pair.

The difficult problem is the evaluation of X_{01}. It has been the subject of many theories but here we shall use the simplest approach assuming that the difference of energies $w_{01} - (w_{00} + w_{11})/2$ is small compared to $k_B T$ and that the molecules are randomly distributed. If we consider that one molecule of type 0 has s neighbours the probability for one of these neighbours to be of type 1 is $zN_1/N_T = \varphi$. Therefore the number of pairs 0–1 it forms is $s\varphi$. Since there are N_0 molecules of this type the total number of pairs is $sX_{01} = sN_0\,\varphi$ and the energy (or enthalpy) of mixing is

$$\Delta U = [w_{01} - \tfrac{1}{2}(w_{00} + w_{11})]sN_0\varphi = [w_{01} - \tfrac{1}{2}(w_{00} + w_{11})]s\varphi(1-\varphi)N_T. \tag{A2.49}$$

Adding the entropy of mixing of ideal solutions to the energy of mixing which has just been evaluated leads to what is called the regular solution. Its free energy of mixing is given by

$$g_c = k_B T[(1-\varphi)\log(1-\varphi) + \varphi\log\varphi] + [w_{01} - \tfrac{1}{2}(w_{00} + w_{11})]s\varphi(1-\varphi). \tag{A2.50}$$

Following Flory, we introduce the parameter χ called the interaction parameter and defined by

$$\chi = \frac{s}{k_B T}[w_{01} - \tfrac{1}{2}(w_{00} + w_{11})] \tag{A2.51}$$

obtaining for regular solutions

$$g_c = k_B T[(1-\varphi)\log(1-\varphi) + \varphi\log\varphi + \chi\varphi(1-\varphi)]. \tag{A2.52}$$

A2.6 FREE ENERGY OF MIXING OF POLYMER SOLUTIONS AND MIXTURES

The simultaneous discovery of a general formula giving the free energy of mixing of polymer mixtures by Flory (1942) and Huggins (1942) has been an important step in our understanding of the properties of macromolecular solutions and mixtures. The theories (ideal and regular solutions) which we developed in the last paragraph cannot be applied to macromolecular solutions. They fail since the molecules have very different volumes and the entropy of mixing is not given by eqn (A2.40). The extension of the calculation of the entropy of mixing to mixtures of molecules of different size and shape is still an unsolved problem and approximations are necessary. Flory divided the macromolecules into z units of volume v_0, and assumed that each unit occupies a cell of the lattice and evaluated the number of possible ways of putting N_1 macromolecules of solute on this lattice. For this purpose he was obliged to assume that the probability of finding a free lattice cell is governed by the average occupation of the lattice by macromolecular segments. This is equivalent to the assumption of mean field theories since it neglects the concentration fluctuations. This calculation is not difficult but rather long and we refer to Flory's book for details, giving here only the result, which is for the entropy of mixing per cell

$$\Delta s_c = -k_B\left[(1-\varphi)\log(1-\varphi) + \frac{\varphi}{z}\log\varphi\right] \tag{A2.53}$$

where φ is the volume fraction of the polymer which has z as degree of polymerization. Since a mean field approximation is used for the entropy it can also be used for the energy. The energy term of regular solutions is not modified and the final result is

$$g_c = k_B T\left[(1-\varphi)\log(1-\varphi) + \frac{\varphi}{z}\log\varphi + \chi\varphi(1-\varphi)\right]. \tag{A2.54}$$

This equation is easily generalized to the case of multicomponent systems. If we assume that we have the solvent plus p species numbered from 1 to p we write

$$\frac{g_c}{k_B T} = \varphi_0\log\varphi_0 + \sum_{i=1}^{p}\frac{\varphi_i}{z_i}\log\varphi_i + \sum_{i=1}^{p}\chi_{0i}\varphi_0\varphi_i + \sum_{1}^{p}\sum_{<j}^{p}\chi_{ij}\varphi_i\varphi_j. \tag{A2.55}$$

It is understood that the χ_{ij} are the interaction parameters between the monomeric units i and j, the monomeric unit being by definition the part of the polymer having the same volume as a molecule of solvent. From these equations the different quantities used in scattering theory can be evaluated as

$$\frac{1}{k_B T}\mu_0 = \log\varphi_0 + (1-\varphi_0) - \sum_{i=1}^{p}\frac{1}{z_i}\varphi_i + (1-\varphi_0)\sum_{i=1}^{p}\chi_{0i}\varphi_i - \sum_{i}^{p}\sum_{<j}^{p}\chi_{ij}\varphi_i\varphi_j \tag{A2.56}$$

$$\frac{1}{k_B T}\bar{\mu}_i = -\log\varphi_0 + \frac{1}{z_i}\log\varphi_i + \frac{1}{z_i} - 1 + \chi_{0i}\varphi_0 - \sum_{i=1}^{p}\chi_{0i}\varphi_i + \sum_{k=1}^{p}\chi_{ik}\varphi_k \tag{A2.57}$$

$$\frac{1}{k_B T}\frac{\partial^2 g_c}{\partial\varphi_i\partial\varphi_j} = \frac{1}{\varphi_0} - \chi_{0i} - \chi_{0j} + \chi_{ij} \tag{A2.58}$$

$$\frac{1}{k_B T}\frac{\partial^2 g_c}{\partial\varphi_i^2} = \frac{1}{z_i\varphi_i} + \frac{1}{\varphi_0} - 2\chi_{0i}. \tag{A2.59}$$

(If one wants these second derivatives of g_c in terms of excluded volume parameters v_{ij} one has to use eqn (7.65).)

When differentiating g_c or $\bar{\mu}$ with respect to φ_i, φ_0 has to be written as $1 - \Sigma\varphi_k$ since the variables are N_T, φ_1, φ_2, . . ., φ_p. If the system is a mixture of polymers without solvent we have to transform the expression for g_c. If we simply make $\varphi_0 = 0$ we obtain for eqn (A2.55)

$$\frac{g_c}{k_B T} = \sum_{i=1}^{p}\frac{\varphi_i}{z_i}\log\varphi_i + \sum_{i}^{p}\sum_{<j}^{p}\chi_{ij}\varphi_i\varphi_j \tag{A2.60}$$

but we must remember that all the φ_i are not independent since $\sum_{i=1}^{p}\varphi_i = 1$.

One of the species has to replace the solvent and becomes the reference. We give it the index zero, writing for the system with p components

$$\frac{g_c}{k_B T} = \frac{\varphi_0}{z_0}\log\varphi_0 + \sum_{i=1}^{p-1}\frac{\varphi_i}{z_i}\log\varphi_i + \sum_{i=1}^{p-1}\chi_{0i}\varphi_0\varphi_i + \sum_{i}^{p-1}\sum_{<j}^{p-1}\chi_{ij}\varphi_i\varphi_j. \tag{A2.61}$$

The derivatives have again to be taken assuming that φ_0 is equal to $1 - \sum_{i=1}^{p-1}\varphi_i$. In fact the only modification which has been effected has been to change φ_0 into φ_0/z_0 in the first term. This could have been done without decreasing the number of species by one, by saying that we use one of the polymers as a solvent. There is still a problem for the choice of the values of the parameter z. The choice of its dimensions is arbitrary, but it does affect the magnitude of the coefficient χ. Therefore, if one wants to give numerical values for χ, one has to define precisely what has been used for the definition of a 'monomer'.

A2.7 PHASE EQUILIBRIA IN FLORY–HUGGINS THEORY

Let us consider a mixture of two polymers and draw the free energy per cell g_c as function φ, the volume fraction occupied by polymer 1 for dif-

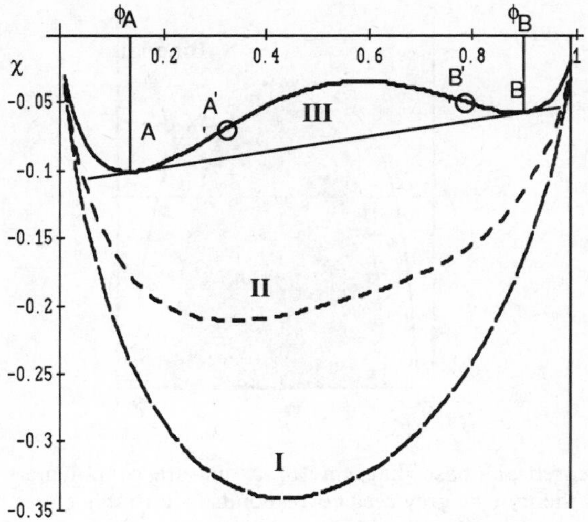

Fig. A2.1 The quantity g_c for a binary mixture as a function of the composition φ for different values of the interaction parameter χ. (— — —) $\chi = 1$; (---) $\chi = 1.65$; (—) $\chi = 2$.

ferent values of χ. Depending on the value of χ we obtain one of the three types of curves in the Fig. A2.1. The curve I corresponds to low values of χ and shows that the two species are miscible in all proportions. If χ is higher one obtains the curve III. This curve has a double tangent touching the curve at the points A and B. This means that between φ_A and φ_B the two species 0 and 1 are not miscible and that the system is made of two phases of respective compositions φ_A and φ_B. If, starting from pure A one adds B, we have one phase until the composition in polymer 1 reaches φ_A. When we increase still further the concentration of polymer, a second phase appears as soon as φ is larger than φ_A and the system is a mixture of two phases of composition φ_A and φ_B. When the total composition is larger than φ_B we again have one phase. If now we decrease χ the distance AB decreases and reaches zero for a given value of χ called χ_c or χ critical. The phases which are in equilibrium have the same composition φ_c which corresponds to what is called the critical point. There are two other points of interest on this diagram: the points A′ and B′, inflection points of g_c as a function of φ. At these points the composition fluctuations and therefore the scattering intensity becomes infinite and for $\varphi_{A'} < \varphi < \varphi_{B'}$ the mixture is absolutely unstable. Between $\varphi_{A'}$ and $\varphi_{B'}$ it is possible to have a mixture in contrast to what happens in the domains $\varphi_A < \varphi < \varphi_{A'}$ and $\varphi_{B'} < \varphi < \varphi_B$ where the system can exist in a metastable equilibrium.

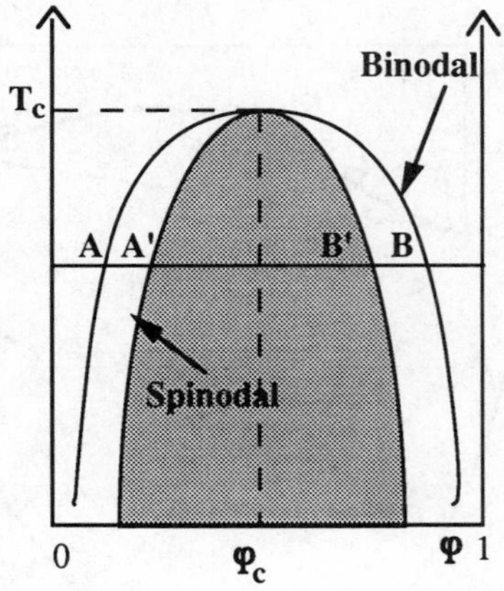

Fig. A2.2 Theoretical phase diagram for a mixture of polymers following the Flory–Huggins theory (the grey area corresponds to unstable states of the system).

We can now draw on a diagram the position of the points φ_A, φ_B, and $\varphi_{A'}$, $\varphi_{B'}$ vs χ (or T since in the simple expression of the Flory–Huggins theory it is assumed that χ depends only on T through the eqn (A2.51) where w is constant). The two curves of Fig. (A2.2) are obtained. The locus of the points A, B is called the binodal and that of the points A′, B′ the spinodal. The equation of the binodal cannot be written as an analytical expression and has been drawn approximately. The equation of the spinodal is, on the contrary, simple to write since it is given by $\partial^2 g_c/\partial\varphi^2 = 0$ or applying eqn (A2.60) to the case of a binary mixture by

$$\frac{1}{z_0(1-\varphi)} + \frac{1}{z_1\varphi} - 2\chi = 0 \tag{A2.62}$$

At the critical point, which is the point where both spinodal and binodal have a common horizontal tangent we have also

$$\frac{\partial^3 g_c}{\partial\varphi^3} = 0 \quad \text{or} \quad \frac{1}{z_0(1-\varphi)^2} - \frac{1}{z_1\varphi^2} = 0.$$

Solving these equations leads to the result

$$\varphi_c = \frac{\sqrt{z_0}}{\sqrt{z_0}+\sqrt{z_1}} \quad \text{and} \quad 2\chi_c = \frac{(\sqrt{z_0}+\sqrt{z_1})^2}{z_0 z_1}. \tag{A2.63}$$

This is simplified into

$$\varphi_c = \frac{1}{1 + \sqrt{z_1}}$$

and

$$\chi_c = \tfrac{1}{2} + \frac{1}{\sqrt{z_1}} + \frac{1}{2z_1} \qquad \text{(A2.64)}$$

if one deals with a solution. If the two molecules have the same size ($z = z_0 = z_1$)

$$\varphi_c = \tfrac{1}{2} \quad \text{and} \quad z\chi_c = 2. \qquad \text{(A2.65)}$$

From this last equation we see that, if the molecular weight increases, it is sufficient to have a small positive value of χ in order to reach phase separation.

A2.8 REAL POLYMER MIXTURES

The Flory–Huggins theory is a very good tool for studying and understanding the properties of polymer solutions and mixtures. It allows us to interpret qualitatively the quasitotality of the phenomena observed and, even now, it is the most frequently used tool for describing the effects which are observed in many circumstances. One could compare it to the Van der Waals equation of state for pure gases and liquids. This equation explains very well the observed phenomena but gives a very poor agreement if one wants to exploit the data quantitatively. Flory and many others have tried to improve the situation but if these improvements are efficient on some problems they do not have the generality and the elegance of the original Flory theory. In order to discuss this point let us concentrate our attention on the parameter χ. It is, in fact, the only parameter of the theory. From its definition (eqn (A2.51)) it should be proportional to T^{-1} since the u_{ij} are supposed to be constant. Flory himself introduced a more general expression for χ, writing

$$\tfrac{1}{2} - \chi = \Psi\left(1 - \frac{\theta}{T}\right)$$

where Ψ is an entropic term and $\Psi\theta$ an enthalpic term. This introduces the well-known θ temperature for which the second virial coefficient and the excluded volume parameter are zero and the chains obey Gaussian statistics. This equation leaves χ independent of concentration and many experimental results show that this is not the case. There is an interesting

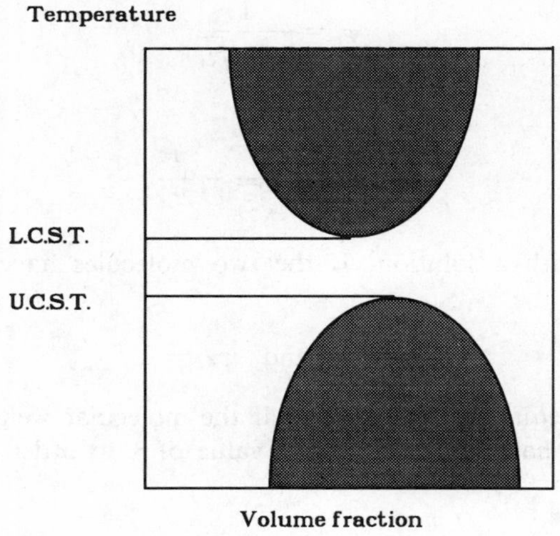

Fig. A2.3 Schematic representation of the phase diagram of a mixture of two polymers and definition of the LCST and the UCST. The grey area corresponds to phase separation.

case which shows that the Flory theory does not explain all the properties of solutions. We have shown that the phase diagram of a binary mixture of two homopolymers was given by curves similar to the cuves represented in Fig. A2.1.

At high temperature the two polymers are miscible in all proportions; when the temperature decreases one observes phase separation; what is called an upper critical solution temperature (UCST). A few decades ago it was shown that one can obtain the reverse situation; the polymers are not miscible at high temperature and become compatible when the temperature decreases; in this case one has what is called an lower critical solution temperature (LCST) (Fig. A2.3). In many cases one observes both phenomena: when temperature increases, first there is phase separation and again complete miscibility. This fact cannot be explained in the framework of the Flory theory and Flory (1965) himself proposed a new theory, taking into account the difference in compressibility of the species present in the mixture, allowing him to explain these experimental results. Intensive studies have been devoted to this problem. Moreover, as we already said, the Flory theory is a 'mean field theory' forgetting the fluctuations in the system. In recent years considerable effort has been made in this field in order to introduce these fluctuations. New concepts have been introduced and a systematic use has been made of what is called scaling arguments and renormalization theories. These theories introduce non-classical critical exponents and they present a considerable improvement specially in the vicinity of the critical points and in the dilute solution range. We shall not discuss these points here and, in order to avoid difficulties we shall take the following point of view. Regardless of the real free energy of mixing, we shall say that the free energy can always be represented by an equation like eqn (A2.55) or eqn (A2.61), putting all that is unknown in the definition of the χ parameter. Thus χ becomes a purely experimental quantity which can be used for the description of the mixture and depends in an unknown way on all the thermodynamic variables. This allows us to interpret all scattering data using this parameter and to leave the discussion of its interpretation to the specialists in the field.

SUGGESTED FURTHER READING

De Gennes (1979).
Flory (1953).
Hill (1956).
Kurata (1982).
Tompa (1956).
Yamakawa (1971).

Appendix 3

Some remarks about the use of theoretical formulae for the interpretation of experimental data; comparison with light and X-ray scattering

A3.1 INTRODUCTION

Throughout this book we have tended to use the volume fraction and the scattering length of what we called 'monomers' but which were in fact units having the same volume as a molecule of solvent. The coherent scattering length of these units was called b. For mixtures of polymers the situation was not very different. We took for a reference one type of monomer and, as in the case of solutions, we defined the 'monomers' of the other polymers as having the same volume. These units are very convenient for writing equations but are in fact not convenient at all for experimenters since their values change from one system to the other and they cannot be found in the literature. It is therefore absolutely necessary to be able to change units. This is in fact very trivial but so many people have difficulties in doing so that it is perhaps useful to emphasize this point. We start with the case of solutions. The attentive reader has perhaps noticed that in eqn (7.18) we introduced $b_{\mathrm{app}} = \mathscr{b}_1 - b_0 V_1/v_0$ and in eqn (7.56) $\bar{b} = \mathscr{b}_1/z_1 - b_0$ with $z_1 = V_1/v_0$. ($\mathscr{b}_1$ is the coherent scattering length of the polymer). These two expressions are not contradictory and need some explanation. In the first, we take the dissolved polymer as reference and subtract from the scattering length of the polymer the scattering length of the same volume of solvent. In the second, we take the solvent as reference and consider the scattering length of a part of the polymer which has the same volume as the solvent. These two expressions have been used because in Section 7.1.3 we were considering only one polymer and focused our attention on the scattering by this polymer whereas in Section 7.3.2, since we had different types of polymers, we were obliged to consider the molecules of the solvent as the reference. Clearly the result one obtains has to be the same and this is the point on which we would like to dwell. For this purpose we will first discuss the zero angle scattering where all the problems reside, and the dilute solution approximation. Subsequently, we will discuss the other cases such as concentrated solutions and the angular dependence of the scattering. Moreover, since all equations which we have been using (except those in Chapter 5 which is devoted to the effect of isotopic substitution) can be applied to light and X-ray scattering, it is interesting to explain the changes which have to be made to the results in order to switch from neutron equations to X-ray and light scattering. All these calculations do not involve any theoretical difficulty but, in the experience of the authors, they can be very frustrating and time-consuming. This appendix has been written in the hope that it might be helpful to experimenters facing this type of difficulty.

A3.1.1 The problem of the contrast factor

It has been shown (eqn (7.17)) that what we have called $I(0)$ or $\partial\sigma/\partial\Omega$ at zero angle can be written as

$$I(0) = b_{\text{app}}^2 \frac{k_B T \varphi}{\partial\pi/\partial\varphi} \frac{V}{V_1^2} \tag{A3.1}$$

or starting from the general eqn (7.94″)

$$\frac{\bar{b}^2}{i(0)} = \frac{1}{\varphi z} + v. \tag{A3.2}$$

Let us consider first the problem of the contrast factor. Until now, if we limited ourselves to the case of coherent scattering, we used four different quantities which were called $\mathscr{b}_1$, b_1, b_{app}, and $\bar{b}$. Let us recall their definitions: $\mathscr{b}_1$ is the coherent scattering length of the polymer and b_1 of the 'monomer'. b_{app} is the quantity $(\mathscr{b}_1 - b_0 V_1/v_0)$ where b_0 is the scattering length of the solvent and V_1/v_0 the ratio of the volume of one of the scattering molecules to the volume of the solvent molecule. Dividing by V_1 one obtains for b_{app}

$$b_{\text{app}} = V_1\left(\frac{\mathscr{b}_1}{V_1} - \frac{b_0}{v_0}\right). \tag{A3.3}$$

This last quantity is the difference between the coherent scattering length per unit volume of the macromolecule and that of the solvent. If the macromolecule is a polymer made of repeating units the quantity $\mathscr{b}_1/V_1$ does not depend on the size of the polymer. Both quantities are proportional to the degree of polymerization and instead of using the letters $\mathscr{b}_1$ and V_1, which refer to the polymer, we shall use the letters b_1 and v_1 which refer to the monomer units. In what follows no distinction will be made between the polymer and the monomer since the quantity b_1 appears only as the ratio b_1/v_1 which is independent of the size of the macromolecule and is the scattering length per unit volume. The same argument gives for $\bar{b}$

$$\bar{b} = \frac{\mathscr{B}_1}{z_1} - b_0 = v_0\left(\frac{\mathscr{B}_1}{V_1} - \frac{b_0}{v_0}\right) = v_0\left(\frac{b_1}{v_1} - \frac{b_0}{v_0}\right). \tag{A3.4}$$

This leads us to introduce the quantity $(b_1/v_1 - b_0/v_0)$ which we shall call b_v

$$b_v = \left(\frac{b_1}{v_1} - \frac{b_0}{v_0}\right). \tag{A3.5}$$

One can, from the values of the coherent scattering length collected in tables and knowing the chemical structure of the polymer, obtain the values of b_1 and b_0 per monomer or molecule. In order to calculate eqn (A3.2) the volume of these molecules in the system under study is required and this quantity is usually not available. In the majority of the cases it can be calculated by assuming that the densities of the constituents in the mixture are equal to the bulk density of the components. This approximation means that one assumes that there is no change of volume during mixing and is used in all the theoretical models we have considered (except for the discussion of the thermodynamical effect of isotopic substitution). Let us assume that we know the specific volumes v_1^s and v_0^s of both polymer and solvent. The mass m_1 of the monomer and m_0 of the solvent are known from the chemical formulae; therefore their volumes should be $v_1 = m_1 v_1^s$ and $v_0 = m_0 v_0^s$ and

$$b_{ap} = \frac{M_1}{N_A} v_1^s \left(\frac{b_1}{m_1 v_1^s} - \frac{b_0}{m_0 v_0^s}\right) = \frac{M_1}{N_A} v_1^s b_v \tag{A3.3$'$}$$

$$\bar{b} = m_0 v_0^s \left(\frac{b_1}{m_1 v_1^s} - \frac{b_0}{m_0 v_0^s}\right) = m_0 v_0^s b_v \tag{A3.4$'$}$$

or, if we use the specific masses ρ_i, we can define the scattering length density ρ_{b_i} as $b_i \rho_i / m_i$. This allows us to write

$$b_v = \left(\frac{b_1 \rho_1}{m_1} - \frac{b_0 \rho_0}{m_0}\right) = \rho_{b_1} - \rho_{b_0} \tag{A3.5$'$}$$

where everything is known. (All the quantities are supposed to be expressed in standard units except for the molecular weight of the polymer which is written in daltons.) As pointed out already neither the quantity inside the bracket nor $\bar{b}$ itself depends on the size of the macromolecules. Thus this parameter b_v is a very useful quantity and from now on we shall use it extensively. In order to evaluate $\bar{b}$ one has to multiply b_v by the reference volume v_0. If one wants b_{app} one multiplies by the volume V_1 of the polymer. In Table A3.1 we have collected values of ρ_{b_i} for some common solvents and polymers using the formula

$$\rho_b = \frac{\sum \rho_i b_i}{\sum m_i 1.66 \times 10^{-24}}$$

Table A3.1 Neutron scattering length densities (ρ_b and contrast factors (b_v^2) for some common polymers and solvents. Note that a few H–H mixtures have a small but finite contrast (the rest are two orders of magnitude smaller) because of the large difference in hydrogen content

	ρ_b → / ↓ ($\times 10^5$ Å^{-2})	H-PS	D-PS	H-PMMA	D-PMMA	H-PE	D-PE
ρ_b ($\times 10^5$ Å^{-2})		0.14	0.642	0.1097	0.679	0.032	0.826
H-PS	0.14		0.252		0.290	0.029	0.470
D-PS	0.692	0.252		0.283		0.454	
H-PMMA	0.1097		0.283		0.324	0.02	0.513
D-PMMA	0.679	0.290		0.324		0.505	
H-PE	0.032	0.029	0.454	0.02	0.505		0.736
D-PE	0.826	0.47		0.513		0.736	
C_6H_6	0.1184		0.274		0.314	0.022	0.707
C_6D_6	0.544	0.163		0.188		0.331	
C_6H_{12}	0.0277	0.028	0.448	0.019	0.499		0.728
C_6D_{12}	0.0672	0.283		0.316		0.496	

ρ_b column: $\times 10^5$ Å^{-2}; table entries: $b_v^2 \times 10^{10}$ Å^{-4}

$$\rho_b = \frac{\sum b \times \rho}{\sum m \times 1.66 \times 10^{-24}} \times 10^{-16} \text{ Å}^{-2}$$

$$b_v = \rho_b^{(1)} - \rho_b^{(2)}$$

b values from Table 3.1, values of the density ρ as below:

Solvents from literature.
Polystyrene from Rawiso *et al.* (1987).
Polymethylmethacrylate from Fernández *et al.* (1990).
Polyethylene (HDPE) from the *Polymer Handbook*.
D-polyethylene assuming same molar volume as H-PE.

and the values of $b_v^2 = (\rho_{b_i} - \rho_{b_j})^2$ for the corresponding solutions. The values for the solvents are taken from the literature. We have used the results of Rawiso *et al.* (1987) for polystyrene, Fernández *et al.* (1990) for polymethacrylate and from Bandrupt and Immergut (1975) for the D-polyethylene assuming that it has the same molar volume as H-PE. (See Table A3.1.)

If we work on a mixture of polymers the tradition is to take as 'reference volume' the volume of one of the monomers but, unless the value of the χ parameter is needed, this choice is completely arbitrary.

A3.2 THE SCATTERING AT INFINITE DILUTION

A3.2.1 Infinitely dilute solution at zero angle

We have to rewrite the first term of eqn (A3.2). In this equation $i(0)$ is the scattering from the volume v_0. The experimental quantity is the scattering per unit volume $I(q)/V$ which is evidently equal to $\dfrac{i(q)}{v_0}$.

Using the fact that we are at infinite dilution and zero angle we write

$$\frac{I(0)}{V} = v_0^2 b_v^2 \varphi z \frac{1}{v_0} = b_v^2 \varphi V_1 \tag{A3.6}$$

since $z = V_1/v_0$ and is a kind of 'degree of polymerization'. If we introduce the molecular weight M_1 of the polymer via $V_1 = M_1/N_A v_1^s$ (N_A is Avogadro's number), we obtain

$$\frac{I(0)}{V} = b_v^2 \varphi \frac{M_1}{N_A} v_1^s. \tag{A3.7}$$

Alternatively, using the concentration c instead of the volume fraction

$$\frac{I(0)}{V} = b_v^2 c \frac{M_1}{N_A} (v_1^s)^2 \tag{A3.8}$$

since concentration and volume fraction are simply related by $cv_1^s = \varphi$. We have recovered the classical result that the scattering is proportional to the product cM or φM and all the coefficients can be measured.

A3.2.2 The equation for dilute solutions at any q

In order to introduce the angular dependence it suffices to multiply eqn (A3.8) by the form factor $P_1(q)$ which is by definition a dimensionless quantity and one obtains:

$$\frac{I(q)}{V} = b_v^2 \varphi \frac{M_1}{N_A} v_1^s P_1(q) = \bar{b}^2 \varphi V_1 P_1(q). \tag{A3.9}$$

(As with the rest of the book all quantities are expressed in molecular units, except for the molecular weights which are expressed in daltons.)

A3.3 THE GENERAL CASE

A3.3.1 Polymer solution

We start from eqn (A3.2) which gives the scattering by a polymer solution and we replace $i(q)$ by $I(q)/N_T$, ($N_T = V/v_0$). In the inverse form the equation becomes

$$\frac{V}{I(q)} = \frac{v_0}{i(q)} = \frac{v_0}{\bar{b}^2}\left\{\frac{1}{\varphi z P_1(q)} + v\right\}. \tag{A3.10}$$

Replacing $\bar{b}$ and z by $v_0\, b_v$ and V_1/v_0 respectively gives

$$\frac{V}{I(q)} = \frac{1}{v_0 b_v^2}\left\{\frac{v_0}{\varphi V_1 P_1(q)} + v\right\} \tag{A3.11}$$

or introducing the molecular weight

$$\frac{V}{I(q)} = \frac{1}{b_v^2}\left\{\frac{1}{\varphi \dfrac{M}{N_A} v_1^s P_1(q)} + \frac{v}{v_0}\right\}. \tag{A3.12}$$

Now if we use the Flory–Huggins theory (eqn (7.86)), for the calculation of the excluded volume parameter v we obtain

$$\frac{V}{I(q)} = \frac{1}{b_v^2}\left\{\frac{1}{\varphi \dfrac{M}{N_A} v_1^s P_1(q)} + \frac{1}{(1-\varphi) m_0 v_0^s} - \frac{2\chi}{v_0}\right\}. \tag{A3.13}$$

A3.3.2 Mixture of two polymers

We want to replace the solvent by a second polymer and to interpret the de Gennes formula in terms of measurable quantities. We identify the polymers as 1 and 2 and fix an arbitrary reference volume as v_0. Following our definition we have to write

$$\bar{b} = \frac{\ell_2}{z_2} - \frac{\ell_1}{z_1} \text{ with } z_1 = \frac{V_1}{v_0} \text{ and } z_2 = \frac{V_2}{v_0} \tag{A3.14}$$

or

$$\bar{b} = v_0 \left\{ \frac{b_1}{v_1} - \frac{b_2}{v_2} \right\} = v_0 b_v . \tag{A3.15}$$

We note that the term in the bracket is, as previously, the difference between the scattering length per unit volume of polymers 1 and 2 and independent of the choice of v_0.

Putting this value in the de Gennes equation

$$\frac{\bar{b}^2}{i(q)} = \frac{1}{\varphi_1 z_1 P_1(q)} + \frac{1}{\varphi_2 z_2 P_2(q)} - 2\chi \tag{A3.16}$$

gives for the inverse of the intensity scattered per unit volume

$$\frac{V}{I(q)} = \frac{v_0}{i(q)} = \frac{v_0}{\bar{b}^2} \left\{ \frac{1}{\varphi_1 z_1 P_1(q)} + \frac{1}{\varphi_2 z_2 P_2(q)} - 2\chi \right\}$$

$$= \frac{1}{b_v^2 v_0} \left\{ \frac{v_0}{\varphi_1 V_1 P_1(q)} + \frac{v_0}{\varphi_2 V_2 P(q)} - 2\chi \right\}$$

$$\frac{V}{I(q)} = \frac{1}{b_v^2} \left\{ \frac{1}{\varphi_1 \frac{M_1}{N_A} P_1(q) v_1^s} + \frac{1}{\varphi_2 \frac{M_2}{N_A} P_2(q) v_2^s} - \frac{2\chi}{v_0} \right\} \tag{A3.17}$$

with $\varphi_1 + \varphi_2 = 1$.

This formula is perfectly symmetric and shows clearly that the value of the parameter χ which can be obtained experimentally depends on the value of the reference volume which is arbitrary and is usually fixed as the volume of one of the monomers of the mixture. One can also express the result using the concentrations

$$\frac{V}{I(q)} = \frac{1}{b_v^2} \left\{ \frac{1}{c_1 \frac{M_1}{N_A} P_1(q) (v_1^s)^2} + \frac{1}{c_2 \frac{M_2}{N_A} P_2(q) (v_2^s)^2} - \frac{2\chi}{v_0} \right\} \tag{A3.18}$$

with the condition $c_1 v_1^s + c_2 v_2^s = 1$.

A3.4 CALCULATION OF THE SCATTERING AT ZERO ANGLE USING OSMOTIC PRESSURE AND VIRIAL COEFFICIENTS

A3.4.1 Osmotic pressure and scattering

We now want to start from eqn (A3.1), which has been used frequently to relate scattering and osmotic compressibility and to show that this gives exactly the same result. Using the definition of b_{app} (eqn (A3.4) or (7.18)) we have

$$\frac{I(0)}{V} = \left(\frac{b_1}{v_1} - \frac{b_0}{v_0}\right)^2 \frac{k_B T\varphi}{(\partial\pi/\partial\varphi)} = b_v^2 \frac{k_B T\varphi}{(\partial\pi/\partial\varphi)} . \tag{A3.19}$$

Again we have, as contrast factor, the quantity b_v, which depends only on the difference of the scattering lengths per unit volume, and the volume v_0 disappears from the final result. In order to show that this equation is identical to eqn (A3.12) we have to calculate the osmotic compressibility $(\partial\pi/\partial\varphi)$ as function of the quantities defined in Appendix 2. From classical text books the osmotic pressure is $-v_0(\mu_0 - \mu_0^0)$ where $(\mu_0 - \mu_0^0)$ is the difference between the actual chemical potential and its value in the pure solvent and v_0 is the volume of one molecule of solvent. Differentiating this expression with respect to φ gives

$$\frac{\partial\pi}{\partial\varphi} = -v_0 \frac{\partial\mu_0}{\partial\varphi} . \tag{A3.20}$$

In order to show the equivalence between this equation and eqn (A3.12) we shall use, as was done in Appendix 1, the exchange chemical potential

$$\bar{\mu}_1 = \frac{\mu_1}{z_1} - \mu_0 . \tag{A3.21}$$

Differentiating with respect to φ and assuming that z is independent of φ gives

$$\frac{\partial\bar{\mu}_1}{\partial\varphi} = \frac{1}{z}\frac{\partial\mu_1}{\partial\varphi} - \frac{\partial\mu_0}{\partial\varphi} . \tag{A3.22}$$

The Gibbs–Duhem relation $N_1 d\mu_1 + N_0 d\mu_0 = 0$, on differentiation with respect to φ gives

$$\frac{\varphi}{z}\frac{\partial\mu_1}{\partial\varphi} + (1-\varphi)\frac{\partial\mu_0}{\partial\varphi} = 0 \tag{A3.23}$$

or

$$\frac{\partial \mu_1}{\partial \varphi} = -\frac{1-\varphi}{\varphi} z \frac{\partial \mu_0}{\partial \varphi}. \tag{A3.24}$$

Using these relationships one can eliminate $\partial\mu_1/\partial\varphi$ between eqns (A3.22) and (A3.23) obtaining

$$\frac{\partial \mu_0}{\partial \varphi} = -\varphi \frac{\partial \bar{\mu}_1}{\partial \varphi}. \tag{A3.25}$$

Since, from eqn (A2.32), $\partial\bar{\mu}_1/\partial\varphi = \partial^2 g_c/\partial\varphi^2$, replacing $\partial\mu_0/\partial\varphi$ by its value in eqn (A3.20) gives:

$$\frac{I(0)}{V} = v_0 b_v^2 \frac{k_B T}{\dfrac{\partial^2 g_c}{\partial \varphi^2}} = \frac{1}{v_0} \bar{b}^2 \frac{k_B T}{\dfrac{\partial^2 g_c}{\partial \varphi^2}}. \tag{A3.26}$$

This is $i(0)/v_0$ and this result could have been obtained directly from eqn (7.63).

A3.4.2 Scattering and virial coefficients

The virial coefficients are defined by expanding the osmotic pressure as function of the concentration following the formula

$$\frac{\pi}{c} = RT\left(\frac{1}{M_1} + A_2 c + A_3 c^2 + \dots\right) \tag{A3.27}$$

where M_1 is the molecular weight of the polymer, R the gas constant and the A_is the virial coefficients. Since eqn (A3.1) (which at zero angle is equivalent is to eqn (A3.2)) gives the result as a function of $\partial\mu_0/\partial\varphi$ we shall start from this quantity and write from the definition of π

$$\frac{\partial \pi}{\partial \varphi} = \frac{\partial \pi}{\partial c}\frac{\mathrm{d}c}{\mathrm{d}\varphi} = \frac{RT}{v_1^s}\left\{\frac{1}{M_1} + 2A_2 c + 3A_3 c^2 + \dots\right\} \tag{A3.28}$$

which gives for the scattered intensity

$$\frac{I(0)}{V} = \left(\frac{b_1}{v_1} - \frac{b_2}{v_0}\right)^2 \frac{c(v_1^s)^2}{N_A\left\{\dfrac{1}{M_1} + 2A_2 c + 3A_3 c^2 + \dots\right\}}. \tag{A3.29}$$

The inverse quantity is usually used and we obtain, neglecting the higher terms

$$\frac{cV}{I(0)} = \frac{1}{b_v^2}\frac{N_A}{(v_1^s)^2}\left\{\frac{1}{M_1} - 2A_2 c + \dots\right\}. \tag{A3.30}$$

In order to generalize this equation for non-zero angle it suffices to introduce the form factor $P_1(q)$ as was done in Section 7.3. This gives what is generally known as the Zimm equation

$$\frac{cV}{I(q)} = \frac{1}{b_v^2}\frac{N_A}{(v_1^s)^2}\left\{\frac{1}{M_1 P_1(q)} - 2A_2 c + \ldots\right\}. \tag{A3.30'}$$

A3.5 REMARKS ABOUT CONTINUOUS MEDIA

A3.5.1 Introduction

Frequently the notion of a statistical element with volume v_1 is not useful (for example for colloidal particles) and it is preferable to introduce the total volume of the particle, the total length for rods or the lotal surface for discs. In the relevant chapters this has been done automatically for the form factors, since there we allowed the number of units to go to infinity and this gave a simple result since, by definition, the form factor is equal to 1 for $q = 0$. To be complete it is perhaps useful to see what happens for the scattering length term. The easiest way is to go back to the basic formula (eqn (6.2)) and write, for a dilute solution, neglecting interactions

$$I(q) = Nb_v^2 v_0^2 z^2 P_1(q) = N\bar{b}_v^2 V_1^2 P_1(q)$$

and

$$\frac{I(q)}{V} = b_v^2 \varphi V_1 P_1(q) \tag{A3.31}$$

where V_1 is the volume of the particles and is independent of the number of units which has been chosen for the calculation. This equation is identical to eqn (A3.9). From this remark follows the simple rule; when we replace the number of units by the total volume of the particles we have to replace $\bar{b}$ by b_v.

A3.5.2 Definition of the scattering length density

Three-dimensional objects

In some cases, especially for colloidal particles or even reflection studies (see Chapter 10) it is interesting to introduce a scattering length or contrast factor density ρ_b in order to characterize the scattering medium. One very general approach could be to start from the general formula, the basis of the definition of the correlation function, established in Chapter 4 (eqn (4.30)) for systems made of only one constituent

$$\left(\frac{\partial \sigma}{\partial \Omega}\right) = b^2 \iint_{V, V'} \exp(-\mathrm{i} \cdot \boldsymbol{q}(\boldsymbol{r} - \boldsymbol{r}')) \langle n(\boldsymbol{r}) n(\boldsymbol{r}') \rangle \, \mathrm{d}\boldsymbol{r} \, \mathrm{d}\boldsymbol{r}' \quad \text{(A3.32)}$$

where the double summation is extended to the whole scattering volume. If we replace the local scattering density $n(r)$ by the quantity $bn(r) = \rho_b(r)$ thus, introducing the coherent scattering length density instead of the particle density we obtain

$$\left(\frac{\partial \sigma}{\partial \Omega}\right)_{\text{coh}} = \iint_{V, V'} \exp(-\mathrm{i} \cdot \boldsymbol{q}(\boldsymbol{r} - \boldsymbol{r}')) \langle \rho_b(\boldsymbol{r}) \rho_b(\boldsymbol{r}') \rangle \, \mathrm{d}\boldsymbol{r} \, \mathrm{d}\boldsymbol{r}' \quad \text{(A3.32}')$$

which makes the formulae simpler and more general since this formalism makes it possible to consider complex media. In order to take as an example a more concrete case let us consider a two-component system, let us say a solution of independent particles. We have to replace b by b_v and define ρ_b as $b_v n(r)$, obtaining:

$$\frac{I(q)}{V} = \frac{N}{V} \iint_{V, V'} \exp(-\mathrm{i} \cdot \boldsymbol{q}(\boldsymbol{r} - \boldsymbol{r}')) \langle \rho_b(\boldsymbol{r}) \rho_b(\boldsymbol{r}') \rangle \, \mathrm{d}\boldsymbol{r} \, \mathrm{d}\boldsymbol{r}' \quad \text{(A3.33)}$$

where the integration has to be performed over one particle. (The use of b_v is obvious since we are in a continuous medium.) As a more practical example let us first consider a sphere of constant density for which ρ_b is independent of r

$$\frac{I(q)}{V} = \frac{N}{V} \rho_b^2 \left[\int_0^R \frac{\sin qr}{qr} 4\pi r^2 \mathrm{d}r \right]^2 = \frac{N}{V} \rho_b^2 V_1^2 P_1(q). \quad \text{(A3.34)}$$

This formula has been obtained using the procedure developed in Section (6.3). If the sphere has a concentration profile of scattering material $\rho_b(r)$ depending on the distance from its centre this result is easily generalized into:

$$\frac{I(q)}{V} = \frac{N}{V} \left[\int_0^R \rho_b(r) \frac{\sin qr}{qr} 4\pi r^2 \mathrm{d}r \right]^2. \quad \text{(A3.35)}$$

In this case, the notion of a geometric factor $P_1(q)$ disappears since the result depends strongly on the profile of $\rho(r)$. This result has been used for spheres coated with polymers where ρ_b is divided in two parts the colloidal sphere and the polymer film (see eqn (6.98) for an application). Note that the use of this formula is only valid if the density depends only on r; if there are fluctuations in $\rho_b(r)$ depending for instance on the coating density or the concentration in a mixture of H and D polymers an extra term has to be introduced, a term which, in practice, is difficult to evaluate.

Surfaces

If we have a surface of thickness e small compared to $1/q$ the form factor depends only on the geometry of the surface and not on its thickness e. One can therefore extract this parameter from the integral and write the result for independent particles as

$$\frac{I(q)}{V} = \frac{N}{V}\rho_b^2 e^2 \iint_{S,S'} \langle \exp(-\mathrm{i}\cdot \boldsymbol{q}(\boldsymbol{r}-\boldsymbol{r}'))\rangle \,\mathrm{d}^2\boldsymbol{r}\,\mathrm{d}^2\boldsymbol{r}' \qquad \text{(A3.36)}$$

where $\boldsymbol{r}$ is the vector characterizing the position of a point on the surface S. This leads to the definition of a new quantity $\rho_s = \rho_b e$, the scattering density per unit surface, where e is the thickness of the lamellae. This allows us to write for a dilute solution without interactions

$$\frac{I(q)}{V} = \frac{N}{V}\rho_s^2 S_1^2 P(q). \qquad \text{(A3.37)}$$

If the density of points in the surface is not constant perpendicular to the surface $\rho_s = \rho_b e$ can be replaced by $\langle\rho_s\rangle = \int_0^e \rho_b(x)\,\mathrm{d}x$ as long as $qe \ll 1$.

Filaments

The same procedure can be used for linear molecules, (for example Porod–Kratky chains) having a uniform finite cross section σ. We introduce the linear density by the relation

$$\rho_l = \rho_b \sigma \qquad \text{(A3.38)}$$

where σ is the surface area of the cross section of the chain and we obtain for the scattering from dilute solutions when $\sigma \gg q^{-1}$:

$$\frac{I(q)}{V} = \frac{N}{V}\rho_l^2 L^2 P(q). \qquad \text{(A3.39)}$$

Again if the density changes with the distance r to the axis of the filament we have to take an average values of ρ_l and replace it by:

$$\langle\rho_l\rangle = \int \rho_b(r)^2\, 2\pi r \mathrm{d}r.$$

A3.6 EXTENSION TO LIGHT-SCATTERING PROBLEMS

A3.6.1 General formula

All the expressions derived and discussed in this book have been developed for their applications to neutron scattering problems. Since there is a rigorous correspondence between neutron and light scattering, it is interesting to show how one can apply this formalism to light scattering problems.

Since both problems are scattering problems one can use exactly the same equations; the only point which has to be discussed is the physical meaning of the scattering length b when one is using light instead of neutrons. This is equivalent to the evaluation of the normalization constants relating the scattered intensity to the intensity of the incident beam. The easiest method is to look at the equivalent of eqn (7.27)

$$\mathcal{B} = \sum_{k=1}^{p} b_k N_k \tag{A3.40}$$

which gives for the scattered intensity

$$I(q = 0) = cst\, \overline{\Delta \mathcal{B}^2} \tag{A3.41}$$

where the constant has to be evaluated

The classical Einstein relation for light scattering can be written, for isotropic molecules, as

$$I(0) = \frac{\pi^2}{2\lambda^4} \overline{\Delta \varepsilon^2}(1 + \cos^2\theta) \tag{A3.42}$$

where λ is the wavelength of the incident beam, θ the scattering angle, and ε the relative dielectric constant of the scattering medium. In fact knowing this relation allows us to make all the necessary transformations and we could stop after this definition but it is interesting to discuss in a very simple way the reasons for these changes. Moreover, light scattering was developed before neutron scattering, especially for solutions (since measurements in bulk are difficult if not impossible). This means that the variables used in this domain of physics are practically always the concentration c_i and not the volume fractions φ_i. Using the Maxwell relation $\varepsilon = \nu^2$, which relates the refractive index to the dielectric constant we can write, differentiating the dielectric constant ε and introducing the partial quantities $(\partial \nu / \partial c_i)$, at constant c_j.

$$\Delta \varepsilon = 2\nu \Delta \nu = 2\nu \sum_{i=1}^{p} \frac{\partial \nu}{\partial c_i} \Delta c_i . \tag{A3.43}$$

This transforms $\overline{\Delta \varepsilon^2}$ into

$$\overline{\Delta \varepsilon^2} = 4\nu^2 \sum_{i=1}^{p} \sum_{k=1}^{p} \left(\frac{\partial \nu}{\partial c_i}\right) \left(\frac{\partial \nu}{\partial c_k}\right) \overline{\Delta c_i \Delta c_k} \tag{A3.44}$$

and eqn (A3.39) into

$$I(0) = \frac{2\pi^2}{\lambda^4}(1 + \cos^2\theta)\,\nu^2 \sum_{i=1}^{p}\sum_{k=1}^{p}\left(\frac{\partial\nu}{\partial c_i}\right)\left(\frac{\partial\nu}{\partial c_k}\right)\overline{\Delta c_i \Delta c_k}. \tag{A3.45}$$

The quantity $\partial\nu/\partial c_k$ is called the refractive index increment. This quantity is easy to measure since it suffices, while keeping the concentration of the species $j \neq k$ constant, to determine the refractive index of the solution as a function of c_k. The initial slope of the curve gives the required numerical value. Many of these values have been tabulated. Luckily the curve $\partial\nu/\partial c = f(c)$ is almost a straight line in a broad domain of concentrations which means that, even if one is working at high concentration or in polymer mixtures one can use the values measured at low concentrations.

With no further problems we can extrapolate this result to the angular dependence of the scattered intensity which means that we can now use everything derived for neutrons for the case of light scattering. This way of presenting the problem is very elegant but does not immediately create an expert in light scattering from the neutron specialist. We shall therefore take a less abstract route and explain in simple terms how one can obtain the light scattering results directly and the origin of this mysterious $2\pi^2/\lambda^4$ coefficient which has been called the λ^{-4} Rayleigh law.

A3.6.2 Scattering by a perfect gas

Let us now assume that we have a plane wave of light hitting a molecule placed at the origin O and that we look at the scattered light at a distance r in the direction θ. What happens physically is that the molecule situated at the origin is excited by the electric field $\boldsymbol{E}$ of the incident beam which produces an induced electric dipole, $\boldsymbol{p} = \alpha\boldsymbol{E}$. Since we are interested only in the isotropic scattering we can assume that α is a number (in the general case it is a tensor). This dipole oscillates at the frequency ω of the incident wave and emits radiation which, for large r, can be assumed to be a spherical wave centred in O. The magnitude of the electrical field at the distance r is thus given by the Hertz formula

$$\boldsymbol{E}' = -\frac{1}{c^2 r}\frac{\partial^2 \boldsymbol{p}}{\partial t^2} = \frac{\omega^2}{c^2 r}\boldsymbol{p}\sin\gamma \tag{A3.46}$$

ω being the circular frequency of the incident beam and γ the angle between the dipole and the direction $\boldsymbol{OM}$ (see Fig. A3.1). Since ω is equal to $2\pi/T$ where T is the period

$$\boldsymbol{E}' = \frac{4\pi^2}{r\lambda^2}\boldsymbol{p} = \frac{1}{r}\frac{4\pi^2}{\lambda^2}\alpha\sin\gamma\boldsymbol{E}. \tag{A3.47}$$

The analogy between this problem and the neutron scattering problem requires that we write

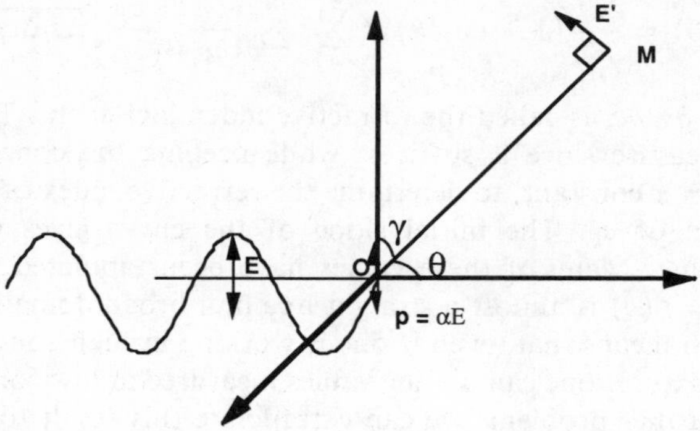

Fig. A3.1 Diagram showing the geometry of the electrical vector of the wave scattered in M by a molecule in O. The incident wave is a plane wave, vertically polarized, going from left to right.

$$b^2 = \frac{16\pi^4}{\lambda^4}\alpha^2 \sin^2\gamma \tag{A3.48}$$

but there is a difficulty due to the transverse nature of the electrical field of the electromagnetic wave which is responsible for the existence of the factor $\sin\gamma$. In Fig. A3.1 we are considering a vertically polarized wave and we should consider an unpolarized vibration. Without entering into the details of the calculation we claim, without demonstration, that considering an unpolarized wave transforms eqn (A3.48) into

$$b^2 = \frac{8\pi^4}{\lambda^4}\alpha^2(1 + \cos^2\theta) \tag{A3.49}$$

and thus we obtain for a perfect gas

$$\frac{\partial\sigma}{\partial\Omega} = Nb^2 = N\frac{8\pi^4}{\lambda^4}\alpha^2(1 + \cos^2\theta). \tag{A3.50}$$

There is a difference between this result and the neutron formula due to the presence of the factor $(1 + \cos^2\theta)$ sometimes called by X-ray specialists the Thomson factor. It is not important when one works at low angle since it is practically equal to 2 but has to be taken into account at higher angles.

Relation between polarizability and refractive index

The last problem we have to solve in order to compare neutron and light scattering is to establish the relationship between polarizability and refractive index. Using Gaussian units which are more convenient for this problem we write, remembering that for a perfect gas ν is of the order of unity

$$\varepsilon - 1 = \nu^2 - 1 \approx 2(\nu - 1) = 4\pi n\alpha \tag{A3.51}$$

n is the number of molecules per unit volume and ε the dielectric permittivity. This gives what light scatterers call the Rayleigh ratio and is equal to the quantity $\partial\sigma/\partial\Omega$ per unit volume

$$\mathcal{R} = \frac{1}{V}\frac{\partial\sigma}{\partial\Omega} = \frac{2\pi^2}{\lambda^4}(\nu - 1)^2\frac{1 + \cos^2\theta}{n}. \tag{A3.52}$$

(This result could have been obtained directly applying eqn (A3.45) to a one-constituent system.)

A3.6.3 Ideal solutions of small molecules

If the solution is sufficiently dilute to be considered as ideal the last formula can be used replacing the vacuum by a solvent and the polarizability by an apparent polarizability $\bar{\alpha}$, i.e. the increment per molecule of the dielectric constant of the medium. This can be done using the following approach; we call ε_0 the dielectric permittivity of the solvent and ε that of the solution and write

$$\varepsilon_0 - 1 = 4\pi\alpha_0$$

$$\varepsilon - 1 = 4\pi\alpha \tag{A3.53}$$

where α_0 and α are the polarization of the solvent and the solution respectively. The difference $\alpha - \alpha_0$ is due to the N/V dipole increment due to the presence of the solute and can be written as $N/V\bar{\alpha}$ defining $\bar{\alpha}$ by the relationship

$$\bar{\alpha} = (\alpha - \alpha_0) = \frac{\varepsilon - \varepsilon_0}{4\pi n} = \frac{\nu^2 - \nu_0^2}{4\pi n} \approx \nu_0\left(\frac{\nu - \nu_0}{2\pi n}\right). \tag{A3.54}$$

Replacing α in eqn (3.50) by $\bar{\alpha}$ gives

$$\mathcal{R} = \frac{4\pi^2}{n\lambda^4}\nu_0^2(\nu - \nu_0)^2(1 + \cos^2\theta).$$

From the definition of the experimental quantity $d\nu/dc$ we write

$$\frac{d\nu}{dc} = \frac{(\nu - \nu_0)}{c} \tag{A3.55}$$

obtaining the final form

$$\mathcal{R} = \frac{2\pi^2}{\lambda^4}\,\nu_0^2\left(\frac{\mathrm{d}\nu}{\mathrm{d}c}\right)^2\frac{c^2}{n}\,(1+\cos^2\theta) \tag{A3.56}$$

replacing n by the concentration $c = nM_1/N_A$ gives

$$\mathcal{R} = \frac{2\pi^2}{\lambda^4}\,\nu_0^2\left(\frac{\mathrm{d}\nu}{\mathrm{d}c}\right)^2\frac{cM}{N_A}\,(1+\cos^2\theta). \tag{A3.57}$$

(As already explained the factor $1 + \cos^2\theta$ does not depend on the size of the molecules.)

A3.6.4 Non-ideal solutions

We now want to generalize the result obtained but this time starting from eqn (A3.42), assuming that the compressibility can be neglected and considering only one constituent. The only variable is the volume fraction which we replace by the concentration writing

$$I(0) = (1+\cos^2\theta)\,\frac{2\pi^2}{\lambda^4}\,\nu_0^2\left(\frac{\mathrm{d}\nu}{\mathrm{d}c}\right)^2\overline{\Delta c^2} \tag{A3.58}$$

since $c = \varphi/v_1^s$, $\overline{\Delta c^2} = 1/(v_1^s)^2\overline{\Delta\varphi^2}$.

Considering the unit volume and using eqn (7.62) with $N_T = v_0^{-1}$ (since we discuss the scattering per unit volume)

$$\overline{\Delta\varphi^2} = N_T\frac{k_BT}{\dfrac{\partial^2 g_c}{\partial\varphi^2}} = \frac{1}{v_0}\frac{k_BT}{\dfrac{\partial^2 g_c}{\partial\varphi^2}} \tag{A3.59}$$

since from eqns (A3.20) and (A3.25) $\partial\pi/\partial\varphi = -v_0\,\partial\mu_0/\partial\varphi = \varphi v_0\,\partial^2 g/\partial\varphi^2$

$$\overline{\Delta c^2} = \frac{1}{(v_1^s)^2}\frac{k_BT\varphi}{(\partial\pi/\partial\varphi)} \tag{A3.60}$$

which gives for the Rayleigh factor of small molecules

$$\mathcal{R} = (1+\cos^2\theta)\,\frac{2\pi^2}{\lambda^4}\,\nu_0^2\left(\frac{\mathrm{d}\nu}{\mathrm{d}c}\right)^2\frac{k_BTc}{\partial\pi/\partial c} \tag{A3.61}$$

or expanding the osmotic pressure as function of concentration and taking into account the angular dependence

$$\mathcal{R} = (1 + \cos^2\theta)\,\frac{2\pi^2}{\lambda^4}\,\frac{\nu_0^2}{N_A}\left(\frac{d\nu}{dc}\right)^2 \frac{c}{\dfrac{1}{M_1 P_1(q)} + 2A_2 c + \ldots} \tag{A3.62}$$

which is the classical Zimm formula. (For the definition of q one has to use the wavelength λ' of the light in the solvent and not *in vacuo*: $\lambda' = \lambda/\nu$.)

One can easily translate all the formulae which have been written in Chapter 7 into the light scattering language. The method is the following: first introduce the factor $(1 + \cos^2\theta)\ 2\pi^2/\lambda^4\nu_0^2$ and second replace the quantity b_v by the quantity $\partial\nu/\partial\varphi_i$ or $1/v_1^s\,(\partial\nu/\partial c_i)$. The introduction of the angular dependence is not a problem since, knowing the normalization for $q = 0$ one has just to add the corresponding form factor or structure factor as we did for neutron scattering. It is useful to verify for these formulae that the dimensions are correct. The Rayleigh constant or the ratio $1/V(\partial\sigma/\partial\Omega)$ has the dimensions of the inverse of a length as it has been shown in Chapter 3.

A3.7 COMPARISON BETWEEN NEUTRON AND X-RAY SCATTERING

This case is especially simple: every electron, struck by an X-ray beam behaves (except in the case of heavy atoms) as if it were free. Therefore all the electrons have the same scattering length b which is given by the Thomson formula:

$$b_0 = \left(\frac{1 + \cos^2\theta}{2}\right) 2.8 \times 10^{-13}\,\text{cm}. \tag{A3.63}$$

The scattering by one atom is the sum of the scattering by all its electrons and is therefore Zb_0, where Z is the atomic number. This allows us to evaluate directly the factor b corresponding to atoms and molecules by summing the number of electrons and multiplying this sum by the constant given in the preceding equation. It seems, therefore, at first sight that the problem is extremely simple. There is, however, a difficulty which comes from the fact that atoms are not very small compared to the wavelength of the X-rays which is of the order of one angstrom. One has therefore to take into account the form factor of the atoms. Since they can be considered as spherically symmetric our discussion in Chapter 6 shows that one has just to multiply the product $b_i b_j$ by the product of the q-dependent amplitude scattered by each atom.

Another effect which we have neglected in the case of neutron scattering is due to the fact that, when the energy of the photons is close to an allowed transition of the scatterers, the laws which we have described are no longer valid, leading to what is called anomalous scattering. More precisely, if the photon energy is in the immediate vicinity, of an absorption edge the scattering length is no longer purely a real term but has an imaginary component and is equal to $b_0[Z + f'(E) + if''(E)]$ where $f'(E)$ corresponds to the real part and $f''(E)$ to the imaginary part of the added scattering length. These two parts which are related by the Kramers–Kronig relation show a strong energy dependence which is illustrated in the Fig. A3.2.

The quantity $f'(E)$ can reach large negative values which can correspond to the scattering of a few free electrons. It is therefore evident that by working in this energy range one can modify the scattering length of one type of atoms by a slight change of wavelength and therefore use all the discussions on the effect of labelling on the coefficients of the different partial structure factors (see for instance eqn (5.15)). The big advantage is that it is not necessary to use chemical labelling techniques and that, for the same sample, one can have different forms of the scattered intensity for the different energies of the incident beam allowing, in principle, deter-

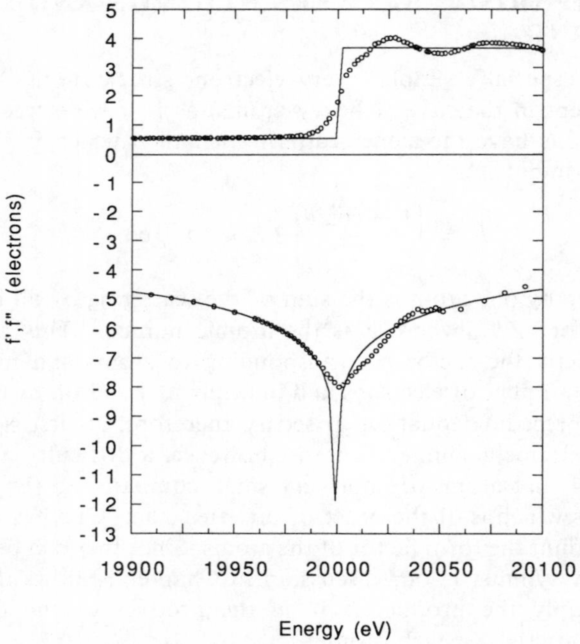

Fig. A3.2 Real and imaginary parts of the scattering length (expressed in electron units) as function of the energy of the incident beam near the absorption *K* edge of Mo (Tonnerre 1989). The full line is a theoretical curve, the points correspond to experiments.

mination of the partial structure factors. However, two major difficulties prevent the systematic use of this technique. The first one is that, in order to work in the region of absorption of photons, one has to work at very high energies unless one uses heavy atoms and these do not very frequently occur in polymers. The second one is that, as soon as $f(E)$ begins to be different from zero, the imaginary part $f''(E)$ cannot be neglected; there is absorption of the beam by the specimen and this makes the interpretation of the data more difficult. Until now, very few results have been published in this field but it is possible that, with the forthcoming improvements of the synchrotron radiation sources, this technique may become available in the near future.

SUGGESTED FURTHER READING

Abramowitz and Stegun (1970).
Bracewell (1965).
Champeney (1973).
Cotton (1991).
Huglin (1972).
Stacey (1956).
Williams (1991).

Appendix 4

Chemical formulae of some of the polymers used in this book

Polybutadiene 1–4 $-\!\!(CH_2\text{–}CH\text{=}CH\text{–}CH_2)\!\!-_N$

Polybutadiene 1–2

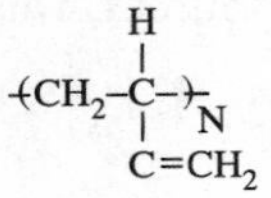

Polycarbonate

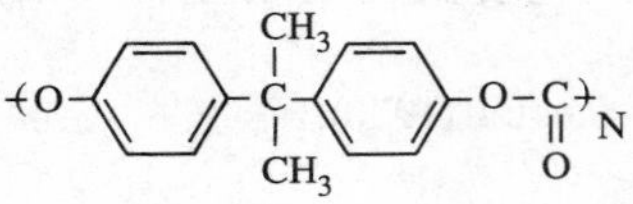

PDMS: Polydimethylsiloxane

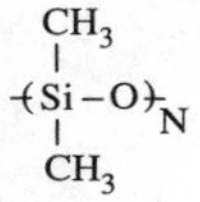

PEO: Polyethylene oxide $-\!\!(CH_2\text{–}CH_2\text{–}O)\!\!-_N$

Polyisoprene 1–4 *cis*

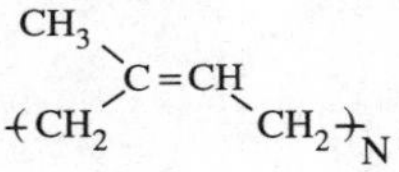

Polyisoprene 1–4 *trans*

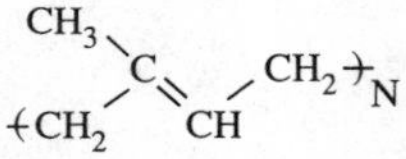

Polyisoprene 1–2

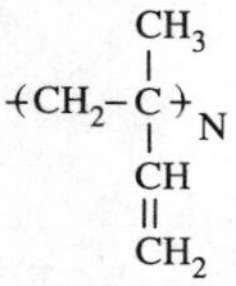

Polyisoprene 3–4

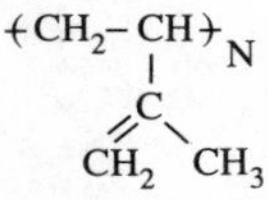

PMMA: Polymethylmethacrylate

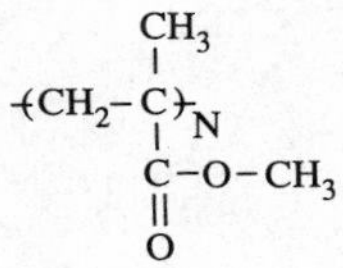

Polypropylene

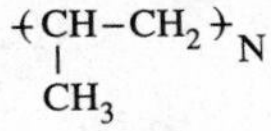

PS: Polystyrene

PS - H

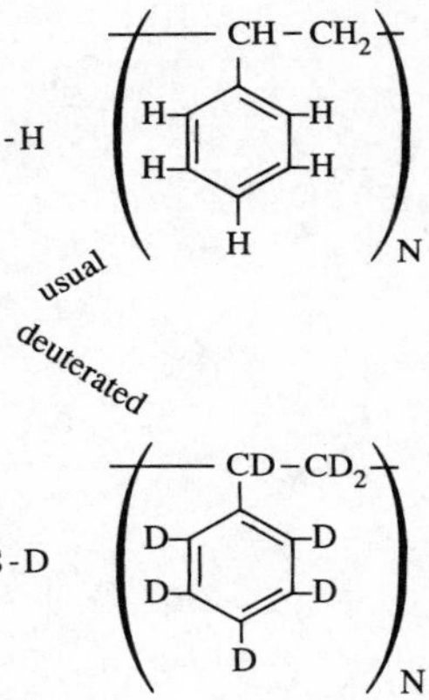

PS - D

PTHF: Polytetrahydrofuran $-\!\!(CH_2\text{–}CH_2\text{–}CH_2\text{–}CH_2\text{–}O)\!\!-_N$

PVME: Polyvinylmethylether

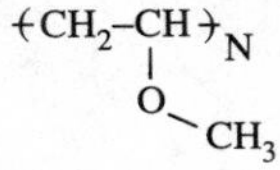

References

Abramowitz, M., and Stegun, I. A (1970). *Handbook of mathematical functions*, (9th edn). Dover Publications, New York.

Akcasu, A. Z., Benmouna, M., and Han, C. C. (1980*a*). *Polymer*, **13**, 409.

Akcasu, A., Summerfield, G. C., Jahshan, S. N., Han, C. C., Kim, C. Y., and Yu, H. (1980*b*). *Journal of Polymer Science* (*Polymer Physics*), **18**, 863.

Allegra, G., Higgins, J. S., Ganazzoli, F., Lucchelli, E., and Bruckner, S. (1984). *Macromolecules*, **17**, 1253.

Allen, G., Brier, P. N., and Higgins, J. S. (1972). *Polymer*, **13**, 157.

Allen, G., Ghosh, R. E., Heidemann, A., Higgins, J. S., and Howells, W. S. (1974). *Chemical Physics Letters*, **27**, 308.

Allen, G., Wright, C. J., and Higgins, J. S. (1975). *Polymer*, **15**, 319.

Allen, G., Ghosh, R., Higgins, J. S., Cotton, J.-P., Farnoux, B., Jannink, G. *et al.* (1976). *Chemical Physics Letters*, **38**, 575.

Amaral, L. W., Vinhas, L. A., and Herdade, S. B. (1976). *Journal of Polymer Science* (*Polymer Physics*), **14**, 1077.

Anastasiadis, S. H., Russell, T. P., Satija, K. K., and Majkrzak, C. F. (1990). *Journal of Chemical Physics*, **92**, 5677.

Ausserre, D., Hervet, H., and Rondelez, F. (1985). *Physical Review Letters*, **59**, 1948.

Auvray, L. and Cotton, J.-P. (1987). *Macromolecules*, **20**, 1890.

Bacon, G. E. (1975). *Neutron diffraction*, (3rd edn). Oxford University Press, Oxford.

Bandrupt, J. and Immergut, E. H. (1975). *Polymer handbook*, Chap. IV, pp. 34–51. Wiley, New York.

Bartsch, E., Fujara, F., Kiebel, M., and Sillescu, H. (1989). *Bericht Bunsengesellschaft Physikalische Chemie*, **93**, 1252.

Bastide, J. and Benoît, H. C. (1990). *Initiation à la chimie et à la physicochimie macromoléculaires*, Vol. 8: *Structure des Polymères et Méthodes d'Études*, Chap. VIII Groupe François des Polymères, Grenoble.

Bastide, J., Duplessix, R., Picot, C., and Candau, S. (1984). *Macromolecules*, **17**, 83.

Bastide, J., Leibler, L., and Prost, J. (1990). *Macromolecules*, **23**, 1821.

Bates, F. S. (1985) *Macromolecules*, **18**, 525.

Bates, F. S., Berney, C. V., and Cohen, R. E. (1983). *Polymer*, **24**, 519.

Bates, F. S., Wignall, G. D., and Koehler, W. C. (1985). *Phys. Rev. Lett.*, **55**, 2425.

Bates, F. S., Wignall, G. D., and Koehler, W. C. (1986). *Macromolecules*, **19**, 932.

Bée, M. (1988). *Quasi-elastic neutron scattering*. Adam Hilger, Bristol.

Benmouna, M. and Akcasu, A. Z. (1980). *Macromolecules*, **13**, 409.

Benoît, H. (1953). *Journal of Polymer Science*, **11**, 507.

Benoît, H. (1991). *Polymer*, **32**, 579.

Benoît, H. and Benmouna, M. (1984). *Polymer*, **25**, 1059.

Benoît, H. and Doty, P. (1953). *Journal of Physical Chemistry*, **57**, 958.
Benoît, H. and Hadziioannou, G. (1988). *Macromolecules*, **21**, 1449.
Benoît, H. and Wippler, C. (1960). *Journal de Chimie Physique*, **57**, 524.
Benoît, H., Koberstein, J., and Leibler, L. (1981). *Die Makromolekulare Chemie Suppl.* **4**, 85.
Benoît, H., Picot, C., and Benmouna, M. (1984). *Journal of Polymer Sciences (Polymer Physics)*, **22**, 1545.
Benoît, H., Wu, W., Benmouna, M., Mozer, B., Bauer, B., and Lapp, A. (1985). *Macromolecules*, **18**, 986.
Benoît, H., Benmouna, M., and Wu, W. (1990). *Macromolecules*, **23**, 1511.
Benoît, H., Benmouna, M., and Vilgis T. (1991). *Compte Rendus de l'Académie des Sciences Paris*, **313**, Série II 869.
Berne, B. J. and Pecora, R. (1976). *Dynamic light scattering* Wiley, New York.
Boué, F., Daoud, M., Nierlich, M., Williams, C., Cotton J.-P., Farnoux, B., *et al.* (1978). *Neutron inelastic scattering 1977*, Vol. 1, p. 563. International Atomic Energy Agency, Vienna.
Boué, F., Nierlich, M., Jannink, G., and Ball, R. (1982). *Journal de Physique*, **43**, 137.
Bracewell, R. (1965). *The Fourier transform and its applications*. McGraw–Hill, New York.
Brereton, M. G., Fischer, E. W., Herkt-Maetzky, Ch., and Mortensen, K. (1987). *Journal of Physical Chemistry*, **87**, 6144.
Buckingham, A. D. and Hentschel, H. G. E. (1980). *Journal of Polymer Science (Polymer Physics)*, **18**, 853.
Cabane, B. (1987). Small angle scattering methods. In *Surfactant solutions new methods of investigation*, (ed. R. Zana). Marcel Dekker, New York.
Cabannes, J. (1929). *La diffusion moléculaire de la lumiére*. Paris Presses Universitaires, Paris.
Cahn, J. W. (1986). *Transactions of the Metallurgical Society AIME*, **242**, 169.
Carlile, C. J. and Adams, M. (1992). *Physica B*, **182**, 431.
Cassasa, E. (1965). *Journal of Polymer Science A*, **3**, 605.
Cebula, D. J., Goodwin, J. W., Jeffrey, G. C., Ottewill, R. H., Parentich, A., and Richardson, R. A. (1963). *Discussions of the Faraday Society*, **76**, 37.
Cebula, D. J., Myers, D. Y., and Ottewill, R. H. (1982). *Colloid and Polymer Science*, **206**, 95.
Champeney, D. C. (1973). *Fourier transforms and their physical applications*. Academic Press, London.
Clark, J. N., Higgins, J. S., and Peiffer, D. G. (1991). *Polymer Engineering and Science*, **32**, 49.
Collins, M. F. (1989). *Magnetic critical scattering*. Oxford University Press, Oxford.
Cosgrove, T., Heath, T. G., Ryan, K., and Crowley, T. L. (1987). *Macromolecules*, **20**, 2879.
Cosgrove, T., Crowley, T. L., Ryan, K., and Webster, J. R. P. (1990). *Colloids & Surfaces*, **51**, 255.
Cotton, J.-P. (1974). *Introduction à la spectrométrie neutronique diffusion aux petits angles: cours disponible au service de spectrométrie neutronique*. Orme des Merisiers, Gif-sur-Yvette.
Cotton, J.-P. (1991). In *Neutron, X-ray and light scattering: introduction to an investigative tool for colloidal and polymeric systems*, (ed. P. Lindner and Th. Zemb). North Holland, Amsterdam.

Cotton, J.-P., Decker, D., Benoît, H., Farnoux, B., Higgins, J., Jannink, G. *et al.* (1974). *Macromolecules*, **7**, 863.

Cotton, J.-P., Nierlich, M., Boué, F., Daoud, M., Farnoux, B., Jannink *et al.* (1976). *Journal of Chemical Physics*, **65**, 1101.

Csiba, T., Jannink, G., Durand, D., Papoular, R., Lapp, A., Auvray L. *et al.* (1991). *Journale de Physique II*, **1**, 381.

Dagleish, P., Hayter, J. B., and Mezei, F., (1980). In *Neutron spin-echo*, (ed. F. Mezei), Lecture Notes in Physics, Vol. 128. Springer, Berlin.

Daoud, M., Cotton, J.-P., Farnoux, B., Jannink, G., Sarma, G., Benoît, H. *et al.* (1975). *Macromolecules*, **8**, 804.

Debye, P. (1944). *Journal of Applied Physics*, **15**, 338.

Debye, P. (1946). *Journal of Chemical Physics*, **14**, 636.

Debye, P. and Bueche, A. M. (1949). *Journal of Applied Physics*, **20**, 518.

Debye, P., Anderson, H. R., and Brumberger, H. (1957). *Journal of Applied Physics*, **28**, 679.

de Gennes, P. G. (1967). *Physics*, **3**, 37.

de Gennes, P. G. (1970). *Journale de Physique*, **31**, 235.

de Gennes, P. G. (1971). *Journal of Chemical Physics*, **55**, 572.

de Gennes, P. G. (1980). *Journal de Physique*, **42**, 735.

de Gennes, P. G. (1979). *Scaling concepts in polymer physics*. Cornell University Press, New York.

des Cloizeaux, J. (1973). *Macromolecules*, **6**, 403.

des Cloizeaux, J. and Jannink, G. (1980). *Physica*, **102A**, 1206.

des Cloizeaux, J. and Jannink, G. (1987). *Les polymères en solution: leur modélisation et leur structure*. Les Éditions de Physique, Les Ulis, France.

Dettenmaier, M., Maconnachie, A., Higgins, J. S., Kausch, H. H., and Nguyen, T. Q. (1986). *Macromolecules*, **19**, 773.

Doi, M. and Edwards, S. F. (1979). *Journal of the Chemical Society*, *Faraday Transactions*, 2, **74**, 1789, 1818.

du Bois Violette, E. and de Gennes, P. G. (1976). *Physics*, **3**, 181.

Duval, M., Duplessix, R., Picot, C., Decker, D., Rempp, P., Benoît, H., *et al.* (1976). *Journal of Polymer Science* (*Polymer Letters*) **14**, 588.

Duval, M., Picot, C., Benoît, H., Borsali, R., Benmouna, M., and Lartigue, C. (1991). *Macromolecules*, **24**, 3185.

Eastoe, J. and Penfold, J. (1992). Rutherford Appleton Report RAL-92-05.

Edwards, S. F. (1966). *Proceedings of the Physical Society* (London), **88**, 266.

Egelstaff, P. A. (ed.) (1965). *Thermal neutron scattering*. Academic Press, New York.

Einstein, A. (1910). *Annalen der Physik*, **33**, 1910.

Felcher, G. P. and Russell, T. P. (ed.) (1991). *Physica B: Condensed Matter*, **173**, Nos 1 and 2.

Fermi, E. and Zinn, W. (1946). *Physical Review*, **70**, 103.

Fernández, M. L., Higgins, J. S., Penfold, J., and Shackleton, C. S. (1990). *Polymer Communications*, **31**, 124.

Fernández, M. L., Higgins, J. S., Penfold, J., and Shackleton, C. (1991). *Journal of the Chemical Society*, *Faraday Transactions*, **87**, 2055.

Flory, P. J. (1942). *Journal of Chemical Physics*, **10**, 51.

Flory, P. J. (1953). *Principles of polymer chemistry*. Cornell University Press, New York.

Flory, P. J. (1965). *Journal of the American Chemical Society*, **86**, 1833.

Frick, B., Richter, D., Petry, W., and Buchenau, U. (1988). *Zeitschrift für Physik B.*, **70**, 73.

Gabryś, D., Higgins, J. S., Ma, K. T., and Roots, J. E. (1984). *Macromolecules*, **17**, 560.

Gabryś, B., Higgins, J. S., and Scharpf, O. (1986). *Journal of the Chemical Society, Faraday Transactions*, **82**, 1929.

Gawrisch, W., Brereton, M. G., and Fischer, E. W. (1981). *Polymer Bulletin*, **4**, 687.

Glatter, O. and Kratky, O. (1982). *Small angle X-ray scattering*. Academic Press, London.

Goyal, P. S., King, J. S., and Summerfield, G. C. (1983). *Polymer*, **24**, 131.

Gradshtein, I. S. and Ryzhik, I. M. (1965). *Table of integrals series and products*, (Translation ed. A. Jeffrey). Academic Press, New York.

Guinier, A. (1956). *Théorie et technique de la radiocristallographie*. Dunod, Paris.

Guinier, A. and Fournet, G. (1955). *Small angle scattering of X-rays*. Wiley, New York.

Hadziiognnou, G., Cotts, P. M., ten Brinke, G., Han, C. C., Lutz, P., Strazielle, C. *et al*. (1987). *Macromolecules*, **20**, 493.

Hansen, J. P. and McDonald, I. R. (1986). *Theory of simple liquids*, (2nd edn). Academic Press, London.

Hayashi, H., Flory, P. J., and Wignall, G. D. (1983). *Macromolecules*, **16**, 1328.

Hayter, J. B. (1978). In *Neutron diffraction*, (ed. H. Dachs). Springer, Berlin.

Hayter, J. B. (1980). In *Neutron spin-echo*, (ed. F. Mezei), Lecture Notes in Physics, Vol. 128. Springer. Berlin.

Hayter, J. B. and Penfold, J. (1983). *Colloid and Polymer Science*, **261**, 1022.

Heidemann, A. (1980). In *Neutron spin-echo*, (ed. F. Mezei), Lecture Notes in Physics, Vol. 128. Springer, Berlin.

Henderson, J. A., Richards, R. W., Penfold, J., and Thomas, R. K. (1993). *Macromolecules*, **26**, 65.

Henry, A. W. and Safford, G. J. (1969). *Journal of Polymer Science* 7, 433.

Higgins, J. S. (1984). In *Developments in polymer characterisation*, Vol. 4, (ed. J. V. Dawkins). Applied Science, Barking England.

Higgins, J. S. and Maconnachie, A. M. (1987). In *Methods in experimental physics*, Vol. 23, Part c, p. 287. Academic Press, New York.

Higgins, J. S. and Roots, J. E. (1985). *Journal of the Chemical Society, Faraday Transactions*, **81**, 757.

Higgins, J. S., Ghosh, R. E., Howells, W. S., and Allen, G. (1977). *Journal of the Chemical Society, Faraday Transactions*, **73**, 40.

Higgins, J. S., Ma, K. T., Nicholson, K., Hayter, J. B., Dodgson, K., and Semlyen, J. A. (1983). *Polymer*, **24**, 793.

Higgins, J. S., Blake, S., Tomlins, P. E., Ross-Murphy, S. B., Staples, E., Penfold, J. *et al*. (1988). *Polymer*, **29**, 1968.

Higgins, J. S., Fruitwala, H., and Tomlins, P. E. (1989*a*). *British Polymer Journal*, **21**, 247.

Higgins, J. S., Fruitwala, H., and Tomlins, P. E. (1989*b*). *Macromolecules*, **22**, 3674.

Hill, T. L. (1956). *Statistical mechanics*. McGraw-Hill, New York.

Holtzer, A. M. (1955). *Journal of Polymer Science*, **17**, 432.

Horton, J. C., Squires, G. L., Boothroyd, A. T., Fetters, L. J., Rennie, A. R., Glinka, G. J. *et al.* (1989). *Macromolecules*, **22**, 681.

Huggins, M. L. (1942). *Journal of Physical Chemistry*, **46**, 151.

Huglin, B. (ed.) (1972). *Light scattering from polymer solutions*. Academic Press, London.

ILL (1992). *Guide to neutron research facilities at the I.L.L.* Institut Laue–Langevin, Grenoble, France.

Jacrot, B. (1976). *Reports on Progress in Physics*, **39**, 911.

Joanny, J.-F. and Candau, S. (1989). In *Comprehensive polymer science*, Vol. 2, Chap. 7, (ed. G. Allen, J. C. Bevington, C. Booth, and C. Price). Pergamon Press, Oxford.

Kaji, K., Urakawa, H., Kanaya, T., and Kitimaru, R. (1984). *Macromolecules*, **17**, 1835.

Kanaya, T., Kaji, K., and Inone, K. (1991). *Macromolecules*, **24**, 1826.

King, J., Boyer, W., Wignall, G. D., and Ullman, R. (1985). *Macromolecules*, **18**, 709.

Kirkwood, J. G. and Goldberg, R. J. (1950). *Journal of Chemical Physics*, **18**, 54.

Kirste, R., Kruse, W. A., and Schelten, J. (1972). *Makromoleculare Chemie*, **162**, 299.

Kirste, R., Kruse, W. A., and Ibel, K. (1975). *Polymer*, **16**, 120.

Koester, L., Rauch, R., and Seymann, E. (1991). *Atomic Data and Nuclear Data Tables*, **49**, 65.

Kratky, O. and Porod, G. (1949*a*). *Journal of Colloid and Interface Science*, **4**, 35.

Kratky, O. and Porod, G. (1949*b*). *Recueil des Travaux Chimiques des Pays-Bas*, **68**, 1106.

Kuhn, W. (1934). *Kolloid Zeitschrift*, **68**, 2.

Kurata, M. (1982). *Thermodynamics of polymer solutions*. Harwood, London.

Leibler, L. (1980). *Macromolecules*, **13**, 1602.

Leibler, L. and Benoît, H. (1981). *Polymer*, **22**, 195.

Lekner, J. (1987). *Theory of reflection of electromagnetic and particle waves*. Martinus Nijhoff, Dordrecht.

Lindner, P. and Zemb, Th. (ed.) (1991). *Neutron, X-ray and light scattering: introduction to an investigative tool for colloidal and polymeric systems*. North Holland, Amsterdam.

Lovesey, S. W. (1984). *Theory of neutron scattering from condensed matter*, Vol. 1. Oxford University Press, Oxford.

Ma, K. T. (1981). Ph.D. thesis, Imperial College, University of London.

Macdonald, W. A., McLenaghan, A. D. W., McLean, J. G., and Richards, R. W. (1991). *Macromolecules*, **24**, 6164.

Maconnachie, A. (1984). *Polymer*, **25**, 1068.

Maconnachie, A., Allen, G., and Richards, R. W. (1981). *Polymer*, **22**, 1157.

Maeda, T. and Fujime, S. (1984). *Macromolecules*, **17**, 1157.

Maier-Leibnitz, H. and Springer, T. (1963). *Reactor Science and Technology*, **17**, 217.

Mandlebrot, B. (1982). *The fractal geometry of nature*. B. W. H. Freeman, New York.

May, R. P., Ibel, K., and Haas, J. (1982). *Journal of Applied Crystallography*, **15**, 15.

Meier, G., Fujara, F., and Petry, W. (1989). *Macromolecules*, **22**, 4421.

Mezei, F. (1972). *Zeitschrift Physick*, **255**, 146.

Myers, W., Summerfield, G. C., and King, J. S. (1966). *Journal of Chemical Physics*, **44**, 189.

Nicholson, L. K. (1981). *Contemporary Physics*, **22**, 451.

Nicholson, L. K., Higgins, J. S., and Hayter, J. B. (1981). *Macromolecules*, **14**, 836.

Neugebauer, T. (1943). *Annalen der Physik*, **42**, 509.

Oeser, R., Picot, C., and Herz, J. (1987). In *Polymer motion in dense systems*, (ed. D. Richter and T. Springer), Springer Proceedings in Physics, Vol. 29. Springer, Berlin.

Ono, M., Okamoto, S., Kanaya, T., Nishida, K., Urakawa, H., Kaji, K. *et al.* (1986). *Physica*, **138B**, 49.

Ornstein, L. S. and Zernike, F. (1914). *Proceedings Academy Sciences, Amsterdam*, **17**, 793.

Pecora, R. (1968). *Journal of Chemical Physics*, **49**, 1032.

Pellow, C. (1978). Private communication.

Penfold, J. and Thomas, R. K. (1990). *Journal of Physics of Condensed Matter*, **2**, 1369.

Penfold, J., Ward, R. C., and Williams, W. G. (1987). *Journal of Physics E Scientific Instruments*, **20**, 1411.

Phoon, C.-L. (1992). Ph.D. thesis, Imperial College, University of London.

Placzek, G. (1954). *Physical Reviews*, **93**, 985.

Porod, G. (1951). *Kolloid Zeitschrift*, **124**, 83.

Proceedings of a workshop on methods of analysis and interpretation of neutron reflection data, *Physica B (Condensed Matter)*, **173**, (1991).

Putzeys, P. and Brosteraux, J., (1935). *Transactions of the Faraday Society*, **31**, 1374.

Rawiso, M., Duplessix, R., and Picot, C. (1987). *Macromolecules*, **20**, 630.

Rayleigh, Lord (1914), *Proceeding of the Royal Society*, **A 90**, 219.

Rennie, A. R., Crawford, R. J., Lee, E. N., Thomas, R. K., Crowley, T. L., Roberts, S. *et al.* (1989). *Macromolecules*, **22**, 3466.

Richards, R. W. and Thomason, J. L. (1983). *Macromolecules*, **16**, 982.

Richter, D., Baumgartner, A., Binder, K., Ewen, B., and Hayter, J. B. (1981). *Physical Review Letters*, **55**, 109.

Richter, D., Farago, B., Fetters, L. J., Huang, J. S., Ewen, B., and Lartigue, C. (1990). *Physical Review Letters*, **64**, 1389.

Ronca, G. (1983). *Journal of Chemical Physics*, **79**, 1031.

Rouse, P., Jr (1953). *Journal of Chemical Physics*, **21**, 1272.

Sadler, D. (1984). In *Crystalline polymers*, (ed. I. H. Hall). Elsevier, Amsterdam.

Safford, G. J., Danner, H. R., Boutin, H., and Berger, M. (1964). *Journal of Chemical Physics*, **40**, 426.

Schaefer, D. W. (1987). In *Scattering, deformation, and fracture in polymers materials*, Research Society Symposia Proceedings, Vol. 79, (ed. G. B. Wignall, B. Crist, T. P. Russell, and E. L. Thomas). Materials Research Society, Pittsburgh, USA.

Schelten, J. and Stamm, M. (1981). *Macromolecules*, **14**, 818.

Schelten, J., Kruse, W. A., and Kirste, R. G. (1973). *Kolloid Zeitschrift Polymer*, **251**, 919.

Schelten, J., Ballard, D. G. H., Wignall, G. D., Longman, G., and Schmatz, W. (1976). *Polymer*, **17**, 751.

Schultz, G. V. (1939). *Zeitschrift Physikalische Chemie Abteilung*, **B 43**, 25.

Schwahn, D., Janssen, S., and Springer, T. (1992). *Journal of Chemical Physics*, **97**, 8775.

Sears, V. F. (1992). *Neutron News*, **3**, 26.

Sears, V. F. (1984). *Thermal neutron scattering lengths for condensed matter research*. Chalk River Nuclear Laboratories, Chalk River, Ontario, Canada.

Sivia, D. S., Carlile, C. J. Howells, W. S., and Konig, S. (1992). *Physical B*, **182**, 341.

Springer, T. (1972). *Quasielastic neutron scattering for the investigation of diffusive motions in solids and liquids*, (ed. H. Höhler). Springer, Berlin.

Squires, G. L. (1978). *Introduction to the theory of thermal neutron scattering*. Cambridge University Press, Cambridge.

Stacey, K. A. (1956). *Light-scattering in physical chemistry*. Butterworths, London.

Stanley, E. (1971). *Introduction to phase transitions and critical phenomena*. Oxford University Press, New York.

Strazielle, C. and Benoît, H. (1975). *Macromolecules*, **8**, 203.

Stockmayer, W. H. (1950). *Journal of Chemical Physics*, **18**, 58.

Sun, X., Bouchaud, E., Lapp, A., Farnoux, B., Daoud, M., and Jannink, G. (1988). *Europhysics Letters*, **6**, 207.

Takeuchi, H., Higgins, J. S., Hill, A., Maconnachie, A., Allen, G., and Stirling, G. C. (1982). *Polymer*, **23**, 499.

Tidswell, I. M., Odeo, B. M., Pershan, P. S., Wasserman, G. R., Whitesides, G. M., and Axe, J. D. (1990). *Physical Review B*, **41**, 1111.

Tompa, H. (1956). *Polymer solutions*. Butterworths, London.

Tonnerre, J. M. (1989). Ph.D. thesis, University Paris Sud Orsay.

Twisleton, J. F., and White, J. W. (1972). *Polymer*, **13**, 40.

Twisleton, J. F., White, J. W., and Reynolds, P. W. (1982). *Polymer*, **23**, 578.

Van Hove, L. (1954). *Physical Review*, **95**, 249.

Wignall, G. D. (1987). In *Encyclopedia of polymer science and engineering*, Vol. 12, p. 112. Wiley, New York.

Wignall, G. D. and Bates, F. S. (1987). *Journal of Applied Crystallography*, **20**, 28.

Wignall, G. D., Christen, D. K., and Ramakrishnan, R. (1988). *Journal of Applied Crystallography*, **21**, 438.

Williams, C. E. (1991). In *Neutron, X-ray and light scattering*, (ed. P. Lindner and Th. Zemb). Elsevier, Amsterdam.

Williams, C. E., Nierlich, M., Cotton, J.-P. Jannink, G., Boué, F., Daoud, M. *et al.* (1979). *Journal of Polymer Sciences* (*Polymer Letters*), **17**, 379.

Willis, B. T. M. (ed.) (1973). *Chemical applications of thermal neutron scattering*. Oxford University Press, Oxford.

Windsor, C. G. (1981). *Pulsed neutron scattering*. Taylor & Francis, London.

Wu, W.-L., Bauer, B. J., and Su, W. J. (1989). *Polymer*, **30**, 1384.

Yamakawa, H. (1971). *Modern theory of polymer solutions*. Harper & Row, New York.

Yamakawa, H. (1977). *Macromolecules*, **10**, 692.

Yang, H., Hadziioannou, G., and Stein, R. S. (1983). *Journal of Polymer Science* (*Polymer Physics*), **21**, 159.

Yerukhimovich, Y. (1979). *Polymer Science USSR*, **20**, 470.
Yoon, D. Y. and Flory, P. J. (1981). *Polymer Bulletin*, **4**, 693.
Zerbi, G. and Piseri, L. (1968). *Journal of Chemical Physics*, **49**, 3840.
Zimm, B. (1948*a*). *Journal of Chemical Physics*, **16**, 1093.
Zimm, B. H. (1948*b*). *Journal of Chemical Physics*, **16**, 1099.
Zimm, B. (1956). *Journal of Chemical Physics*, **24**, 269.

Name index

Subject index